Distributed Energy Systems

This book provides the insight of various topology and control algorithms used for power control in distributed energy power conversion systems such as solar, wind, and other power sources. It covers traditional and advanced control algorithms of power filtering including modelling and simulations, and hybrid power generation systems. The adaptive control, model predictive control, fuzzy-based controllers, Artificial Intelligence-based control algorithm, and optimization techniques application for estimating the error regulator gains are discussed.

Features of this book include the following:

- Covers the schemes for power quality enhancement, and voltage and frequency control.
- Provides complete mathematical modelling and simulation results of the various configurations of the renewable energy-based distribution systems.
- Includes design, control, and experimental results.
- Discusses mathematical modelling of classical and adaptive control techniques.
- Explores recent application of control algorithm and power conversion.

This book is aimed at researchers, professionals, and graduate students in power electronics, distributed power generation systems, control engineering, Artificial Intelligence-based control algorithms, optimization techniques, and renewable energy systems.

Control Theory and Applications

Series Editor: Dipankar Deb, *Institute of Infrastructure, Technology, Research and Management (IITRAM)*

This book series is envisaged to add to the scholarly discourse on high-quality books in all areas related to control theory and applications. The book series provides a forum for the control scientists and engineers to exchange related knowledge and experience on contemporary research and development in control and automation. This includes aircraft control, adaptive control, sliding mode control, evolutionary control, fuzzy theory and control, robotic manipulators, and even control applications in areas such as the Internet of Things and Big Data. The scope includes all aspects of control engineering needed to implement practical control systems, from analysis and design, through simulation and hardware, with a special emphasis on bridging the gap between theory and practice. It aims to explore the latest research findings and provide attention to emerging topics in control theory and its applications to diverse domains of engineering and technology, to expand the knowledge base and applications of this rapidly evolving and interdisciplinary field. The series will include textbooks, references, handbooks, and short-form books.

Control Strategies of Permanent Magnet Synchronous Motor Drive for Electric Vehicles
Chiranjit Sain, Atanu Banerjee, and Pabitra Kumar Biswas

Distributed Energy Systems
Design, Modeling, and Control
Edited by Ashutosh K. Giri, Sabha Raj Arya, and Dmitri Vinnikov

For more information about this series, please visit: www.routledge.com/Control-Theory-and-Applications/book-series/CRCCTA

Distributed Energy Systems

Design, Modeling, and Control

Edited by Ashutosh K. Giri, Sabha Raj Arya, and Dmitri Vinnikov

CRC Press
Taylor & Francis Group
Boca Raton London New York

CRC Press is an imprint of the
Taylor & Francis Group, an **informa** business

First edition published 2023
by CRC Press
6000 Broken Sound Parkway NW, Suite 300, Boca Raton, FL 33487–2742

and by CRC Press
4 Park Square, Milton Park, Abingdon, Oxon, OX14 4RN

CRC Press is an imprint of Taylor & Francis Group, LLC

ISBN: 978-1-032-13424-6 (hbk)
ISBN: 978-1-032-13425-3 (pbk)
ISBN: 978-1-003-22912-4 (ebk)

DOI: 10.1201/9781003229124

Typeset in Times
by Apex CoVantage, LLC

Contents

Editor Biographies

Ashutosh K. Giri received his bachelor of engineering in electrical engineering from College of Technology, G. B. Pant University of Agriculture and Technology, Pantnagar Uttarakhand, India, in 2002; M.E in electrical engineering with specialization in power systems from the L.D. College of Engineering, Ahmedabad, Gujarat, India, in 2005; and PhD from the Department of Electrical Engineering, Sardar Vallabh National Institute of Technology, Surat, India in 2020. In June 2005, he joined Department of Electrical Engineering, PIET Limda (now deemed university) Vadodara as a lecturer. In May 2011, he joined the Department of Electrical Engineering, Government Engineering College, Bharuch, as an assistant professor. He has authored and co-authored more than 30 research papers in reputed journals such as *IEEE Transactions* and *IET Journal*, and books published by Wiley, Taylor & Francis, Springer, etc. He has also co-authored one book which is published in Springer Publication, Singapore. He is working on another book on *Power Quality: Infrastructure and Control* with Springer Nature Publication, Singapore. He has received the sponsored research grant of Rs. 25 Lakhs from the director of technical education, Gujarat state under STEM Scheme. He is working as a principal investigator in this project. He has coordinated five short-term training programme including the recently concluded All-India Council for Technical Education (AICTE) Training and Learning faculty development program (ATAL FDP). He has qualified GATE Examination three times with reputed percentile scores. He has delivered national and international level expert talk at various reputed universities across India, including SVNIT, SURAT, and Delhi Technical University, New Delhi. He has received a Research Opportunity Week (ROW) scholarship and visited Technical University of Munich (Germany) in March–April 2019. He has also served as a Technical Programme Committee member in conferences organized at Scientific Federation of Dallas, USA and Kenitra, Morocco. He has also chaired Technical Session in IEEE-sponsored international conference at IIITDM Kancheepuram, Chennai. He organized a special session at the world-reputed IEEE-ECCE2020 Conference at Detroit, USA. He also organized a special session at ECCE-ASIA 2021 at Singapore. He is serving as a potential reviewer in various reputed journals such as *IEEE Transactions of Power Electronics*, *IEEE Transactions on Industrial Electronics*, *IET*, Wiley etc. He was nominated best researcher of the year 2019 at electrical department, SVNIT, Surat in 2019. He is also serving as an honorary advisor (technical services) to GMDT Marine and Industrial Engineering Pvt. Ltd, Ahmedabad, India, and is serving as an honorary advisor (technical services) in Cubatics Industries Pvt. Ltd, Surat, India. He is life member of the Indian Society of Technical Education. Recently, he was included on an advisory board as a member in Society of Power Engineers, Vadodara Chapter.

Sabha Raj Arya (M 2012, SM 2015) received a bachelor of engineering degree in electrical engineering from Government Engineering College Jabalpur, in 2002; Master of Technology in Power Electronics from Motilal National Institute of Technology, Allahabad, in 2004; and PhD degree in electrical engineering from Indian Institute of Technology (IIT) Delhi, New Delhi, India, in 2014. He is an assistant professor in the Department of Electrical Engineering, Sardar Vallabhbhai National Institute of Technology, Surat. In January 2019, he was promoted as associate professor in same institute. His fields of interest include power electronics, power quality, design of power filters, and distributed power generation. He received two national awards, namely the INAE Young Engineer Award from Indian National Academy of Engineering and the POSOCO Power System Award from Power Grid Corporation of India in the year of 2014 for his research work. He is also received AmitGarg Memorial Research Award in 2014 from IIT Delhi from the high-impact publication in a quality journal during the session 2013–2014. At present, he has published more than a hundred research paper in internal national journals and conferences in the field of electrical power quality. He also serves as an associate editor for the journal *IET Renewable Power Generation.*

Dmitri Vinnikov (M 2007, SM 2011) received the Dipl. Eng., M.Sc., and Dr.Sc.Techn. degrees in electrical engineering from the Tallinn University of Technology, Tallinn, Estonia, in 1999, 2001, and 2005, respectively. He is currently the head of the Power Electronics Group, Department of Electrical Engineering, Tallinn University of Technology, and a guest researcher at the Institute of Industrial Electronics and Electrical Engineering, Riga Technical University, Riga, Latvia. He has authored more than 150 published papers on power converter design and development, and is the holder of numerous patents and utility models in this field. His research interests include switch-mode power converters, modeling and simulation of power systems, applied design of power converters and control systems, and application and development of energy-storage systems.

Contributors

Prof. B. K. Panigrahi and Prof. Bhim Singh
Department of Electrical Engineering, IIT Delhi

Dimosthenis Peftitsis
NTNU, Norway

Geeta Pathak
G.B. Pant University of Agriculture and Technology, Pantnagar, Uttarakhand

Oleksandr Husev
TalTech University, Eastonia

Rakesh Maurya
SVNIT, Surat

Mr. Athul Vijay P. K., Prof. (Dr.) Varsha A. Shah, Mr. Ujjval B. Vyas, Dr. Nikunj Patel[1,2,3]
Department of Electrical Engineering, Sardar Vallabhbhai National Institute of Technology, Surat, Gujarat, India.[4]
Department of Electrical Engineering, Shree Swami Atmanand Saraswathi, Institute of Technology, Surat, Gujarat, India

Dr. Ashutosh K. Giri
Government Engineering College, Bharuch

Dr. Sabha Raj Arya
Sardar Vallabhbhai National Institute of Technology, Surat

Prof. (Dr.) Madhusudan Singh
Delhi Technological University, New Delhi

Mr. Sombir
Delhi Technological University, New Delhi

Mr. Vamja Rajan Vinodray
Sardar Vallabhbhai National Institute of Technology, Surat, Gujarat, India

Miss. Uliya Mitra
Maulana Azad National Institute of Technology, Bhopal

Dr. Anoop Arya
Maulana Azad National Institute of Technology, Bhopal

Dr. Sushma Gupta
Maulana Azad National Institute of Technology, Bhopal

Shweta Mehroliya
Maulana Azad National Institute of Technology, Bhopal

Shilpi Tomar
SATI Vidisha-India

Karuna Nikum
Thakur College of Engineering & Technology, Mumbai

Rishi Kumar Singh
Maulana Azad National Institute of Technology, Bhopal

Prateek Mundra
Maulana Azad National Institute of Technology, Bhopal

Dr. Kalpana R.
Department of Electrical and Electronics Engg., National Institute of Technology Surathkal

Mr. Ujjval B. Vyas, Prof. (Dr.) Varsha A. Shah, Mr. Athul Vijay P. K.
Department of Electrical Engineering, Sardar Vallabhbhai National Institute of Technology, Surat, Gujarat, India

Dr. J. Saikrishna Goud
Electrical & Electronics Engg., Manipal Institute of Technology, Manipal, Manipal Academy of Higher Education, Manipal

Dr. Anjana Jain
Amrita School of Engineering, Amrita Vishwa Vidyapeetham, Bengaluru, Karnataka

Dr. Saravanakumar Rajendran
University of Santiago of Chile, Electrical Engineering Department, Santiago, Chile

Dr. K. Deepa
Amrita School of Engineering, Amrita Vishwa Vidyapeetham, Bengaluru, Karnataka

Dr. Prabhakaran K. K.
National Institute of Technology, Surathkal, Karnataka

Mr. Biswajeet Rout
Kansas State University, USA

Dr. Vuddanti Sandeep
Department of Electrical Engineering, National Institute of Technology (NIT) Andhra Pradesh, Tadepalligudem

Ms. Pagidipala Sravanthi
Department of Electrical Engineering, National Institute of Technology (NIT) Andhra Pradesh, Tadepalligudem

Vadim Sidorov
Department of electrical power engineering and mechatronics, Tallinn University of Technology, Estonia

Abualkasim Bakeer
Department of electrical power engineering and mechatronics, Tallinn University of Technology, Estonia

Hamed Mashinchi Maheri
Department of electrical power engineering and mechatronics, Tallinn University of Technology, Estonia

Naser Hassanpour
Department of electrical power engineering and mechatronics, Tallinn University of Technology, Estonia

Showrov Rahman
Department of electrical power engineering and mechatronics, Tallinn University of Technology, Estonia

Andrii Chub
Department of electrical power engineering and mechatronics, Tallinn University of Technology, Estonia

Oleksandr Matiushkin
Department of electrical power engineering and mechatronics, Tallinn University of Technology, Estonia

Dr. Mahmadasraf A. Mulla
Department of Electrical Engineering, Sardar Vallabhbhai National Institute of Technology, Surat. Gujarat. India

Dr. Payal Patel
Department of Electrical Engineering, Sardar Vallabhbhai National Institute of Technology, Surat, Gujarat, India

Dr. Shailendra Kumar and Dr. More Raju
Maulana Azad National Institute of Technology, Bhopal

Dr. Roya Ahmadiahangar
Department of electrical power engineering and mechatronics, Tallinn University of Technology, Estonia

Dr. Andrei Blinov
Department of electrical power engineering and mechatronics, Tallinn University of Technology, Estonia

Preface

The main aim of the book is to present a novel idea and concepts developed by the researchers/academia and practicing engineers in the power electronics industry. The distributed power generation based on wind, solar, hydro, and many other types is increased day by day due to advancements in power electronics and processor technology. The importance of renewable energy–based distributed power generation systems has also increased in recent years. Power quality (PQ) issues in distributed power generation systems are of major concern to customers. The role of power electronics–based custom power devices such as voltage source converters is to mitigate the PQ issues and make the system more reliable and practically feasible. The operation of voltage source converters is dependent on the control algorithms used for estimating the modulating signal followed by pulse width modulation techniques that are used for generation of gate pulses. Therefore, this creates awareness among teachers, research students, and industry persons about better utilization of distributed power generation systems by making them more efficient with the use of power electronics.

The main features of this book are the following.

1. The application of various converter topologies in power generation by solar, wind, hydro, diesel engines, and other sources, as well as enegy conversion, are the thrust areas of this book.
2. This is intended to cover all the traditional, as well as advanced, control algorithms of power filtering and their control, including modeling and simulations and hybrid power generation systems.
3. The application of adaptive control, model predictive control, fuzzy-based controllers, Artificial Intelligence–based control algorithm and optimization techniques for estimating the error regulator gains are the focus area for estimating the reference current and system protection issues.
4. The authors can submit their book chapter mentioned above, however there is no restriction for some innovative applications used in the domain of wind, solar and others based distributed power generation system.

The readers will be able to utilize the concepts presented in this book for doing additional research in this domain. This book will provide the insight of various topology, as well as control algorithms used for power control in distributed enegy power conversion systems such as solar, wind, and others.

The entire book has been arranged in nineteen chapters. Chapter 1 is the book's introduction, with the adopted system and its basic structure discussed in brief. The overview of one such distributed generation system based on wind energy is also provided here. Chapter 2 provides greater use of active power filters, passive filters, and hybrid filters for power quality improvements. In Chapter 3, some DIgSILENT power simulation results of wind energy–based systems is presented. In Chapter 4, performance review of some control algorithms used in single phase distributed power generation system is explored. The design control and modelling of wind and hydro combined generation systems is discussed in Chapter 5. The modelling of energy sources, along with loads for distributed power generation systems, is covered in Chapter 6. The focus of this book is equally given on solar photo voltaic system modelling, design and control. In Chapter 7, detail design technique of solar thin film have been discussed. In Chapter 8, design, control, and operation aspects are covered for solar grid interface. The interesting results with MATLAB simulation, followed by hardware experimentation, is also provided in the same. The other application of solar photovoltaic energy systems such as design and control of solar water pumping systems is provided in Chapter 9. In Chapter 10, solar wind hybrid system design and control is covered for better understanding by the reader. The application of some control algorithms in solar wind microgrid system is provided with MATLAB simulation results in Chapter 11. The application of new control algorithms such as sliding mode controller for controlling the operating conditions in wind energy systems is provided in Chapter 12. Chapter 13, Chapter 14, and Chapter 15 belong to design and modelling of energy storage technology such as lithium-ion battery and ultra-capacitor. Chapter 16 presents the energy scheduling for green buildings and methods for quantification of energy flexibility. The modelling and design of direct current (DC) motor–based wind emulators is presented in Chapter 17. The requirements of filter always exist in solar and wind energy conversion system, and therefore, Chapter 18 deals the LCL filter design and its applications.

The authors have tried their level best to provide a complete package in this book related to wind- and solar-based distributed power generation systems. The simulation results and experimental studies are an added advantage for researchers to get proper ideas in this work domain. The authors hope that readers will find the contents most suitable for further application in their respective working domains.

1 Introduction

Ashutosh K. Giri, Sabha Raj Arya and Rakesh Maurya

CONTENTS

1.1 DISTRIBUTED POWER GENERATION SYSTEM

Smaller generating units are used for power generation at different points of location where renewable energy is abundant, known as a distributed power generation system (DPGS) [1]. There are two types of distributed generations systems available, first one that is connected or integrated from the available AC grid and second is one working in off-grid mode connected directly through the consumers. The distributed power can be generated from the nonconventional sources of energy such as wind, solar, tidal, biomass, etc. The generated power ranges less than a kilowatt (kW) to tens of megawatts (MW). The potential of DPGS can meet entire or some major part of a consumer's needs. It can be installed on rooftops for excess power generation on office building premises, also generating the local revenue. Distributed power generation has engrossed a lot of popularity worldwide [2–4]. By employing the DGPS, the dependency on conventional fossil fuels will be reduced that results the reduction in emission of greenhouse gases (GHGs). The policies of government largely affect the development of DGPS and its operation either off grid or grid integrated.

1.2 NEED OF DISTRIBUTED POWER GENERATION SYSTEM

Isolated locations are deprived from economic growth due to non-availability of power grid. For such locations and populations, the renewable energy–based distributed power generation system can supply electricity with adequate quality and reliability. It will raise the living standard of the entire population and will feed power at quite reasonable rates. Generally, such generation and distribution systems are highly efficient due to absence of long transmission lines. Voltage regulation problems and frequency regulation problems can be easily solved using local control centres. The complexity of the system is much less as compared to the interconnected grid. Intercommunication between the different controllers and converters is also easy. These days, wind-based mini- and micro-generation is getting much attention in distributed generation systems (DGs). The intrinsic short circuit protection features with low capital investments makes the induction generator an apparent choice for this case. The voltage and frequency regulation with poor power factor are major limitations of self-excited induction generator. The technical issues in distributed power generation system are discussed in the rest of this chapter.

1.3 TECHNICAL CHALLENGES IN DISTRIBUTED POWER GENERATION SYSTEMS

Basically, distributed power generation is dependent on natural resources of energy. These resources provide power with some degree of randomness. Maintaining voltage and frequency with such input power variations in most of the system except solar is a challenging task. Moreover, due to popular use of nonlinear loads on the generation system, harmonics are also present in the supply current. This creates waveform distortions and it results power factor problem at the supply end. Load balancing creates additional problems in three-phase four-wire type distributed power generation systems.

The standalone distributed power generation system suffers from the aforesaid problems. In addition to it, it also suffers from voltage and frequency variations under either load variations or mechanical input variation [5]. For effective utilisation, need arises for a voltage source converter (VSC) which controls the power flow to mitigate the power quality issues as per the guideline of IEEE standards-519. Power quality standards are discussed in the following section.

DOI: 10.1201/9781003229124-1

1.4 POWER QUALITY STANDARDS

An ample amount of information related to power quality issues has been reported in research papers which is very useful to give the complete basics about voltage- and current-based power quality issues. The monitoring of power quality problems, causes and remedies of voltage and current quality are discussed in detail. So many standards are specified about the level of harmonics and monitoring the power quality of the distributed power generation system. IEEE Std.519–2014 [6] on recommended practices and the requirements for harmonic control in distribution systems is one of them. It provides guidance to design the distributed power systems with nonlinear loads. The other IEEE Std.1531–2003 [7, 8] suggests the guidelines for control of harmonic filters and other power quality problems. The harmonic filter is basically governed by the rules and guidelines of these standards, although most of the components used in harmonic filter are now also covered. The major application includes utility medium-voltage systems, industrial low-voltage facilities, induction furnace installations, etc.

1.5 VSC USED AS AN ACTIVE POWER FILTER

A VSC can be used as an active power filter in wind-based distributed power generation systems. The usefulness of VSC is explained in various sources as following. Akagi [10], Singh et al. [11] and Valdez et al. [12] have explained since the beginning of 1983 the basics of the active filters and control strategies based on unity-power factor control for the distorted supply voltage and current. They have suggested that supply current can be made sinusoidal by connecting active filter (AF) at the point of common coupling. It also presents a comprehensive review of active filter configurations, selection of components, other related economic and technical considerations, and their selection for specific applications. The operation of a VSC is dependent on the utilisation of effective control techniques used for gate pulse generation. The various control techniques are reported in the literature as following. Chaer et al. [13] have presented the current applications of advanced Artificial Intelligence techniques in power quality analysis. Applications of some advanced mathematical tools in general, and wavelet transform in particular, are also provided. Rao et al. [14], Singh et al. [15] and Luo et al. [16] have presented many harmonic extraction techniques and comprehensive review of active power filters control. These reviews give the design guide lines of filters for power quality improvement. Zou et al. [17] have proposed the control methodology of shunt active power filters based on frequency-adaptive fractional-order repetitive control for three-phase or single-phase application. Moreover, the definition of voltage and current in steady state is also provided on the basis of the average value concept. Further, for three-phase power application, the definition of instantaneous reactive power in transient state for arbitrary voltage and current waveforms is also provided on the basis of the instantaneous value concept. Sreeraj et al. [18] have presented a power active filter based on one-cycle control that permits a very good compensation of the higher current harmonics generated in the mains by a controlled or uncontrolled three-phase rectifier and also explain static frequency converter with direct current link and active power filter. Kanjiya et al. [19], Arya et al. [20], Yazdani et al. [21–22], Cirrincione et al. [23], Tey et al. [24] have proposed the active power filter implementation, application issues in different consumer loads. Huang et al. [25], Tolbert et al. [26] and Cardenas et al. [27] have proposed the identification idea of instantaneous current components, reduction of neutral currents in three-phase circuit, new phase shift control circuit for active filter, design of active power filters under non-ideal mains voltages and harmonic cancellation by mixing single and three-phase nonlinear loads, respectively. In conclusion, the authors have explored the usefulness of VSC as an active power filter and its further application in the area of distributed power generation systems, as discussed in the following subsections.

1.6 APPLICATION OF VSC IN THREE-PHASE FOUR-WIRE DISTRIBUTED POWER GENERATION SYSTEMS

The problems in the most of the distributed power generation system are its randomness in power generation. The reliability and power quality is a major issue. Therefore, the use of VSC for power control is necessary. The utilization of VSC in distributed power generation system for power quality improvement is provided in the some of the literatures. Chilipi et al. [28], Kasal et al. [29] and Ouazeneet et al. [30] have presented a survey of different topology of distributed generation with neutral currents compensation in three-phase four-wire systems, explaining that the imbalance of three-phase currents are the reason behind neutral current flow in three-phase four-wire power systems. However, in some nonlinear loads such as computer systems and diode-based rectifier type loads, it is observed that a very high neutral current even load is balanced in phases. To determine the extent of the neutral current problem, some results from a sample of computer power systems in the United States are presented. Alolah et al. [31] have explained an Optimisation-based steady state analysis of three-phase self-excited induction generator. The compensation characteristics illustrate that the proposed scheme works well not only with harmonic voltage source loads but also with harmonic current-source loads. Jabri and Alolah [32] have explained critical overview on zero-sequence compensation in distorted and unbalanced three-phase four-wire systems. The generalised theory of instantaneous powers has been used in power systems application for compensation of instantaneous powers and harmonic cancellation. Results are presented for a three-phase four-wire system that includes significant zero-sequence current. Murthy et al. [33] have presented a novel and comprehensive performance analysis of a single-phase two-winding self-excited induction generator. For keeping the power factor at unity and currents fully balanced for the power utility, the VSC provides compensation currents under distorted loading conditions. Singh et al. [34] have presented analysis and control of a battery-based isolated induction generator which can suppress zero-sequence currents in the neutral wire of three-phase four-wire systems for harmonic-current source loads, as well as in harmonic voltage source loads. The reduced VSC volt ampere rating is required rather than the conventional neutral-current–suppressing methods. With the controlled voltage of DC capacitors, the requirement of rectifier part of the neutral harmonic suppressor has been eliminated. Singh et al. [35–38] have presented a topology of static distribution compensator distribution in three-phase four-wire systems reducing the volt ampere rating of VSC by using a zig-zag transformer. This topology is used for the reduction in the source harmonic current, compensation of reactive power, elimination of neutral current and load balancing. In addition, the voltage and frequency regulation can also be achieved at the point of common interface. Some applications of control algorithm implementation are discussed throughout the references [39–45] for distributed power generation and its control.

1.7 CONFIGURATIONS OF VSC AS AN ACTIVE FILTER IN DISTRIBUTED POWER GENERATION SYSTEMS

This can be classified based on the type of its use and supply system. The classification based on supply system is single phase or three phase. Three-phase systems are further classified into three-phase three-wire and three-phase four-wire VSC.

The development of different configurations is reported in the various literatures. The basic single-phase configuration for VSC-based distributed power generation system is shown in Figure 1.1. Three-phase three-wire VSC [31–33] may be classified in isolated and non-isolated mode with topologies. A three-leg VSC-based topology is used as a compensator, as shown in Figure 1.2. This configuration has six IGBTs, three AC inductors and a BESS (battery energy storage system) unit. The required compensation to be provided by the VSC decides the rating of the VSC components.

The rating of the solid state switches is based on the voltage and current rating of the compensation system.

1.8 CONTROL ALGORITHMS FOR VSC IN DISTRIBUTED POWER GENERATION SYSTEMS

The standalone distributed power generation system is effectively utilised with the help of a shunt-connected power filter—widely known as VSC—at the distribution end. For operating the VSC to meet the power quality requirement, the control algorithms are needed to generate the reference source current. Generally, classical and adaptive control algorithms are more popular due to their accuracy and speed under dynamic conditions. The classical control techniques are mostly dependent on phase-locked loops (PLLs) for phase angle and unit magnitude signal estimation. The key demerits of such control algorithms are erroneous output under unbalanced and distorted input

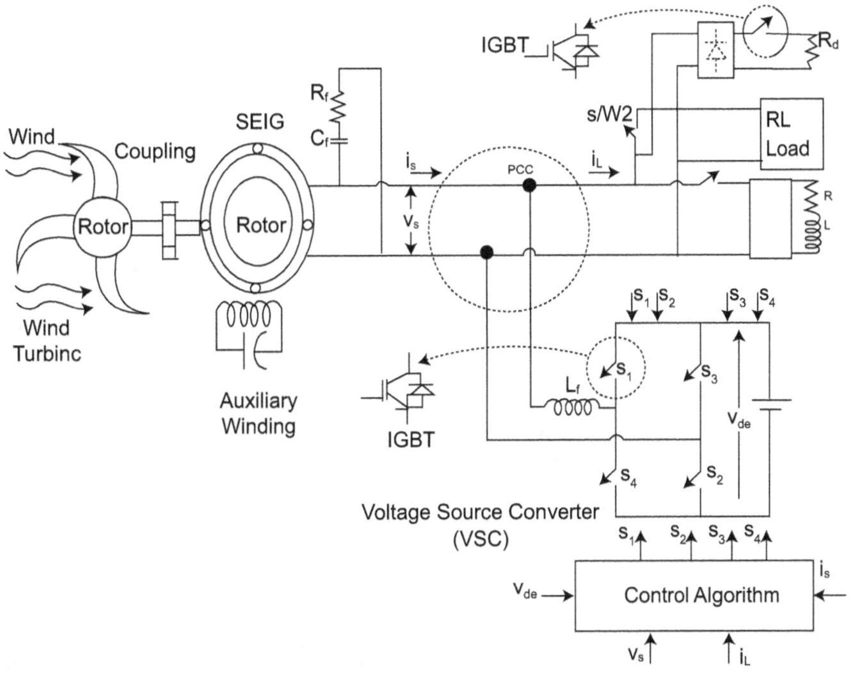

FIGURE 1.1 Single-phase two-wire distributed power generation system.

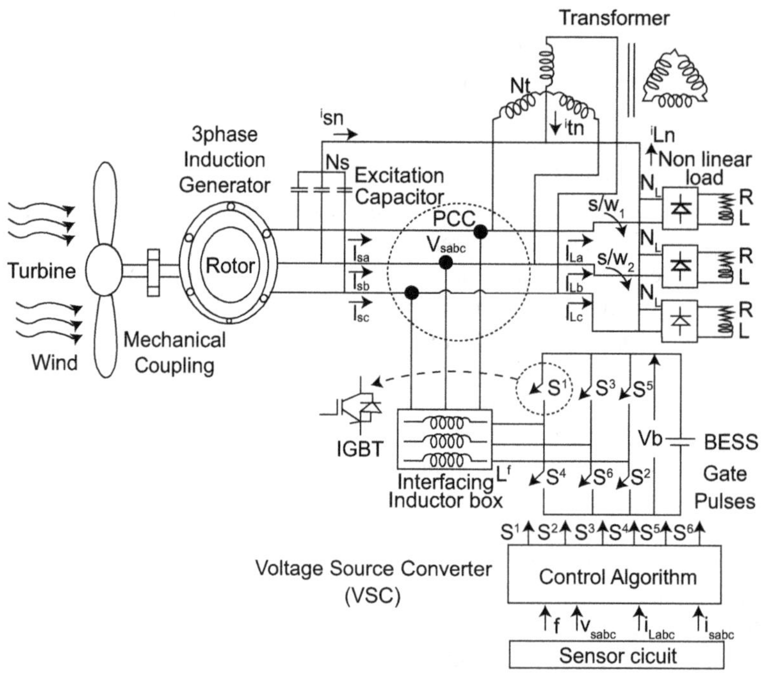

FIGURE 1.2 Three-leg VSC-based three-phase four-wire distributed power generation system.

conditions. Therefore, authors in this book have decided to implement adaptive control techniques in wind-based distributed power generation system where changes are random in nature. These techniques are elaborated in coming chapters of this book.

1.9 ORGANISATION OF THE BOOK

The entire book has been arranged in nineteen chapters including this chapter. The first chapter belongs to the book introduction means that the adopted system and its basic structure are discussed in brief. The overview of one such distributed generation system based on wind energy is also provided here. The second chapter provides greater use of active power filter, passive filter and hybrid filter for power quality improvements. In Chapter 3, some DigSILENT power simulation results of wind energy based system is presented. In Chapter 4, performance review of some control algorithms used in single phase distributed power generation system is explored. The design control and modelling of wind and hydro combined generation system is discussed in Chapter 5. The modelling of energy sources along with load for distributed power generation system is covered in Chapter 6. The focus of this book is equally given on solar photo voltaic system modelling, design and control. In Chapter 7, detail design technique of solar thin film have been discussed. In Chapter 8, design, control and operation aspect is covered for solar grid interface. The interesting results with MATLAB simulation followed by hardware experimentation is also provided in the same. The other application of solar photovoltaic energy system such as design and control of solar water pumping system is provided in Chapter 9. In Chapter 10, solar wind hybrid system design and control is covered for better understanding of reader. The application of some control algorithm in solar wind microgrid system is provided with MATLAB Simulation results in Chapter 11. The application of new control algorithms such as sliding mode controller for controlling the operating conditions in wind energy system is provided in Chapter 12. The Chapters 13, 14 and 15 belongs to design and modelling of energy storage technology such as lithium ion battery and ultra-capacitor. The Chapter 16 presents the energy scheduling for green building and methods for quantification of energy flexibility is presented. The modelling and design of DC motor based wind emulator is presented in Chapter 17. The requirement of filter is always there in solar and wind energy conversion system. Therefore, Chapter 18 dealt the LCL filter design and its applications. The protection features of distributed power generation system are also taken up and presented in Chapter 19.

Author have tried their level best to provide complete package in this book related to wind and solar based distributed power generation system. The simulation results and experimental study are added advantage for researchers to get proper idea in this work domain. Authors hope that readers will find contents most suitable for further application in their respective working domain.

REFERENCES

[1] M. Godoy Simoes and Felix A. Farret, *Alternate Energy Systems, Design and Analysis with Induction Generators*, Second edition, CRC Press, Tailor and Fransis Group, London, 2008.

[2] L. L. Lai and T. F. Chan, *Distributed Generation: Induction and Permanent Magnet Generators*, John-Wiley and Sons, New York, 2007.

[3] Brendan Fox and Damian Flynn, *Wind Power Integration Connection and System Operational Aspects*, The Institution of Engineering and Technology, London, 2007.

[4] Jean Marc Chapallaz, Jacques Dos Ghali, Peter Eichenberger and Gerhard Fischer, *Manual on Induction Motors Used as Generators*, Springer Vieweg, Singapore, 1992.

[5] M. Stiebler, *Wind Energy Systems for Electric Power Generation. Green Energy and Technology*, Springer-Verlag, Berlin and Heidelberg, Singapore, 2008.

[6] IEEE Standards 519–1992, *IEEE recommended practices and requirements for harmonic control in electric power systems.*

[7] Ann Marie Borbely and Jan F. Kreider, *Distributed Generation: The Power Paradigm for the New Millennium*, CRC Press, Boca Raton, FL, 2001.

[8] A. R. Jha, *Wind Turbine Technology*, CRC Press, Taylor and Francis Group, London, 2011.

[9] Ion Boldea, *Synchronous Generators Hand Book*, CRC Press, Tailor and Fransis Group, London, 2016.

[10] H. Akagi, Y. Kanazawa and A. Nabae, "Generalized theory of the instantaneous reactive power in three-phase circuits," in *Proc. IEEE/JIEE Int. Power Electronics Conference (IPEC'83)*, 1983, pp. 821–827.

[11] B. Singh and V. Verma, "Selective compensation of power-quality problems through active power filter by current decomposition," *IEEE Transactions on Power Delivery*, vol. 23, no. 2, pp. 792–799, Apr. 2008.

[12] A. A. Valdez-Fernández, P. R. Martínez- Rodríguez, G. Escobar, C. A. Limones-Pozos and J. M. Sosa, "A model-based controller for the cascade H-bridge multilevel converter used as a shunt active filter," *IEEE Transaction on Industrial Electronics*, vol. 60, no. 11, pp. 5019–5028, Nov. 2013.

[13] T. Chaer, J. P. Gaubert, L. Rambault and M. Najjar, "Linear feedback control of a parallel active harmonic conditioner in power systems," *IEEE Transaction on Power Electronics*, vol. 24, no. 3, pp. 641–653, Mar. 2009.

[14] U. K. Rao, M. K. Mishra and A. Ghosh, "Control strategies for load compensation using instantaneous symmetrical component theory under different supply voltages," *IEEE Transactions on Power Delivery*, vol. 23, no. 4, pp. 2310–2317, Oct. 2008.

[15] B. Singh and S. Arya, "Back-propagation control algorithm for power quality improvement using VSC," *IEEE Transaction on Industrial Electronics*, vol. 61, no. 3, pp. 1204–1212, Mar. 2014.

[16] A. Luo, X. Xu, L. Fang, H. Fang, J. Wu and C. Wu, "Feedback feed forward PI-type iterative learning control strategy for hybrid active power filter with injection circuit," *IEEE Transaction on Industrial Electronics*, vol. 57, no. 11, pp. 3767–3779, Nov. 2010.

[17] Z. Zou, K. Zhou, Z. Wang and M. Cheng, "Frequency-adaptive fractional-order repetitive control of shunt active power filters," *IEEE Transactions on Industrial Electronics*, vol. 62, no. 3, pp. 1659–1668, Mar. 2015.

[18] E. S. Sreeraj, E. K. K. Prejith, K. Chatterjee and S. Bandyopadhyay, "An active harmonic filter based on one-cycle control," *IEEE Transactions on Industrial Electronics*, vol. 61, no. 8, pp. 3799–3909, Aug. 2014.

[19] P. Kanjiya, V. Khadkikar and H. H. Zeineldin, "A non iterative optimized algorithm for shunt active power filter under distorted and unbalanced supply voltages," *IEEE Transactions on Industrial Electronics*, vol. 60, no. 12, pp. 5376–5390, Dec. 2013.

[20] S. Arya and B. Singh, "Composite observer-based control algorithm for distribution static compensator in four-wire supply system," *IET Power Electronics*, vol. 6, no. 2, pp. 251–260, Feb. 2013.

[21] Davood Yazdani, Alireza Bakhshai and Geza Joos, "A real-time sequence components decomposition for transient analysis in grid-connected distributed generation systems," in *Proc. IEEE Symposium on Industrial Electronics*, pp, 1651–1656, 2008.

[22] Davood Yazdani, Mohsen Mojiri, Alireza Bakhshai and GezaJoos, "A fast and accurate synchronization technique for extraction of symmetrical components," *IEEE Transactions on Power Electronics*, vol. 24, no. 3, pp. 674–684, Mar. 2009.

[23] M. Cirrincione, M. Pucci, G. Vitale and A. Miraoui, "Current harmonic compensation by a single-phase shunt active power filter controlled by adaptive neural filtering," *IEEE Transaction on Industrial Electronics*, vol. 56, no. 8, Aug. 2009.

[24] L. H. Tey, P. L. So and Y. C. Chu, "Improvement of power quality using adaptive shunt active filter," *IEEE Transactions on Power Delivery*, vol. 20, no. 2, pp. 1558–1568, Apr. 2005.

[25] S. Huang and J. Wu, "A control algorithm for three-phase threewiredactive power filters under non-ideal mains voltage," *IEEE Trans. Power Electronics*, vol. 14, no. 4, pp. 753–760, 1999.

[26] L. M. Tolbert and T. G. Hablter, "Comparison of time-based non-activepower definitions for active filtering," in *Proc. CIEP, Acapulco*, Mexico, pp. 73–79, 2000.

[27] V. Cardenas, L. Moran, A. Bahamondes and J. Dixon, "Comparative study of real time reference generation techniques forfour-wire shunt active power filters," *IEEE Proc. (PESC), Acapulco*, Mexico, pp. 791–796, 2003.

[28] Rajasekhara Reddy Chilipi, Bhim Singh and S. S. Murthy, "Performance of a self-excited induction generator with VSC-DTC drive-based voltage and frequency controller," *IEEE Transactions on Energy Conversion*, vol. 29, no. 3, pp. 545–557, Sept. 2014.

[29] Gaurav Kumar Kasal and Bhim Singh, "Voltage and frequency controllers for an asynchronous generator-based isolated wind energy conversion system," *IEEE Transactions on Energy Conversion*, vol. 26, no. 2, pp. 402–416, June 2011.

[30] L. Ouazene and G. McPherson Jr., "Analysis of the isolated induction generator," *IEEE Transactions on Power Apparatus and Systems*, vol. PAS-102, no. 8, pp. 2793–2798, Aug. 1983.

[31] A. I. Alolah and M. A. Alkanhal, "Optimization-based steady state analysis of three-phase self excited induction generator," *IEEE Transactions on Energy Conversion*, vol. 15, no. 1, pp. 61–65, Mar. 2000.

[32] A. K. Jabri and A. I. Alolah, "Capacitance requirement for isolated self-excited induction generator," *IEE Proceedings, PB*, vol. 137, no. 3, pp. 154–159, May 1990.

[33] S. S. Murthy, Bhim Singh and Vuddanti Sandeep, "A Novel and comprehensive performance analysis of a single-phase two-winding self-excited induction generator," *IEEE Transactions on Energy Conversion*, vol. 27, no. 1, pp 117–127, Mar. 2012.

[34] Bhim Singh, Gaurav Kasal, Ambrish Charndra and kamal-Al Haddad, "Battery based voltage and frequency controller for parallel operated isolated asynchronous generators,"

in *Proc. of IEEE International Symposium on Industrial Electronics (ISIE-2007)*, Vigo, Spain, June 4–7, 2007.

[35] V. Rajagopal, B. Singh and G. K. Kasal, "Electronic load controller with power quality improvement of isolated induction generator for small hydro power generation," *IET Renewable Power Generation*, vol. 5, no. 2, pp. 202–213, 2011.

[36] Ujjwal Kumar Kalla, Bhim Singh and S. S. Murthy, "Normalised adaptive linear element-based control of single-phase self excited induction generator feeding fluctuating loads," *IET Power Electronics*, vol. 7, no. 8, pp. 2151–2160, 2014.

[37] B. Singh, S. S. Murthy and Sushma Gupta, "A voltage and frequencycontroller for self-excited induction generators," *Journal of Electrical Power Components and Systems*, vol. 34, pp. 141–157, 2006.

[38] Bhim Singh, S. S. Murthy and Sushma Gupta, "An electronic voltage and frequency controller for single-phase self-excited induction generators for pico hydro applications," in *Proc. of IEEE Power Electronics Drives System Conf.*, vol. 1, pp. 240–245, 2005.

[39] Ashutosh K. Giri, Amin Qureshi, Sabha Raj Arya, Rakesh Maurya and B. Chitti Babu, "Features of power quality in single phase distributed power generation using adaptive nature vectorial filter," *IEEE Transactions on Power Electronics*, vol. 33, no. 11, pp. 9482–9495, Nov. 2018.

[40] Ashutosh K. Giri, Sabha Raj Arya, and Rakesh Maurya, "Compensation of power quality problems in wind based renewable energy system for small consumer as isolated loads," *IEEE Transactions on Industrial Electronics*, vol. 66, no. 11, pp. 9023–9031, Nov. 2019.

[41] Ashutosh K. Giri, Sabha Raj Arya, Rakesh Maurya and B. Chitti Babu, "Power quality improvement in stand-alone SEIG based distributed generation system using lorentzian norm adaptive filter," *IEEE Transactions on Industry Applications*, vol. 54, no. 5, pp. 5256–5266, Sep./Oct. 2018.

[42] Ashutosh K. Giri, Sabha Raj Arya, Senior Member IEEE, Rakesh Maurya, Member IEEE and R. K. Mehar, "Variable learning adaptive gradient based control algorithm for VSC in distributed generation," *IET Renewable Power Generation*, vol. 12, no. 16, pp. 1883–1892, 2018.

[43] Ashutosh K. Giri, Sabha Raj Arya, Rakesh Maurya and B. Chitti Babu, "VCO less PLL control based VSC for power quality improvement in distributed generation system," *IET Electrical Power Application*, vol. 13, no. 8, pp. 1114–1124, 2019.

[44] Ashutosh K. Giri, Sabha Raj Arya, Rakesh Maurya and B. Chitti Babu, "Control of VSC for enhancement of power quality in off-grid distributed power generation," *IET Renewable Power Generation*, vol. 14, no. 5, pp. 771–778, 2020.

[45] Ashutosh K. Giri, Sabha Raj Arya and Rakesh Maurya, "Mitigation of power quality problems in permanent magnet synchronous generator using quasi-newton based algorithm," *International Transactions of Electric Power and Energy System*, vol. 29, no. 11, pp. 1–26, Wiley Publication, 2019.

2 Design and Performance Analysis of Grid Connected Hybrid-Distributed Generation Using DIgSILENT PowerFactory

Anoop Arya, Shweta Mehroliya, Uliya Mitra, Sushma Gupta, Rishi Kumar Singh, and Prateek Mundra

CONTENTS

2.1 INTRODUCTION

Renewable sources are the sources by which one can generate power using natural resources like solar, wind, water and heat generated by the earth. In the present scenario, distributed generation has become a very interesting and popular topic for power system engineers. In recent years, distributed generation has expanded very rapidly because of its several advantages; it gives an alternate economical solution towards load growth. There are many locations for fixing the distributed generation, but the location should be optimized and power losses are minimum and it must have proper size [1]. Introducing distribution resources (DR) in the distribution network possesses benefits of reduced losses, improved power quality and better system reliability. In the power system era, many of the latest trends are adopting as the demand and requirement of power or energy are increasing. Thereby, moving towards the next century, many new technologies of generating power are becoming noticeable that will change the requirements of energy generation, transmission and distribution. A micro-grid consists of distributed energy resources (DER), mainly small-scale renewable energy sources (RES), and many loads that can be operated with larger utility grid autonomously [2]. According to need, it can be connected with the main grid or can be disconnected from the grid. Grid interconnected DC micro-grids for residential houses and data centre are developed. Different schemes are suggested for the hybrid supplies like wind power, solar power and fuel cells [3, 4]. A hybrid type of distributed generation consists of renewable generation like solar radiation–based PV cells, wind turbine models and battery-operated systems attached to DC grid side and AC loads are connected at the AC grid side [5, 6]. Various hybrid micro-grid structures are discussed in and various power controlling techniques are used in AC/DC micro-grids [7]. The integration of micro-grids into a power system improves reliability and flexibility of the system. Power quality can also be improved in hybrid micro-grid by submerging AC and DC sub-grids [8–10]. Different energy management schemes have been analysed for their different modes of operation. Droop control schemes are presents with complete survey [11]. The system power balance is obtained by keeping the constant DC link voltage through charging and discharging of batteries in energy management systems (EMSs) [12–14]. Distributed generation can generate power at small scale and its technologies are based on combustion and non-combustion type. Distributed generation scenario: The total renewable power contribution is 175 GW and in India and in the next ten years, it can have increased up to 227 GW [15]. The distributed generation (Figure 2.1) can be a solar or photovoltaic array, wind turbine–based, fuel cells or natural gas. Hybrid-distributed generation also uses the help of different technologies. A fundamental frequency type doubly fed induction generator (DFIG) has been modelled, and various operating conditions were discussed [16].

The digital simulation has done for understanding the transient behaviour of wind turbines, and the wind turbine model was developed with the help of EPRI programmes [17]. The performance of induction machines, speed control of induction generators and solar PV array designing are discussed in [18]. The penetration of wind turbines

DOI: 10.1201/9781003229124-2

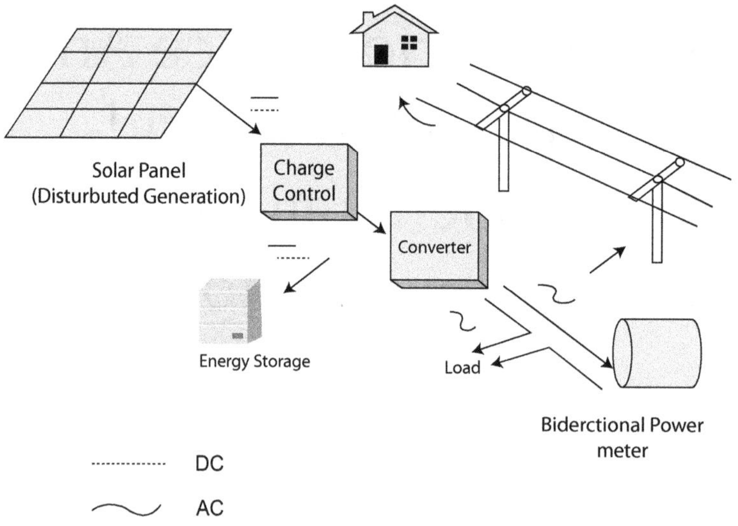

FIGURE 2.1 Distributed generation.

is increasing day by day due to their popularity, the synchronous machines dominating power systems and having a experience of change in operational and dynamic characteristics. The transient and small signal stability analysis is done in [19] for doubly fed induction generator based wind turbines. Sensitivity analysis has determined for Eigen value, and this has achieved through conversion of DFIG into synchronous machines with round rotors. Transient stability analysis with solar PV integration has been achieved through participation factors and damping ratios; various load conditions are considered in [20]. In DIgSILENT PowerFactory, a new model has developed for doubly fed induction generator and its principle is based on voltage source controller [21]. Various challenges and techniques for performance and control of micro-grids has been discussed in [22]. A comparative study of three levels of control—that is: primary, tertiary and secondary—has been done. The potential impact of distributed generation on distribution network and how this distributed generation, especially solar generation, will enhance distribution systems has been studied. The complete study has done in DIgSILENT and open DSS (open distribution simulation software) [23]. Over- and under-frequency methods were used to detect islanding phenomena [24]. By taking detection time along with different conditions, systems are analysed using the DIgSILENT PowerFactory simulation tool. To perform switching operation tests of vacuum circuit breakers (VCB) and to get the dielectric parameter characteristics. The DIgSILENT is used to carry out the modelling and simulation of DG systems [25, 26].

2.1.1 Wind Power Generation

For producing power or electricity, wind has become one of the prominent and easily available renewable sources. Its concept is simple and based on transforming kinetic energy into mechanical energy, and then again, this mechanical energy is transformed into electrical energy. The integration of wind turbine generators (WTGs) can be done through power transformer (Figure 2.2). Generally, wind turbine speed varies from 2–13m/s [27, 28]. The wind power generation has the following characteristics.

1. Carbon footprints can be minimized.
2. Wind power plants can be installed anywhere, as wind is a natural source.
3. Operation and maintenance of equipment is easy.

As per the National Energy Administration (NEA), the primary frequency regulation for wind turbines are as follows.

- If the turbine output power exceeds 20% of its rated power, it can convert output power for control of primary frequency.
- The time of response should not greater than five seconds.
- As the grid frequency decreases to certain value, the turbine has to increase 8–10% of its rated power.
- Wind turbines have to balance increased power output up to ten seconds.

2.1.2 Solar Radiation Power Generation

To produce power form the radiation of the sun, what is known as "solar PV generation" is one of the clean and environmental friendly power generation methods; its working depends on the concept of turning solar radiation energy into useful electrical energy. It is one of the most popular sources of producing electricity. Sun provides light

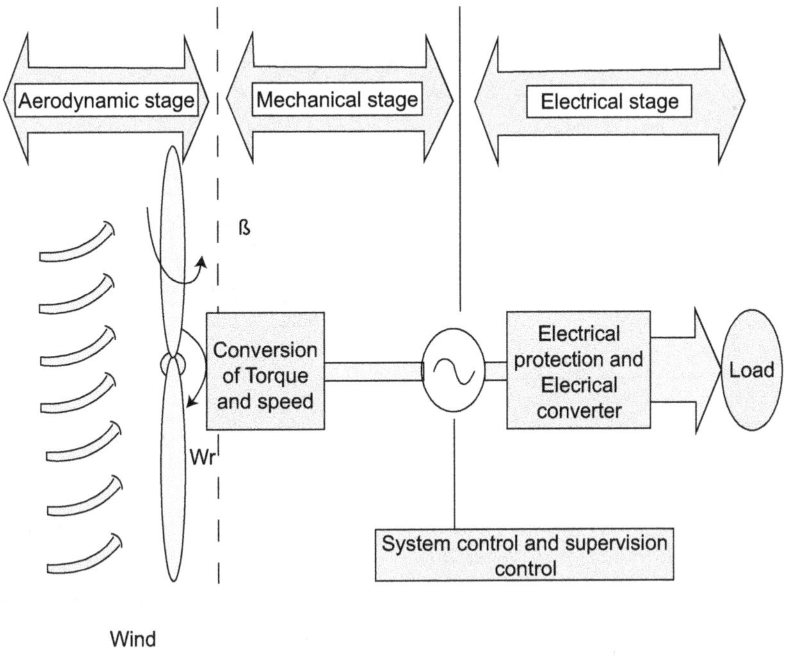

FIGURE 2.2 Wind turbine-based distributed generation.

everywhere without any cost; therefore, solar energy can be provided very easily. In solar power generation, photovoltaic arrays are used. Photovoltaic cells are made up of silicon material; after silicon, Cadmium telluride (CdTe) material is used. Figure 2.3 shows the principle of energy conversion in solar cells: In a cell, a junction is formed in the cell (P and N) [29, 30]. Cells made from silicon have high reliability, low cost and long life. By using these photovoltaic arrays, the solar energy is converted into electrical energy.

As today's world is based on the smart technologies, solar energy enables any individual to produce electricity on their own [31–35]. These technologies can be very helpful in the developed and developing countries like India. In India, the total solar energy production is about 100 MW. Many researchers are working in this area with their new technology. Therefore, this solar energy is very helpful in small villages where there is no facility to provide electricity. By using solar power generation, a micro-grid can be formed. Everyone can be benefited from these technologies. Solar energy is also very useful in hilly areas where the installation of transmission towers is such a difficult task for the workers. Therefore, this type of energy is useful in areas where there is no energy grid. It does not take any heavy maintenance during servicing period. It also has no moving parts, so there will be no wear and tear [36]. Therefore, keeping all factors of solar energy, it is concluded that it has following advantages.

1. Carbon footprints can be minimized by solar energy.
2. Electricity bill can be reduced by using it.
3. It has diverse applications.
4. Its maintenance cost is less.

Solar energy has some limitations, also.

1. Its capital cost is high.
2. It is dependent on weather; it cannot work without sun.
3. It requires lot of space.
4. It requires regular maintenance due to dust particles on PV arrays.

2.1.3 Small Hydropower Generation

In today's century, as population are increasing, demand for power is also increasing. To fulfil demand of power, researchers and scientists are continuously working towards generation of maximum power by renewable sources. Generation of power from water is one of the most easily available and natural sources [37, 38]. It is also one of the oldest renewable sources. Hydropower is the method of generating power with the help of moving water; kinetic energy is used to produce power by converting it into useful power. In developing countries like India, rural populations generally face problems of not getting electricity. Therefore, renewable generation of power like by hydropower generation is proven to be a most prominent method. The scale of power generation is different, as follows.

- Small-scale hydropower generation.
- Large-scale hydropower generation.
- Micro-scale hydropower generation.

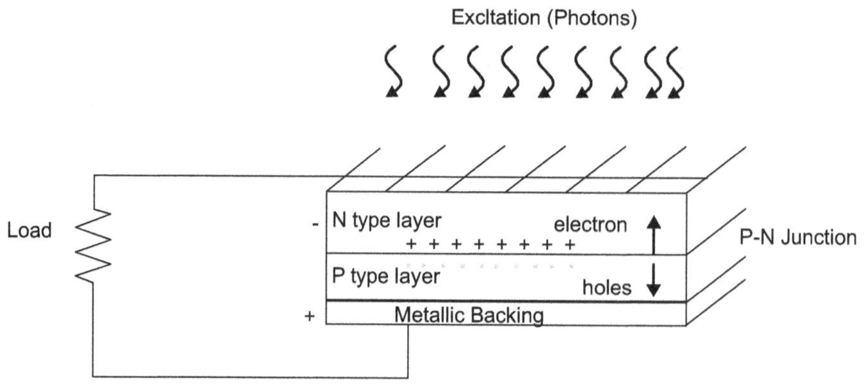

FIGURE 2.3 Construction of a solar cell.

Small-scale power generation is easily available; it works on the concept of head and flow. Head is height of water flow and flow is the volume of water. An isolated region is used for producing power in a small hydropower generation. A setup of small hydropower generation is installed near a river.

2.1.4 Introduction to DIgSILENT Simulation Tool

DIgSILENT is power system simulation software used to solve various problems associated with power systems and electrical networks. DIgSILENT is a short form of digital simulation and electrical networks. It includes all the function of power system analysis [39, 40]. PowerFactory is also used for transmission system operation and control. The features and functions of DIgSILENT simulation software are as follows.

- Load flow sensitivities.
- Network analysis for low voltage.
- Analysis for protection.
- Network reduction.
- Wind turbine model.
- Probabilistic analysis.
- Unit commitment.
- Management and planning for outage.
- Tools for transmission network.
- Tools for distribution network.
- Short circuit analysis.
- Sensitivity factor.
- VP/VQ curve calculation.

Figure 2.4 shows how one can operate DIgSILENT software. This chapter is divided into two parts. In the first part, wind turbine is modelled using DIgSILENT software; and in the second part, a transient stability analysis has done. In this software, system models already exist for analysis, but for those wanting to build their own models, this software provides facility for doing just that. It enables real-time simulation and has wide application in the area of industries, businesses and research. This software is also used for monitoring of power systems.

2.2 MODELLING OF WIND TURBINE AND HYBRID-DISTRIBUTED GENERATION IN DIGSILENT POWER SIMULATION TOOL

2.2.1 Modelling of Wind Turbine

The wind turbine with doubly fed induction motor consists of induction generator, wind turbine drive train, pitch controller and converters on the rotor side and at the grid side. The stator winding and rotor windings are coupled to the grid and power electronics converter, respectively. The rotor frequency of both electrical and mechanical system is decoupled. This is done by using power converters and injects a rotor current so that it can offer compensation between electrical and mechanical systems [41–43]. With the help of doubly fed induction generators, reactive power and active power can be controlled by controlling the rotor excitation current independently. The need of slip rings is the main drawback.

The speed controllers are used for adjusting generator torque, but this leads to overloading of rotor converter as well as generator during heavy wind speeds. To overcome this problem, a pitch controller is used.

2.1.1.1 Dynamic Simulation Language (DSL) in Wind Turbine

The study and analysis of following systems are necessary in the modelling of the wind turbine.

- Aero-dynamic system.
- Mechanical system.
- Wind speed control system.
- Storage system.

The DSL enables dynamic modelling of both nonlinear and linear systems, as it is the feature of DIgSILENT. Thus apart from modelling non-electrical wind components, it helps in designing and implementing different type of controllers.

In electrical models, the first step is to model the load flow; the DSL models should be started carefully. Overall system performance depends on the initialization of the model under normal and faulty conditions. There are some

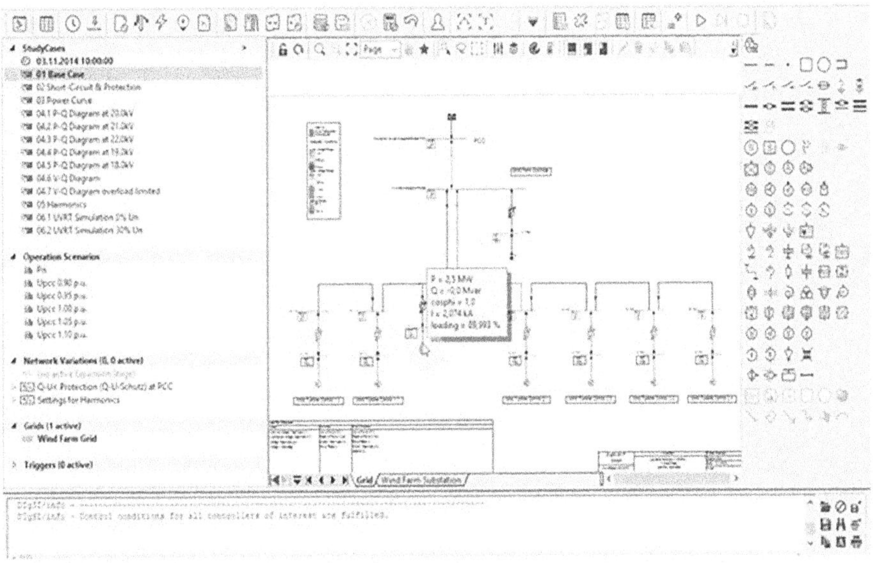

FIGURE 2.4 DIgSILENT software.

issues associated with the initialization of the wind turbine in DSL model. The integration of a new model into power system simulation model needs proper initialization because the whole system performance is dependent on this procedure. Power system simulation starts with load flow parameters; to obtain the parameters, the operator has to assign input data. For a DFIG, the necessary load flow data are stator active power, reactive power and slip. Once the electrical parameter initialization is over, the mechanical component model starts. The grid connected turbine model is initialized in a sequential manner. The difficulty is to find the order of sequence of model, which they are following. The solution of simulation equations are carried out by fixing time derivative to zero, load flow results and mechanical input data. It is required to point out the input, output and states, which is to be initialized.

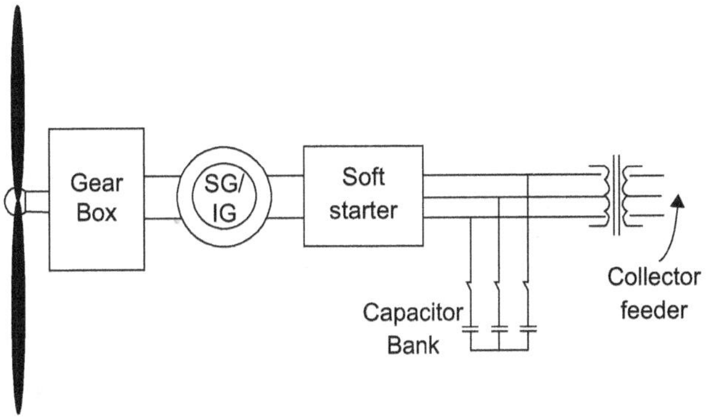

FIGURE 2.5 Fixed-speed induction generator.

2.2.1.2 Various Types of Wind-Turbine Generators

2.2.1.2.1 Fixed-Speed Induction Generator

This wind turbine concept (Figure 2.5) is based on the principle of a squirrel-cage induction generator [44]. Turbine speed is kept fixed with electrical grid frequency. While generating active power, the induction generator will absorb VARs (volt-ampere reactives, a measurement of reactive power).

2.2.1.2.2 Variable-Speed Induction Generator

These are very similar to a fixed-speed induction generator. A variable-speed induction generator (Figure 2.6) has variable resistance in the rotor circuit. The stator circuit is same as the fixed-speed type of induction generator.

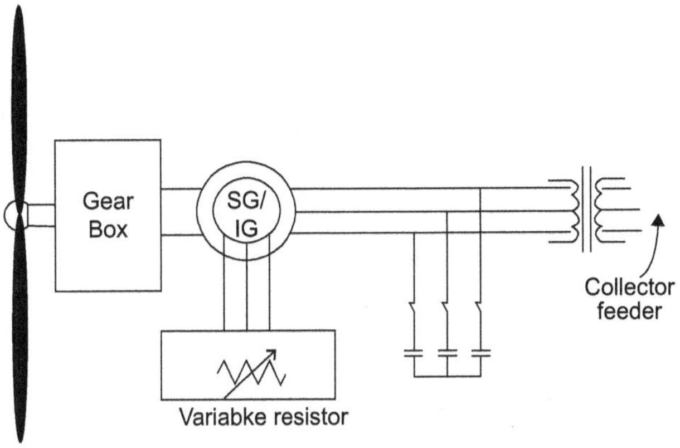

FIGURE 2.6 Variable-speed induction generator

2.2.1.2.3 DFIG (Doubly Fed Induction Generator)

These are the combination of fixed-speed and variable-speed types of induction generator. In the rotor circuit, AC excitation for variable frequency is added. A small current is incorporated in the rotor circuit so that a large control of power is achieved in the stator circuit. DFIG (Figure 2.7) enables fast recovery of voltage.

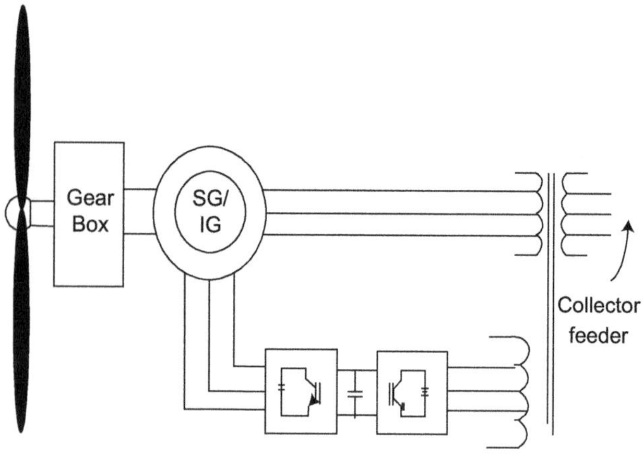

FIGURE 2.7 Doubly fed induction generator.

2.2.1.2.4 Full Power Converter Type Wind-Turbine Generators

Back-to-back frequency converter plays a vital role in this type of generator (Figure 2.8), the generator directly coupled with the grid. All power output is obtained to the grid via converters. A generator can be a permanent magnet or a synchronous generator with a wound rotor. Wider range of reactive power, speed variation and voltage control capability is achieved.

2.2.1.3 Mathematical Formulation

As per Newton's second law of motion,

$$F = ma \tag{2.1}$$

The kinetic energy is given by

$$E = m*a*s \tag{2.2}$$

From the equation of kinetics of solid motion,

$$v^2 = u^2 + 2*a*s \tag{2.3}$$

where, u is the initial velocity of the object.

Let the initial velocity of the object be zero, then:

$$E = \frac{1}{2} m * v^2 \tag{2.4}$$

Power (P) is represented as the rate of change of kinetic energy:

$$P = \frac{dE}{dt} = \frac{1}{2}\frac{dm}{dt} v_w{}^2 \tag{2.5}$$

V_w is the velocity of wind $\frac{dm}{dt} = \rho A v_w$ and ρ is the density of air.

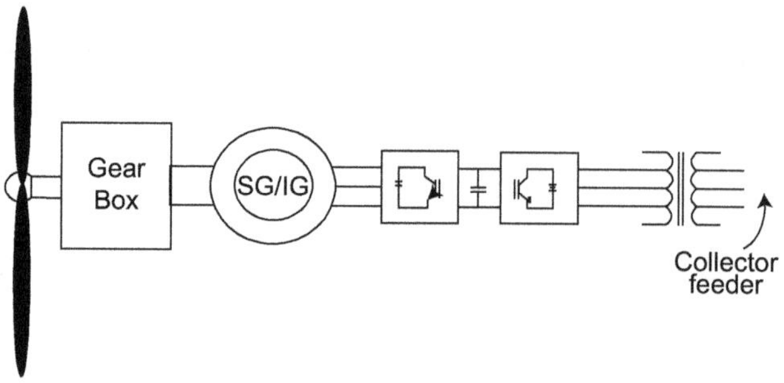

FIGURE 2.8 Full power converter type wind turbine generator.

From Eq. (2.5), power becomes

$$P = \frac{1}{2} \rho A v_w^3 \tag{2.6}$$

The actual mechanical power is determined by the difference of upstream and downstream wind power:

$$P_w = \frac{1}{2} \rho A v_w \left(v_u^2 - v_d^2\right) \tag{2.7}$$

v_u: Upstream velocity of wind of rotor blades at its entrance in m/s

v_d: Downstream velocity of wind of rotor blades at its exit in m/s

These two velocities give a boost to blade tip speed ratio. Mass flow rate can be written as:

$$\rho A v_w = \frac{\rho A \left(v_u + v_d\right)}{2} \tag{2.8}$$

Therefore, Eq. (2.7) can be written as:

$$P_w = \frac{1}{2} \rho A \left(v_u^2 - v_d^2\right)\left(\frac{v_u + v_d}{2}\right) \tag{2.9}$$

$$P_w = \frac{1}{2}\left[\rho A \left\{\frac{v_u}{2}\left(v_u{}^2 - v_d{}^2\right) + \frac{v_d}{2}\left(v_u^2 - v_d^2\right)\right\}\right] \tag{2.10}$$

$$P_w = \frac{1}{2}\left[\rho A \frac{\left\{v_u^3 - v_u v_d^2 + v_u v_d^2 - v_d^3\right\}}{2}\right] \tag{2.11}$$

$$P_w = \frac{1}{2}\left\{\rho A v_u^3 \left[\frac{1 - \left(\frac{v_d}{v_u}\right)^2 + \frac{v_d}{v_u} - \left(\frac{v_d}{v_u}\right)^3}{2}\right]\right\} \tag{2.12}$$

Alternatively, it can be written as:

$$P_w = \frac{1}{2}\rho A v_u^3 C_p \quad (2.13)$$

where, $C_p = \dfrac{1-\left(\dfrac{v_d}{v_u}\right)^2+\dfrac{v_d}{v_u}-\left(\dfrac{v_d}{v_u}\right)^3}{2}$

The value of C_p is proportional to upstream and downstream wind power, and it is sometimes called the "Betz limit" after famous scientist Albert Betz. The ratio of upstream wind speed and downstream wind speed is called its coefficient λ [45, 46]:

$$\lambda = \frac{v_d}{v_u} \quad (2.14)$$

or it can be written as;

$$\lambda = \frac{Blade\ tip\ speed}{Wind\ speed}$$

$$\lambda = \frac{\omega * R}{v} \quad (2.15)$$

The blade tip speed extracted from the rotational speed of the turbine and the blade length of turbine.

$$\text{Blade tip speed} = \frac{angular\ speed\ of\ turbine(w) * R}{wind\ speed} \quad (2.16)$$

w is angular speed of turbine measured in radian per second.

By substituting λ in equation of C_p, we get:

$$C_p = \frac{(1+\lambda)(1-\lambda^2)}{2} \quad (2.17)$$

Now to get the maximum value of C_p, differentiate it with respect to λ.

$$\frac{dC_p}{d\lambda} = \frac{(1+\lambda)(-2\lambda)+(1-\lambda^2)1}{2} \quad (2.18)$$

The values of λ are −1 and 1/3. Therefore the maximum value of C_p is 16/27.

The slow motion of the rotor causes the wind to pass through the blades opening and there is some power extraction. On the other side, the fast movement of the rotor causes the solid wall to obstruct the wind flow. Higher value of λ enables better and efficient operation of the generator, but there are the follwing drawbacks arises with higher value of λ.

1. Higher value of λ causes lower efficiency.
2. Due to its higher value, vibration and noise occurs.
3. Due to dust and sand particles, there is an erosion of the leading edge of the blade.

2.2.1.4 Analysis of Power Coefficient

For power regulation, the power coefficient plays a significant role. It is nonlinear function and is dependent on the tip ratio and blade pitch angle. Every turbine manufacturer prepares a table of C_p for proper operation. The C_p can be formulate as in terms of its speed tip ratio and blade pitch angle is given by the following formula;

$$C_p(\lambda,\theta) = C1\left(C_2\frac{1}{\beta} - C_3\beta\theta - C_4\theta^x - C_5\right)e^{-C_6\frac{1}{\beta}} \quad (2.19)$$

The value of $C_1, C_2, C_3, C_4, C_5, C_6$ and x are dependent on type of turbine. For example, for a particular turbine type, let us take different value of C_1–C_6.

β is represented in terms of λ and θ as

$$\frac{1}{\beta} = \frac{1}{\lambda + 0.08\theta} - \frac{0.035}{1+\theta^3} \quad (2.20)$$

The graph shown in Figure 2.9 is plotted between power coefficient and tip speed ratio for and angle $\theta = 0°$.

2.2.1.5 Machine Equations

For DFIG modelling, the following equations are taken into consideration.

$$v_{ds} = -R_s i_{ds} - \omega_s \phi_{qs} + \frac{d\phi_{ds}}{dt} \quad (2.21)$$

$$v_{qs} = -R_s i_{qs} + \omega_s \phi_{ds} + \frac{d\phi_{qs}}{dt} \quad (2.22)$$

$$v_{dr} = -R_r i_{dr} - s\omega_s \phi_{qr} + \frac{d\phi_{dr}}{dt} \quad (2.23)$$

TABLE 2.1
Different Values of C

C_1	C_2	C_3	C_4	C_5	C_6
0.6	118	0.4	0	5	21

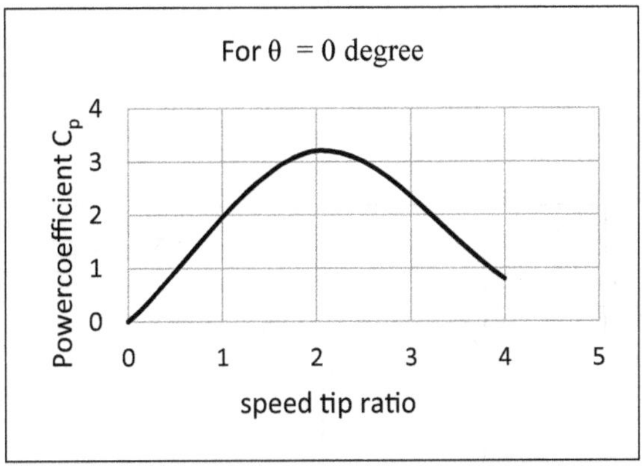

FIGURE 2.9 Power curve.

$$v_{ds} = -R_r i_{qr} + s\omega_s \phi_{dr} + \frac{d\phi_{qr}}{dt} \quad (2.24)$$

Direct and quadrature axis components have represented as d and q, respectively; s and r represent stator and rotor quantities, respectively. ψ and s are flux linkage in V_s and rotor slip, respectively; i stands for current in ampere, v stands for voltage in volts and R stands for resistance in ohms. The stator and rotor have active and reactive power, which can be calculated by following equations.

$$P_s = u_{ds} i_{ds} + u_{qs} i_{qs} \quad (2.25)$$

$$Q_s = u_{qs} i_{ds} - u_{ds} i_{qs} \quad (2.26)$$

$$P_r = u_{ds} r + u_{qr} i_{qr} \quad (2.27)$$

$$Q_r = u_{qr} i_{dr} - u_{dr} i_{qr} \quad (2.28)$$

In doubly fed Induction generator (DFIG)

$$P = P_s + P_r = u_{ds} i_{ds} + u_{qs} i_{qs} + u_{dr} i_{dr} + u_{qr} i_{qr} \quad (2.29)$$

Similarly,

$$Q = Q_s + Q_r = u_{qs} i_{ds} - u_{ds} i_{qs} + u_{qr} i_{dr} - u_{dr} i_{dr} \quad (2.30)$$

2.2.2 Modelling of Solar Cells

A photovoltaic cell consists of a semiconductor diode with p–n junction. The photovoltaic cell concept is based on the photoelectric effect. Monocrystalline and polycrystalline silicon cells have become popular for use on a commercial basis [47]. Figure 2.10 represents the equivalent circuit of a photovoltaic module without a load. The three main points of photo voltaic graph are as follows.

1) Maximum power point.
2) Open circuit (OC) voltage.
3) Short circuit (SC) current.

The relation for voltage and current in solar cell is given by:

$$I = I_L - I_0\left(e^{\frac{V+R_s I}{\alpha V_T}} - 1\right) - \frac{V + IR_s}{R_{sh}} \quad (2.31)$$

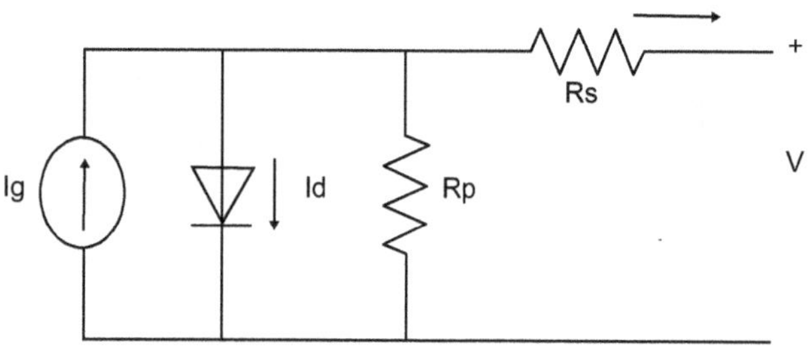

FIGURE 2.10 Single diode model of theoretical PV cell.

The photocurrent can simply be calculated using Eq. (2.31). This equation is a function of temperature and irradiance values [48].

$$I_L = \frac{I_{scref}}{G_{Tcref}} G_{Tc} - \alpha(T_c - T_{cref}) \quad (2.32)$$

I_L: Photocurrent (A)	*T_c: Condition Temperature*
I_o: Saturation current of diode (A)	*V_T: Thermal voltage equivalent*
R_s: Series Resistor (Ω)	*C_{TC}: Irradiance (W/m^2)*
R_{sh}: Shunt Resistor (Ω)	*G_{Tcref}: Reference*
α: Ideality factor of diode	*I_{scref}: Short circuit current for reference condition*
a: Curve fitting parameter	*T_{cref}: Reference Temperature*

The diode current I_0 can be calculated by the following formula:

$$I_o = \frac{(R_{sh} + R_s)I_{sc} - V_{oc}}{R_{sh} e^{(\frac{V_{oc}}{\alpha V_T})}} \quad (2.33)$$

$$V_T = \frac{n_s kT}{q} \quad (2.34)$$

The multiplication of operating voltage and output current represents output power. This output power is dependent on irradiance and temperature.

2.2.3 Modelling of SHP (Small Hydropower)

The power derived from water current is the multiplication of head and flow rate [49]. The formula is given by:

$$P_{hyd} = \rho g Q H \quad (2.35)$$

P_{hyd} is the mechanical power developed at turbine shaft. ρ is water density in kg/m^3. The value of ρ is 1000kg/m^3. g represents gravity and has value of 9.82 m/s^2. Q represents water flow rate in m/s^3 and H represents effective pressure head in metres.

The waterpower is converted into mechanical shaft power by hydro turbine. Relation can give the mechanical and hydraulic power:

$$P_n = n_h P_{Hyd} \quad (2.36)$$

$$Q = Av \quad (2.37)$$

The Bernoulli's equation is given by

$$\frac{v^2}{2g} + h + \frac{p}{\rho g} = \frac{P_{hyd}}{\rho g Q} \quad (2.38)$$

v represents water flow speed in m/s, A and p are the area of cross section in m^2 and pressure of water in m/s^2, respectively.

2.3 METHODOLOGY

A capacity of 1.6 MW (DFIG) wind-based turbine has to develop by using simulation tool. A method of parameter identification is used in developing the turbine. The modelling error is reduced by comparing the utilizing factor of wind energy listed in table of DIgSILENT software with the actual wind turbine generator, the complete paper is divided into the following two sections.

2.3.1 Methodology for Modelling of Wind Turbine Using Parameter Identification in DIgSILENT Simulation Software

The modelling of the wind turbine has been done in DIgSILENT software. The primary function of the model is to identify component and constructing a new utilization factor for wind energy. DIgSILENT has its own model of DFIG-based wind turbine. An actual data of wind farm is taken for consideration. The parameter value of actual generator differs from the parameter value of simulated module. Therefore, both will have different output characteristics. The characteristics will be in the form of output voltage (V), current (I), active (MW) and reactive power (MVAR). We have to check the consistency of output characteristics for both the simulation model and actual data. For balancing the output characteristics of simulation and actual generator, parameter identification is required. The process of parameter identification has explained in Figure 2.12. In Figure 2.11, there are three modules are used, i.e. measurement file module, component module, comparator module and optimizer module. The comparator module will find the error between actual data and simulation data, and it will be the objective function. In optimizer module, all the objective function will be summed up here.

The objective function is given as:

$$F = \sum_{i=1}^{n} \left[(M_i - S_i)\omega_i \right]^P \qquad (2.39)$$

F represents the objective function and the objective to minimize the error. M_i and S_i are the measured data and simulation data. ω_i is the weighting factor and it can be changed as per the situation. The value of p ranges from 1–10. Here, the value of p is 2. The process of identification of parameter starts with the simulation and the measured data is transferred at the comparator module and power data at simulation output. Now the desired function is calculated. In the next step, the controller parameter is modified until the minimized function is obtained. Our requirement is to minimize the function F. The error is less between the actual data and simulation data. The reading of actual active power of generator is taken by measurement file module. It is then passed to the comparator module block, and in the last process, power measurement module deliver data to comparator module and the objective function is determined. The parameter identification module can understand by Figure 2.12. The default value of the software is considered as an initial value to perform parameter identification. Table 2.2 shows the initial values. In DIgSILENT, the input data has to be saved in text documents. By using measurement file module, the data can be read out in the simulation model. After this, parameter identification is achieved. The parameter identification module will read the data of wind speed and the simulation programme can run successfully, showing parameters modified.

2.3.2 Methodology for Analysis of Transient Stability of Hybrid-Distributed Generation

The methodology for study and analysis of transient stability for hybrid-distributed generation is dependent on the following modes of operation:

1. **Import mode of operation:** Apart from the power generated, the grid will also supply active and reactive power.
2. **Export mode of operation:** Power, in excess amount, is transferred or export to the grid.
3. **Load = generation mode of operation:** The generated power of hybrid-distributed generation is sufficient to meet the load demand. The penetration level is defined as:

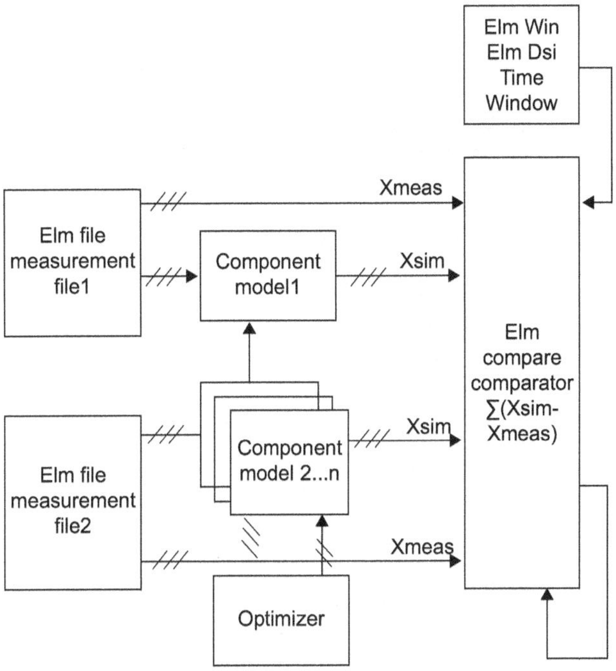

FIGURE 2.11 Parameter identification process.

$$\% P_{LDG} = \frac{P_{DG}}{P_{DG} + P_{CG}} * 100 \qquad (2.40)$$

where, P_{LDG} represents percentage of penetration level, P_{DG} represents the power generated by two DG, and P_{CG} represents grid generated power.

The simulation setup of the problem has been modelled in the DIgSILENT PowerFactory. A single line diagram of the system is shown in Figure 2.13. The modelling is done with the single line diagram, which is having a synchronous generator of capacity 80 MW and 40 MVAR. A transformer is connected between bus numbers 1 and 2. Two loads of 80 MW and 40 MVAR are present. The grid voltage is maintained at 230 KV. The lines 1 and 2 are 630 km in length, and line 3 is 330 km. The doubly fed induction generator has rating of 8 MW, 4 MVAR, 0.8 pf lagging. The modelling of solar PV is done in DIgSILENT software as a DC current source with 8 MW of real power, and there is no reactive power; this integration is done through pulse width modulation (PWM) inverter. The hydropower has capacity of 8 MW and 4 MVAR with active and reactive power, respectively. The three-phase short circuit fault takes place at line 2, and by disconnecting the line, the issue of fault is resolved after 200 ms. The system possess the following order of penetration.

1. Import mode (P_{LDG} = 40%).
2. Export mode (P_{LDG} =120%).
3. Load = generation (P_{LDG} = 50%).

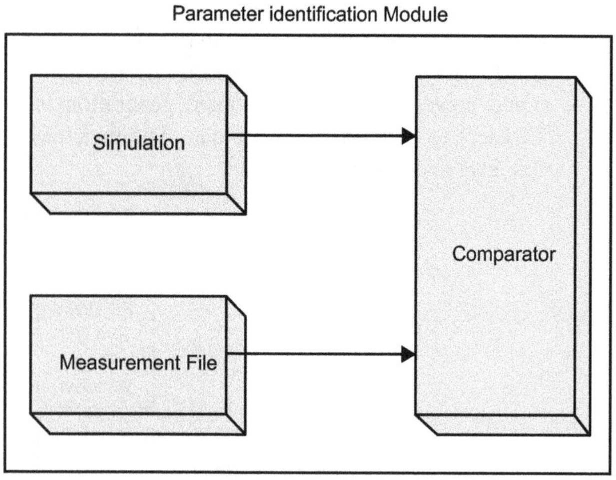

FIGURE 2.12 Parameter identification module.

2.4 RESULT AND DISCUSSION

Parameter identification method has successfully applied in DIgSILENT PowerFactory simulation tool. For processing sample data, measurement data is required. The input data should be saved as text documents. By using measurement file module, data can be read in to simulation model. For one second of interval data, the generated power output waveform and wind speed will show consistency for about ten seconds. The timeline is compressed and the interval is set at ten seconds. After modifying the parameter, the identification module read the data and perform simulation. Table 2.3 shows identified parameters. Figure 2.19 compares the results of waveform after the parameters are identified.

Figure 2.14, Figure 2.15, Figure 2.16 and Figure 2.17 show that two sets of wind data are taken, i.e. the wind speed (dynamic) and wind speed (steady). This data is sufficient to check the performance of the parameter. In practical operation, doubly fed induction generator–based turbine has some losses, which were not taken in simulation model. In Figure 2.19, the green-coloured waveform is showing actual active power and red-coloured waveform is showing the output active power. Power variation is almost same in both active powers. After getting simulation results, it is understood that both actual and simulated results have same nature of tendency.

Table 2.3 has shown the critical clearing time for solar. By comparing Table 2.4 with Table 2.5, Table 2.6, and Table 2.7, it is seen that penetration increases as critical clearing time drops. By a comparing study of Table 2.6 and Table 2.7 for export mode of operation, it is seen that integration of two distributed generations performs better than three distributed generations. By comparing Table 2.6 and Table 2.7 for export mode column, the critical clearing time is greater in Table 2.6, i.e. 330 ms. But for import mode, three

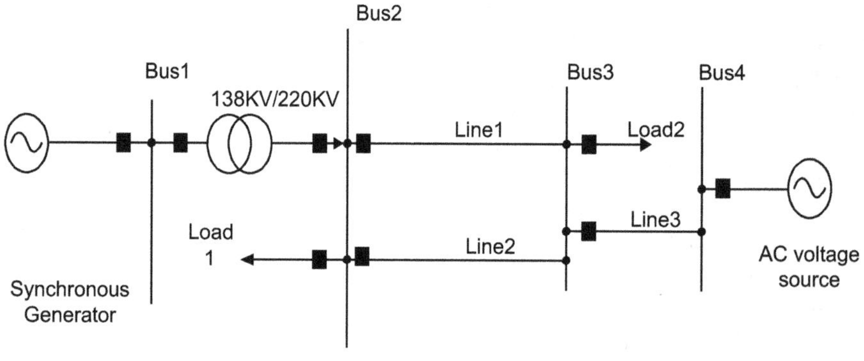

FIGURE 2.13 DIgSILENT simulation setup.

TABLE 2.2
Initial Parameters of the Controller

Power Controller Parameter

T_{tr}	K_p	T_p	K_q	T_q
0.0012	1	0.1	1	0.1

Parameter for Current Controller

K_q	T_q	T_d
0.2	0.01	0.01

Parameter for Shaft Module

J-turb	*K-shaft*	*D-shaft*
6200000	8300000	1400000

TABLE 2.3
Parameter after Identification

Power Controller Parameter

T_{tr}	K_p	T_p	K_q	T_q
0.1	58.7043	0.00756	0.07041	1.3103

Current Controller Parameter

K_q	T_q	T_d
0.0822	0.1542	0.00975

Shaft Module Parameter

J_turb	*K_shaft*	*D_shaft*
1202228	145667228	1578965

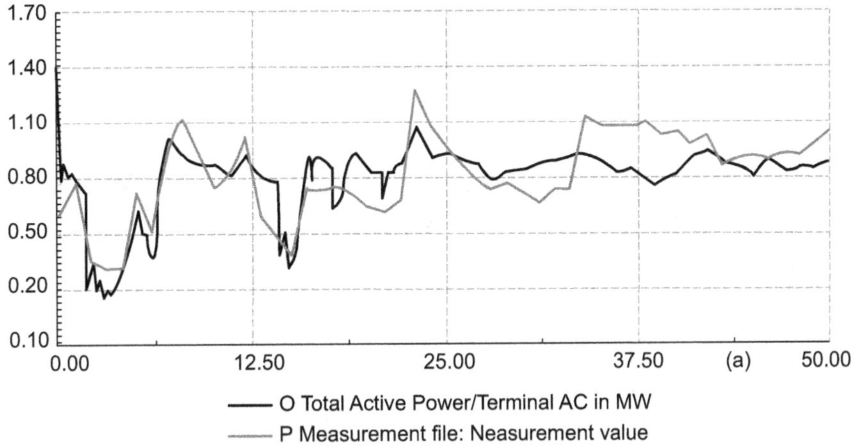

FIGURE 2.14 Comparison analysis of simulation results after parameter identification and actual results.

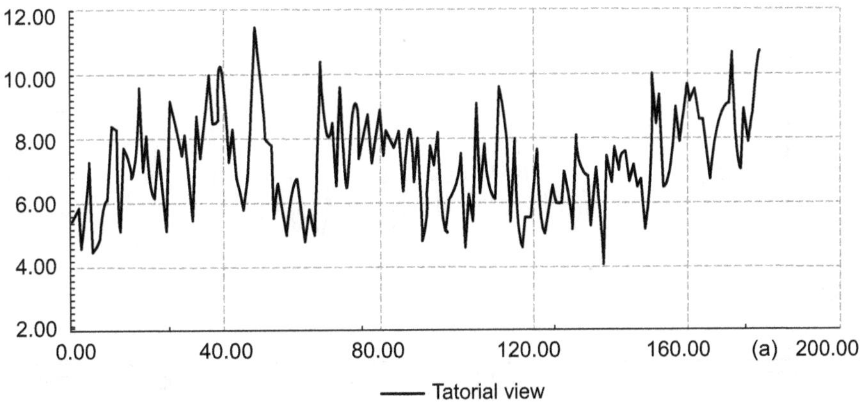

FIGURE 2.15 Dynamic wind speed.

distributed generations perform better than two distributed generation.

2.5 CONCLUSION

This chapter gives the complete analysis of modelling of wind turbine generator and transient analysis of transient or rotor angle stability in hybrid-distributed generation. A capacity of 1.6 MW is successfully modelled using DIgSILENT PowerFactory software. A new utilization factor table is reformed and model parameters are introduced. A set of parameters for turbine is successfully identified. On the other hand, a hybrid-distributed generation has taken, and transient stability analysis is considered. For import, export and load equals generation mode, different rotor angle analysis waveforms are obtained using the same

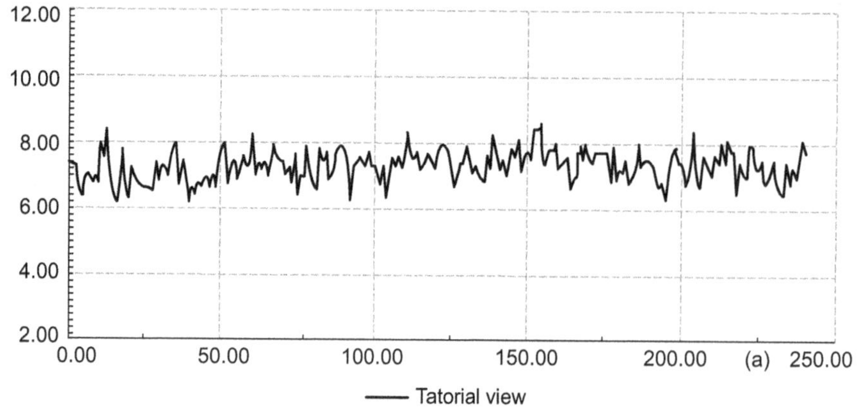

FIGURE 2.16 Steady wind speed.

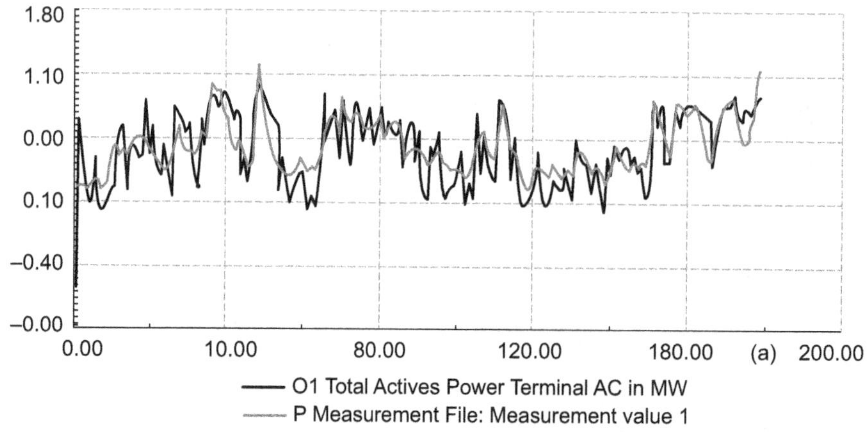

FIGURE 2.17 Comparison analysis of output power and actual output power (during steady wind).

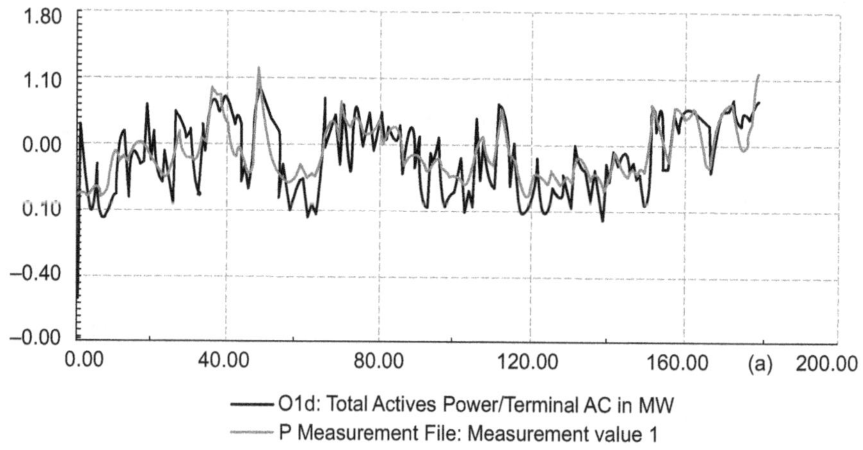

FIGURE 2.18 Comparison analysis of simulation power output and actual power output during dynamic wind speed.

software. By using critical clearing time, the integration of hybrid-distributed generation has applied on single machine infinite system. The system has analysed by critical clearing time and duration of oscillation. The waveforms of the system shows that the stress of the system is more as we increase the number of generators. The various modes of operation like import export and load equals generation mode are analysed. Therefore, based on the above discussion, the following can be concluded.

- The doubly fed induction generator has been modelled in DIgSILENT software.

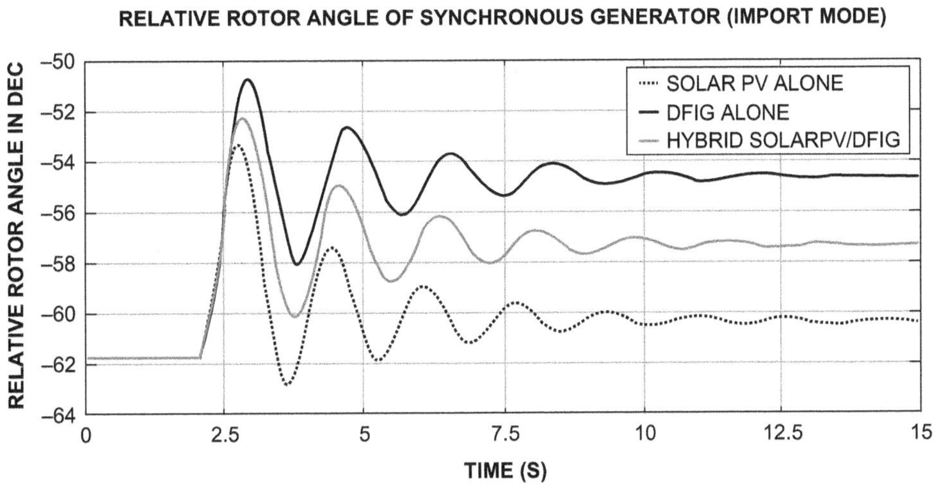

FIGURE 2.19 Rotor angle comparison between solar, DFIG and hybrid solar/DFIG (import mode of operation).

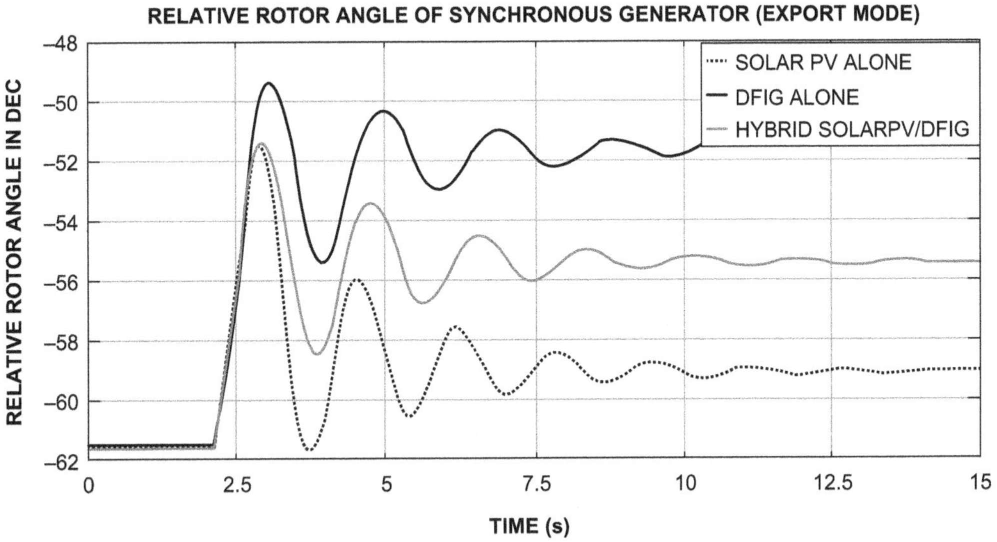

FIGURE 2.20 Rotor angle comparison between solar, DFIG and hybrid solar/DFIG (export mode of operation).

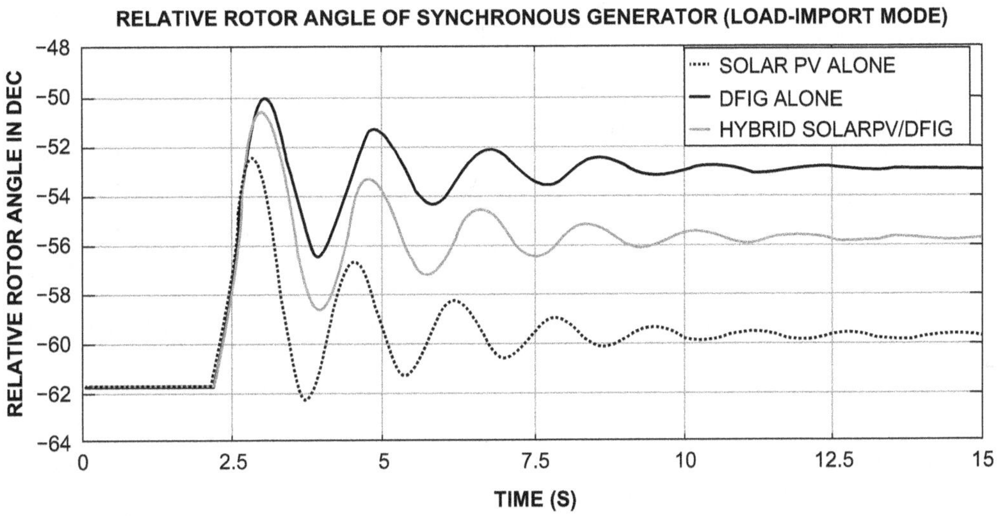

FIGURE 2.21 Rotor angle comparison between solar, DFIG and hybrid solar/DFIG (load = generation mode of operation).

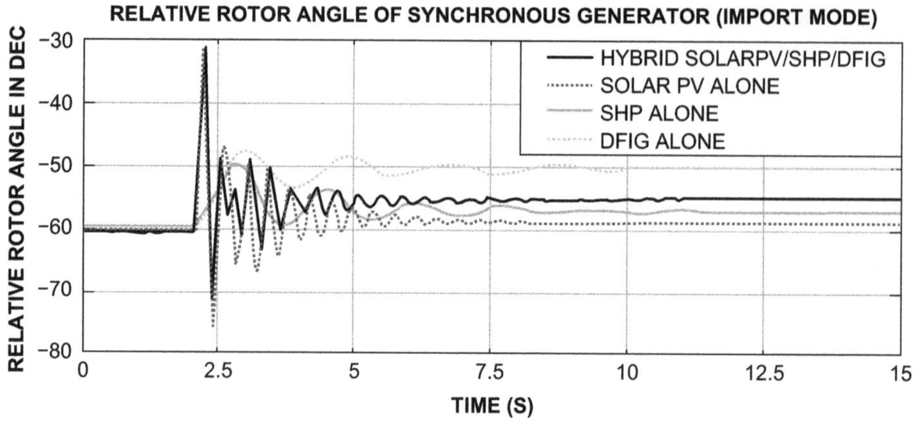

FIGURE 2.22 Rotor angle comparison analysis between solar, DFIG, (SHP) small hydropower and hybrid solar/DFIG/SHP for import mode of operation.

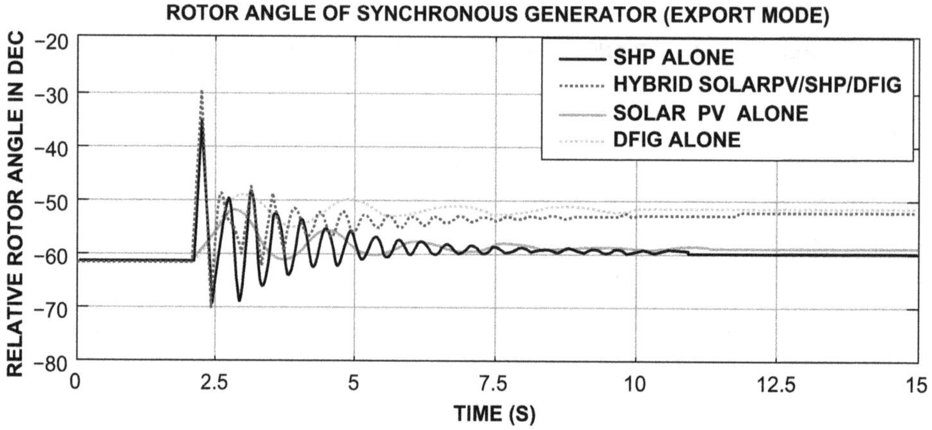

FIGURE 2.23 Rotor angle comparison analysis between solar, DFIG, (SHP) small hydropower and hybrid solar/DFIG/SHP for export mode of operation.

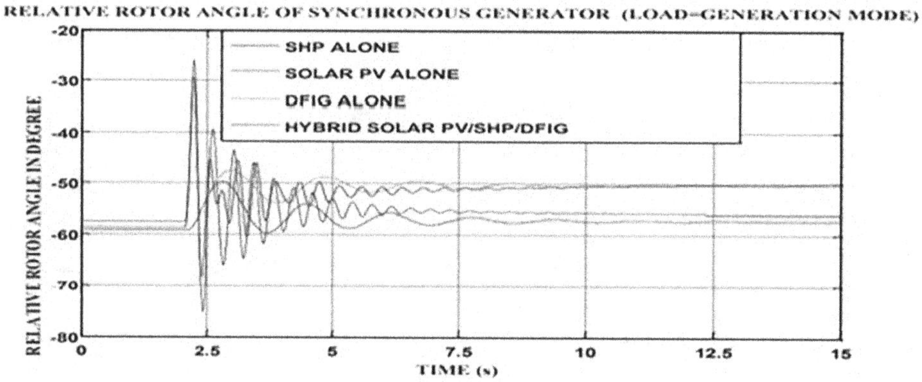

FIGURE 2.24 Rotor angle comparison analysis between solar, DFIG, (SHP) small hydropower and hybrid solar/DFIG/SHP for load = generation mode of operation.

- A wind utilization factor table is formed for the parameter identification method.
- Rotor angle stability is analysed for different modes of operation: import mode, export mode and load equals generation mode.

2.6 NUMERICALS

Q1. If the wind speed is 20 m/s and blade length is 45 m, calculate the power in the wind (air density = 1.23kg/m^3).

Solution: Given that wind speed (v) = 20m/s,

TABLE 2.4
Table for Critical Clearing Time for Synchronous Generator

Fault Location (%)	Fault Location (km)	Import (ms)	Export (ms)	Load = Generation (ms)
0	0	340	330	335
20	124	440	418	430
40	248	540	500	520
60	372	601	550	570
80	437	560	500	530
100	620	430	400	430

TABLE 2.5
Table for Critical Clearing Time of Synchronous Generator along with Hybrid DG (Solar/Small Hydropower [SHP])

Fault Location (%)	Fault Location (km)	Import (ms)	Export (ms)	Load = Generation (ms)
0	0	350	340	430
20	124	450	430	440
40	248	640	590	600
60	372	800	690	730
80	436	770	670	700
100	620	560	510	530

TABLE 2.6
Table for Critical Clearing Time of Synchronous Generator along with Hybrid DG (Solar/DFIG)

Fault Location (%)	Fault Location (km)	Import (ms)	Export (ms)	Load = Generation (ms)
0	0	340	330	330
20	124	430	420	420
40	248	530	500	510
60	372	580	550	560
80	436	550	510	520
100	620	430	420	410

TABLE 2.7
Table for Critical Clearing Time of Synchronous Generator along with Hybrid DG (Solar/SHP/DFIG)

Fault Location (%)	Fault Location (km)	Import (ms)	Export (ms)	Load = Generation (ms)
0	0	340	320	330
20	124	470	400	420
40	248	630	480	510
60	372	790	510	550
80	436	750	480	520
100	620	550	390	410

Blade length (r) = 45m
Wind power is given by Eq. (2.6)

$$P = \frac{1}{2}\rho A v^3$$

where, $A = \pi r^2$; ρ = air density; v = wind speed

$$A = 3.14 * (45)^2 = 6358.5\text{m}^2$$

$$P = \frac{1}{2} * 1.23 * 6358.5 * 20^3$$

$$P = 31.2 \text{ MW}$$

Q2. A wind turbine has a wind speed of 20m/s and blade has a length of 25m on a day having air density of 1.335kg/m³. Calculate the power produced by wind turbine if it has 35% efficiency.

Solution. Given that wind speed (v) = 20m/s,

Length of blade = 25m
Efficiency (P_{max}) = 35%

As we know that $P_{max=} \frac{1}{2}\rho A v^3$

Therefore, $P_{max} = \frac{1}{2} * 1.335 * \pi (25)^2 * (20)^3$ W

P_{max} = 10.47 MW

As $efficiency(\eta) = \frac{P_e}{P_{max}}$

Therefore; $P_e = \eta * P_{max}$

$P_e = 0.3 * P_{max} = 3.14$ MW

PRACTICE PROBLEMS

Q1. A windmill has cross sectional area and wind speed are 25m² and 6m/s. What will be the power generated by the windmill in the Betz limit?

(Answer: 2.064 KW)

Q2. The following data is given:

Density of water	996 kg/m³
Efficiency of water pump	0.6
Efficiency of transmission from rotor to pump	0.9
Tip speed ratio (λ)	0.75
Density of air	1.2 kg/m³
Power coefficient (C_p)	0.31

Calculate the rotor radius and rotor speed for multi-blade wind machine operating at a design speed of 25km/hr. The machine operates a water pump having capacity of 5.1m³/hr and lift of 9m.

(Answer: rotor radius = 1.087m, rotor speed = 45.74rpm)

Q3. At a hydroelectric power plant, the water flow is $100m^3s^{-1}$ and water pressure head is at a height at 300m. If the turbine generator efficiency is 60%, find the electric power available from the plant ($g = 9.8ms^{-2}$).

(Ans: 176.4 MW)

REFERENCES

[1] Wang, C., M. H. Nehrir, Analytical approaches for optimal placement of distributed generation sources in Power Systems. *IEEE Transactions on Power Systems*, vol. 19, no. 4, pp. 2068–2076, Nov. 2004. https://doi.org/10.1109/TPWRS.2004.836189.

[2] Celli, G., E. Ghaiani, S. Mocci, F. Pilo, A multiobjective evolutionary algorithm for the sizing and sitting of distributed generation. *IEEE Transactions on Power Systems*, vol. 20, no. 2, pp. 750–757, May 2005. https://doi.org/10.1109/TPWRS.2005.846219.

[3] Li, Weixing, Peng Wang, Zhimin Li, Yingchun Liu, Reliability evaluation of complex radial distribution systems considering restoration sequence and network constraints. *IEEE Transactions on Power Delivery*, vol. 19, no. 2, Apr. 2004. https://doi.org/10.1109/TPWRD.2003.822960.

[4] Arya, A., S. Verma, S. Mehroliya, S. Tomar, C. S. Rajeshwari, Optimal placement of distributed generators in power system using sensitivity analysis. In: Bansal R. C., Agarwal A., Jadoun V. K. (eds) *Advances in Energy Technology. Lecture Notes in Electrical Engineering*, vol. 766. Springer, Singapore, 2022. https://doi.org/10.1007/978-981-16-1476-7_67

[5] Deshmukh, M. K., S. S. Deshmukh, Modeling of hybrid renewable energy systems. *Renewable and Sustainable Energy Reviews*, vol. 12, no. 1, pp. 235–249, 2008. https://doi.org/10.1016/j.rser.2006.07.011.

[6] Sainz, E., A. Llombart, J. J. Guerrero, Robust filtering for the characterization of wind turbines: improving its operation and maintenance. *Energy Conversion and Management*, vol. 50, no. 9, pp. 2136–2147, 2009. https://doi.org/10.1016/j.enconman.2009.04.036.

[7] Villanueva, D., A. E. Feijóo, Reformulation of parameters of the logistic function applied to power curves of wind turbines. *Electric Power Systems Research*, vol. 137, pp. 51–58, 2016. http://doi.org/10.1016/j.epsr.2016.03.045.

[8] Yang, H., L. Lu, W. Zhou, A novel optimization sizing model for hybrid solar-wind power generation system. *Solar Energy*, vol. 81, no. 1, pp. 76–84, 2007. https://doi.org/10.1016/j.solener.2006.06.010.

[9] Lydia, M., S. Suresh Kumar, A. Immanuel Selvakumar, G. Edwin Prem Kumar, Wind resource estimation using wind speed and power curve models. *Renewable Energy*, vol. 83, pp. 425–434, 2015. http://doi.org/10.1016/j.renene.2015.04.045.

[10] Hansen, A. D., C. Jauch, P. Sørensen, F. Iov, F. Blaabjerg, *Dynamic wind turbine models in power system simulation tool DIgSILENT.* Risø National Laboratory, Roskilde, 2003. www.osti.gov/etdeweb/servlets/purl/20437623.

[11] Jain, R., A. Arya, A comprehensive review on micro grid operation, challenges and control strategies. *Proceedings of the 2015 ACM Sixth International Conference on Future Energy Systems*, Bangalore, pp. 295–300, 2015. http://doi.org/10.1145/2768510.2768514.

[12] Atwa, M., E. F. El-Saadany, M. M. A. Salama, R. Seethapathy, M. Assam, S. Conti, Adequacy evaluation of distribution system including wind/solar DG during different modes of operation. *IEEE Transactions on Power Systems*, vol. 26, no. 4, pp. 1945–1952, 2011. https://doi.org/10.1109/TPWRS.2011.2112783.

[13] Akhmatov, V., Variable-speed wind turbines with doubly-fed induction generators: Part I: Modelling in dynamic simulation tools. *Wind Eng*, vol. 26(2), pp. 85–108, 2002. doi:10.1260/030952402761699278.

[14] Lei, Y., A. Mullane, G. Lightbody, R. Yacamini, Modelling of the wind turbine with a doubly fed induction generator for grid integration studies. *IEEE Trans. Energy Convers*, vol. 21, pp. 257–264, 2006. https://doi.org/10.1109/TEC.2005.847958.

[15] Central Electricity Authority, *Load Generation and Balance Report.* Ministry of Power, Government of India, 2019–20. https://cea.nic.in/l-g-b-r-report.

[16] Poller, M., Doubly fed induction machine models for stability assessment of wind farms. *Power Tech. Conf.,* Bologna, 2003. https://doi.org/10.1109/PTC.2003.1304462.

[17] Anderson, P. M., B. Anjan, Stability simulation of wind turbine systems. *IEEE Transactions on Power Operators and Systems*, vol. 12, pp. 3791–3795, 1983. https://doi.org/10.1109/TPAS.1983.317873.

[18] Mukund, R. P., *Wind and Solar Power Systems.* CRC Press, London, 1999. www.routledge.com/Wind-and-Solar-Power-Systems-Design-Analysis-and-Operation-Second-Edition/Patel/p/book/9780849315701.

[19] Gautam, D., V. Vittal, T. Harbour, Impact of increased penetration of DFIG-based wind turbine generators on transient and small signal stability of power systems. *IEEE Transactions on Power Systems*, vol. 24, no. 3, pp. 1426–1434, 2009. https://doi.org/10.1109/TPWRS.2009.2021234.

[20] Mehta, B., P. Bhatt, V. Pandya, Small signal stability analysis of power systems with DFIG based wind power penetration. *International Journal of Electrical Power & Energy Systems*, vol. 58, pp. 64–74, 2014. https://doi.org/10.1016/j.ijepes.2014.01.005.

[21] Alejandro, R., L. Alvaro, G. V. Azquez, D. Aguilar, G. Azevedo, Modeling of a variable wind speed turbine with a permanent magnet synchronous generator. *IEEE Intenational Symposium on Industrial Electronics (ISIE)*, pp. 734–739, 2009. https://doi.org/10.1109/ISIE.2009.5218120.

[22] Singh, P., P. Paliwal, A. Arya, A review on challenges and techniques for secondary control of microgrid. *IOP Conf. Series: Materials Science and Engineering*, vol. 561, pp. 1–13, 2019. doi:10.1088/1757-899X/561/1/012075.

[23] Insu Kim, K., R. Regassa, R. G. Harley, The modelling of distribution feeders enhanced by distributed generation in DIgSILENT. *IEEE 42nd Conference on Photovoltaic Specialists (PVSC)*, pp. 1–5. doi:10.1109/PVSC.2015.7356246.

[24] Angulo, N., J. F. Medina, J. Cidrás, C. Carrillo, A. Feijóo, Analysis of the behaviour of asynchronous wind turbines under network frequency variation. *European Union Wind Energy Conference*, 2003. http://jcidras.webs.uvigo.es/publicaciones/2003_EUWEC_FreqVariations.pdf

[25] Rui-min, Z., L. Jian-hua, L. Zuo-hong, Modeling of large-scale wind farms in the probabilistic power flow analysis considering wake effect. *Journal of Xi'an Jiao Tong University*, vol. 12, no. 42, pp. 1515–1520, 2008.

[26] Vidyanandan, K., N. Senroy, Primary frequency regulation by deloaded wind turbines using variable droop. *IEEE Transactions on Power Systems*, vol. 28, no. 2, pp. 837–846, 2012. https://doi.org/10.1109/TPWRS.2012.2208233.

[27] Akhmatov, V., An aggregated model of a large wind-farm with variable speed wind turbines equipped with doubly-fed induction generators. *Wind Engineering*, vol. 4, no. 28, pp. 479–488, 2004. https://doi.org/10.1260%2F0309524042886423.

[28] Li, P., W. Hu, R. Hu, Q. Huang, J. Yao, Z. Chen, Strategy for wind power plant contribution to frequency control under variable wind speed. *Renewable Energy*, vol. 130, pp. 1226–1236, 2019. https://doi.org/10.1016/j.renene.2017.12.046.

[29] Mehroliya, S., A. Arya, U. Mitra, P. Paliwal, P. Mundra, Comparative analysis of conventional technologies and emerging trends in wind turbine generator. *2021 IEEE 2nd International Conference on Electrical Power and Energy Systems (ICEPES)*, Bhopal, pp. 1–6, 2021. doi:10.1109/ICEPES52894.2021.9699538.

[30] Döşoğlu, M. K., A new approach for low voltage ride through capability in DFIG based wind farm. *International Journal of Electrical Power & Energy Systems*, vol. 83, pp. 251–258, 2016. https://doi.org/10.1016/j.ijepes.2016.04.027.

[31] Liang, J., W. Qiao, R. G. Harley, Direct transient control of wind turbine driven DFIG for low voltage ride-through. *IEEE Power Electronics and Machines in Wind Applications*, vol. 3, pp. 1–7, 2009. https://doi.org/10.1109/PEMWA.2009.5208403.

[32] Kumar, M., S. K. Morla, R. N. Mahanty, Modeling and simulation of a Micro-grid connected with PV solar cell & its protection strategy. *4th International Conference on Recent Trends on Electronics, Information, Communication & Technology (RTEICT)*, pp. 146–150, 2019. http://doi.org/10.1109/RTEICT46194.2019.9016924.

[32] Krpan, M., I. Kuzle, Dynamic characteristics of virtual inertial response provision by DFIG-based wind turbines. *Electric Power Systems Research*, vol. 178, p. 106005, 2020. https://doi.org/10.1016/j.epsr.2019.106005.

[33] Krupke, C., J. Wang, J. Clarke, X. Luo, Modeling and experimental study of a wind turbine system in hybrid connection with compressed air energy storage. *IEEE Transactions on Energy Conversion*, vol. 32, no. 1, pp. 137–145, 2016. https://doi.org/10.1109/TEC.2016.2594285.

[34] Margaris, I. D., S. A. Papathanassiou, N. D. Hatziargyriou, A. D. Hansen, P. Sorensen, Frequency control in autonomous power systems with high wind power penetration. *IEEE Transactions on Sustainable Energy*, vol. 3, no. 2, pp. 189–199, 2012. https://doi.org/10.1109/TSTE.2011.2174660.

[35] Kumar, R. Senthil, N. Puja Priyadharshini, E. Natarajan, Experimental and numerical analysis of photovoltaic solar panel using thermoelectric cooling. *Indian Journal of Science and Technology*, vol. 8, no. 36, pp. 1–9, 2015. http://doi.org/10.17485/ijst/2015/v8i36/87646.

[36] Kim, M. K., Optimal control and operation strategy for wind turbines contributing to grid primary frequency regulation. *Applied Sciences*, vol. 7, no. 9, p. 927, 2017. http://doi.org/10.3390/app7090927.

[37] Maina, D. K., M. J. Sanjari, N. C. Nair, Voltage and frequency response of small hydro power plant in grid connected and islanded mode. *Australasian Universities Power Engineering Conference (AUPEC)*, pp. 1–7, 2018. doi:10.1109/AUPEC.2018.8757944.

[38] Wijesinghe, A., L. L. Lai, Small hydro power plant analysis and development, *2011 4th International Conference on Electric Utility Deregulation and Restructuring and Power Technologies (DRPT)*, pp. 25–30, 2011. doi:10.1109/DRPT.2011.5993857.

[39] DIgSILENT GmbH, *Manuals, Version 14.0*, DIgSILENT PowerFactory, 2008. www.digsilent.me/dme/attachment.php?attachmentid=288&d=1355070647.

[40] DIgSILENT GmbH, *DigSILENT PowerFactory V13—User Manual*, 2002. www.digsilent.de/en/downloads.html.

[41] Kong, S., R. Bansal, Z. Dong, Comparative small-signal stability analyses of PMSG-, DFIG-and SCIG-based wind farms. *International Journal of Ambient Energy*, vol. 33, no. 2, pp. 87–97, 2012. http://doi.org/10.1080/01430750.2011.640802.

[42] Lei, Y., A. Mullane, G. Lightbody, R. Yacamini, Modelling of the wind turbine with a doubly fed induction generator for grid integration studies. *IEEE Trans. Energy Converters*, vol. 21, pp. 257–264. https://doi.org/10.1109/TEC.2005.847958.

[43] Hansen, A. D., P. Sorensen, F. Iov, F. Blaabjerg, Initialisation of grid connected wind turbine models in power-system simulations. *Wind Engineering*, vol. 27, no. 1, pp. 21–38, 2003. http://doi.org/10.1260/030952403321833734.

[44] Yangfei, Z., Y. Yue, C. Xiaohu, H. Jian, W. Bowen, Analysis on wind turbine parameters identifiability. *Automation of Electric Power System*, vol. 33, pp. 86–89, 2009.

[45] Momoh, J., G. D. Boswell, Improving power grid efficiency using distributed generation. *Proceedings of the 7th International Conference on Power System Operation and Planning*, pp. 11–17, 2007. www.ijrte.org/wp-content/uploads/papers/v8i5/E6164018520.pdf.

[46] Karki, R., P. Hu, R. Billinton, Reliability evaluation considering wind and hydro power coordination. *IEEE Transactions on Power Systems*, vol. 25, no. 2, pp. 685–693, 2010. https://doi.org/10.1109/TPWRS.2009.2032758.

[47] Echiu, I., H. Camblong, G. Papia, B. Dakyo, C. Nichita, Dynamic simulation model of a hybrid power system: performance analysis. *International Journal of Automotive Technology*, vol. 7, no. 19, 2007. www.researchgate.net/publication/228731923_Dynamic_Simulation_Model_of_a_Hybrid_Power_System_Performance_Analysis.

[48] Fraile-Ardanuy, J., J. R. Wilhelmi, J. J. Fraile-Mora, and J. I. Perez, Variable speed hydro generation: Operational aspects and control. *IEEE Transactions on Energy Conversion*, vol. 21, no. 2, 2006. https://doi.org/10.1109/TEC.2005.858084.

[49] Sahoo, A. K., V. T. Chitra Kanagapriya, Load frequency control for a distributed grid system involving wind and hydro power generation. *2nd International Conference on Power, Control and Embedded Systems*, 2012. doi: https://doi.org/10.1109/ICPCES.2012.6508090.

3 Active, Passive and Hybrid Filters in DG Systems for Power Quality Improvement

Anoop Arya, Uliya Mitra, Shweta Mehroliya, Sushma Gupta, Shilpi Tomar and Karuna Nikum

CONTENTS

3.1 INTRODUCTION

Advanced technology and changing life styles have added many complications in our day-to-day routine which make our life easy and trouble free. These complications from their manufacture to use, consumes a lot of electrical energy, which increases power consumption. This could be understood by a report issued by CEA of India, which stated that the electrical energy consumption of India in 1947 was 482 GWH [1] that further increases to 1291494 GWH in 2020. The continuous increase in demand of electrical energy has made the power industry to search and move to alternative generating sources to go along with the prevailing ones. Rising energy demand has always been a major concern for the governments of the country. There are certain other factors also which prove to be add-ons: increased fuel costs, limited supply of coal and water, congestion in transmission lines and changes in climatic conditions due to pollution which increases our concerns for environmental impacts. This paved the way for the electrical systems that are smarter. This made us to opt for renewable or non-conventional energy systems. The mix of power generation due to conventional and non-conventional sources will help in achieving targets. Since the last decade, renewable energy sources are inculcated in the system tremendously. This has been appreciated by governments by providing them incentives.

Distributed generation (DG) systems mean on-site or local generation which is decentralized in form, unlike the prevailing one that is centralized. Not only generation but storage of excess electric power can also be achieved. The power flow direction in case of decentralized local generation is bidirectional, which differs from the conventional centralized generating stations which have been unidirectional. The range of DG systems varies from 1 kilowatt to 100 megawatts [2]. A lot of research work is carried out in the areas of DG technologies, their sizing and location, their stability and protection, financial and economic analyses of integration, etc. (Figure 3.1, Figure 3.2). The DG generation is also known as dispersed or embedded generation. Popular distributed generation technologies are listed as follows [3]:

- Photovoltaic systems (PV systems)
- Fuel cells
- Wind turbines
- Micro-turbines
- Natural gas or reciprocating diesel engines
- Combustion gas turbine

The various trends, challenges, benefits and applications are discussed in details in reference [4]. Apart from generating clean energy, integration of distributed power generation systems (DPGS) with the utility and power grid raises certain issues and challenges for the prevailing system. These issues are discussed briefly in the next section of this chapter. DPGS systems are always fluctuating and time varying, affecting the stability of the network. The DPGS systems must withstand any abnormal interruption. Thus, the operators of transmission systems and distribution systems, along with stakeholders, legalized certain interconnection codes that guide the operation and commissioning process of DPGS [5]. Due to these standards, there is an

DOI: 10.1201/9781003229124-3

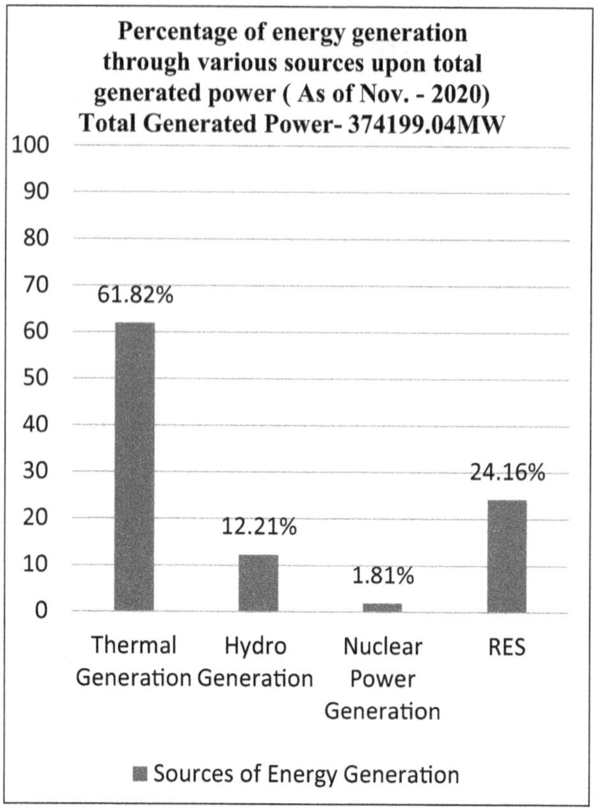

FIGURE 3.1 Graph showing percentage contribution of various methods of generation on total generated power until 30 November 2020.

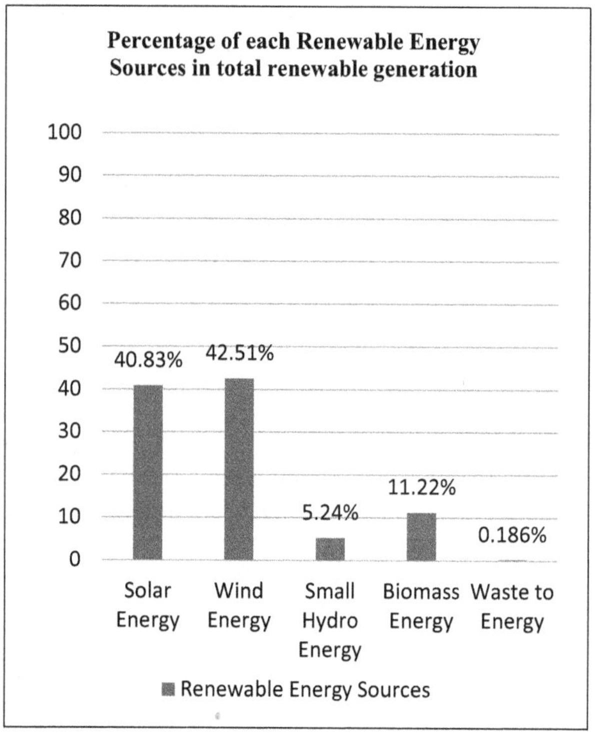

FIGURE 3.2 Graph showing percentage contribution of generation by each renewable energy source as compared to total energy generated by all renewable sources together until 30 November 2020.

increment of DPGS penetration in our prevalent system, as this are referred for planning and designing purposes of DPGS. Power electronics has been proven a boon for DPGS development. These developments have been further enhanced by advanced power converters. Classification of DG systems has been done in the following [6].

1. Micro distribution generation: ranges from 1 W to 5 KW
2. Small distributed generation: ranges from 5 KW to 5 MW
3. Medium distributed generation: ranges from 5 MW to 50 MW
4. Large distributed generation: ranges from 50 MW to 300 MW

This chapter focuses on such a DG system. The application of various filters in PV systems for reduction of harmonics are to be discussed in this chapter. But, the question arises: Why have we chosen PV system only?

India lies in the torrid zone, which is near to the equator. As a result, it is normally warm amidst a tropical monsoon climate. The annual energy received per year in India from sun is 5,000 trillion kWh. The installed solar power generation capacity in India through 31 January 2021 stands to be 38794 MW which includes rooftop installation of 4233 MW and 34561 MW of ground installation. India stands in fifth position globally in solar power capacity. India is also the head of International Solar Alliance. The photovoltaic cell technology is developing quickly, as constantly falling prices of PV modules and the continuously lowering cost of PV modules and advanced power electronics are the major reasons for inculcating a large number of PV systems. With the continuous declining cost of photovoltaic technology, makes it more competitive and increasingly adopted as compared to other non-conventional energy systems. It is due to these mentioned reasons that today we can find small PV installations in buildings, farms, factories, etc. This is increasing the future scope of photovoltaic DPGS in country. PV DPGS generates the power by conversion of energy resources from solar to electrical. The power generation is achieved by exploiting the photovoltaic effect that converts solar energy to electrical energy. Power electronics plays a key role in this energy conversion, making it more reliable and efficient. Solar energy highly depends on environmental factors. Photovoltaic converters—when properly controlled—help us to extract maximum power from photovoltaic panels in various conditions such as anti-islanding, weather conditions and grid faults.

The development of PV DPGS is rapid and is being widely used. Countries like China, the United States, Japan, Germany and India each generate a large percentage of electricity from this system. This, however, meets the huge energy demand of the world. Due to the variable and non-dispatchable nature of photovoltaic systems, the economical operation and stability of the distributed grids are affected. For safer operation of distributed grids, it is necessary that solar energy fed to PV cells and further to grids should be efficient, less harmful and more reliable, which can be achieved by some mandatory requirements. One method to perform these functions is by using smart inverters along with power converters [7]. PV DPGS have certain demands which are further sectionalized in three zones. These three zones are generator side, photovoltaic converter side and grid side. Distributed grid behaviour matches with PV DPGS power characteristics, because of which its requirements are easier to satisfy as compared to wind distributed power generation. For capturing maximum energy from the sun, there should be a control on voltage or current of PV panels. Maximum power point tracking (MPPT) control is done for every PV system, regardless of their power ratings. There is an oscillation in MPP of the PV array during whole day with an irregularity in atmospheric scenarios [8]. Therefore, it is recommended to implement MPPT technique in PV systems for their optimal utilization. For PV DPGS, the configuration is determined by power rating. Proper diagnosis, as well as monitoring at panel level, is required as they develop some defects during operation. Power quality at the grid side should be maintained satisfactorily.

Generation of power through PV DPGS is low. The energy cost is high currently, and to bring it down to a decent amount, an efficient conversion of power is required. As we know, power electronic converters play a specific role; similarly, the photovoltaic power converters have a great significance for PV DPGS. These converters are aided by the extension of total energy production and cost reduction. Although photovoltaic power converters are being placed in a small chamber, there characteristics should be more insensitive to temperature failing, which would result in increase in degradation. Thus, the temperature should be managed and check properly for this system. Proper communication and coordinated control is required for proper operation of the system. These all can be achieved by inverters also. The power produced per generating unit by wind technology is more than that of PV DPGS (this refers to power generation by a single panel or string). Thus, to increase the power output, various PV panels are connected in parallel. The centre inverters and string inverters are used for interfacing with the grid. The exclusive use of centre inverters takes place in DG systems as they are capable of extracting DC power at a low construction cost from photovoltaic panels. Nonlinearity in the system due to various loads affects efficiency and power factor; there is a reduction in both. Due to non-sinusoidal current, there is a change in impedance which is the effect of linear and non-linear charges [9]. Also due to this, non-sinusoidal currents harmonics are introduced in the system and voltage is distorted which influences both sides of the system. Here, this PV system is employed with various types of filters. With use of power filters, MPPT and the harmonic currents damping in the grid is easily done. In case of active filters, compensating currents used are being injected at each phase equal in value and opposite in sign of load distortion, resulting in harmonic cancellation [10].

3.1.1 Modelling of PV System

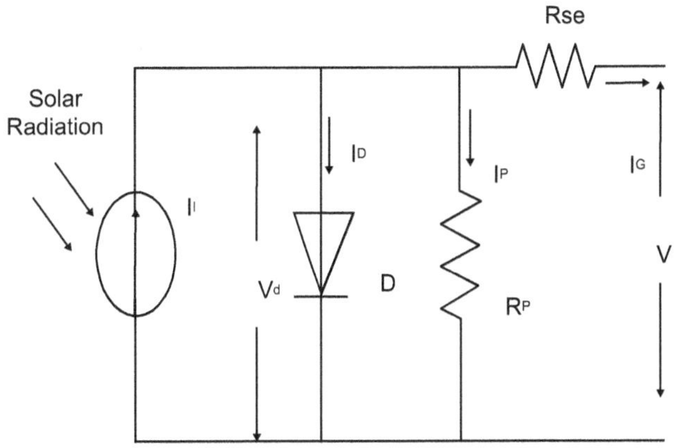

FIGURE 3.3 Photovoltaic cell (PV cell) model.

Figure 3.3 is a model of a PV cell [11]. If we apply Kirchhoff's Current Law (KCL) at the node of equivalent solar cell circuit, then we will get the value of final current generated (Figure 3.4) as:

$$I_G = I_I - I_D - I_P \tag{3.1}$$

where, I_I = Current due to solar irradiance.

I_D = Diode current dependent on temperature of solar cell (Reverse saturation current).

I_P = Current flowing through the parallel resistance of PV cell.

$$V_d = V + R_{se} I_G \tag{3.2}$$

$$I_P = \frac{V + R_{se} I_G}{R_P} \tag{3.3}$$

$$I_D = I_S \left[\exp\left(\frac{qV_d}{nKT} \right) - 1 \right] \tag{3.4}$$

n = Ideality factor of the cell (=1.92)
V_d = Diode voltage
q = Electron charge value
K = Boltzmann's Constant in Joule per degree Kelvin
T = Solar temperature measured at Kelvin scale
I_S = Value of rated short-circuit current of the solar cell

The voltage obtained at the output of PV cell is generally small. By achieving PV cells in series, we can achieve an output voltage of increased magnitude. Similarly, when PV cells have parallel connection, they enhance the current rating and also form a PV module. The voltage we obtained at the output of PV cell is due to photo current which depends on level of solar insolation.

$$V = \frac{nKT_C}{q} \ln\left(\frac{I_I + I_D - I_G}{I_D} \right) - R_{se} I_G \tag{3.5}$$

where, V = Output voltage of PV cell
I_G = Output current of cell
T_C = Temperature of cell at STC (standard test conditions) in degrees Celsius

The value of solar power (P) (in watts) obtained at any given area can be calculated by the given formula:

$$P = G * A \tag{3.6}$$

where, G = Irradiation
A = Area of the collector or object

Temperature coefficient of a solar cell (k) (in $°C / W / m^2$) is found by the formula:

$$k = \frac{NOCT - 20}{800} \tag{3.7}$$

where, NOCT = normal operating cell temperature.

The value of cell temperature of the module (T_{cell}) is given by the formula:

$$T_{cell} = T_a + K * G \tag{3.8}$$

where, T_a—Ambient temperature of cell.

3.1.2 Maximum Power Point Tracking (MPPT)

In any solar or PV cell array, a large number of solar or PV cells are connected either in a series or in a parallel manner. The array and single solar or PV cell have similar I-V (current voltage) characteristics. These characteristics depends on changes in isolation. The point at which the product of current and voltage of the array is maximized is known as maximum power point. It is desirable to operate the array at maximum power point, and thus, the tracking of maximum power point is needed [12]. This method is not always used, nor is it cost effective. The point of maximum power depends on changes in temperature and the age of the array. The analysis of this point could result in 5–20% increase in energy output annually, which is the consequence of location and design of PV cell system along with tracker's efficiency. Here, DC voltage is needed, and thus, DC/DC converters or choppers are adopted for voltage transformation of AC to DC.

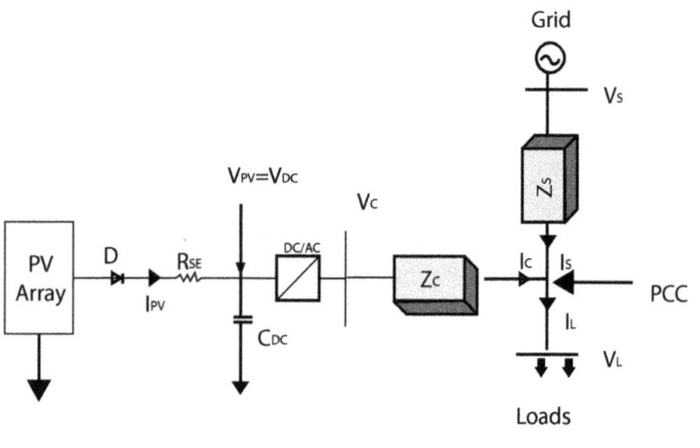

FIGURE 3.4 SLG of an interfacing of PV system with grid.

The point of maximum power in a PV or solar array is gained by increasing the output power to the load. To achieve or gain the maximum power, errors should be minimized between the reference and the operating power. The calculation of maximum power is to be done for variations in temperature and level of solar irradiation before doing its comparison with operating power. The PV DPGS system is linked with the system or utility grid by using voltage source inverters. This inverter is given a DC input of maximum power that is obtained at the PV array's output. The converters like boost/buck is used to extract maximum power available at the output of a photovoltaic array. There are various algorithms for MPPT such as:

1. P&O (perturbation and observation) method:
   a. The conventional perturbation and observation method
   b. Method of incremental conductance
2. Methods based on linearity:
   a. SCC (short-circuit current) method
   b. Method of open circuit voltage
3. Artificial Intelligence (AI) methods:
   a. FLC (fuzzy-logic control) method
   b. Neural network control

The most commonly used MPPT technique is the P&O method [13].

3.2 Issues in Distributed Generation Systems

Distributed generation systems, as we know, are the most discussed and useful of the last two or three decades. A lot of advantages of these systems—such as PV systems, wind energy conversion systems (WECS), gas turbines, Biomass, geothermal energy, etc.—have been seen. These systems offer a lot of advantages such as fulfilling energy demands, low pollution, less cost and local generation; further, they also improve power quality and voltage profiles, and corrects power factors. This decentralized generated power when injected into the grid system in a large scale introduces many problems or issues for the system. These issues are further classified as technical and non-technical issues [14]. They are further classified as described in the following subsections.

3.2.1 Technical Issues

1. The integration of distributed generation systems with grids results in affecting the power quality. The various factors affecting power quality are

also discussed in what follows. As this integration introduces harmonics in the system, there is fluctuation in the voltage and frequency of the grid.

2. Issues of power fluctuations are there in the system. These fluctuations can be for small or long time periods, depending on loads.
3. Storage of the excess energy produced more than what is required is also of major concern.
4. The major issue of all the system is protection of the system and regaining the system stability from faults.
5. The location or placement of the distributed generation systems is among the priorities.
6. Finally, islanding of the system is also to be take care of.

These mentioned issues are discussed in what follows.

3.2.1.1 Harmonics

In distributed systems like PV and WECS, various converters are used. These basically convert DC to AC and introduce harmonics into the system which are further reduced by use of filters and inverters in the systems. Also harmonics are caused because of components from additional frequency signals present in current or voltage mains. This further becomes the integral multiples of main frequency. The use of various non-linear loads also introduces harmonics in the system. This results in harmonic distortion. The reduction of harmonics with the use of filters is discussed in this chapter. Harmonics basically are maximum in the system during starting and stopping times [15]. The levels of harmonic distortion are described and analysed in terms of magnitude and phase angles of every individual harmonic of harmonic spectrum. Total harmonic distortion (THD) is a single quantity used for the measuring harmonic distortions effectively.

3.2.1.2 Voltage Related Issues

1. **Voltage flickering:** It is seen that when voltage fluctuates, there is a decrease in a lamp's brightness. In DG systems, this occurs due to fast changes in output [16]. It is one of the basic problem arises by integrating DG systems with the grid, which results in improper function and reduced usage time of the equipment as stated by Jain et al. (2008). This issue could be resolved by use of voltage regulators, by injection of reactive powers and by controlling the loads connected to the network.
2. **Unbalanced voltage:** Unbalance in voltage occurs due to the addition of various renewable systems of single phase in nature with the grid [17]. When these decentralized DPGS are added in the network in large numbers, the profile of voltage becomes unstable as the flow of power is bidirectional in nature. This creates problems in controlling the voltage, and further various contingencies occur.
3. **Reactive power compensation:** Various loads such as irrigation pumps consumes power of reactive nature type as stated by Yamujala et al. (2014). Not only that various DPGS technology uses synchronous generator for generation which in turn delivers reactive power to the system. Also, various flexible AC transmission systems (FACTS) devices are used for injecting reactive power.

3.2.1.3 Frequency of the System

The stability of frequency and balance of the network depends on machines equipped such as synchronous generators and speed governors [18]. Various micro sources having power rating constant if penetrated in large number in the system increases the stability of the frequency of grid.

3.2.1.4 Power Fluctuations

If we consider a PV DPGS, there is a fluctuation in power in the output side for a particular day [19]. Due to fluctuations in power, there is a risk for security and stability of the grid. Due to this, the power generated is more than required; thus, installation of a storage system becomes must.

3.2.1.5 Energy Storage

If we consider that the primary generation will be from renewable sources, we should have a storage system for this energy. DGs connected to microgrid because of their intermittency and power reliability require systems for energy storage [20]. Absence of storage systems during disturbances in network and fault could cause the system to operate out of the stability zone [21], but there are certain problems while employing a storage system for generated energy such as its increased costs and increased system complexity due to its design for a particular network, which adds to the total cost. As for a particular network, the storage system for generated energy is manufactured from various materials; therefore, the rate of working will be different for each system or network. Also, its flexibility is a matter of great concern. Thus, to retain the power quality while supplying energy to load from the storage system, examination of design and working of storage element should be done before its use [22]. Also, solar water pumps could be installed in place of storage systems.

3.2.1.6 Protection of the System

These systems affect the protection scheme of the existing distribution grid or system. Due to their power ratings, placement of DG systems, operational modes, applied technology and configurations of network of distributed systems, the protection of the system from overcurrent may vary [23]. During normal condition or operation, the protection scheme of the system will remain same, but during fault condition, the distributed generation systems must be disconnected. This demand calls for some new solutions. If there is a branch of low voltage supplied by a distributed energy system of low energy, suppose a fault occurs; then the current due to fault in transformer will result in operating transformer protection with the fault current being supplied by the unit (distributed generation). As the impedance of the system may be high, it will be difficult to operate the protection system of the DG set. Thus, it is recommended that the DG system should be disconnected from the system as soon as the fault occurs. Supervisory control and data acquisition (SCADA) systems are employed nowadays in distribution networks, making them automated. The use of SCADA will design and coordinate the protection scheme. This issue can also be defined as not adequate fault current extent. The main generation in distribution systems includes that from solar and wind energy resources. Due to this, a large number of power electronics converters are used. This—when used at low voltage—limits the level of fault current. The over-current (OC) relay (conventional type) is somewhat incapable to operate, as the extent in which fault current is to be feed is adequate [24]. Thus, the distributed generation systems contribute less fault current to the relays of networks or feeders of power systems. More use of power electronics devices in system results in curtailing fault current, as these devices have the ability to sustain thermally, which is a major drawback.

3.2.1.7 Location or Placement

Various methods are there to suggest the optimal location for DGs with specific objectives. The process named as analytical hierarchical is used for this purpose, and it does select the location on the basis of priority of the objectives. In the present scenario of power systems, there are certain changes in structure and regulatory systems, and also, the network is expanding and the load is concentrated; these factors conclude that the method suggested is good and satisfactory regarding the present scenario—but the location which is optimal now will not be same forever. The optimal location will change with the increasing penetration of DG sets [25]. This requires a new study and planning for finding the optimal location. Also, the fuel availability will affect the location of DG's in near future.

3.2.1.8 Islanding

Due to loss in grid connectivity, some of the parts of system might work in isolation as stated by Chahyal et al. (2017). As a result, complications occur in reclosing procedures, which leads to equipment damage. Thus, there is a risk with power quality expressing swell or sag in voltage. If there is a fault in a grid and it is repaired without disconnecting it from the system or de-energizing, then it causes severe threat to the safety of the personnel working on it. For a system having auto-reclosing mode, it is assumed that units of distributed generations are to be disconnected before reclosing. These problems are not properly resolved if conventional elements of protection scheme are used, as they are unable to sense the variations in frequency and voltage of power generation at low levels. Thus, it is very much necessary to plan a strong protection strategy for operation in case of islanding mode as stated by Wang et al. (2011).

3.2.2 Non-Technical Issues

There are certain non-technical issues in case of DG systems. This are described as follows:

3.2.2.1 Economic and Regulatory Factors

The capital cost of installed power in per-kilowatt term is comparatively higher than that of central plants, which is a major problem [26]. Though capital cost differs according to the type of DG system, this cost could be reduced by technological advancements, so the major concern is its economic sustainability in the future. Their survival depends on the agreement of energy supply to the load in the future and commitment from the government side. Expansion of distributed generation is possible only if proper return is obtained from the market. A clear and proper policy with regulatory instruments should be there for proper treatment of various distributed generation systems, as DG networks have been operated—and still are being developed as passive networks—for a long time. This would also help to integrate the distributed generation systems into grids or networks.

3.2.2.2 Less Technically Skilled Personnel

Anees (2012) states that, for installation and maintenance of DG systems, technically skilled personnel or manpower is required. If it is not, there will be connectivity issues in local generation which could damage the equipment, and also problems with grid integration which would result in reduction in power quality, stability and reliability.

3.3 FILTER AND ITS TOPOLOGIES

The power quality of a system is basically evaluated by the percentage of harmonics present in the system. Harmonics are present in both voltage and current. These harmonics occur in the system because of usage loads which are nonlinear in nature. Non-linear loads can be present in both domestic and industrial fields, in equipment such as televisions, computers, florescent lamps ballast and power electronics based on switch mode. For industrial and tertiary applications, devices like welding equipment, arc furnaces and electric drives introduce non-linearity in the system. These loads of non-linear nature draw non-sinusoidal current from the system, which results in harmonic contents in the system on supply voltage side with reference to fundamental component of the supply voltage. The harmonics in the system could cause inappropriate operation of sensitive loads such as measuring and monitoring devices, process controllers, etc. Also, if the harmonic content of the AC line to which devices are connected is higher, there are chances of failure of equipment. The electrical machines, transmission lines and protection schemes also could not perform well. The effects on voltage at supply side due to load can be correctly measured at the point of common coupling (PCC). Two types of harmonics are produced by the load which can be measured and identified at bus bar. These are current-type and voltage-type harmonics. Both harmonics types have specific characteristics and dual properties. Capacitors are used for filtering. Basically, the current harmonics are due to the phase-controlled SCRs and diode rectifiers feeding inductive loads. The loads responsible for voltage harmonics are diode rectifiers that feed to DC capacitors used for filtering purposes. Based on their specific properties, single or both type of harmonics-producing loads require specific filter configurations. To remove these harmonics, various filters are being used. They are as the following.

1. Active filters
2. Passive filters
3. Hybrid filters

Each type of filter is explained in this chapter, along with the configurations used. Also, harmonics can be suppressed by reducing harmonic injection in the grid by various devices of power electronics, improvement of sources of harmonics and using pulse width modulation converters, and also, the increase in the count of converter pulses [27] could help to achieve this objective.

3.3.1 Active Filters

Active power filters are used efficiently for mitigating harmonics, balancing the load, compensating reactive powers, regulating the voltage and reducing voltage. These filters are basically analog-type electronic filters. These filters are used in variable and high voltage supply. They are basically complex and costly, but are preferred over passive filters due to issues with those.

The systems of active filter designed for DC/DC converters serve the following two purposes:

1. Removal of high-frequency electromagnetic interference (EMI) from converter input current
2. Removal of ripples from converter output voltage

TABLE 3.1
CLASSIFICATION OF ACTIVE FILTERS

S. No.	Basis of Classification	Types of Active Filters
1.	On the basis of converter used	a) Voltage source b) Current source
2.	On basis of phases	a) Single phase b) Three phase (for three-wire or four-wire systems)
3.	On the basis of their topologies	a) Shunt active power filter b) Series active power filter c) Hybrid power filters d) Unified power quality conditioner

3.3.1.1 Voltage Source Active Filters

When a large capacitor is employed and tied with the DC bus at converter side, it acts as a voltage source. These filters are lighter and cheaper, and are easily controlled as compared to that of active filter of current-source type. Here, the element used for storage is capacitor.

3.3.1.2 Current Source Active Filters

When an inductor is employed as a storage device of DC energy, it is termed as an active filter of current-source type. The same inductive element is used as a storage device for both single- and three-phase current-source type shunt active filters and capacitive elements for voltage-source types. The method used for controlling is pulse width modulation.

3.3.1.3 Shunt Active Power Filter

Shunt active power filters (SAPFs) are connected to electrical networks or systems in a parallel manner (Figure 3.5). This filter reduces distortions in current, improves reactive power in the system, balances load of the system and improves neutral current. It is operated at harmonic current of magnitude same as that due to non-linear load, and with different phases is being injected in the utility network or system so that at PCC, a sinusoidal current is maintained. The main aim of this filter is to improve the power factor by compensating for harmonic currents. It finds difficulty in compensating for non-linear loads of voltage source forms. Thus, it can be said that being connected to PCC, it is basically efficient for reducing reactive power in the system from the front end and mitigating current harmonics [28]. They also work as static VAR compensators (SVCs) in the system so as to compensate other disturbances like flickering and unbalancing.

Advantages:

1. Due to the inductance present in source side, the ability of SAPF to compensate harmonics is not affected.
2. It is of reasonable cost for industrial loads ranging from low to medium KVA.
3. In case of a distribution feeder, harmonic propagations can be damped.

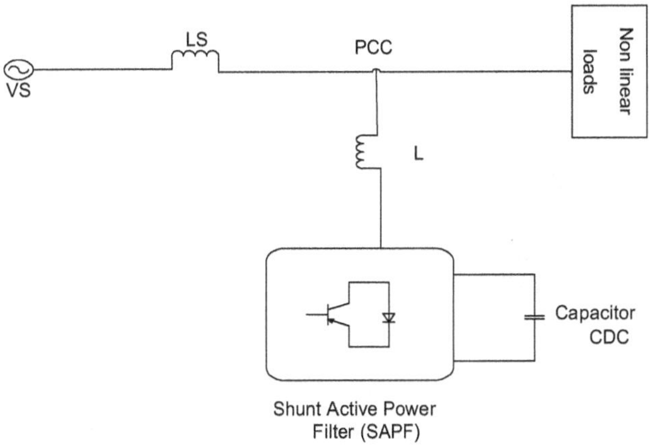

FIGURE 3.5 Configuration of a shunt active power filter.

Disadvantages:

1. KVA rating of higher value of power electronics based inverters for industrial loads of high power, as the converter should withstand the supplied harmonic current, utility voltage and line frequency.
2. The load voltage harmonics is also not compensated in this case.

3.3.1.4 Series Active Power Filter

When the active power filter is connected in series with the network or utility system, it is known as a series active power filter. With the help of a matching transformer, a barrier is provided for the harmonic currents so that they are unable to reach the supply system. In this case, voltage injection is needed for compensating voltage in the harmonics and distortion in voltage on load side. The controlled method is designed in such a manner so that at PCC, there is zero impedance in response to fundamental frequency [29]. The harmonic propagation resulted due to resonance of shunt passive filters and line impedance are damped by these filters. Compensation of loads of non-linear voltage source type are effectively done by this type of filter. Switched-mode power supply (SMPS) circuits, uninterruptible power supplies (UPS), frequency converters, etc., have capacitors of larger rating fixed in a filter in the rectifier circuit's DC side. These loads mentioned are basically non-linear loads of voltage source type. The main function of this filter is isolating the harmonics of current between load and source then compensating current harmonics at the load side. These filters are used for mitigating the harmonics and compensating the reactive power at PCC. This PCC in micro grid is near to load bus. The design and development of series active filters are of controllable voltage-source type or controllable current-source type; the previous one is converter being fed by voltage and the later one is current fed. The disadvantage of this filter is its lagging performance in case of direct compensation for current harmonics. Also, suppressing neutral current, reactive power compensation and load current balancing is difficult for this type of filter. The advantage of these filters is that they can carry the full load current and can also withstand larger ratings of power by employing a suitable transformer.

3.3.1.5 Unified Power Quality Conditioner (UPQC)

The unified power quality conditioner (Figure 3.6) is a combination of shunt and series active power filter connected in back to back configuration having a self-supporting DC common to both. This has an advantage of both filters. The main purpose of a UPQC is to compensate for voltage imbalance, reactive power, negative-sequence current and harmonics. It is a main link. Voltage compensation in this case is done by series filter. The shunt filter is used for compensating harmonic currents. This UPQC is costly and complex to control. Various concerns of power quality such as voltage flicker, voltage harmonics, unbalanced voltage, current harmonics, voltage swell and sag, unbalanced current, neutral current and reactive power of load side are also compensated by using this UPQC [30]. Hence, its drawback is its high cost and complexity of control. There is a growing interest to use this as its performance is efficient and superior as compared to others.

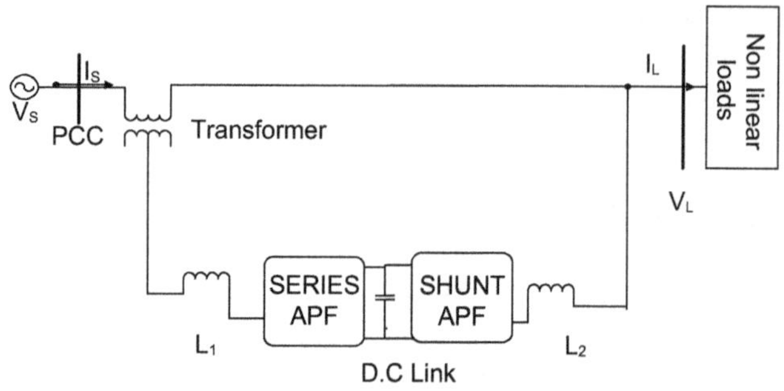

FIGURE 3.6 Configuration of a UPQC (unified power quality controller).

3.3.2 Passive Filters

Filters consisting of passive elements such as inductors and capacitors are known as passive filters. Externally connected power sources are not required in these types of filters. The cost of these filters are very low and they offer a path of minimum impedance to the unwanted harmonics [31]. This filter type is used in bypassing power supplies, power distribution systems, discrete circuits, etc. It is basically a series tuned resonant circuit having low value of impedance. These filters are basically good at compensating current harmonics.

3.3.2.1 Band Pass Filter

A band pass filter passes signals of frequencies lying within a particular range and rejects all other frequencies other than this range. This frequency range is the difference between high cut-off frequency and lower cut-off frequency. Its types are described in what follows.

TABLE 3.2
CLASSIFICATION OF PASSIVE FILTERS

S. No.	Filter Type	Sub Type
A.	Band pass filter	i) Single-tuned filter ii) Double-tuned filter
B.	High pass filter	i) First-order filter ii) Second-order filter iii) Third-order filter iv) C-type filter
C.	Composite filter	–

3.3.2.1.1 Single-Tuned Filter

When a capacitor, small damped resistor and an inductor is connected in series, this is known as single-tuned filter (Figure 3.7). This filter is further connected parallel with loads of non-linear type. The loads previously mentioned provide a path of low impedance for distinguished harmonic frequencies, which leads to absorption of harmonic currents that flow through the load. At operating frequency of the system, the reactive power is compensated. The impedance and frequency relation of this system follows:

$$Z(s) = \frac{\left(1 + RCs + LCs^2\right)}{Cs} \quad (3.9)$$

3.3.2.1.2 Double-Tuned Filter

When two single-tuned filters are connected in parallel, it is known as a double-tuned filter (Figure 3.8). It is further attached with load of non-linear type and is further tuned to two harmonic frequencies. In a double-tuned filter, a resistor is connected in parallel with inductance and their combination is connected in series with capacitor. The value given to the inductance and capacitance of the filter is less such that the impedance of filter is negligible at a particular harmonic frequency. A damping resistor can be connected in series to the circuit so as to adjust tuning sharpness. Capacitance and inductance in this case will be calculated in the same way as in single tuned filters. The value of resistance follows [32].

$$R = \sqrt{\frac{L}{C}} * Q \quad (3.10)$$

where, the quality factor lies between the range of 0.5–5.

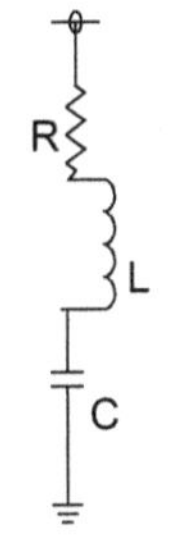

FIGURE 3.7 Circuit diagram of a single-tuned filter.

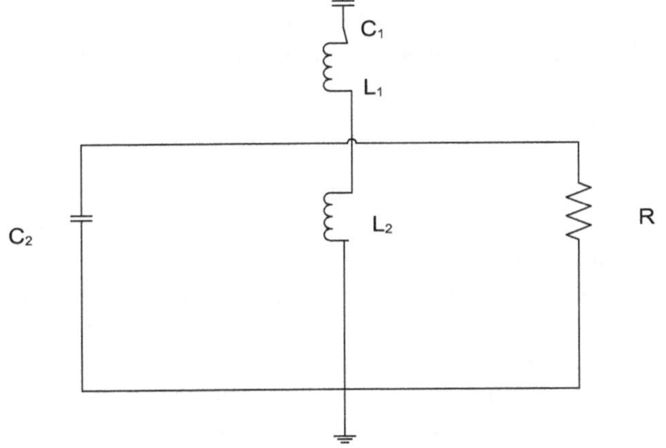

FIGURE 3.8 Circuit diagram of a double-tuned filter.

3.3.2.2 High-Pass Filter

In case of a higher-order filter, parallel connection is adopted for connecting the inductance and resistance. It is basically a single-tuned filter. This filter is of wide band nature, with high frequency impedance which is limited by a resistance. The following equation gives inductor and capacitor value.

The value of total impedance:

$$Z(s) = \frac{R + Ls + RLCs^2}{RCs + LCs^2} \quad (3.11)$$

At a particular quality factor, resistance value is given by:

$$R = QLw,\ 0.5 < Q < 5 \quad (3.12)$$

For minimum power loss in system, value of R should be minimized.

3.3.2.2.1 L Filter

This is a type of first order filter with an attenuation of 20 dB/decade as mentioned by Jianjun et al. (2002). Such a filter's transfer function is given by:

$$G(s) = \frac{1}{sL} \quad (3.13)$$

3.3.2.2.2 LC Filter

An LC filter (Figure 3.9) is a type of second-order filter [33]. In this filter, the inductance is in series connection and capacitance is in parallel connection with former connected with inverter and later connected with the grid. The value of inductance can be reduced by this parallel capacitance. The losses and cost of this filter are less as compared to an L filter. If a large value of capacitance is used, this results in large inrush current due to charging of capacitor

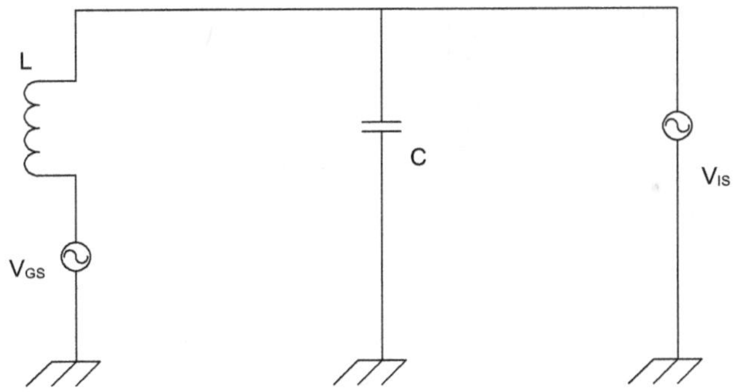

FIGURE 3.9 Circuit diagram of an LC filter connected to a grid.

at fundamental frequency and also attenuation of harmonics will occur. A resistance is placed in series with capacitor of the filter as it overcomes the dynamics which are unstable in nature. This functions as a damping resistance in the filter.

The transfer function of voltage of both input side of inverter and grid side of LC filter is given by [34]:

$$G(s) = \frac{V_{GS}}{V_{IS}} \tag{3.14}$$

$$G(s) = \frac{1}{s^2 LC + 1} \tag{3.15}$$

3.3.2.2.3 LCL Filter

This is a third-order filter. It consists of an inductor on the grid side, inductance on a load/grid side and a capacitor, as shown in Figure 3.10. This filter is used for high power rating of 100 KVA with the use of small inductors and capacitors. An LCL filter is better compared to L filter, as it decouples between the power grid and filter [35]. In the LCL filter, a damping resistance is added. It is used to reduce the damping and attenuation at resonance frequency.

The transfer function of this filter is given by:

$$G(s) = \frac{I_2}{V_{IS}} \tag{3.16}$$

$$G(s) = \frac{1}{s^3 L_1 L_2 C + (L_1 + L_2)s} \tag{3.17}$$

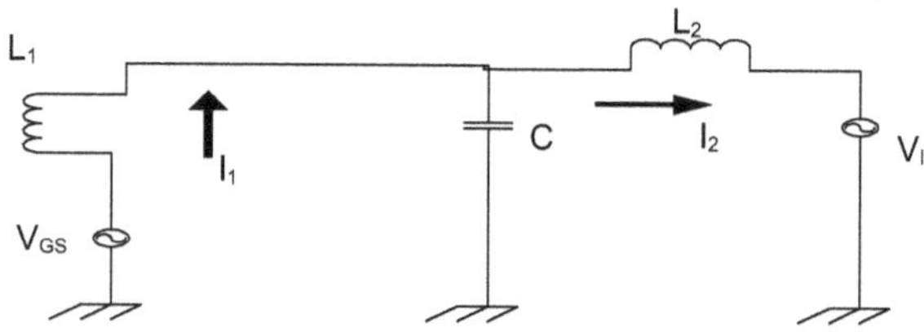

FIGURE 3.10 Circuit diagram of an LCL filter connected to a grid.

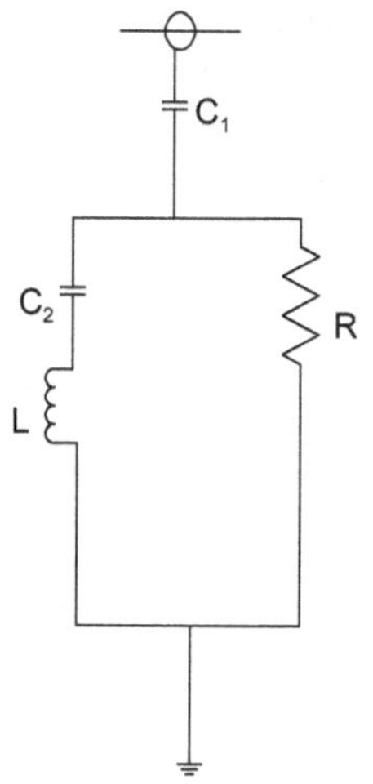

FIGURE 3.11 Circuit diagram of C-type filter.

LCL filter with series damping resistance:

$$G(s) = \frac{1 + sRC}{L_1 L_2 C s^3 + RC(L_1 + L_2)s + (L_1 + L_2)s} \tag{3.18}$$

LCL filter with parallel damping resistance:

$$G(s) = \frac{R}{L_1 L_2 R C s^3 + L_1 L_2 s^2 + R(L_1 + L_2)s} \tag{3.19}$$

3.3.2.2.4 C-Type Filter

In a C-type filter (Figure 3.11), the inductor and capacitor are connected in series and their combination is connected in parallel with the resistor and is tuned at required frequency [36]. The inductance and capacitance branch and at tuning frequency the resistive branch will be bypassed. Thus, at fundamental frequency, the filter works as a capacitive branch such that no current flows in resistor and the power loss of the resistor becomes minimum. The value of inductive reactance is large at frequency of higher value, thus making the current to flow from resistive path of the filter such as it behaves like a high pass filter of first order.

3.3.2.3 Series Passive Filters

These filters are made of resonant branches connected in parallel. These filters are further connected in series with the loads of non-linear type. These filters have three tuned filters as mentioned in Power systems-filtering techniques for power quality improvement (2021). High impedance has been offered to harmonic currents; as a result, they do not flow in the power system. In rectifier circuit, ripples of current are reduced with the help of this filters. These filters are easy for implementation and low cost, but harmonics caused due to large loads are limited. This filter operates in lagging power factor. These filters are basically used for reduction in harmonics of load from voltage-fed types.

3.3.3 Hybrid Filters

Hybrid filters (Figure 3.12) are combination of both active filters and passive filters. These filters have different structures. These filters are basically used to remove problems of passive filters. As passive filters suffer from the non-dynamic response and resonance probability, the high and increased active filter's cost promotes the use of hybrid filter. This filter uses the advantage of both active and passive filter and also it is of low cost [37]. The passive filters here not only compensate harmonics of higher order and reactive power, but also absorb the reactive power. Also, the active power filter in hybrid filter configuration results in an increase in the higher order harmonics damping performance [38]. Active filters compensate dynamic harmonics, and also compensates low frequency and surge reactive powers.

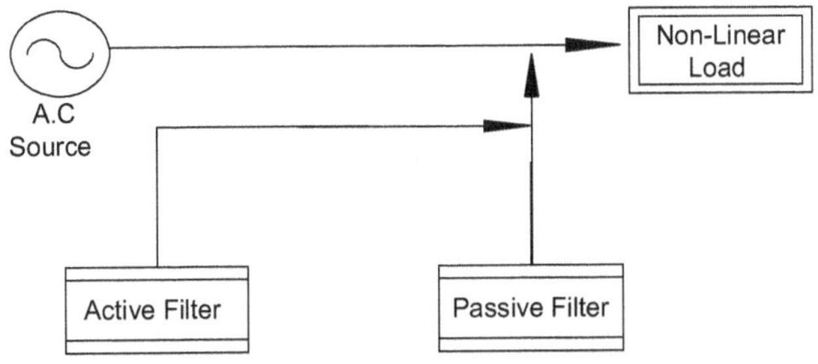

FIGURE 3.12 Hybrid filter circuit configuration.

3.4 REDUCTION OF HARMONICS IN PV SYSTEMS BY ACTIVE POWER FILTERS

The shunt active power filter topology is basically preferred for harmonic reduction in PV system as compared to the series one. For designing a SAPF, measurement of line voltage and current is necessary, along with calculation of THD and harmonic currents. Also for SAPF, the value of inductor in the filter, capacitor of DC link and the power rating can be selected with the help of iteration process [39]. SAPF mitigates the harmonics present in the line current by injecting a harmonic current that is negatively generated at the point of common coupling. This filter can be used for both single- and three-phase supply, but in case of a three-phase active power filter, a phase-locked loop (PLL) is required, depending on the control method also used. PLL concept contributes in generation of reference currents and compensating currents. The PLL is used for sensing the load current and further by varying the converter's firing angle at which this current is supplied to compensate current as stated by Delfanti et al. (2014). Similarly, for a single-phase APF, a zero-crossing detector (ZCD) is used for tracking frequency and phase angle of corresponding frequency. Here, a three-phase SAPF, along with voltage source inverter configuration, has been used due to its simplicity and economical design [40]. A conditioning device is employed by a controller which results in a smooth profile of injected current making it the duplicate of those harmonic currents which needs to be mitigated. The value obtained by VSI is controlled so as to generate required compensating current. Various control strategies are listed in what follows [41].

1. Instantaneous reactive power theory
2. Synchronous reference frame
3. Indirect power control mode
4. Space vector pulse width modulation (SPWM)
5. Hysteresis current control
6. Carrier phase-shifted SPWM
7. Repetitive control
8. One-cycle control
9. Dead-beat control
10. Fuzzy-logic control
11. Artificial neural network
12. Sliding-mode control

The SAPF has been analysed by dividing it into the following two important sections.

- CCVSI (current-controlled voltage source inverter)
- Control block

First, we will discuss the CCVSI, then we will discuss the controlling techniques used.

3.4.1 CCVSI

This is basically three phase-two level voltage source inverters (VSIs) fitted with insulated-gate bipolar transistors (IGBTs). Here, hysteresis controllers are also being used for closing current loops. Sampling of current obtained at the output is done at fixed frequency so that the inverter's switching frequency is limited [42]; the trade between filtered high frequencies and high change in flow of current with respect to time in inductor for damping of harmonic currents decides the coupling inductor's value. The second-order filters connected in parallel further filter all those switching frequencies. Capacitance connected at DC side is used for minimizing the voltage ripple. Here, a proportional controller is used to control the DC voltage level. This proportional controller is used to modify the active power given as a reference to the converter. In order to generate the compensation current, the active power and reactive power needs to be sensed by a rotating equivalent DC having two dimensions. This is possible by a two-stage transforming process. First, the parameters of the grid will be converted into reference frame stationary in nature and then through proper transformation again, they will be converted into a rotating system (Figure 3.13).

3.4.2 Control Strategy

There are different control strategies available, but here we are using the concept of instantaneous reactive power theory or p–q theory [43], as it is most widely commercially applied. This theory basically focuses on three-phase power systems. It can be applied in both steady-state operations and transient operations. In this theory, there is a transformation of reference frame known as Clark's transformation

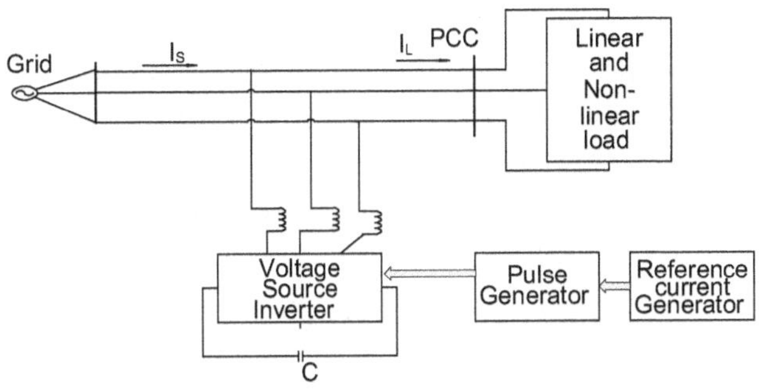

FIGURE 3.13 Circuit diagram.

to transform the voltage and current signals from reference frame of *abc* to *αβ0*, and further using them for calculation of instantaneous power [44]. A regulator is used for calculating active currents and reactive currents represented by i_p and i_q, respectively. Both the currents are instantaneous in nature and are used as references fed to the converter. The following equations help us to calculate the instantaneous current and voltage value.

$$\begin{bmatrix} v_\alpha \\ v_\beta \end{bmatrix} = [\mathrm{A}] \begin{bmatrix} v_a \\ v_b \\ v_c \end{bmatrix} \tag{3.17}$$

$$\begin{bmatrix} i_\alpha \\ i_\beta \end{bmatrix} = [\mathrm{A}] \begin{bmatrix} i_a \\ i_b \\ i_c \end{bmatrix} \tag{3.18}$$

Here, A is the Clark's transformation [45] and is given by:

$$\mathrm{A} = \sqrt{\frac{2}{3}} \begin{bmatrix} 1 & \frac{-1}{2} & \frac{-1}{2} \\ 0 & \frac{\sqrt{3}}{2} & \frac{-\sqrt{3}}{2} \end{bmatrix} \tag{3.19}$$

The zero-sequence phase components of voltage signals and current signals are neglected for simplicity. Thus, the instantaneous real or active power and imaginary or reactive power, according to Babu et al. (2016), can be defined as:

$$\begin{bmatrix} p \\ q \end{bmatrix} = \begin{bmatrix} V_\alpha & V_\beta \\ -V_\beta & V_\alpha \end{bmatrix} \begin{bmatrix} i_\alpha \\ i_\beta \end{bmatrix} \tag{3.20}$$

$$\begin{bmatrix} i_\alpha \\ i_\beta \end{bmatrix} = \frac{1}{V_\alpha^2 V_\beta^2} \begin{bmatrix} V_\alpha & V_\beta \\ -V_\beta & V_\alpha \end{bmatrix} \begin{bmatrix} p \\ q \end{bmatrix} \tag{3.21}$$

The compensating currents obtained can be given as:

$$\begin{bmatrix} i^*_{c\alpha} \\ i^*_{c\beta} \end{bmatrix} = \frac{1}{V^2_{\alpha+V_\beta^2}} \begin{bmatrix} V_\alpha & V_\beta \\ -V_\beta & V_\alpha \end{bmatrix} \begin{bmatrix} p \\ q \end{bmatrix} \tag{3.22}$$

This compensating current can further be transformed in to abc reference frame using Clark's Transformation

$$\begin{bmatrix} i^*_{ca} \\ i^*_{cb} \\ i^*_{cc} \end{bmatrix} = \sqrt{\frac{2}{3}} \begin{bmatrix} 1 & 0 \\ \frac{-1}{2} & \frac{\sqrt{3}}{2} \\ \frac{-1}{2} & \frac{-\sqrt{3}}{2} \end{bmatrix} \begin{bmatrix} i^*_{c\alpha} \\ i^*_{c\beta} \end{bmatrix} \tag{3.23}$$

If we include the zero sequence components of currents in abc frame then it will result in following compensating currents:

$$\begin{bmatrix} i^*_{ca} \\ i^*_{cb} \\ i^*_{cc} \end{bmatrix} = \sqrt{\frac{2}{3}} \begin{bmatrix} \frac{1}{\sqrt{2}} & 1 & 0 \\ \frac{1}{\sqrt{2}} & \frac{-1}{2} & \frac{\sqrt{3}}{2} \\ \frac{1}{\sqrt{2}} & \frac{-1}{2} & \frac{-\sqrt{3}}{2} \end{bmatrix} \begin{bmatrix} -i_0 \\ i^*_{c\alpha} \\ i^*_{c\beta} \end{bmatrix} \tag{3.24}$$

Here, i_0 is the zero-sequence component of current and is given by:

$$i_0 = \frac{1}{\sqrt{3}}\left(i_a + i_b + i_c\right) \tag{3.25}$$

Since we have considered three-phase three-wire systems, we can say that $i_0 = 0$.

3.5 REDUCTION OF HARMONICS IN PV SYSTEMS BY PASSIVE POWER FILTERS

Passive filters are the cheapest method to reduce harmonics. Being least expensive, they are basically used in low-power and medium-power systems.

Figure 3.14 shows block diagram of connection of passive filter in the system. From the system, we can see that solar radiance from the PV array is fed to a boost converter. The output of the converter is further fed to an inverter and then to a non-linear load. Here, the passive filter is connected in parallel to load.

3.5.1 BOOST CONVERTER (DC/DC) AND MPPT

The voltage available at the output of a PV array is low and is of DC nature. Therefore, a DC/DC boost converter is used. This converter is a switching power supply used for converting low voltage to high voltage DC, which is further fed to inverter. Here, MPPT have been used for tracking the power and controlling the system with the help of a source adapter.

3.5.2 MODEL OF DC/DC BOOST CONVERTER

The circuit diagram of a step up DC/DC boost converter is presented in Figure 3.15. The converter consists of an input

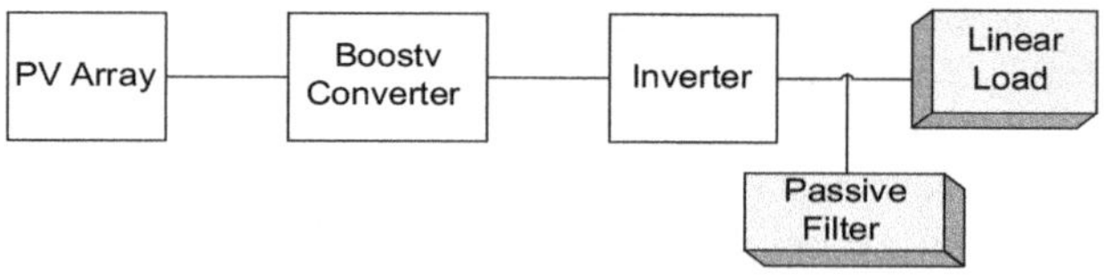

FIGURE 3.14 Block diagram of a circuit using a passive filter.

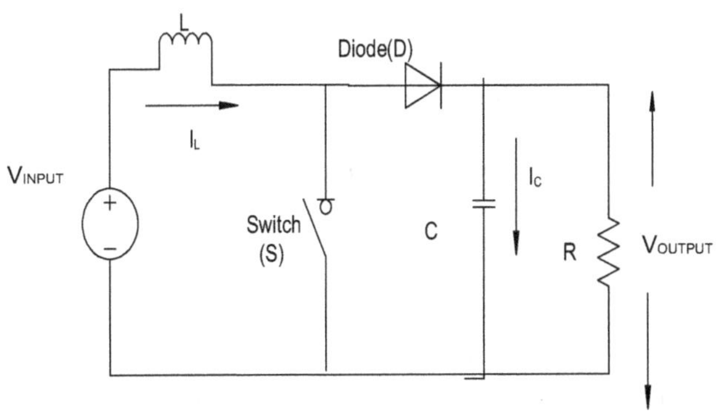

FIGURE 3.15 Circuit diagram of a boost converter.

voltage which is DC in nature (V_{INPUT}), an inductor L, diode D, capacitor C, controlled switch S and a load resistance R. Here, V_{OUTPUT} is the output voltage of the DC/DC boost converter. The value of V_{OUTPUT} is given by:

$$\frac{V_{OUTPUT}}{V_{INPUT}} = \frac{1}{1-d} \tag{3.26}$$

where, d is the duty cycle of the boost converter

The value of inductor depends upon the current flowing through it, as the current should be continuous. Jayaraman et al. (2013) stated that the ripple factor of the current (ΔI) compared to the output current is 5% at maximum times and ripple factor of the voltage (ΔV) compared to the output voltage is 3% generally. The value of inductor and capacitor is given by:

$$L = \frac{1}{8}\frac{V_{INPUT}d}{\Delta I f_s} \tag{3.27}$$

$$C = \frac{I_0 d}{f_s \Delta V} \tag{3.28}$$

Here, f_s = Boost converter's switching frequency.

3.5.3 DC/AC Voltage Inverter

The DC voltage which we have obtained at the output of boost converter is fed to inverter circuit. Here, a three-phase inverter converts the DC output voltage into AC voltage in line with the network Chtouki et al. (2016). Here, pulse width modulation (PWM) technique is used for controlling the switches. PWM technique can improve the spectral quality of inverter output. This is because it eliminates lower order harmonics and filters higher order harmonics as mentioned by Sharma et al. (2014).

3.5.4 LCL Filter

We have various type of passive filters that we have studied in Section 3.3. Here, high-pass filter is considered for studying harmonic effects in this section. Among various type of high-pass filters, we will be using LCL filters here. These filters are very popular nowadays in renewable energy industries. These filters are very economical and efficient in improving the power quality that is fed into an isolated load or the grid from PV systems. LCL filters have better performance than L, LC and LLCC filters, with an attenuation of 60 Db/decade ahead of the resonant frequency Chtouki et al. (2016). This filter has also been chosen because it has a fine current ripple attenuation for small values of inductance.

3.5.5 Designing of LCL Filter

As shown in Figure 3.9, the circuit of LCL filter consists of L_1 and L_2 in the inverter and load side, respectively, and a capacitor C. The transfer function of LCL filter is given by:

$$G(s) = \frac{1 + Z_{load}sC}{s^3 L_1 L_2 C + s^2 (L_1 + L_2) C Z_{load} + s(L_1 + L_2)} \tag{3.29}$$

where, $s = jw$

Z_{load} = Load impedance connected to the PV generation system (standalone)

For designing the parameters of an LCL filter, we should also consider inverter's power rating, frequency of the line and switching frequency of inverter switches. The range of resonant frequency of this filter is $10\omega_0 \le \omega_{res} \le \left(\frac{\omega_{switch}}{2}\right)$ [47]. This helps in ignoring resonance problems.

where, ω_0 = angular frequency of utility measured in rad/s

ω_{res} = resonant angular frequency

ω_{switch} = angular switching frequency in rad/s.

The value of ω_{res} of passive LCL filter is given by Kotturu et al. (2014) and can be obtained by:

$$\omega_{res} = \sqrt{\frac{L_1 + L_2}{L_1 L_2 C}} \tag{3.30}$$

Larger value of C in the circuit results in suppression of higher order harmonics, which in turn increases reactive power demand and current demand from L_1 side, which decreases the overall efficiency of the filter system. Larger value of inductance can mitigate the harmonic requirement, as well as the size of capacitor. Also, it can be said that the absorption of reactive power by the filters during rated condition should also be considered for selecting the value of capacitance. The value of C for LCL filters can be obtained by the given formula:

$$C = \frac{Q}{2\pi f_0 V^2} \tag{3.31}$$

where, Q = Absorbed reactive power
V = Rated voltage of the system
f_0 = Frequency at the output of system.

The value of inductor to be used in designing purpose have to consider the ripples occurring in the current. If we take a large value of inductor, there is a decrease in ripples of current of small value and switching losses. Thus, the value of inductance L_1 is given by:

$$L_1 = \frac{V_{in}}{8hf_{res}} \quad (3.32)$$

where, V_{in} = Voltage fed at input side of inverter
h = Quantity of ripple present in current (0.05 × rated current)

Since we have two inductances in this filter, value of L_2 is selected by the relation:

$$L_1 = aL_2 \quad (3.33)$$

Here, term a used is the ratio factor of inductance. If a > 1 then $L_2 < L_1$. This condition exists for low and medium values of power. This also improves the efficiency of the filter as well as the system.

3.6 REDUCTION OF HARMONICS IN PV SYSTEMS BY HYBRID POWER FILTERS

Hybrid filters, as we have seen earlier, employ both active and passive filters. Here, the passive filter used is the traditional LC filter and the active filter used here comprises of IGBT switches. The active filter will use to reduce harmonics, whereas power capacitor in passive filter will improve the power factor. Thus, both the filters will help in improving power quality. The block diagram shown in Figure 3.16 is the proposed system Sharma et al. (2014). As per the block diagram, the passive filter is series connected and active filter is connected in parallel to the source side and load side. The IGBT diode switching is of six-pulse type.

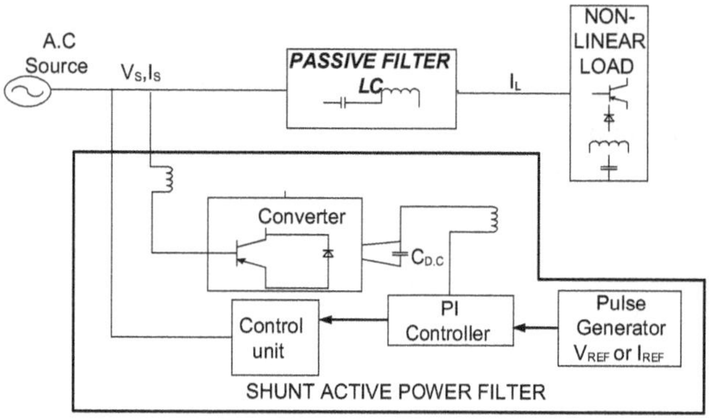

FIGURE 3.16 Design of hybrid filter.

The active filter used here is shunt type. The active filter configuration we can see here has converters and a PI controller and controlling scheme. Active filters were discussed and explained in the previous section. Here, the controlling scheme used is instantaneous reactive power theory or p–q theory [47]. This has been discussed earlier. Here the DC reference voltage or pulses is fed in the active filter with PI controller. The source current generated is compared with the reference current which generates pulse and is fed in converter circuit. The load used here is of non-linear type. Here also, there is a transformation of instantaneous voltage signal and current signal from three-phase to two-phase with the help of Clark's transformation. This modelling was explained previously in Section 3.4.2.

3.7 RESULT AND DISCUSSION

3.7.1 Case 1: When Shunt Active Filter Is Used in PV DPGS

In case of active-power filters, a DC link is used. The value of DC capacitor can be found by:

$$C_{D.C} = \frac{2\Delta E_{D.C}}{\left(V_{DC,Ref} + V_{D.C}\right)\left(V_{DC,Ref} - V_{D.C}\right)} \quad (3.34)$$

where, $V_{D.C}$ is the voltage that can be collected through source
$V_{DC,Ref}$ is the DC reference voltage of capacitor and is given by:

$$V_{DC,Ref} = \frac{2\sqrt{2}}{\sqrt{3}} \frac{V_L}{m} \quad (3.35)$$

V_L = Line voltage obtained at point of common coupling
m = It is defined as the suitability factor.

The variation in load as estimated should be considered for selecting the value of capacitor as stated in Blorfan et al. (2011). Here in eqn. (1) the value of $\Delta E_{D.C}$ can be found out by:

$$\Delta E_{D.C} = \frac{1}{2} C_{D.C}\left(V_{DC,Ref} + V_{D.C}\right)\left(V_{DC,Ref} - V_{D.C}\right) \quad (3.36)$$

The waveform of system's source voltage and system's load current with respect to time has been shown in Figure 3.17 and Figure 3.18, respectively. After inclusion of active filter in the system, the current flowing through the filter is shown through a waveform in Figure 3.19, and the effect of filter on source current can be seen through the waveform in Figure 3.20.

In the preceding graphs, we can see that as soon as load is applied to system, some harmonics occurs in system which causes changes in source current. But as soon as a filter is opened, filter current starts to flow. Thus the waveform of source current becomes continuous. From Figure

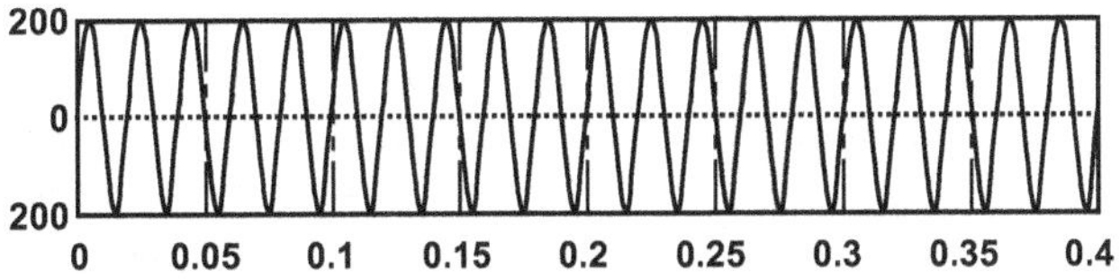

FIGURE 3.17 Waveform of system's source voltage with respect to (w.r.t) time (in secs).

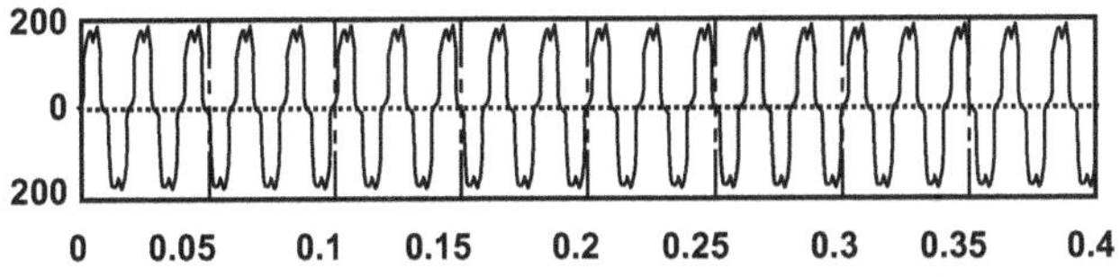

FIGURE 3.18 Waveform of system's load current w.r.t time (in secs).

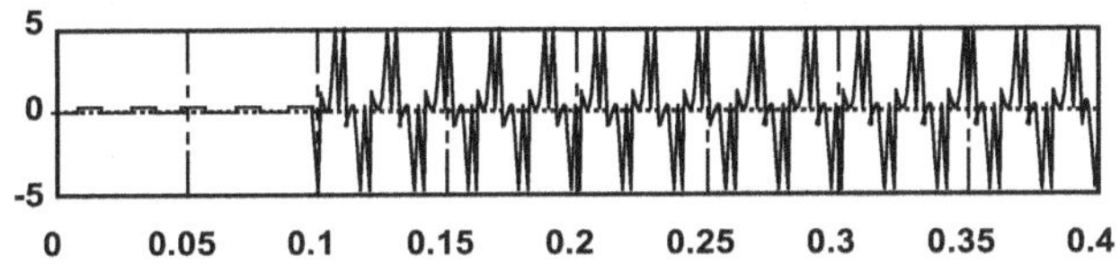

FIGURE 3.19 Waveform of system's filter current w.r.t time (in secs).

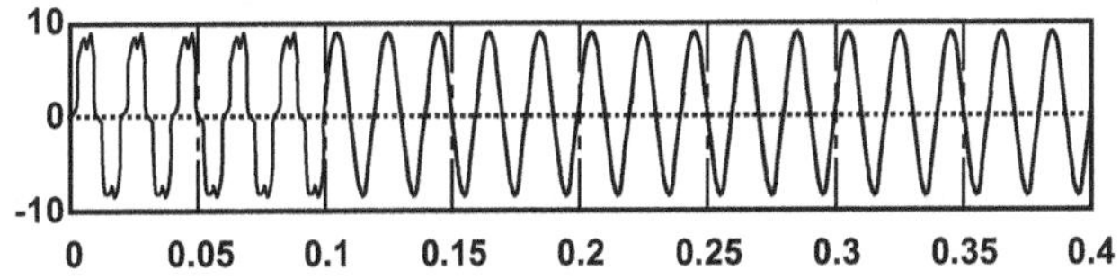

FIGURE 3.20 Waveform of system's source current w.r.t time (in secs).

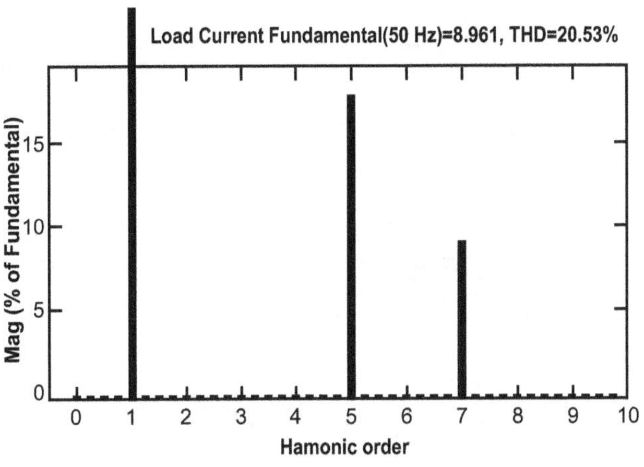

FIGURE 3.21 Total harmonic distortion in load current.

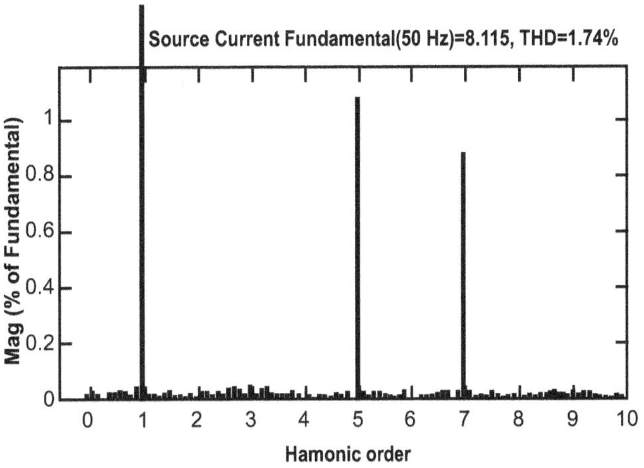

FIGURE 3.22 Total harmonic distortion in source current.

3.21 and Figure 3.22, we can observe that at balanced supply (PQ control):

- Total harmonic distortion of load current is: 20.53%
- Total harmonic distortion at source current is: 1.74%
- Displacement factor: Near unity

3.7.2 Case 2: When Passive Filter Is Used with PV DPGS

Here we have used a LCL filter as a passive filter. The basic block diagram of the whole system is shown in Figure 3.14, and the circuit of converter is shown in Figure 3.15. The waveform of output voltage of PV system before and after using the filter can be seen through the waveforms shown in Figure 3.23 and Figure 3.24, respectively.

From Figure 3.25 and Figure 3.26, it can observe that the voltage we obtain at the inverter output through simulation has reduced the harmonics to 5.44% from 51.83% with the use of LCL filter.

3.7.3 Case 3: When Hybrid Filter Is Used with PV DPGS

MATLAB/Simulink is used for validation of the system. The system is shown in Figure 3.16. Here, a non-linear

TABLE 3.3
Value of Parameters Considered for the System

PV Module's Output Voltage	60 Volts
Converter parameters	L = 12.4 mH C = 160.1 μF Frequency: 25 KHz Output Voltage = 325 V
Inverter's output voltage	230 V (rms)
Passive filter (LCL filter)	L =18.2 mH, C = 14.8 μF

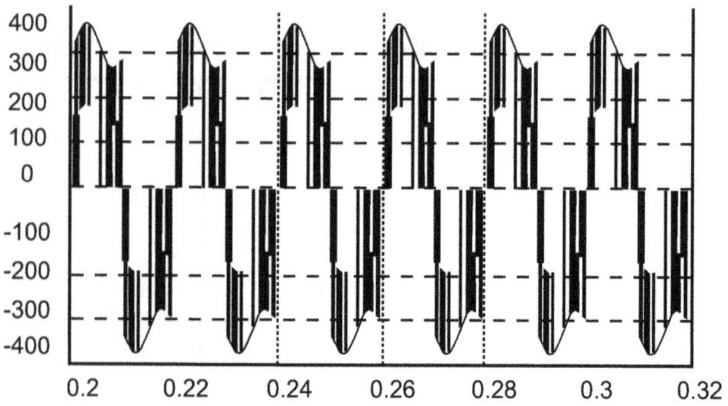

FIGURE 3.23 Output voltage of PV system w.r.t time without any compensation.

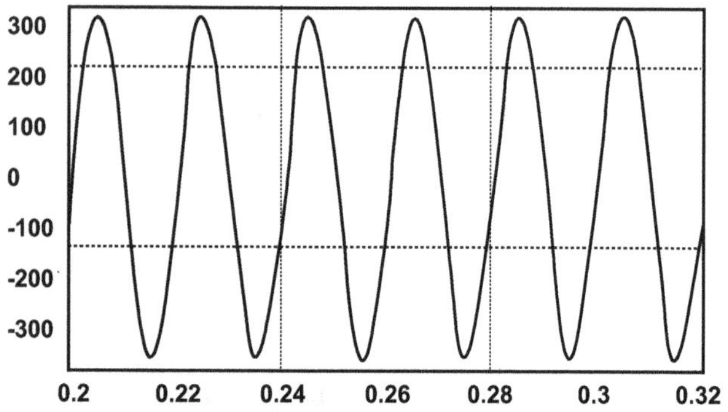

FIGURE 3.24 Output voltage of PV system w.r.t time with compensation.

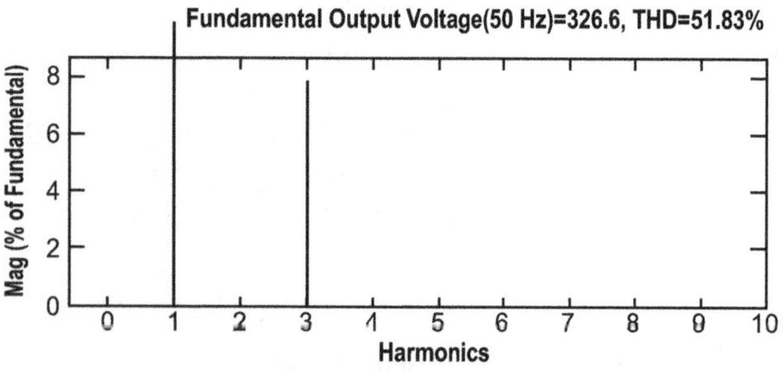

FIGURE 3.25 Spectrum of harmonics of output voltage without compensation.

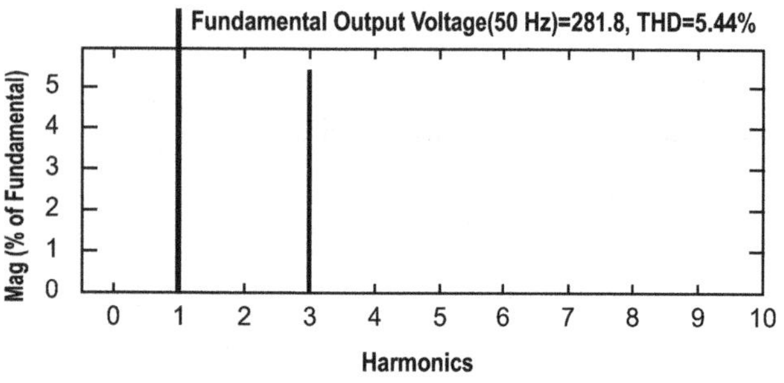

FIGURE 3.26 Spectrum of harmonics of output voltage with compensation.

load has been used because non-linear load harmonics are present in the system. A hybrid filter is used for compensation of harmonics. The simulation time is up to 0.5 seconds. The responses of voltage and current of the system at before compensation and after compensation is shown in Table 3.4. The waveform of source voltage and source current before applying the filters can be seen through Figure 3.27 and Figure 3.28, respectively. After applying filters, the harmonics of source voltage and source current get compensated which is visible from the waveforms shown in Figure 3.29 and Figure 3.30, respectively.

TABLE 3.4
System Parameters

Source	Supply voltage (L-L) = 415 (r.m.s value), frequency = 50 Hz
Shunt active power filter	DC link: i) $V_{D.C}$ =600 volts ii) $C_{D.C}$ = 1200 μF
	Filter values: i) R_f = 6.8 Ω ii) L_f = 4.2 mH
	PI controller: i) K_p = 0.32 ii) K_i = 0.13
Passive filter	L= 1.24 mH, C= 10 μF
Loads	Non-linear load (diode rectifier of three phase)

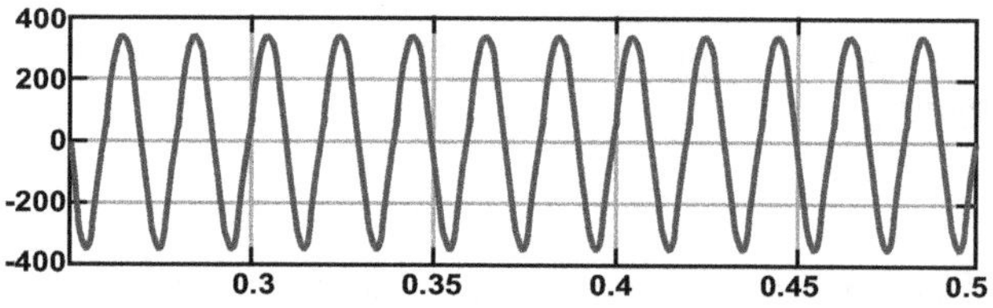

FIGURE 3.27 Source voltage w.r.t time (in seconds) before compensation.

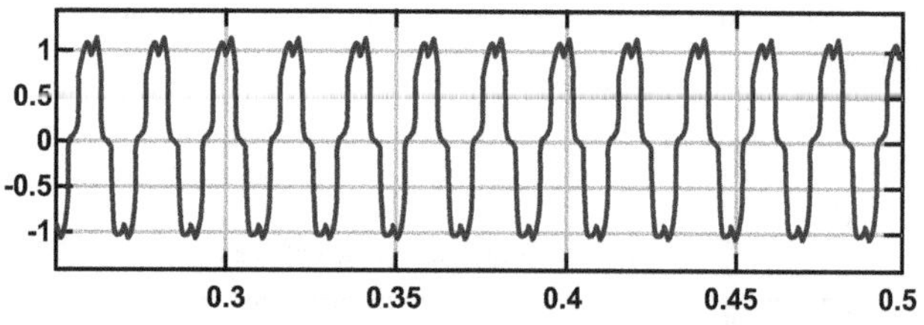

FIGURE 3.28 Source current w.r.t time (in seconds) before compensation.

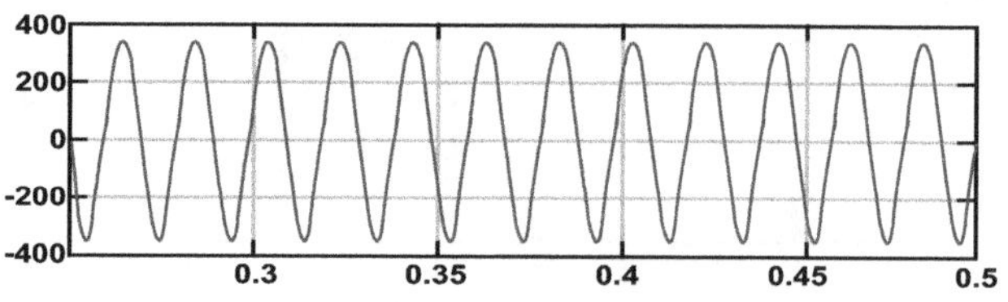

FIGURE 3.29 Source voltage w.r.t time (in seconds) after compensation.

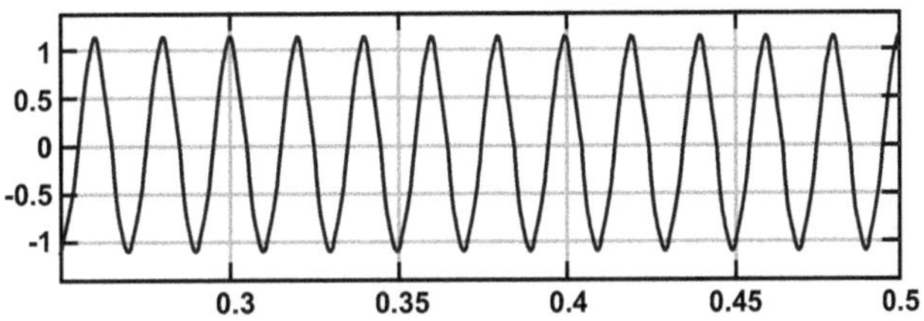

FIGURE 3.30 Source current w.r.t time (in seconds) after compensation.

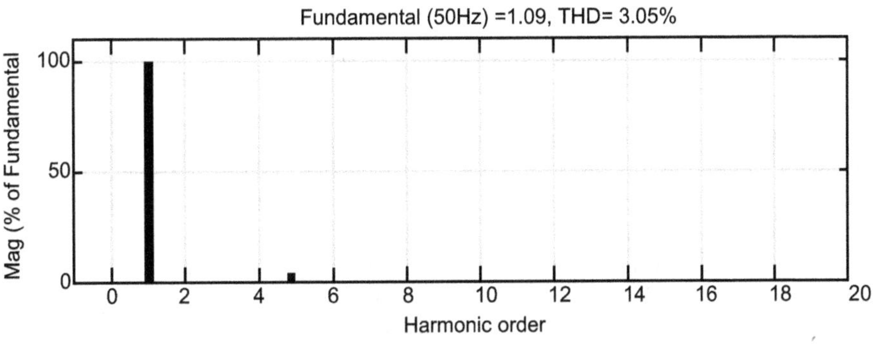

FIGURE 3.31 Spectrum of harmonics before compensation of source voltage.

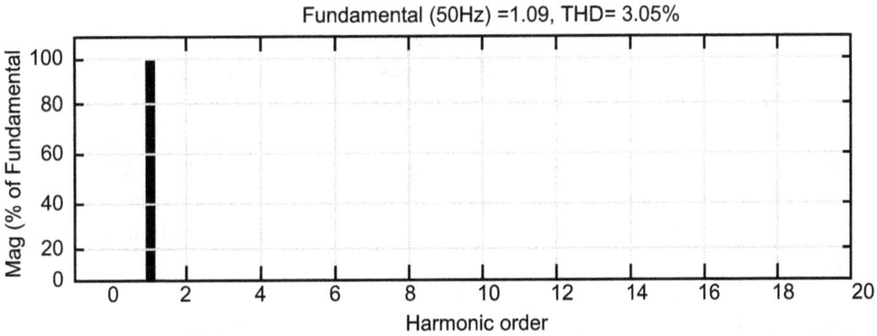

FIGURE 3.32 Spectrum of harmonics after compensation of source voltage.

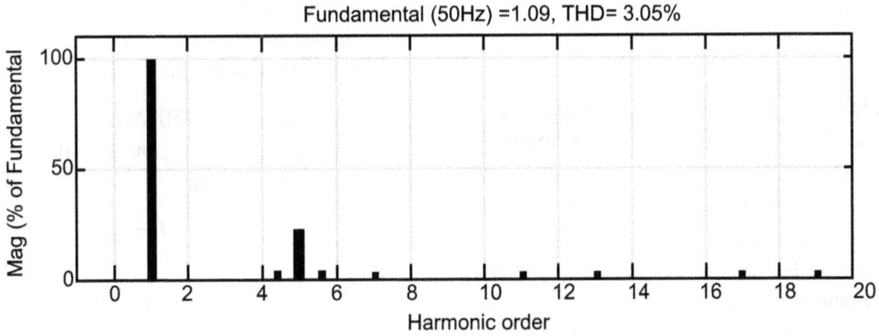

FIGURE 3.33 Spectrum of harmonics before compensation of source current.

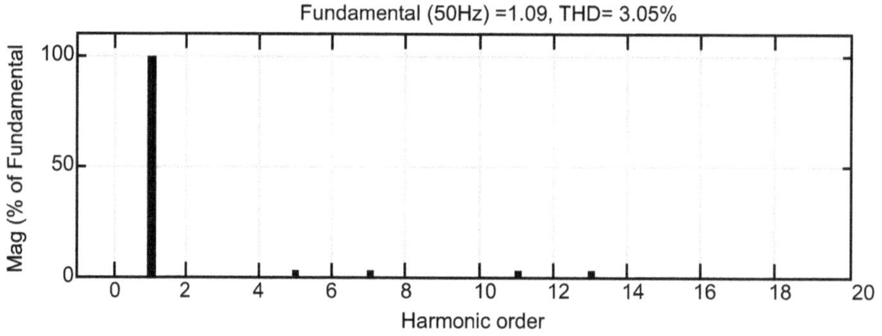

FIGURE 3.34 Spectrum of harmonics after compensation of source current.

From these graphs and FFTs shown in Figure 3.31 and Figure 3.32, it can be said that due to hybrid filtering, the harmonics introduced by non-linear load in voltage side is reduced to 0.43% from 2.81%. The harmonics in current side has been reduced to 3.05% from 25.44%, as seen in Figure 3.33 and Figure 3.34.

3.8 CONCLUSION

A large number of household and industrial loads exist nowadays. These loads require variable supply, meaning that their voltage rating can vary from few volts to kilovolts. Due to this, various power electronics–based devices are used. These power electronics devices are basically non-linear in nature, and therefore they are responsible for injecting harmonics at the point of common coupling. This harmonic current increases heat and reduces efficiency of the system. In this chapter, the mitigation of harmonics in a PV DPGS have been seen. Distributed power generation systems have been studied here. Various issues in distributed power generation systems are listed and discussed. Since this chapter has dealt with PV systems, the equivalent circuit or general structure of PV cell has also been seen. Here, MPPT has been used for extraction of maximum power. The output from PV systems is fed to grids or islanded loads, since harmonic distortion increases due to non-linearity of loads. As it is known that for mitigation of harmonics and reduction of their effects on power quality filters has been used, various filters and their classification, configurations and their applications have been discussed. The harmonic effects on the power quality of the system before and after compensation has been discussed. The harmonic compensation has been done with the configurations of active, passive and hybrid filters. The whole analysis has been done in MATLAB/Simulink.

At first, shunt active filters have been used for compensation. The use of shunt APF mitigated the harmonics at the point of common coupling. Thus maximum power has been transferred with the help of PQ or instantaneous reactive power controlling technique used. After seeing the results, it can be concluded that this filter reduces the total harmonic distortion. As after the introduction of non-linear loads in the system, the THD of the source current due to filter is still 1.74%. It is also observed that at source side, the performance of SAPF is better than that at load side. Second, the performance of the system when LCL passive filter is used has also been discussed. The result shows us that there is a reduction in total harmonics in currents. Here, MPPT controlling theory is applied. Total harmonics in voltages are reduced from 51.83% to 5.44%. At last, hybrid filters were used and their performance in the system is observed. The source current and voltage harmonics is 25.44% and 2.81%, respectively, without filtering stage, and source current harmonics level goes to 3.05% as well as source voltage harmonic becomes 0.43%. After comparing the performance of three filters, it can be concluded that the passive filters are cheap and are used for LV and MV, but SAPF filters are better compared to passive ones. Hybrid filters, as they possess both passive and active configuration, can be used for MV and HV configurations.

3.9 NUMERICALS

Q1. Determine the value of solar power falling on a root area of 20 m^2 for an irradiance of 750 W/m^2 perpendicular to the roof.

Solution: Given that area of the panel, A = 20 m^2

Irradiance, G = 750 W/m^2

The value of solar power will be calculated by the given formula:

$$P = G * A$$

$$P = 750 \times 20$$

$$P = 15{,}000 \text{ Watts} = 15 \text{ KW}$$

Q2. The normal operating cell temperature (NOCT) of a particular solar module is 49°C. Calculate the temperature coefficient of cell. Also determine the value of cell temperature when the irradiance is 650 W/m^2 and the ambient temperature of cell is 40°C.

Solution: Given that irradiance, G = 650 W/m^2

Ambient temperature of cell, T_a = 40°C

Normal operating cell temperature, NOCT = 49°C

Temperature coefficient of cell is found by:

$$k = \frac{NOCT - 20}{800}$$

$$k = \frac{49 - 20}{800}$$

$$k = \frac{29}{800} = 0.03625°C / W / m^2$$

The value of cell temperature of the module is given by:

$$T_{cell} = T_a + K * G$$

$$T_{cell} = 40 + 0.03625 * 650$$

$$T_{cell} = 63.5625°C$$

Q3. A photo diode delivers 3 A short-circuit current under a certain condition. A current of 60 pA is its reverse saturation current. Calculate the open circuit voltage if all the resistances are neglected.
Solution: Given that the short-circuit current, $I_s = 3$ A

The reverse saturation current, $I_D = 60$ pA
Since all the resistances are neglected, therefore the open circuit voltage is calculated by:

$$V_{oc} = \frac{kT}{q} \ln \frac{I_S}{I_D}$$

$$V_{oc} = 0.026 \ln \frac{3}{60 * 10^{-12}}$$

$$V_{oc} = 0.64 \text{ volt.}$$

PRACTICE PROBLEMS

1. For a particular solar module, the NOCT is given as 49°C. Calculate the cell temperature of the module for an ambient temperature of 30°C and irradiance of 500 W/m^2.
   Answer: 48.125 °C
2. Determine the daily amount of solar energy received by a solar water collector having dimensions of 2 m × 2 m, situated in a location receiving 30 MJ/m^2 in five days.
   Answer: 24 MJ

REFERENCES

[1] A report on "Growth of electricity sector in India from 1947–2020" by Central Electricity Authority, Ministry of Power, Government of India, New Delhi, Oct. 2020.

[2] Agarwal, U., Jain, N., Distributed energy resources and supportive methodologies for their optimal planning under modern distribution network: A review. *Technology and Economics of Smart Grids and Sustainable Energy*, vol. 4, no. 10, Jan. 2019. http://doi.org/10.1007/s40866-019-0060-6.

[3] Singh, S. N., Østergaard, J., Jain, N., Distributed generation in power systems: An overview and key issues. *Proceedings of IEC, 24th Indian Engineering Congress*, NIT Surathkal, Kerala, Dec. 2009. www.researchgate.net/publication/228838433_Distributed_Generation_in_Power_Systems_An_Overview_and_Key_Issues.

[4] Jain, N., Singh, S. N., Wen, F., Distribution Generation recent trends and future challenges. *International Journal of Power and Energy Systems*, vol. 23, pp. 53–61, 2008. www.researchgate.net/publication/236965777_Distribution_Generation_recent_trends_and_future_challenges.

[5] Preda, T., Uhlen, K., Nordgård, D. E., An overview of the present grid codes for integration of distributed generation. *CIRED 2012 Workshop: Integration of Renewables into the Distribution Grid*, Portugal, pp. 1–4, May 2012. http://doi.org/10.1049/cp.2012.0716.

[6] Prakash, P., Khatod, D. K., Optimal sizing and siting techniques for distributed generation in distribution systems: A review. *Renewable and Sustainable Energy Reviews*, vol. 57, pp. 111–130, May 2016. http://doi.org/10.1016/j.rser.2015.12.099.

[7] Seuss, J., Reno, M. J., Lave, M., Broderick, R. J., Grijalva, S., Advanced inverter controls to dispatch distributed PV systems. *IEEE 43rd Photovoltaic Specialists Conference (PVSC)*, IEEE, Portland, pp. 1387–1392, Jun. 2016. http://doi.org/10.1109/PVSC.2016.7749842.

[8] Bagi, S. M., Kudchi, F. N., Bagewadi, S., Power quality improvement using a shunt active Power filter for grid connected photovoltaic generation system. *2020 IEEE Bangalore Humanitarian Technology Conference (B-HTC)*, IEEE, BLDE's V P Dr. P G Halakati College of Engineering and Technology, Vijiyapur, Karnataka, pp. 1–4, Oct. 2020. http://doi.org/10.1109/B-HTC50970.2020.9298001.

[9] Kumar, S., Singh, B., Optimum filtering theory based control for grid tied PV-battery microgrid system. *IEEE International Conference on Power Electronics, Drives and Energy Systems (PEDES)*, IEEE, Jaipur, pp. 1–5, Dec. 2018. http://doi.org/10.1109/PEDES.2018.8707626.

[10] Colque, J. C., Ruppert, E., Delgado-Huayta, I., Azcue, J. L., Application of three-phase grid-tied PV system for the electrical grid power factor improved with filtering function. *IEEE XXVI International Conference on Electronics, Electrical Engineering and Computing (INTERCON)*, IEEE, Lima, Peru, pp. 1–4, Aug. 2019. http://doi.org/10.1109/INTERCON.2019.8853553.

[11] Tali, M., Obbadi, A., Elfajri, A., Errami, Y., Passive filter for harmonics mitigation in standalone PV system for nonlinear load. *International Renewable and Sustainable Energy Conference (IRSEC)*, IEEE, Morocco, pp. 499–504, Oct. 2014. http://doi.org/10.1109/IRSEC.2014.7059834.

[12] Delfanti, M., Falabretti, D., Merlo, M., Monfredini, G., Distributed generation integration in the electric grid: Energy storage system for frequency control. *Journal of Applied Mathematics*, Article 198427, pp. 1–14, 2014. http://doi.org/10.1155/2014/198427.

[13] Karaca, M., Mamizadeh, A., Genc, N., Sular, A., Analysis of passive filters for PV inverters under variable irradiances. *8th International Conference on Renewable Energy Research and Applications (ICRERA)*, pp. 680–685, 2019. http://doi.org/10.1109/ICRERA47325.2019.8997111.

[14] Anees, S., Grid integration of renewable energy sources: Challenges, issues and possible solutions. *5th India International Conference on Power Electronics (IICPE)*, IEEE,

Delhi, pp. 1–6, 2012. http://doi.org/10.1109/IICPE.2012.6450514.

[15] Yamujala, S., Fatima, M., Sri Teja, I. A., Lahari, Y., Present scenario of distributed generation in India—technologies, cost analysis & power quality issues. *International Conference on Innovative Applications of Computational Intelligence on Power, Energy and Controls with their Impact on Humanity (CIPECH14)*, Krishna Institute of Engineering and Technology Ghaziabad, India, pp. 417–421, Nov. 2014. http://doi.org/10.1109/CIPECH.2014.7019089.

[16] Naresh, M., Tripathi, R. K., Power flow control and power quality issues in distributed generation system. *IEEE 1st International Conference on Power Electronics, Intelligent Control and Energy Systems (ICPEICES)*, IEEE, New Delhi, pp. 1–5, Jul. 2016. http://doi.org/10.1109/ICPEICES.2016.7853558.

[17] Dulău, L. I., Abrudean, M., Bică, D., Effects of distributed generation on electric power systems. *Procedia Technology*, vol. 12, pp. 681–686, 2014. http://doi.org/10.1016/j.protcy.2013.12.549.

[18] Wang, B., Sun, M., Dong, B., The existed problems and possible solutions of distributed generation microgrid operation. *Asia-Pacific Power and Energy Engineering Conference*, IEEE, Wuhan, China, pp. 1–4, Mar. 2011. http://doi.org/10.1109/APPEEC.2011.5748616.

[19] Driesen, J., Belmans, R., Distributed generation: Challenges and possible solutions. *IEEE Power Engineering Society General Meeting*, vol. 8, 2006. http://doi.org/10.1109/PES.2006.1709099.

[20] Singh, P., Paliwal, P., Arya, A., A review on challenges and techniques for secondary control of microgrid. *IOP Conf. Series: Materials Science and Engineering*, vol. 561, Nov. 2019. http://doi.org/10.1088/1757-899X/561/1/012075.

[21] Jain, R., Arya, A., A comprehensive review on micro grid operation, challenges and control strategies. *Proceedings of the 2015 ACM Sixth International Conference on Future Energy Systems*, Bangalore, 295–300, Jul. 2015. http://doi.org/10.1145/2768510.2768514.

[22] Vineetha, C. P., Babu, C. A., Smart grid challenges, issues and solutions. *International Conference on Intelligent Green Building and Smart Grid (IGBSG)*, IEEE, Taipei, Taiwan, pp. 1–4, Apr. 2014. http://doi.org/10.1109/IGBSG.2014.6835208.

[23] Blaabjerg, F., Yang, Y., Yang, D., Wang, X., Distributed power-generation systems and protection. *Proceedings of the IEEE*, vol. 105, no. 7, pp. 1311–1331, Jul. 2017. http://doi.org/10.1109/JPROC.2017.2696878.

[24] Chahyal, D. M., Kalgunde, M. N., Kalage, A. A., A study on protection issues in presence of distributed generation. *International Conference on Algorithms, Methodology, Models and Applications in Emerging Technologies (ICAMMAET)*, IEEE, Chennai, Tamil Nadu, pp. 1–4, Feb. 2017. http://doi.org/10.1109/ICAMMAET.2017.8186734.

[25] Arya, A., Verma, S., Mehroliya, S., Tomar, S., Rajeshwari, C. S., Optimal placement of distributed generators in power system using sensitivity analysis. In: Bansal, R. C., Agarwal, A., Jadoun, V. K. (eds) *Advances in Energy Technology. Lecture Notes in Electrical Engineering*, vol. 766. Springer, Singapore, 2022. https://doi.org/10.1007/978-981-16-1476-7_67.

[26] Purchala, K., Belmans, R., Leuven, K. U., Exarchakos, L., Hawkes, A., *Distributed Generation and the Grid Integration Issues*, 2006. www.eusustel.be/public/documents_publ/WP/WP3/WP%203.4.1%20Distributed%20generation%20and%20grid%20integration%20issues.pdf.

[27] Chen, H., Sun, Y., Chen, W., Harmonic suppression of grid-connected distributed generation based on novel Hybrid power filter. *4th IEEE Conference on Industrial Electronics and Applications*, IEEE, Xi'an, China, pp. 2914–2918, May 2009. http://doi.org/10.1109/ICIEA.2009.5138742.

[28] Electrical Engineering and Science for Manufacturing Applications, *Power systems-filtering techniques for power quality improvement (partl)*, Jan. 2021. www.industrial-electronics.com/PEaMD2e_38.html.

[29] Smadi, A., Lei, H., Johnson, B. K., Distribution system harmonic mitigation using a PV system with hybrid active filter features. *NorthAmerican Power Symposium (NAPS)*, pp. 1–6, 2019. http://doi.org/10.1109/NAPS46351.2019.9000238.

[30] Jianjun, Gu., Dianguo, Xu., Hankui, Liu., Maozhong, Gong, Unified power quality conditioner (UPQC): The principle, control and application. *Proceedings of the Power Conversion Conference-Osaka (Cat. No.02TH8579)*, vol.1, pp. 80–85, 2002. http://doi.org/10.1109/PCC.2002.998518.

[31] Aswal, J., Pal, Y., Passive and active filter for harmonic mitigation in a 3-phase, 3-wire system. *2nd International Conference on Inventive Systems and Control (ICISC)*, IEEE, Piscataway, NJ, pp. 668–672, 2018. http://doi.org/10.1109/ICISC.2018.8398882.

[32] Jayaraman, M., Sreedevi, V. T., Balakrishnan, R., Analysis and design of passive filters for power quality improvement in standalone PV systems. *International Conference on Engineering (NUiCONE)*, Nirma University, pp. 1–6, 2013. http://doi.org/10.1109/NUiCONE.2013.6780164.

[33] Chtouki, A., Zazi, M., Feddi, M., Rayyam, M., LCL filter with passive damping for PV system connected to the network. *International Renewable and Sustainable Energy Conference (IRSEC)*, IEEE, Morocco, pp. 692–697, Nov.2016. http://doi.org/10.1109/IRSEC.2016.7984020.

[34] Hojabri, M., Toudeshki, A., Third-order passive filter improvement for renewable energy systems to meet IEEE 519–1992 standard limits. *IEEE Conference on Energy Conversion (CENCON)*, IEEE, Malaysia, pp. 199–204, Oct. 2015. http://doi.org/10.1109/CENCON.2015.7409539.

[35] Das, S. S., Panda, A., LCL filter based solar photovoltaic distributed generation system. *IEEE International Conference on Power Electronics, Drives and Energy Systems (PEDES)*, IEEE, Jaipur, pp. 1–6, 2018. http://doi.org/10.1109/PEDES.2018.8707818.

[36] Sanjay, J. S., Misra, B., Power quality improvement for non linear load applications using passive filters. *3rd International Conference on Recent Developments in Control, Automation & Power Engineering (RDCAPE)*, pp. 585–589, 2019. http://doi.org/10.1109/RDCAPE47089.2019.8979035.

[37] Ahrabian, G., Shahnia, F., Haque, M. T., Hybrid filter applications for power quality improvement of power distribution networks utilizing renewable energies. *IEEE International Symposium on Industrial Electronics*, pp. 1161–1165, 2006. http://doi.org/10.1109/ISIE.2006.295801.

[38] Blorfan, A., Wira, P., Flieller, D., Sturtzer, G., Mercklé, J., A three-phase hybrid active power filter with photovoltaic generation and hysteresis current control. *IECON 2011–37th Annual Conference of the IEEE Industrial Electronics Society*, IEEE, Melbourne, Australia, pp. 4316–4321, Nov. 2011. http://doi.org/10.1109/IECON.2011.6120018.

[39] Sriranjani, R., Jayalalitha, S., PV interconnected shunt active filter for power balancing and harmonic mitigation. *International Journal of Pure and Applied Mathematics*, vol. 118, no. 18, pp. 2341–2354, 2018. www.ijpam.eu.

[40] Dash, R., Paikray, P., Swain, S. C., Active power filter for harmonic mitigation in a distributed power generation system. *Innovations in Power and Advanced Computing Technologies (i-PACT)*, IEEE, Vellore, pp. 1–6, Apr. 2017. http://doi.org/10.1109/IPACT.2017.8245204.

[41] Corasaniti, V. F., Barbieri, M. B., Arnera, P. L., Valla, M. I., Comparison of active filters topologies in medium voltage distribution power systems. *IEEE Power and Energy Society General Meeting—Conversion and Delivery of Electrical Energy in the 21st Century*, IEEE, Pittsburgh, PA, USA, pp. 1–8, Jul. 2008. http://doi.org/10.1109/PES.2008.4596665.

[42] Naik Anant Jaivant, *Integration of Grid Connected Photovoltaic System WithActive Power Filtering Functionality.* NIT Surathkal, Karnataka, 2014.https://idr.nitk.ac.in/jspui/bitstream/123456789/14374/1/100682EE10F02.pdf.

[43] Babu, P. Narendra, Kar, B., Halder, B., Comparative analysis of a Hybrid active power filter for power quality improvement using different compensation techniques. *International Conference on Recent Advances and Innovations in Engineering (ICRAIE)*, IEEE, Jaipur India, pp. 1–6, 2016, http://doi.org/10.1109/ICRAIE.2016.7939483.

[44] Kotturu, S., Kotturu, J., Reddy, C. Kumar., Reduction of harmonics in 3-Phase, 3-Wire system by the use of shunt active filter. *International Conference on Circuit, Power and Computing Technologies [ICCPCT]*, Institute of Electrical and Electronics Engineering, Nagercoil, India, pp. 7–12, Mar. 2014. http://doi.org/10.1109/ICCPCT.2014.7054767.

[45] Devi, R. Saradha, Seyezhai, R., Mrudhulaa, P. V., Priyadharshini, K., Mitigation of harmonics in a grid connected photovoltaic inverter. *International Journal of Innovative Technology and Exploring Engineering (IJITEE)*, vol. 8, no. 10, pp. 1166–1172, Aug. 2019. http://doi.org/10.35940/ijitee.J9149.0881019.

[46] Sharma, D., Nagar, Y. K., Agrawal, S., Kumar, M., Performance analysis of hybrid filter to mitigate harmonics. *International Conference on Research Trends in Engineering*, Applied Science and Management (ICRTESM-2017), pp. 403–406, 2017. www.researchgate.net/publication/316038654_Performance_Analysis_of_Hybrid_Filter_to_Mitigate_Harmonics.

[47] Khanna, R., Chacko, S. T., Goel, N., Performance and investigation of hybrid filters for Power Quality Improvement. *5th International Power Engineering and Optimization Conference*, IEEE PES Malaysia, Malaysia, pp. 93–97, Jun. 2011. http://doi.org/10.1109/PEOCO.2011.5970409.

4 A Performance Review of Control Algorithms for VSC to Improve the Power Quality Features in Single-Phase Wind-Based Distributed Generation

Ashutosh K. Giri, Sabha Raj Arya and Rakesh Maurya

CONTENTS

4.1 INTRODUCTION

In this chapter, the authors discuss the power quality problems and their remedies in single-phase SEIG utilized as a wind energy–driven islanded distributed power generation system. This generator is useful for feeding single-phase loads located at very remote places where grid supply is not practical. The operation of single-phase SEIG in isolated mode is vulnerable to the load changes, meaning that frequency and voltage both are variable with the load dynamics [1–4]. Further, nonlinear loading on the generator creates harmonics in the source current, which severely affects the generator performance. For controlled and reliable operation of SEIG, custom power devices like VSC are used across the load at common point of interfacing [5–9]. The operation of the entire system—comprising generator, load and VSC—are dependent on the controlled operation of VSC. For VSC control, fast and robust control algorithms are required to generate the modulating signal to create the firing pulses [10–12]. Some researchers have proposed the controlled operation of single-phase SEIG with a variable capacitor in the excitation [13]. However, its dynamical operation is slow. Moreover, change in the capacitance is possible only in discrete steps. Another class of algorithms called adaptive control algorithms are very effective under the fast-varying conditions of voltage and frequency such as wind-based distributed power generation systems. Therefore, the three adaptive-type control algorithms such as AANF [14–16], AVF [17] and VLGLMS [18] have been implemented to control the operation of SEIG through VSC. The use of some more control algorithm is proposed in references [19–28] for the controlled operation of VSC in this area. The control algorithm application reported is mainly based on PLL or some classical type such as power balance theory, instantaneous reactive power theory and synchronous reference frame theory based control. These control algorithms work with good accuracy and with faster dynamic response under grid supply conditions where frequency and voltage variations are not fast; however, their response is slightly slower under the dynamics of distributed generation system. Detailed mathematical analysis, simulation studies and experimental validation have been presented for each aforementioned control in the following subsections.

4.2 CONFIGURATION OF DISTRIBUTED POWER GENERATION SYSTEMS

The configuration of distributed power generation systems taken under study is depicted in Figure 4.1. A VSC is interfaced in parallel to linear/nonlinear load fed by single-phase SEIG. The prime mover for rotating the SEIG is a horizontal axis unregulated wind turbine. To produce nominal voltage at no load in this system, a capacitor connected to the auxiliary winding is required. An interfacing inductor (L_f) is interfaced between VSC and the common point of interfacing (CPI) to reduce the ripple in the compensating current.

A single unit of first order damped high pass filter (R_f, C_f) is connected at the CPI to filter the VSC's high frequency switching noise. A linear and ballast load is also connected, in addition to the nonlinear load. The ballast load is used only when power mismatch occurs between

DOI: 10.1201/9781003229124-4

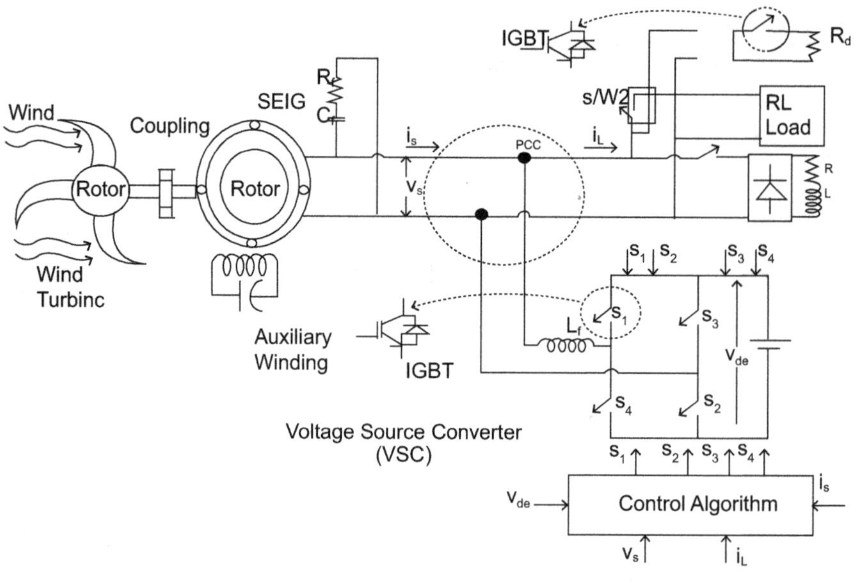

FIGURE 4.1 Proposed system configuration based on single-phase SEIG.

generation and demand. The compensating current is usually supplied from the VSC to compensate for the load current's reactive power and harmonics. As a result, distortion in the source current due to nonlinear generator loading is reduced to the level required by IEEE standard 519–1992. The VSC's switch rating is determined by the amount of harmonic current and reactive power that must be compensated. Appendix A contains the values for the remaining design parameters.

4.3 DESIGN OF SYSTEM COMPONENTS IN DISTRIBUTED POWER GENERATION SYSTEMS

The proposed system under study comprises wind-driven SEIG as a major component. However, choosing an excitation capacitor is a critical task because it affects not only the voltage build-up process at no load, but also electromagnetic torque and isolated system frequency. Each component is created to meet the requirement of the system. The design is described in the subsections that follow.

4.3.1 Selection of the Induction Generator [1, 3, 4]

In a wind-based distributed generation system, a two-winding single-phase induction generator is used to generate electrical power while taking into account the following factors.

1. Wind energy applications appear to benefit from lower maintenance needs and simplified controls.
2. The induction generator is preferred for wind power plants because of its simplicity, robustness, and small size per generated kilowatt.
3. Natural short-circuit protection and low cost compared to other generators.
4. Induction generators accept both constant and variable loads, can start with or without a load, and can operate continuously or intermittently, which is ideal for wind applications with erratic input power.
5. An induction generator with a squirrel-cage rotor can be designed specifically for use with wind or hydro turbines; for example, with a higher slip factor, more convenient deformation in the torque curve, winding sized to support higher saturation current, and so on.

Based on the aforementioned criteria and for targeting the standalone single-phase remotely located loads, the rating of the generator is selected and provided in Appendix A.

4.3.2 Selection of Excitation Capacitor [15–18]

The capacitor in the auxiliary winding should be selected such that the necessary no load voltage build-up across the main winding at specified takes place. The deviation of no-load terminal voltage (*Vs*) of main winding with excitation capacitance is shown in Figure 4.2. It is clear that (*Vs*) increases with capacitor at a known speed. The suitable value of capacitor can be selected depending upon the maximum acceptable voltage across the machine terminals.

In this study, a capacitor of 53μF was found most suitable to build up the terminal voltage of 241V at 1,700 rpm. It is noted that the no-load voltage is susceptible to speed variation and shown in Figure 4.2. Obviously, bigger capacitance is obligatory for increased voltage for the same speed.

4.4 CONTROL ALGORITHMS FOR VSC

The control algorithm of VSC extracts reference supply current from the load current and operates VSC in such

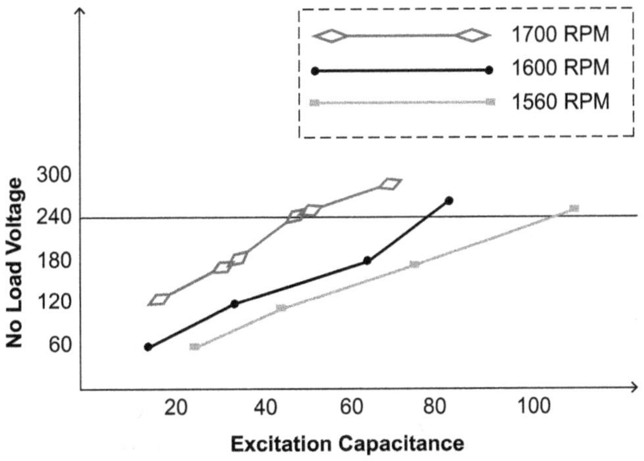

FIGURE 4.2 No-load terminal voltage (*Vs*) excitation capacitor at different speed.

TABLE 4.1
Terminal Voltage Variation Capacitance

Sr. No.	Rotor Speed (RPM)	Value of Capacitance (μF)	No-Load Voltage (Volts)
1	1,700	16	121
		37	170
		53	241
		65	290
2	1,600	18	60
		31	118
		67	172
		82	260
3	1,560	27	50
		42	110
		78	175
		110	245

a manner that it injects compensation current at common point of interfacing (CPI). However, this injection capability mainly depends on the complexity and execution speed of the control scheme. Therefore, fast, accurate and adaptive response is the required features of any control algorithm utilized for reference generation. Many control algorithms based on adaptive control theory have been published in literature for shunt active power filters. In this work, some new adaptive control algorithms have been proposed for VSC which feed single-phase consumer loads.

In this work, three different control approaches are utilized to operate the VSC for improved power quality of the generator. These are explained in what follows.

4.4.1 Amplitude Adaptive Notch Filter (AANF)–Based Control Algorithm [17]

For a VSC-based voltage and frequency regulator of a wind turbine driven islanded single-phase SEIG, an AANF-based control algorithm is proposed. For the sake of power quality and utilization of its installed capacity, the single-phase SEIG is supposed to supply only the fundamental real power component of load current and small losses in various parts of VSC. By adjusting the power consumption between the various parts of the system, the system's frequency is kept constant. The harmonics current and the oscillating component of power demanded by the unbalanced load, as well as SEIG, are supplied by the VSC. This algorithm takes as inputs the voltage at the PCC (V_s), load current (i_L) and source current (i_s). The following are the basic equations for estimating in phase and oscillating constituents of reference source current and total reference source current for real-time monitoring and correction using an AANF-based control algorithm.

4.5 SIMULATION PERFORMANCE

In this section, the performance of the standalone system has been investigated under different operating conditions, and the results are presented. For effective operation of the VSC, it is validated using the MATLAB Simulink tool. The voltage profile control, current profile, frequency control and harmonic eradication functions of the VSC are designed and controlled. Figures 4.3–4.7 portray performance results under various operating conditions.

4.5.1 Performance of SEIG in the Fixed/Change in Wind-Speed Mode Feeding Linear Load

The performance of wind-driven SEIG is shown in Figure 4.3(a) and Figure 4.3(b) under dynamic conditions. Under constant wind velocity mode operation of SEIG, wind velocity is taken to be (v_w = 12 m/s) (normal range of wind velocity is 9–16 m/s [2]), and all results are free of harmonics because either no load or linear load is connected to the generator. The load change is applied twice during the course of simulation. The first load change occurs at (t = 3.5 sec.) and the second load change occurs at (t = 4.1 sec.). During no-load period, it is observed that current is more through compensator path to flow the power towards battery, but while load is applied over the terminal of the generator, power flow towards battery reduces, and consequently, current in compensator circuit also reduces. It is also evident from the results that voltage and frequency are fully controlled throughout the load variations. Now for transient analysis, load is varied from the level (i_L = 9.5A, 0.79pf) to level (i_L = 10A, 0.75pf) at (t = 4.1 sec.). The source current is nearly constant, while the current flowing from the battery rises to meet the increased load demand. Furthermore, the battery's voltage level remains constant with some fluctuations. Because the power flow into the battery is reversed, dump control is disabled in this case. As a result, there is no excess power to be dumped. Other parameters, such as frequency, remain constant around their 50Hz reference value. The results are shown in Figure 4.3(a). Here, change in wind velocity operation of SEIG, velocity of wind is shifted to (v_w = 13.5 m/s) at (t = 4.1 sec.). When the rotor's velocity increases, the frequency rises at the same time. Apart from that, the mutual and self-impedance values of SEIG will increase. As a result, the voltage drop across the PCC terminals increases. This change can be seen from Figure 4.3(b). Now with enhanced speed of the wind at (t = 4.2 sec.), the turbine generates more torque ($T_l \alpha N^2$); hence, output power will also increase. Active power balance at the generator terminal is required to keep the frequency around the reference value. Because of the constant load assumed here, as the battery charging level rises, the terminal voltage rises. For the same load demand, the source current level will rise, and the dump load will be turned on with the help of coordinated control to limit the battery current flow. Frequency and terminal voltage will be maintained as a result of this power balance. The remaining parameters are set to a fixed speed and are kept at their default values. The results are shown in Figure 4.3(b).

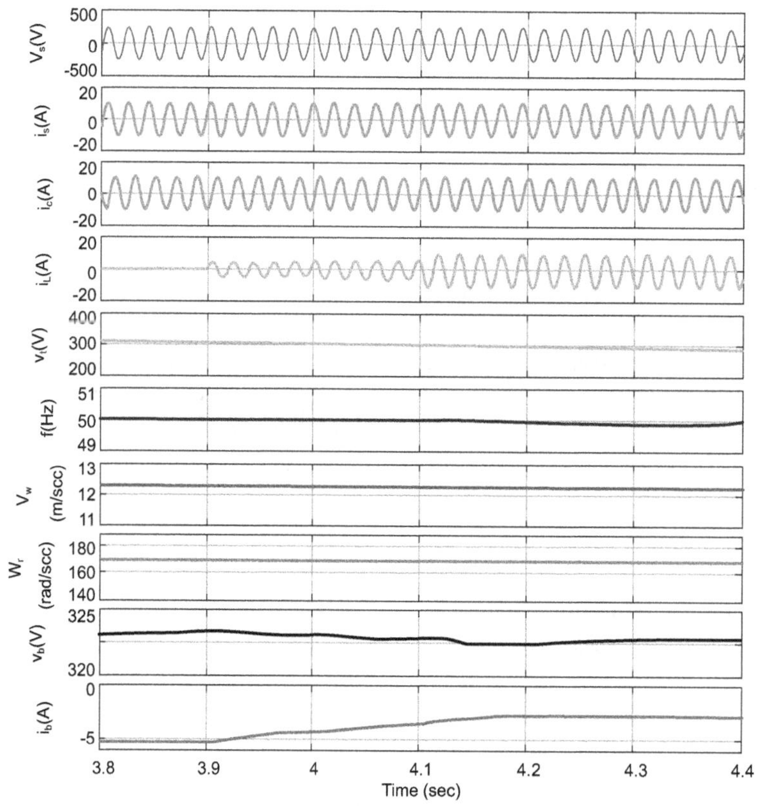

(a)

FIGURE 4.3 Performance of SEIG with: (a) fixed wind speed under linear load; and (b) variable wind speed under linear load.

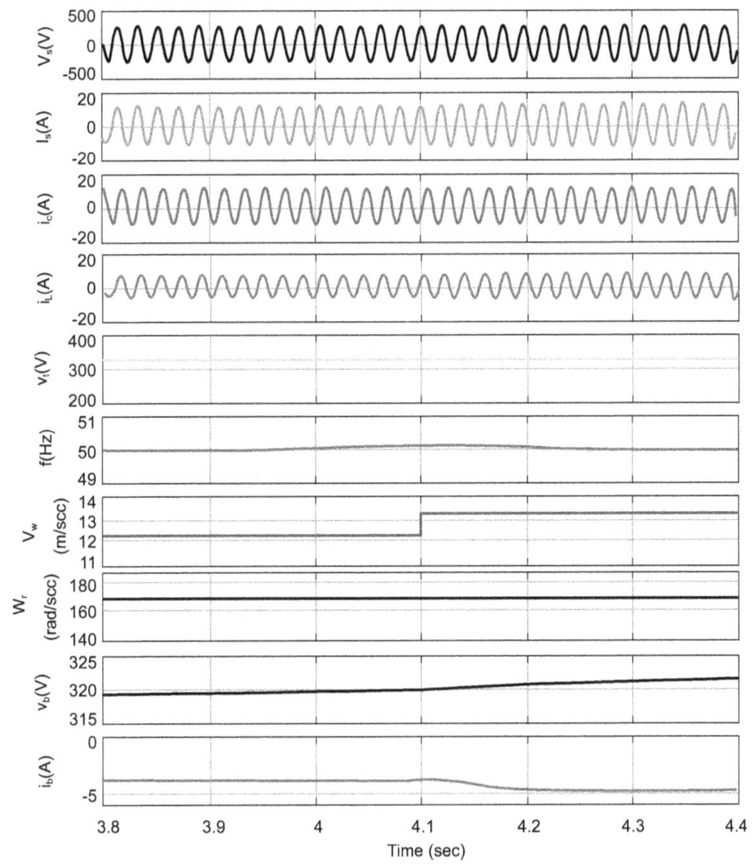

FIGURE 4.3 (Continued)

4.5.2 Performance of SEIG in the Fixed/Variable Wind Speed Mode Under Nonlinear Load

Figures 4.4(a)–(c) show the simulation performance of an SEIG controlled and operated by VSC. As a nonlinear load, a single-phase diode bridge rectifier with an RL load connected on the DC side was used. Except for the diode bridge, the loading condition is the same as in case B. In fixed-speed mode, load is changed at two time instants (at t = 3.9 sec and t = 4.2 sec.) for creating transients. The wind velocity is considered to be (v_w = 12 m/s) under fixed wind velocity of SEIG (normal range of wind velocity is 9–16 m/s [2]), and the voltage amplitude of 330V is observed at CPI. The source current (i_s = 9.8A) is noted down for the connected load (i_L = 12.6A, 0.85pf) at the time instant (t = 4.2 sec. onwards). VSC supplies the reactive and harmonics components of the load current (i_c= 6.8A) in this condition. As can be seen in Figure 4.4(a), SEIG is only supplying the real fundamental component of the load current, because a harmonic component, as well as a reactive component, must be compensated by VSC. The other parameters, such as frequency, are kept at their default value of 50Hz; Figure 4.4(a) depicts them. Under load variation, the battery dynamics are similar to case (b). The wind velocity of the SEIG has been changed to (v_w = 13 m/s) at (t = 4.1 sec.) in variable wind velocity mode. The source current will no longer be sinusoidal due to the presence of non-linearity in the nature of the load. Because the machine is running in standalone mode, this effect will be reflected in the machine parameters. This change can be seen in Figure 4.4(b). Now with enhanced speed of the wind at the time instant (t = 4.1 sec.), the turbine generates more torque ($T_e \alpha N^2$); hence, output power will also increase. Active power balance at the generator terminal is required to keep the frequency around the reference value. For the same load demand, the source current level will rise due to constant terminal voltage, and one coordinated dump controller will be turned on to maintain the frequency and control the charging rate of the battery. Under rated loading conditions with fixed wind speed, total harmonic distortion (THD) of the load current, source current and source voltage is found to be 34.42%, 4.10% and 3.9%, respectively. Figure 4.5(a–c) shows the harmonic spectra, as well as the FFT signal window. Under rated loading conditions and variable wind speed, total harmonic distortion (THD) of the load current, source current, and source voltage is found to be 32.76%, 4.4% and 4.08%, respectively. These figures are well within the IEEE-519–1992 guidelines. Figure 4.6(a–c) shows the harmonic spectra, as well as the FFT signal window.

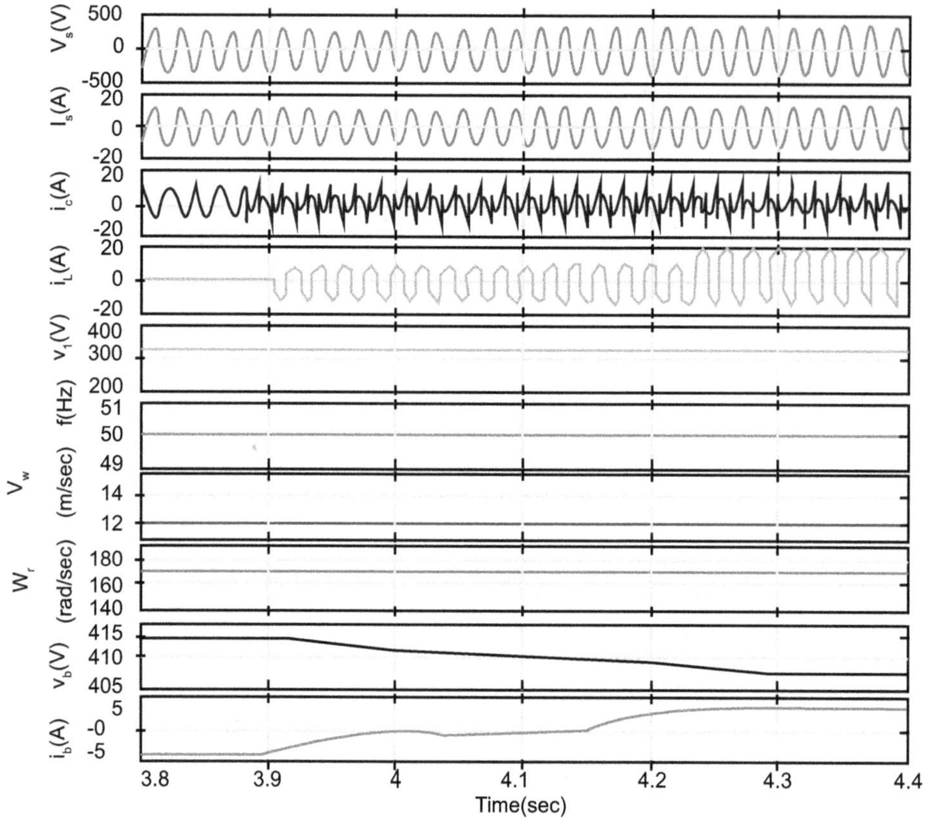

(a)

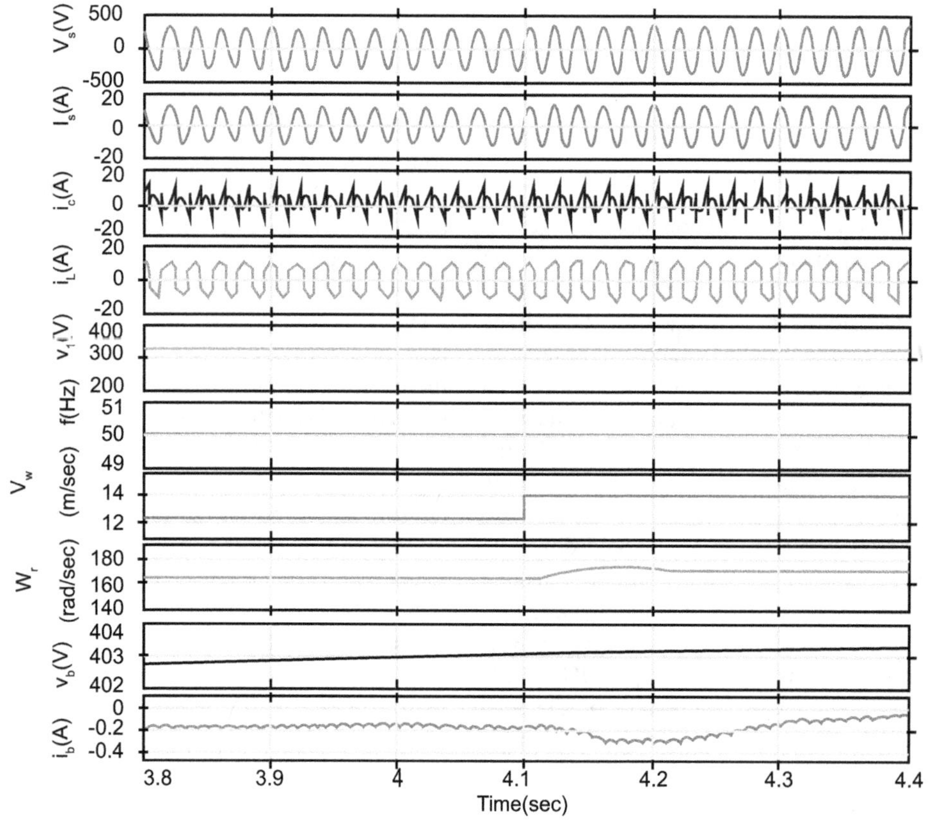

(b)

FIGURE 4.4 Performance of SEIG with: (a) fixed wind speed under nonlinear load; and (b) change in wind speed under nonlinear load.

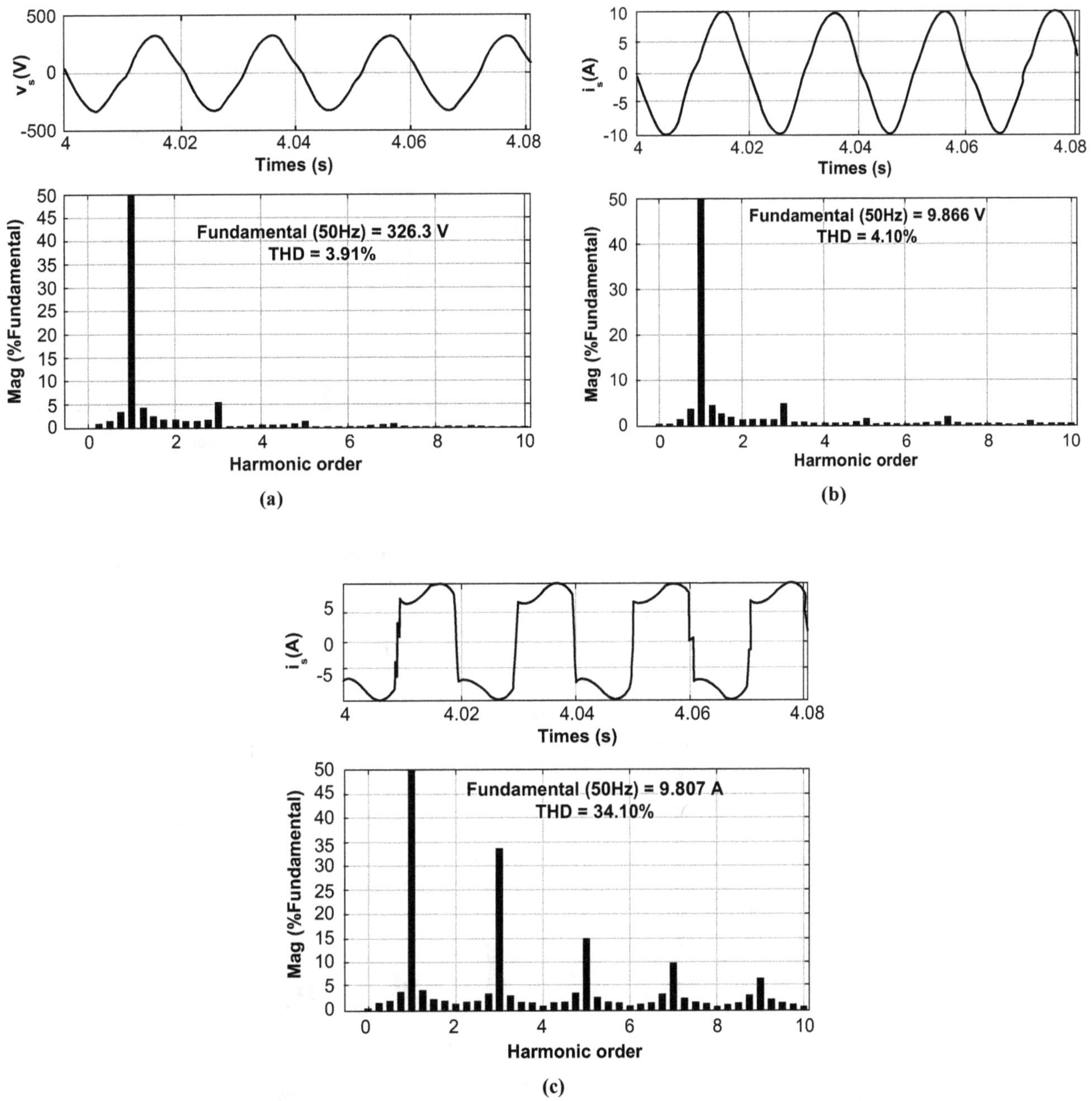

FIGURE 4.5 THD value of: (a) supply voltage; (b) supply current; and (c) load current under nonlinear load with fixed speed.

4.5.3 Power Balance Operation

Figure 4.7 depicts the active power exchange from/to the battery, SEIG, load, and dump load. It is evident from the simulation results that the battery makes the operation balanced either by supplying or absorbing the power mismatch between generator and load. Only when the battery is close to reaching its optimum voltage level does the dump controller intervene. The load is not connected to the PCC of the system in Figure 4.7 (t = 3.4–3.9 sec.), and therefore, all the generated power is stored in the battery. The battery voltage reaches its maximum of 325V. As a result, at the moment (t = 3.6 sec.), some power is diverted to the dump load. When the load is connected at (t = 3.9 sec.), the dump load power decreases, and the battery current flow decreases, as well. Because the generator is overloaded at (t = 4.2 sec.), the battery began discharging to maintain the frequency, causing battery power to flow in the opposite direction.

4.6 EXPERIMENTAL RESULTS

The performance of AANF control algorithm is verified by experimentation. It is tested under dynamics of the system as well as in steady state. The results are shown in the subsequent subsections.

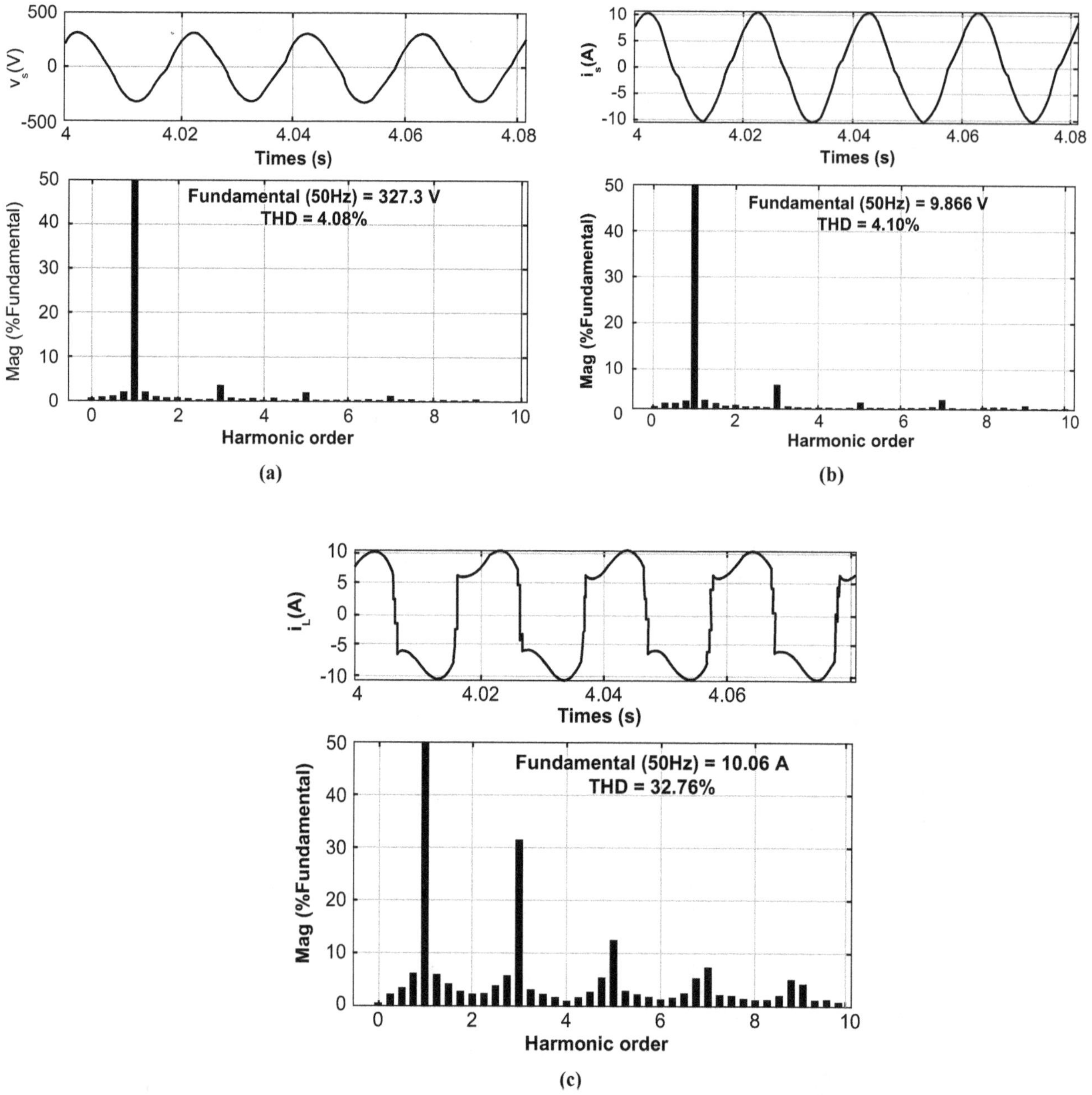

FIGURE 4.6 THD level of: (a) supply voltage; (b) supply current; (c) load current under nonlinear load.

4.6.1 Voltage and Frequency Control of SEIG with Load Dynamics Using AANF Control Algorithm

The dynamics results are taken under load variations and presented in Figure 4.8 (a–d). The objective behind this test is to prove the zero voltage and frequency regulation under any kind of load variation on the terminal of the SEIG. Moreover, the power quality of the current and voltage waveform are also taken into consideration. The load current is changed for introducing the dynamics on the generator load system. From the results shown in Figure 4.8(a), it is clear that the voltage and frequency are intact during load current variation. The compensator current (i_c) ensured the sinusoidal source current (i_s) and supply voltage (v_s) from the generator side though the load current (i_L) is nonlinear in nature. From the waveforms of Figure 4.8(b), it is proved that the dynamics of battery energy storage system (BESS) is taken care to load variation, and hence, frequency of the system remains constant. During the load increment on the generator, battery charging current (i_b) is reduced and thereby supports the generator. Therefore, it can be said that the real power flow is controlled by the AANF algorithm to maintain the frequency. The Figure 4.8(c) depicts the variation in the PCC voltage magnitude (v_t) and frequency (f) under load current variation, and it is found that both are under control. Figure 4.8(d) depicts the relations between various quantities of the generator under the influence of adopted control algorithm.

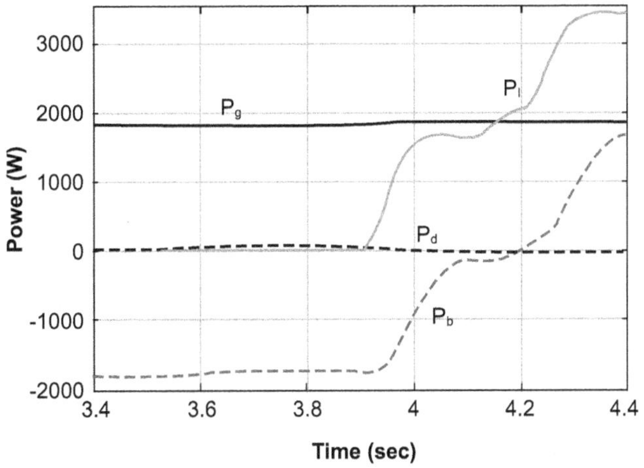

FIGURE 4.7 Power flow balance in the system.

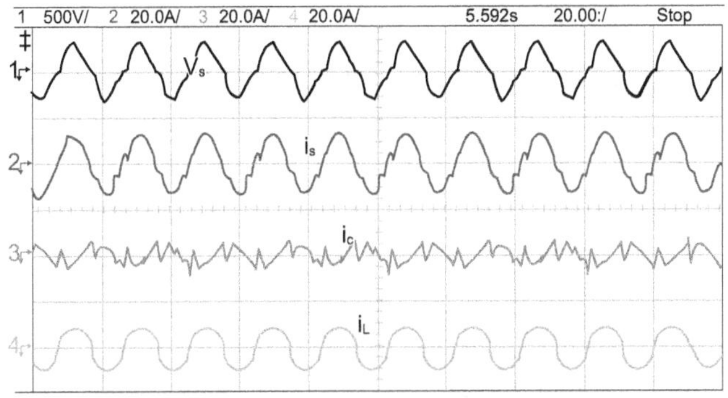

(a) on x axis 20ms/dive and on y axis 1-500v/div,2-20A/div,3-20 A/div, 4-20A/div

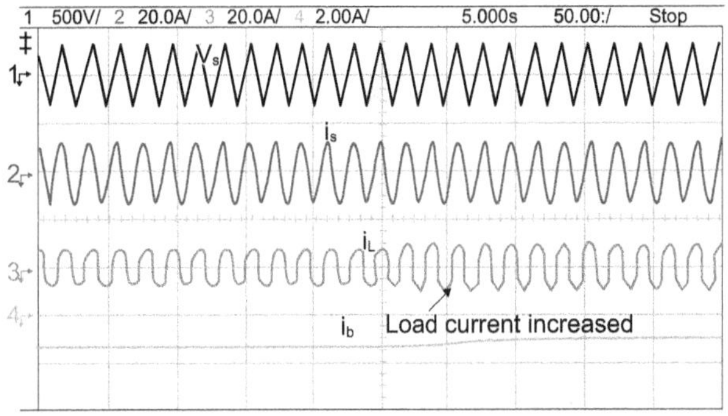

(b) on x axis 50ms/dive and on y axis 1-500v/div,2-20A/div,3-20 A/div,4-20A/div

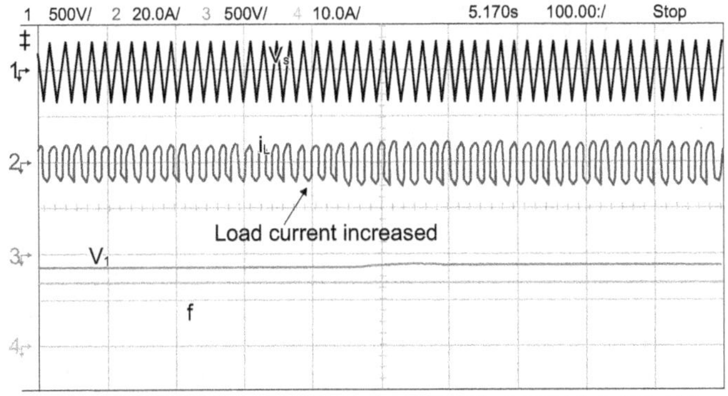

(c) on x axis 100ms/dive and on y axis 1-500v/div,2-20A/div,3-500 A/div, 4-20A/div

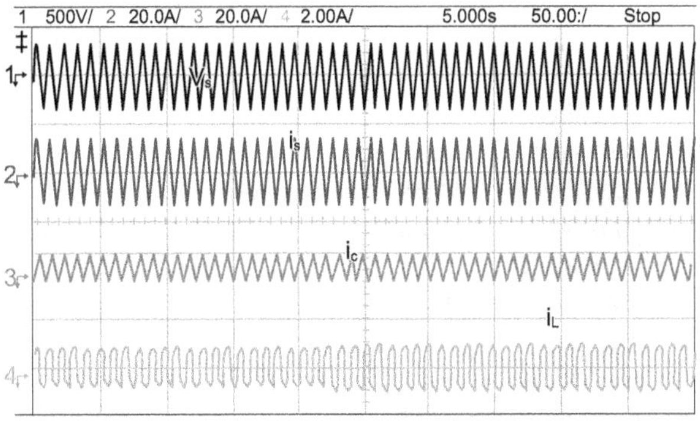

(d) on x axis 100ms/dive and on y axis 1-500v/div,2-20A/div,3-20 A/div,4-20A/div

FIGURE 4.8 Dynamic performance of the SEIG static compensator (STATCOM) with BESS under nonlinear load variation: (a) test results of v_s, i_s, i_c and i_L when load is shifted at some time; (b) test results of v_s, i_s, i_L and i_b when more load is injected; (c) test results of v_s, i_L, v_t and f under load current variation; (d) v_s, i_s, i_c and i_L when load is changed at some time.

4.6.2 Steady-State Operation with AANF Control Algorithm

The steady-state results are also taken and studied to analyze the performance of the adopted control for the system under study. The power quality indices of source voltage (v_s), source current (i_s) and power balance operation are recorded and presented in Figure 4.9. It is evident from the results that the total harmonic distortion in the supply current (i_s) and supply voltage (v_s) are maintained at 4.4% and

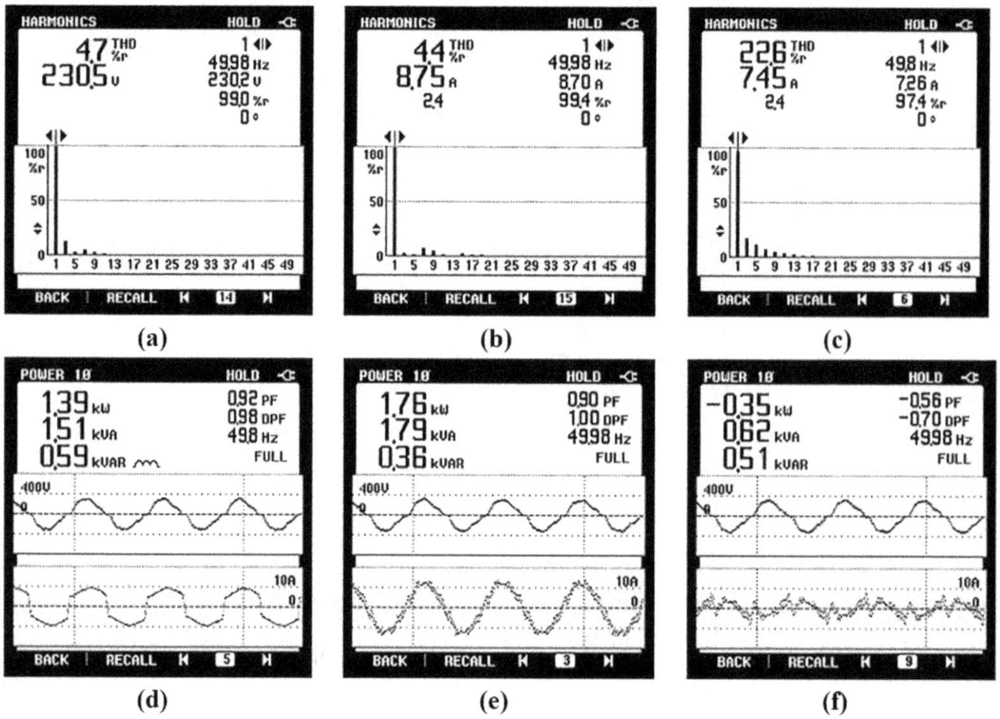

FIGURE 4.9 Steady-state performance of the SEIG STATCOM with BESS under nonlinear load variation: (a) THD of V_s; (b) THD of i_s; (c) THD of i_L; (d) load power; (e) generator power; (f) compensator power

4.7%, respectively, which is below 5% allowable limits as per IEEE-519 standards. The total harmonic distortion in the load current was 22.6%. These results are presented in Figure 4.9(a–c) and clearly indicated that the power quality is maintained at the generator terminal despite the connection of nonlinear load from the generator terminal. The frequency is maintained by the balanced operation of real power between the generator (P_g), load (P_L) and compensator (P_c). These are presented in Figure 4.9(d–f). The real, apparent and reactive power of the load is recorded 1.39 kW, 1.59 kVA and 0.59 kVAR, respectively, while the similar power of the generator is 1.76 kW, 1.79 kVA and 0.36 kVAR, respectively. The compensator power is recorded as 0.35 kW (negative), 0.62 kVA, 0.51 kVAR. These readings of power are clearly indicated that apart from maintaining the power quality, the system maintains the frequency also by balancing the power between the various components of the system.

4.7 CONCLUSION

In this chapter, power quality issues are addressed in detail for single-phase induction generators employed in distributed power generation. Three different adaptive control techniques are implemented to enhance the power quality, as well voltage and frequency control of SEIG working in standalone mode. The performance of the generator is evaluated under different operating conditions. Results in each control are obtained and found to be satisfactory. Under various operating conditions, the proposed controls are capable of generating appropriate control signals, and can also maintain a constant terminal voltage and frequency. These control techniques have improved the power quality features of the induction generator under nonlinear load, according to the experimental results. Table 4.2 presents a summary of the findings. Further, it can be said from the summary of the results that proposed AVF control technique is working relatively more better than other two proposed techniques under the described circumstances. However, power-balanced performance is shown by all and satisfactory.

The proposed control algorithms utilized to control the operation of VSC in single-phase SEIG-fed systems for standalone operation can be compared based on their operational characteristics, though all the controls have been applied in the system taken for study and shown the reliable nature to achieve the objectives. The theoretical comparison of three proposed control algorithm is presented here for better understanding of the reader. The implemented controls are compared based on their characteristics and are presented in Table 4.3.

4.7.1 Comparative Performance of Control Algorithms in Single-Phase SEIG

The comparative analysis for dynamic performance of three control algorithms those are implemented in single-phase distributed power generation system are presented in

Table 4.2
Comparative Performance of AANF, AVF and VLGLMS Control Algorithms

Sr. No.	Proposed Control	Total Harmonic Distortion (THD)			Power Balanced Operation		
		vs	is	iL	Generator	Load	Compensator
1	AANF	4.7%	4.4%	22.6%	1.76 kW, 1.79 kVA, 0.36 kVAR	1.39 kW, 1.5 kVA, 0.59 kVAR	0.35 kW, 0.62 kVA, 0.51 kVAR
2	AVF	4.0%	4.4%	27.1%	1.7 kW, 1.74 kVA, 0.35 kVAR	2.0 kW, 2.25 kVA, 1.01 kVAR	0.31 kW, 0.92 kVA, 0.86 kVAR
3	VLGLMS	4.4%	4.3%	21.0%	1.76 kW, 1.81 kVA, 0.40 kVAR	1.66 kW, 1.42 kVA, 0.65 kVAR	0.15 kW, 3.54 kVA, 3.54 kVAR

Table 4.3
Comparative Performance of AANF, AVF and VLGLMS Control Algorithms

Sr. No	AANF	VLGLMS	AVF
1.	Has a second-order characteristic equation whose stability is sensitive to the selection of damping factor	Stability of the control is dependent on the step size chosen	It is a first-order vectorial nature filter-based control algorithm which has exponential decaying characteristics for any oscillations, and is therefore more suitable to extract fundamental component of load current in oscillatory conditions such as wind variations, etc.
2.	Takes sample time of 80 microseconds for real-time implementation	Takes sample time 60 microseconds for real-time implementation	Takes sample time 45 microseconds for real-time implementation
3.	Computational burden is highest	Computational burden is higher	Computational burden is low
4.	Tuning of filter gain is not able to eliminate the delay in phase angle between fundamental component and reference voltage	There is inherent delay in phase angle between fundamental component and reference voltage	Tuning of filter gain is able to eliminate the delay in phase angle between fundamental component and reference voltage

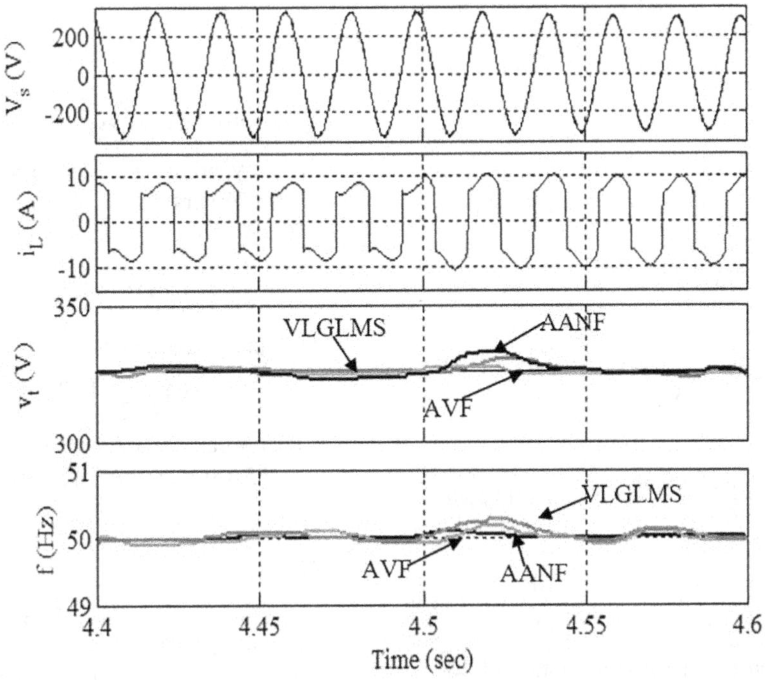

FIGURE 4.10 Comparative analysis of AANF, VLGLMS, and AVF control algorithm implemented in single-phase SEIG.

Figure 4.10. The waveforms of source voltage (v_s), load current (i_L), magnitude of terminal voltage (v_t) and frequency (f) have been observed. The load is changed at the moment (t) is equal to 5 seconds for creating load dynamics. The voltage and frequency response for each control is recorded and plotted on a common axis as shown in Figure 4.10. From the simulation results of load dynamics, it is clear that AVF control algorithm is performing better as compared to AANF and VLGLMS. The reason is its vectorial property and simple structure for fundamental current extraction.

4.8 APPENDIX A

Parameters and Rating of Single-Phase SEIG

Parameters: Winding resistances and inductances: rotor: $R_{mr} = 3.3077\Omega$, $L_{mr} = 0.017544H$, stator: $R_{ms} = 2.93\ \Omega$, $L_{ms} = 0.0267544H$, mutual inductance: $L_{ms} = 0.13814H$, auxiliary winding: stator: $R_{as} = 5.0268\ \Omega$, $L_{as} = 0.024007H$; Inertia = 0.00290763 J (kg/m^2),

Ratings: 2.2 kW, 220V, 50Hz and four-pole.

Compensator parameters: Ls = 5 mH; $R_f = 8\Omega$; $C_f = 12\mu F$; $C_{dc} = 2300\mu F$.

BESS Parameters: lithium-ion type, 400V, 7.5Ah, SOC (10–90%), $r_s = 0.05\Omega$.

Dump load: 220V, 100Ω, 500W.

4.9 SOME EXAMPLES

Q1. A hydroelectric station has to operate with a mean head of 50 m. It utilizes water over a catchment area of 200 km^2. Average annual rainfall is 420 cm with 30% loss due to evaporation. Turbine efficiency is 85% and alternator efficiency is 80%. Evaluate the average power that can be generated.

Answer

As per the question, given 70% of water is used for generation (that is, quantity of water available for utilization is):

$$Q = 200\times10^6\times4.2\times0.7 = 588\times10^6\,m^3$$

h = 50 m

Overall power plant efficiency = 0.8 × 0.85 = 0.68

$$\text{Total energy available} = \frac{9.81QWh\eta\times10^{-3}}{60\times60}$$

$$= \frac{9.81\times588\times10^6\times50\times0.68}{60\times60} = 54.47\times10^6\,kwh$$

$$\text{Average power P} = \frac{Total\ power}{24\times365} = \frac{54.47\times10^6}{24\times365} = 6.22Mw$$

Q2. The peak demand of a generating station of 100 Mw with 65% of load factor, plant capacity factor 50% and plant use factor 75, find the following.

Daily energy generated
Plant storage capacity
Maximum energy that can be produced daily if plants are running all the time
Maximum energy that can be produced daily if plants are fully loaded

Answer

Peak demand (PD) = 100 Mw
Load factor (LF) = 0.65
Plant capacity factor = 0.5
So, average load = MD × LF = 100×0.65 = 65 Mw

Plant use factor = average load × 24 = 65 × 24 = 1560 Mwh

$$\text{Plant Capacity} = \frac{65}{0.5} = 130Mw$$

Plant storage capacity = 130–100 = 30 Mw

Maximum energy that can be produced daily if plants are running all the time $= \dfrac{Actual\ energy\ generated}{Plant\ capacity\ factor}$

$$= \frac{1560}{0.5} = 3120Mwh$$

Maximum energy that can be produced daily if plants are fully loaded $= \dfrac{Actual\ energy\ generated}{Plant\ capacity\ factor} = \dfrac{1560}{0.75} = 2080Mwh$

Q3. The wind turbine is being operated with environmental conditions ($\rho = 1.225$ kg/m^3). It is considered that a turbine efficiency is 0.4. Evaluate the annual energy generation from a wind turbine with a 12 m diameter horizontal axis, operating in a wind regime with an average wind velocity of 8 m/s.

Solutions

The per year energy generation (in kWh) can be evaluated as follows.

Energy production (kWh) = (efficiency) × (*P*) × (8,760 hrs)

Where *P* can be calculated as

$$P = \pi R^2 \frac{1}{2}\rho U^3,$$

for an average wind velocity of 8 m/s:

$$P = \pi(6)^2\times\frac{1}{2}\times(1.225)\times(8)^3 = 35.46kW$$

Therefore, annual energy generation = (0.4) × (35.46) × (8760) = 124,300 kWh

REFERENCES

[1] A. H. Sayed, *Fundamentals of Adaptive Filtering*, Wiley, New York, 2003.

[2] M. Stiebler, *Wind Energy Systems for Electric Power Generation*, Green Energy and Technology, Springer-Verlag, Berlin and Heidelberg, 2008.

[3] Ion Boldea, *Synchronous Generators Hand Book*, CRC Press, Tailor and Fransis Group, London, 2016.

[4] M. Godoy Simoes and Felix A. Farret, *Alternate Energy Systems, Design and Analysis with Induction Generators*, Second edition, CRC Press, Tailor and Fransis Group, London, 2008.

[5] Brendan Fox and Damian Flynn, *Wind Power Integration Connection and System Operational Aspects*, The Institution of Engineering and Technology, London, 2007.

[6] L. L. Lai and T. F. Chan, *Distributed Generation: Induction and Permanent Magnet Generators*, John-Wiley and Sons, New York, 2007.

[7] Ann Marie Borbely and Jan F. Kreider, *Distributed Generation: The Power Paradigm for the New Millennium*, CRC Press, Boca Raton, FL, 2001.

[8] IEEE Standards 519–1992, *IEEE Recommended Practices and Requirements for Harmonic Control in Electric Power Systems*, IEEE, 1993, doi: 10.1109/IEEESTD.1993.114370.

[9] A. A. Valdez-Fernández, P. R. Martínez-Rodríguez, G. Escobar, C. A. Limones-Pozos and J. M. Sosa, "A model-based controller for the cascade H-bridge multilevel converter used as a shunt active filter," *IEEE Transaction on Industrial Electronics*, vol. 60, no. 11, pp. 5019–5028, Nov. 2013.

[10] U. K. Rao, M. K. Mishra and A. Ghosh, "Control strategies for load compensation using instantaneous symmetrical component theory under different supply voltages," *IEEE Transactions on Power Delivery*, vol. 23, no. 4, pp. 2310–2317, Oct. 2008.

[11] A. Bhattacharya, C. Chakraborty and S. Bhattacharya, "Shunt compensation," *IEEE Transactions on Industrial Electronics Magzine*, vol. 3, no. 3, pp. 38–49, 2009.

[12] Dong Wang, Weiming Ma, Fei Xiao, Botao Zhang, Dezhi Liu and An Hu, "A novel stand-alone dual stator-winding induction generator with static excitation regulation," *IEEE Transactions on Energy Conversion*, vol. 20, no. 4, 2005.

[13] W. Chen, Y. Lin, H. Gau and C. Yu, "STATCOM controls for a self-excited induction generator feeding random loads," *IEEE Transactions on Power Delivery*, vol. 23, no. 4, pp. 2207–2215, Oct. 2008.

[14] L. R. Vega, H. Rey, J. Benesty and S. Tressens, "A new robust variable step-size NLMS algorithm," *IEEE Trans. Signal Process.*, vol. 56, no. 5, pp. 1878–1893, May 2008.

[15] D. Yazdani, A. Bakhshai and P. K. Jain, "A three-phase adaptive notch filter-based approach to harmonic/reactive current extraction and harmonic decomposition," *IEEE Transactions on Power Electronics*, vol. 25, no. 4, pp. 914–923, Apr. 2010.

[16] Bhim Singh and ShailendraSharma, "Stand-alone single-phase power generation employing a three-phase isolated asynchronous generator," *IEEE Transactions On Industry Applications*, vol. 48, no. 6, pp. 2414–2423, Nov./Dec. 2012.

[17] M. Z. A. Bhotto and A. Antoniou, "A family of shrinkage adaptive-filtering algorithms," *IEEE Trans. Signal Process*, vol. 61, no. 7, pp. 1689–1697, Apr. 2013.

[18] Sergio Vazquez, Juan A. Sanchez, Manuel R. Reyes, Jose I. Leon and Juan M. Carrasco, "Adaptivevectorial filter for grid synchronization of power converters under unbalanced and/or distorted grid conditions," *IEEE Transactions on Industrial Electronics*, vol. 61, no. 3, pp. 1355–1367, 2014.

[19] Ashutosh K. Giri, Sabha Raj Arya, Rakesh Maurya and B. ChittiBabu, "VCO-less PLL control-based voltage-source converter for power quality improvement in distributed generation system," *IET Electric Power Applications*, vol. 13, no. 8, 2019.

[20] Ashutosh K. Giri, Sabha Raj Arya and RakeshMaurya, "Compensation of power quality problems in wind-based renewable energy system for small consumer as isolated loads," *IEEE Transactions on Industrial Electronics*, vol. 66, no. 11, 2019.

[21] Ashutosh K. Giri, Sabha Raj Arya, Rakesh Maurya and RamakantaMehar, "Variable learning adaptive gradient based control algorithm for voltage source converter in distributed generation," *IET Renewable Power Generation*, vol. 12, no. 16, 2018.

[22] Ashutosh K. Giri, Amin Qureshi, Sabha Raj Arya, Rakesh Maurya and B. Chitti Babu, "Features of power quality in single phase distributed power generation using adaptive nature vectorial filter," *IEEE Transactions on Power Electronics*, vol. 33, no. 11, pp. 9482–9495, Nov. 2018.

[23] Ashutosh K. Giri, Sabha Raj Arya, Rakesh Maurya and B. Chitti Babu, "Power quality improvement in stand-alone SEIG based distributed generation system using lorentzian norm adaptive filter," *IEEE Transactions on Industry Applications*, vol. 54, no. 5, pp. 5256–5266, Sep./Oct. 2018.

[24] Ashutosh K. Giri, Sabha Raj Arya, Senior Member IEEE, Rakesh Maurya, Member IEEE and R. K. Mehar, "Variable learning adaptive gradient based control algorithm for VSC in distributed generation," *IET Renewable Power Generation*, vol. 12, no. 16, pp. 1883–1892, 2018.

[25] Ashutosh K. Giri, Sabha Raj Arya and Rakesh Maurya, "Mitigation of power quality problems in permanent magnet synchronous generator using quasi-newton based algorithm," *International Transactions on Electric Energy System*, vol. 29, no. 11, pp. 1–26, Wiley Publication, 2019.

[26] Sabha Raj Arya, Mittal Patel, Javed Alam, S. Jaideep and Ashutosh K. Giri, "Phase lock loop based algorithms for DSTATCOM to mitigate load created power quality problem," *International Transactions on Electric Energy System*, vol. 30, no. 1, e12161, Wiley Publication, 2019.

[27] Sabha Raj Arya, Rakesh Maurya and Ashutosh K. Giri, "Enhancement of power quality in wind based distributed generation system using adaptive vectorial filter," *International Journal of Emerging Electric Power Systems*, vol. 19, no. 5, 2018.

[28] Sabha Raj Arya, Ashish Patel and Ashutosh K. Giri, "Isolated power generation system using permanent magnet synchronous generator with improved power quality," *Journal of the Institution of Engineers (India): Series B*, vol. 99, no. 3, pp. 281–292, 2018.

5 Variable Speed Wind and Hydro Power-Based Distributed Generation for Remote Electrification

Geeta Pathak, Bhim Singh and B. K. Panigrahi

CONTENTS

5.1 INTRODUCTION

In the modern era, lifestyles and high living standards have increased electrical energy consumption many times that of a few decades before. The exhaustive utilization of conventional resources (coal, diesel, petrol, etc.) has taken them in declining states. As these resources are not sufficient to support energy needs for long time, their prices are taking hikes and they are offering environmental irregularities like global warming and greenhouse gases, and they impose worst effects on health, viz. impairing visibility, dizziness, impaired hearing etc. [1, 2]. The renewable green energy and distributed resources are free of cost, available in abundance, clean and environmentally friendly. Further, power electronic development and advancement have made their growth rates fast and easy. By implementing such solutions and with the development of DGS, it is feasible to replace fossil resources to some extent to create balance between environment and social development [3–5].

On the other hand, it is also true that energy crises are persisting worldwide. Taking the scenario of India, more than 70 million rural households are deprived of electricity access and the rest suffer load shedding of medium to long durations. Metropolitan and Tier 1 cities always face frequent power crises of small duration for few minutes to an hour. It has adverse effect on industry, production and other business. The time to attend and resolve the grid fault in rural areas may take much longer than that in urban areas. About 18,000 Indian villages cannot be easily served through the grid because of their remoteness [6]. Fortunately, these remote villages have good resources of solar, hydro and wind power [7, 8], the reason being most of these villages are located either in hilly or desert areas.

Though wind has been utilized for milling grains, pumping water and mechanical energy generating appliances, etc., for centuries, these have been used commercially for electrical energy production since 1980 onwards. Worldwide wind installation capacity has increased up to 651GW, and in Indian scenario, it will touch to 60GW by 2022. With commercialization and the latest technology, wind energy has become cheaper. Its cost has been reduced from 2.8 to 2.2 Rs. per kWh [9]. In order to harvest wind power, wind turbines together with variety of generator technology as in [10, 11], which can be utilized to convert wind energy into electrical energy. Wind power generation can take place in two ways: 1) constant speed wind power generation; and 2) variable speed wind power generation. Literature on these methods based wind application is available in abundance in [12, 13], with their classification and benefits.

Worldwide hydropower generation is very old technology for electricity generation. Plants with 25MW capacity installations are considered as small hydro power plants; however, in India, nearly one-third of the total identified potential of 20,000MW is yet to be tapped. In the next ten years, it is planned to harness about half of the total potential. Future growth scenarios of small hydropower facilities up to 2050 have been developed using the Vensim DSS software package [14]. Good literature is available in hydropower distributed generation [15, 16]. Its applications with improved power quality are reported in [17, 18]. Current and voltage power quality issues, such as frequency and voltage regulation, harmonics mitigation, reactive power compensation, etc., are being resolved using various small hydro configurations and control techniques.

In [19], a price coordination strategy for wind-hydro station is proposed and internal optimization is used for

DOI: 10.1201/9781003229124-5

problem solving. Similar optimization is reported in [20] for wind, PV and hydro systems. In [21], a model based on a sequential Monte Carlo simulation is considered for calculating the reliability indices, to evaluate how the uncertainty of wind energy production affects system planning, especially with the reduction of capacity of reservoirs associated with hydropower plants. All such studies are useful for distributed generation systems.

In this chapter, for hydropower generation, synchronous reluctance generators (SyRG) are employed to deliver power at nominal voltage and frequency, and reactive power support for generator is provided using a capacitor bank. SyRG are maintenance-free generators due to absence of slip rings and rotor windings. Further, they have reduced rotor losses; as a result, their efficiency is highly enhanced [22, 23]. A permanent magnetic brush-less DC (PMBLDC) generator is used to extract power from the speed wind variation. The average power production of SyRG is higher than the alternators due to its almost quasi-square currents and trapezoidal electromagnetic field (EMF). A P&O technique is implemented for wind power maximum power tracking (MPPT) [24]. The unpredictable and varying behaviour of wind makes it an unreliable resource, and to surmount this problem, a combination of battery power [25, 26] and hydropower, which is always accessible, can be a prudent choice.

This DGS provides power management during variation in load demand and wind fluctuations. A sign least mean square (SLMS) [27] control technique is implemented on DGS VSC which is single and key controlling unit. This algorithm has faster convergence rate and better steady state performance. It supresses harmonics on PCC due to nonlinear loads, reactive power compensation and voltage regulation with demand variation, based on DGS requirements. Power balance among a variety of units like wind and hydro generators, loads and battery system is also achieved through its use.

The chapter is organized as follows: Section 5.2 explains the system topology; Section 5.3 presents modelling of various units in DGS like wind-turbines, hydro generator, battery storage, voltage source converters (VSCs), boost converters, etc.; Section 5.4 describes the control technique of VSC; Section 5.5 discusses wind energy maximum power point technique; Section 5.6 presents the simulated performance of a wind-hydro distributed generation system (DGS) with different power scenarios, i.e. increase in wind speed, load imbalance and variation in load demand—some numerical problems are discussed under this to build up good understanding with the chapter; and finally in Section 5.7, conclusions are made exploring all aspects of the concepts this chapter.

5.2 STRUCTURE OF WIND-HYDRO DISTRIBUTED GENERATION SYSTEMS

A wind and hydro DGS is depicted in Figure 5.1. The SyRG works in constant speed for hydropower generation, and hydropower is directly utilized by the AC loads. Power from variable windspeed is extracted using the PMBLDCG. Wind power is converted into DC power using a diode rectifier, and it is supplied to the DC/DC converter using P&O control for maximum power tracking (MPPT). MPPT power is transferred to the AC loads connected at common coupling point (PCC) of the VSC through interfacing inductors (L_f). The surplus power is kept into the battery, which is coupled at the DC link capacitor of the VSC. Hydro generators (SyRG) and ripple filters are connected to PCC. To provide the reactive power support during voltage buildup across SyRG terminals, a capacitor bank is connected.

5.3 MODELLING OF WIND-HYDRO DGS IN MATLAB ENVIRONMENT

This wind-hydro energy–based DGS is modelled in MATLAB Simulink, and design parameters of VSC and battery bank are also explained.

5.3.1 WIND-TURBINE CHARACTERISTICS

Wind-turbines convert kinetic energy of wind speed in the wind into useful electrical energy. This power is characterized by coefficient of performance (C_p), which is the function of tip-speed ratio (λ) and is defined as:

$$\lambda = \frac{r\omega_t}{v} \tag{5.1}$$

where, r is the radius of the wind-turbine blade in metres, ω_t is turbine shaft speed in rad./sec. and v is the wind velocity in m/sec.

C_p can be calculated in terms of λ and blade pitch angle (β) as:

$$C_p(\lambda) = C_1\left(\frac{C_2}{\lambda_i} - C_3\beta - C_4\right)e^{\frac{-C_5}{\lambda_i}} + C_6\lambda \tag{5.2}$$

where, λ_i is defined as:

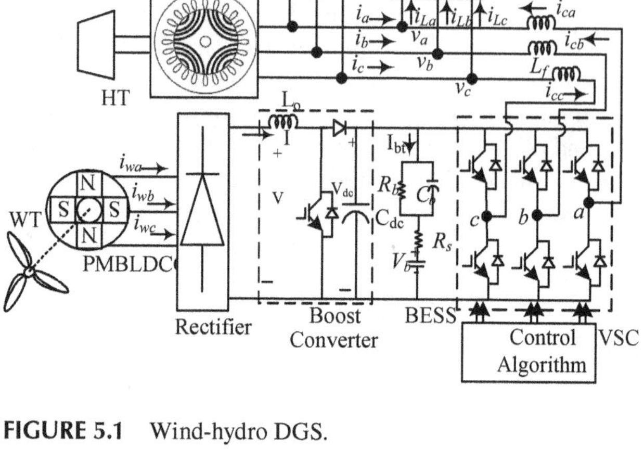

FIGURE 5.1 Wind-hydro DGS.

$$\frac{1}{\lambda_i} = \frac{1}{\lambda + 0.08\beta} - \frac{0.035}{1+\beta^3} \quad (5.3)$$

The mechanical power (P_t) and torque (T_t) made available by the wind-turbine to the PMBLDC is described respectively in equations (5.4 and 5.5) as:

$$P_t = 0.5\rho\pi R^2 C_p v^3 \quad (5.4)$$

$$T_t = \frac{P_t}{\omega_t} \quad (5.5)$$

Wind-turbine power vs. speed curve for variable wind speeds is shown in Figure 5.2, where pitch angle β is zero.

5.3.2 Synchronous Reluctance Generator Model

The SyRG has anisotropic structure due to the design of the magnetic circuit; the reluctance along the flux path depends on the position of the applied magneto-motive force (MMF). The d-(direct) axis of the rotor is along the minimum reluctance path, while the axis q-(quadrature) axis is found along the maximum reluctance. The SyRG has a sinusoidal distributed stator winding; therefore, its model can be derived from the salient-pole wound-rotor synchronous generator, in the rotor reference frame by removing the effect of damping windings.

$$V_{dq} = p\psi_{dq} + \begin{bmatrix} 0 & -\omega_r \\ \omega_r & 0 \end{bmatrix} \psi_{dq} - R_s i_{dq} \quad (5.6)$$

$$P = \frac{3\omega_r}{2}(\psi_f i_q - (L_d - L_q) i_d i_q) \quad (5.7)$$

where,

$$\psi_{dq} = \begin{bmatrix} \psi_f - L_d i_d \\ -L_q i_q \end{bmatrix} \quad (8)$$

The SyRG model is realized by considering the absence of magnetic flux from the rotor ($\psi_f = 0$), then the model appears as:

$$v_d = \omega_r L_q i_q - R_s i_d - L_d \frac{di_d}{dt} \quad (5.9)$$

$$v_q = -\omega_r L_d i_d - R_s i_q - L_q \frac{di_q}{dt} \quad (5.10)$$

$$P = \frac{3\omega_r}{2}(L_d - L_q) i_d i_q \quad (5.11)$$

5.3.3 Voltage Source Converters and Battery Bank Components

VSC components boost converters, and design of battery bank components is considered in detail.

The voltage at the DC link of the VSC is computed as:

$$V_{dc} = \frac{2\sqrt{2}}{\sqrt{3}m} V_L \quad (5.12)$$

where m is the modulation index, V_L is the line rms voltage and V_{dc} is the DC bus voltage.

A Thevénin's theorem for battery storage system is described, whereby C_b and R_b are parallelly connected capacitance and resistances, in series with an ideal voltage source (V_{oc}) having internal resistance (R_s). Equivalent capacitance C_b is computed as:

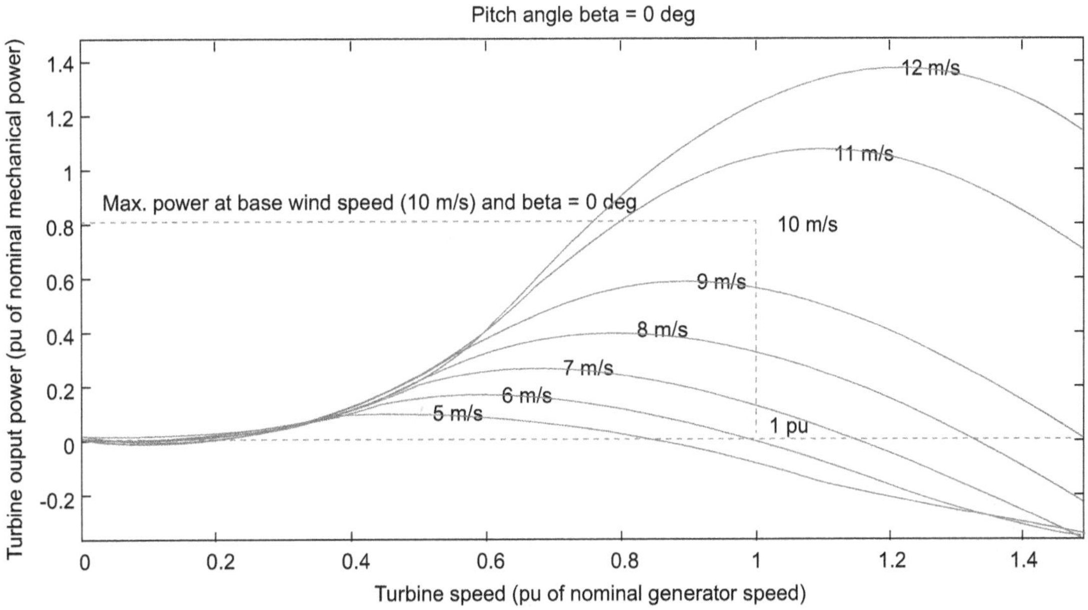

FIGURE 5.2 Wind-turbine power characteristics.

$$C_b = \frac{(kW.h * 3600 * 1000)}{0.5(V_{oc\max}^2 - V_{oc\min}^2)} \quad (5.13)$$

where V_{ocmax} and V_{ocmin} are the maximum and minimum open circuit voltage values under fully charged and discharged states of battery (±10% V_{oc})

5.3.4 Rating of Interfacing Inductors of VSC and RC Filters

Interfacing inductor is formulated as:

$$L_f = \left(\frac{\sqrt{3}}{2}\right) \times \frac{mV_{dc}}{6af_s I_{rpl}} \quad (5.14)$$

V_{dc} and L_f are voltage at DC bus capacitance and interfacing inductance, respectively; $I_{ri.}$ is the ripple current and f_s is switching frequency. The noise in the PCC voltage is filtered out using a high-pass RC filter. This filter filters out the high frequency voltage ripples by providing low impedance path. This offers high impedance to the fundamental frequency values at switching frequency of 10 kHz. RC filter capacitance (c) and resistance (R) are considered 10µF and 5Ω, respectively.

$$Z = \sqrt{\left(R^2 + \left(\frac{1}{2\pi fc}\right)^2\right)} \quad (5.15)$$

5.3.5 Boost Converter Design

Boost converters are employed for second stage solar power transfer to the VSC after maximum power tracking using the incremental conductance (INC) technique. Calculations for input inductor of boost converter to limit ripple current (ΔI_L) with switching frequency f_{ss} is as:

$$L_o = \frac{V_{dc}}{4 \times f_{ss} \times \Delta I_L} \quad (5.16)$$

where V_{dc} is DC bus voltage

5.4 CONTROL ALGORITHM FOR WIND-HYDRO DGS

For accurate and satisfactory performance of the wind-hydro DGS, the control algorithm must be very effective and robust. Performance of DGS with SLMS control technique is depicted in Figure 5.3. The parameters µ (step size), i_{La} (load current for phase a), U_{pa} (in-phase unit template), U_{qa} (quadrature unit template) are used in the control approach.

5.4.1 Calculation for In-Phase and Quadrature Unit Templates of Supply Voltages

The phase voltages (v_{abc}) are calculated from the line voltages as [28]:

$$v_a = \frac{1}{3}(2v_{ab} + v_{bc}) \quad (5.18)$$

$$v_b = \frac{1}{3}(-v_{ab} + v_{bc}) \quad (5.19)$$

$$v_c = \frac{1}{3}(-v_{ab} - 2v_{bc}) \quad (5.20)$$

Voltage V_t across the terminal is calculated as:

$$V_t = \sqrt{\left(2(v_a^2 + v_b^2 + v_c^2)\Big/3\right)} \quad (5.21)$$

In-phase unit templates of all three-phase voltages are calculated as:

$$U_{pa} = v_a / V_t, U_{pb} = v_b / V_t, U_{pc} = v_c / V_t \quad (5.22)$$

The quadrature unit templates of all three-phase voltages are estimated as:

$$U_{qa} = (-U_{pa} + U_{pc}) / \sqrt{3} \quad (5.23)$$

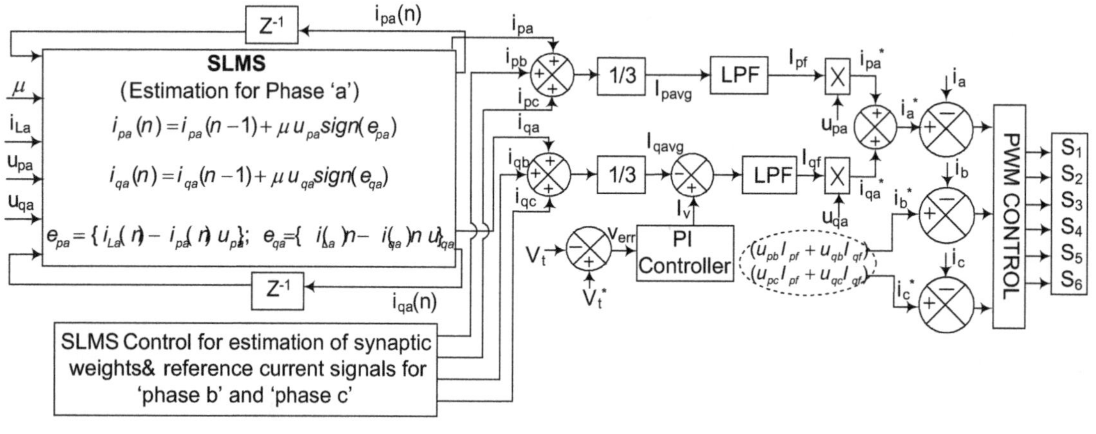

FIGURE 5.3 SLMS control approach.

$$U_{qb} = (3U_{pa} + U_{pb} - U_{pc}) / \sqrt{3} \quad (5.24)$$

$$U_{qc} = (-3U_{pa} + U_{pb} - U_{pc}) / \sqrt{3} \quad (5.25)$$

5.4.2 Extraction of Active and Reactive Components of Load Currents Using SLMS Algorithm

At phase a, the active component of load current is extracted implementing SLMS approach at kth sample as:

$$i_{pa}(k) = i_{pa}(k-1) + \mu u_{pa} sign(e_{pa}) \quad (5.26)$$

The actual and desired outputs difference in terms of error e_{pa} (k) at kth sample is derived as:

$$e_{pa}(k) = \{i_{La}(k) - i_{pa}(k) u_{pa}\} \quad (5.27)$$

Estimation of load active components for phases b–c is made in the same manner as that of phase a using the SLMS control.

At phase a, the reactive component of load current for applying the same approach is computed at kth sample as:

$$i_{qa}(k) = i_{qa}(k-1) + \mu u_{qa} sign(e_{qa}) \quad (5.28)$$

$$e_{qa}(k) = \{i_{La}(k) - i_{qa}(k) u_{qa}\} \quad (5.29)$$

Estimation of active components of the load for phases b–c is made in the similar manner to that of phase a using the same control approach.

The equivalent active component of load is estimated as:

$$I_{pavg} = \frac{i_{pa} + i_{pb} + i_{pc}}{3} \quad (5.30)$$

The equivalent reactive component of the load is estimated as:

$$I_{qavg} = \frac{i_{qa} + i_{qb} + i_{qc}}{3} \quad (5.31)$$

After filtering out through low pass filter (LPF), this component is called I_{pf}.

A PI (proportional and integral) regulator is tuned with k_p (proportional gain) and k_i (integral gain) to maintain common coupling (PCC) voltage. The difference between the magnitudes of sensed and reference voltages V_t and V_t^* is the error signal of the PI regulator, and after regulation, the current signal is obtained as:

$$I_v(k) = I_v(k-1) + k_p\{V_{err}(k) - V_{err}(k-1)\} k_i V_{err}(k) \quad (5.32)$$

where V_{err} is the error signal generated from reference and actual voltage and is shown as:

$$V_{err} = V_t^* - V_t \quad (5.33)$$

Total reactive component of supply current is calculated as:

$$I_{qt} = I_v - I_{qavg} \quad (5.34)$$

This is the reference and actual voltage error, and is shown as I_{fq}.

5.4.3 Computation of Fundamental Active and Reactive Supply Currents Components

The three-phase active and reactive supply reference current components are computed as:

$$i_{pa}^* = I_{pf} u_{pa}\, i_{pb}^* = I_{pf} u_{pb}\, i_{pc}^* = I_{pf} u_{pc} \quad (5.35)$$

$$i_{qa}^* = I_{qf} u_{qa}\, i_{qb}^* = I_{qf} u_{qb}\, i_{qc}^* = I_{qf} u_{qc} \quad (5.36)$$

5.4.4 Derivation of Three-Phase Reference Supply Currents

The three-phase supply reference currents for a, b and c phases are determined as:

$$i_a^* = i_{pa}^* + i_{qa}^*\, i_b^* = i_{pb}^* + i_{qb}^*\, i_c^* = i_{pc}^* + i_{qc}^* \quad (5.37)$$

The PWM pulses are generated from the comparison between reference supply currents (i_a^*, i_b^*, i_c^*) and actual supply currents (i_a, i_b, i_c) for switching to legs of three-phase VSC.

5.4.5 Maximum Power Point Tracking Control Algorithm for Boost Converters

Various MPPT algorithms are available for maximum power extraction from variable speed wind systems. A perturbation and observation (P&O) is a very simple and accurate method used for wind MPPT. It has no requirement of wind speed measurement and speed sensors, as it is a sensor-less hill climb approach. The output voltage (V) and output current (I) of the rectifier are detected and the DC power is calculated using DC power formula $P = I \times V$ where P is a function of current I. Power derivative of MPPT with respect to (wrt) output current, i.e. $\frac{\Delta P}{\Delta I} = 0$, must be satisfied. In the present chapter, a P&O approach [29] is implemented, small perturbation in terms of a constant (d) is made in output current I, and change is observed in terms of power. If with the increase of d in current, power tracks maximum power point (MPP), the same steps are followed in the same direction until the point is towards the MPPT; otherwise, to maintain MPPT, direction of perturbation is changed. A MPPT flowchart is depicted in Figure 5.4(a).

For this purpose, the output reference current is generated according to the rules as:

$$I^* = I + d; \Delta I > 0 \,\&\, \Delta P > 0 \quad (5.38)$$

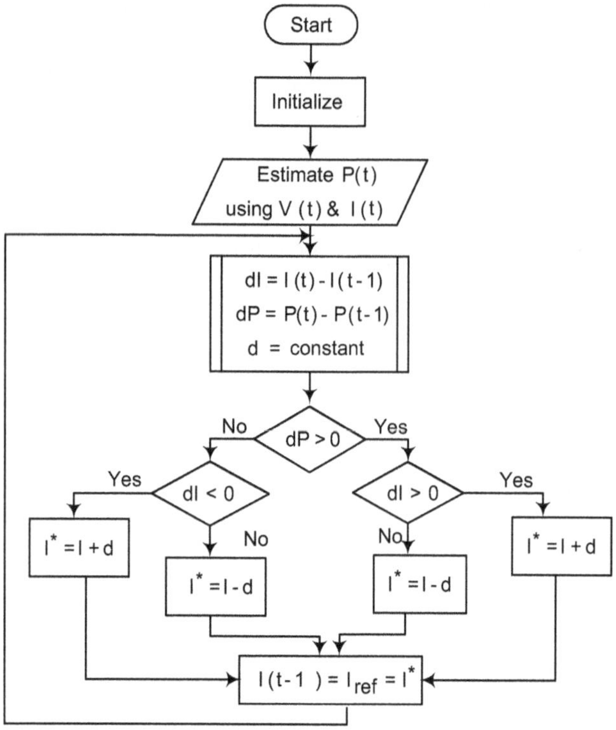

FIGURE 5.4 (a) Flowchart of wind MPPT.

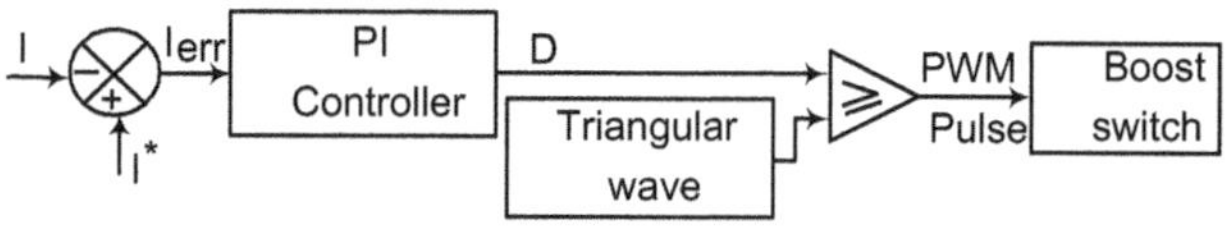

FIGURE 5.4 (b) Switching pulse generation for boost controller.

$$I^* = I + d; \Delta I < 0 \,\&\, \Delta P < 0 \tag{5.39}$$

$$I^* = I - d; \Delta I > 0 \,\&\, \Delta P < 0 \tag{5.40}$$

$$I^* = I - d; \Delta I < 0 \,\&\, \Delta P > 0 \tag{5.41}$$

where P = V × I and δ is any small value

I^* and I are the reference and sensed DC current of the boost inductor

To achieve the duty ratio, MPPT generated reference current and sensed current are compared and current error is fed to the PI regulator.

The equation is presented as:

$$D(n+1) = D(n) + k_{pd}\left\{I_{err}(n+1) - I_{err}(n)\right\} + k_{id} I_{err}(n+1) \tag{5.42}$$

where $I_{err} = I^* - I$ is current error and k_{pd} (proportional gain) and k_{id} (integral gain) of PI regulator

PWM pulse for boost converter is generated from the duty cycle as depicted in Figure 5.4(b).

5.5 RESULTS AND DISCUSSION

Under this section, the simulation results of DGS are described with an SLMS algorithm. The wind MPPT, response of intermediate parameters of control scheme under load imbalance, steady state performance and dynamic behaviour of the wind-hydro system are depicted under nonlinear load.

5.5.1 Wind MPPT

Simulated performance of wind MPPT algorithm is presented in Figure 5.5. The change in wind speed V_w from 10m/s to 12m/s is noticed at t = 2 sec. Increase in wind power is observed with rise in generator current (i_{wa}) simultaneously. The PMBLDC generator voltage ($v_{pm\text{-}bldc}$) is maintained constant. After rectification, the input current (I) and input voltage (V) of the boost converter are also shown. During wind MPPT, the input current (I) of boost converter and extracted input power P = I × V of the converter are increased with the wind variation. These results show the proper working of the P&O control technique.

5.5.2 Intermediate Signal Behaviour of SLMS Control under Nonlinear Loads

Robust and accurate control techniques provide satisfactory and effective performance of wind-hydro DGS. In Figure 5.6, intermediate parameters of the proposed control algorithm are demonstrated under nonlinear load. At t = 1.6s, unbalanced load is realized by throwing the switch of phase a, and at t = 1.8s, the load recovery is done. During unbalanced load, the load current of phase

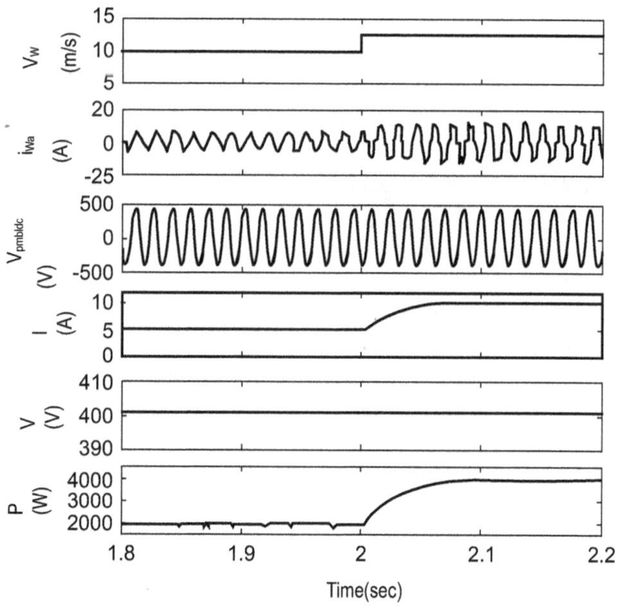

FIGURE 5.5 MPPT of wind generations with wind variation.

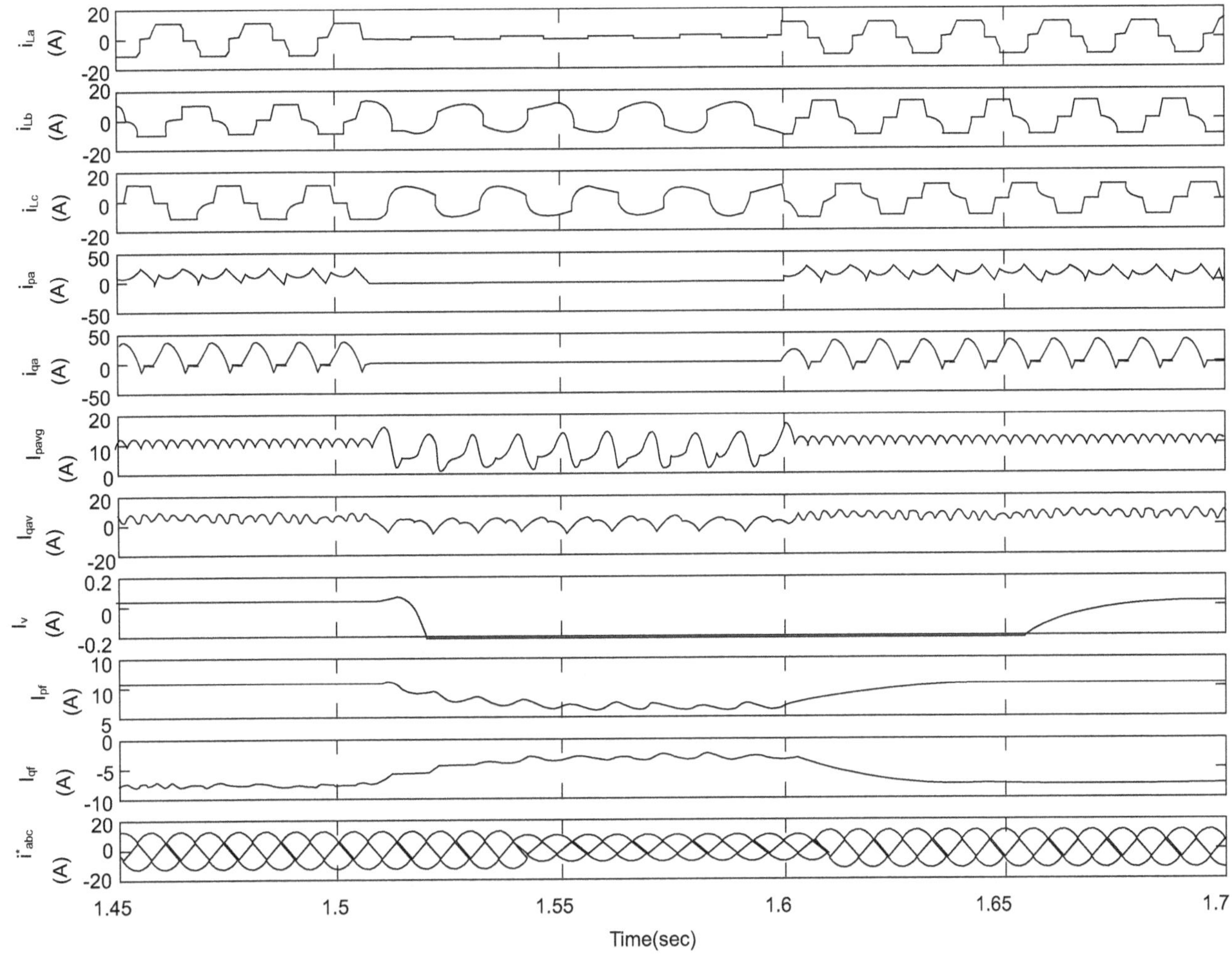

FIGURE 5.6 Intermediate parameters of SLMS control under nonlinear load.

a (i_{La}) is observed to become zero. The active and reactive load components (i_{pa}, i_{qa}) are also reduced to zero simultaneously.

The average active and reactive load components (i_{pavg}, i_{qavg}) are also decreased as the total demand is decreased in absence of load demand of phase a. The PI regulator output current (I_v) provides voltage regulation at point of common coupling. With load variation, reference current active component (I_{pf}) and that of reactive component (I_{qf}) are also changed. Therefore, reduction is observed in i^*_{abc} (reference currents) under disconnection of phase a. The SLMS technique is capable of settling the response under few cycles with adequate performance during dynamics. The response is stable in both the steady states before and after the imbalance.

5.5.3 Steady State Performance of DGS

During normal running conditions, the DGS performance is depicted in Figure 5.7. A sinusoidal PCC voltage (v_a) with 0.07% total harmonic distortion (THD) at phase a has maintained its magnitude at reference value. The THD of the supply current for phase a is 2.45% which keeps it sinusoidal. Both the current and voltages are as per the IEEE-519 limits. However, the very high THD of nonlinear load is 24.98%. After analyzing the results, it is concluded that during steady state condition, voltages and currents at PCC are sinusoidal, balanced and have less harmonics, which indicates the effectiveness of the control algorithm.

5.5.4 Dynamic Performance of DGS

The dynamic response of wind-hydro DGS is shown in Figure 5.8, Figure 5.9 and Figure 5.10 under nonlinear load. To understand the power management of the system, three different scenarios are created. In the first one, wind velocity is changed from 8–10 metres/second at t = 1.4s, as depicted in Figure 5.8.

Therefore, under constant load demand (P_L) and fixed hydro generation, the wind power generation (P_{wind}) is increased. This increase in power generation is accumulated into the battery, it increases battery power (P_{bt}) and battery current I_{bt}. The source current (i_{abc}) and PCC voltages (v_{abc}) remain unaffected, as battery is capable to absorb this excess power.

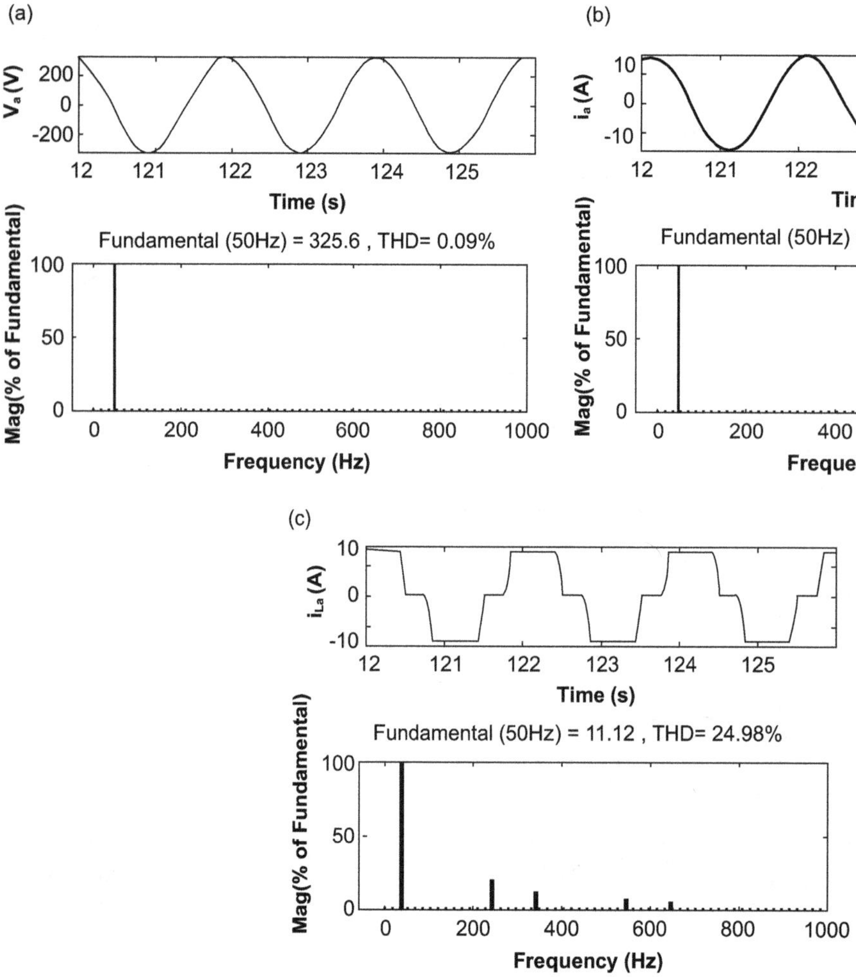

FIGURE 5.7 Steady state performance of the DGS.

In the second scenario: An unbalanced condition is realized in Figure 5.9 by disconnecting the load at phase a. The load behaves as a single phase load but the source side currents (i_{abc}) are balanced and sinusoidal. The PCC voltages are maintained constant and terminal voltage V_t remains near to its reference value. The VSC control is so robust to such contingencies on the load side that it has no impact on system voltage and frequency. The flow of power is also observed at t = 1.5s. The wind power (P_{wind}) is constant with the constant wind speed, and the hydro power supply is almost constant during imbalance. Therefore, that unused power is stored in battery bank through the VSC in terms of P_{bt} and I_{bt} increase. The VSC is delivering less power towards the load during load imbalance. The VSC currents are also changed.

In the third scenario: It is on sudden load increase as shown in Figure 5.10, where the hydro generator is supplying the maximum power to the load. Therefore, the source currents (i_{abc}) are fixed and sinusoidal. Wind generation is also fixed at constant wind speed (V_w). The load demand increases the P_{vsc} power, which is the only power regulating component in the DGS. The load demand is higher than the hydro generator (SyRG) capacity; therefore, the deficit power is delivered by the VSC converting DC power (P_{wind}+ P_{bt}) into AC power to fulfil the load demand. Therefore, the

TABLE 5.1
Wind/Solar DGS Components

S No.	Distributed Generation System Components	Rating
1	3-phase permanent magnet BLDC generator	3.7kW, 13A, 230V,1500rpm
2	Battery bank: V_b	400V
	C_b	1125F
	R_b	10 kΩ
	R_s	0.1Ω
3	PI controller (voltage regulation) gains: k_p	0.4
	k_i	0.08
4	Interfacing inductance (L_f):	4mH
5	RC filter: R	5Ω
	C	10μF
7	DC link: V_o	400 V
	C_{dc}	1600μF

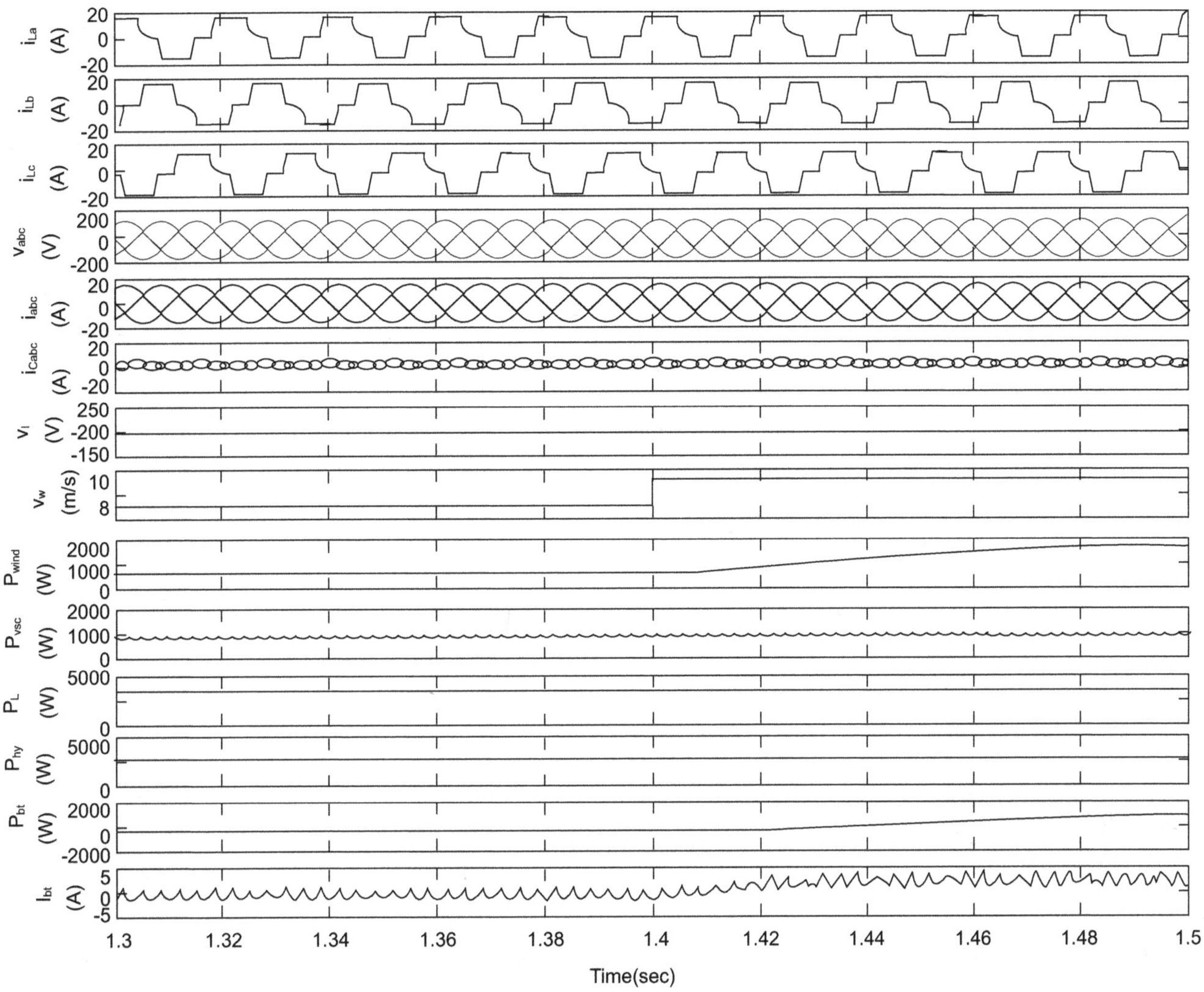

FIGURE 5.8 Dynamic performance of DGS with wind speed variation under nonlinear loads.

battery power (P_{bt}) and the current I_{bt} are reduced, and the VSC currents are increased.

The PCC voltage and system frequencies are maintained throughout the change. These dynamic scenarios are showing the satisfactory operation of VSC for power balancing of the DGS during such contingency conditions. Therefore, the DGS is capable of operating under changing scenarios.

5.6 CONCLUSIONS

In this chapter, wind-hydro renewable DGS is explored for electricity generation. Variable speed wind generation and constant speed hydro power generation is utilized under various power generation and load demand variation scenarios. Brushless generators are implemented due to their cost and maintenance benefits. DG system structure is explained in detail, and modelling and design of wind-turbine, SyRG hydro generator, battery storage system and VSC are described in a systematic and detailed manner. The SLMS control technique is implemented on VSC for the improvement of power quality like voltage regulation, harmonics mitigation, load levelling, etc. The waveforms of the source currents and voltages are sinusoidal and their THDs are also under the limits according to the IEEE standards. The voltage source converter (VSC) control also maintains the power balance among all the energy generation resources such as hydro, wind, battery storage, etc., under changing power generation and demand scenarios. In interior places where the renewable energy is in abundance, energy independence can be provided such systems. Fossil fuel consumption and its affect on human health and environment can also be reduced by such practices. To make the chapter more interactive and to build better understanding, some numerical problems on the covered topics are also included.

5.7 NUMERICAL PROBLEMS

1. A horizontal axis wind-turbine having 14m diameter is operating under average wind speed of 10m/s.

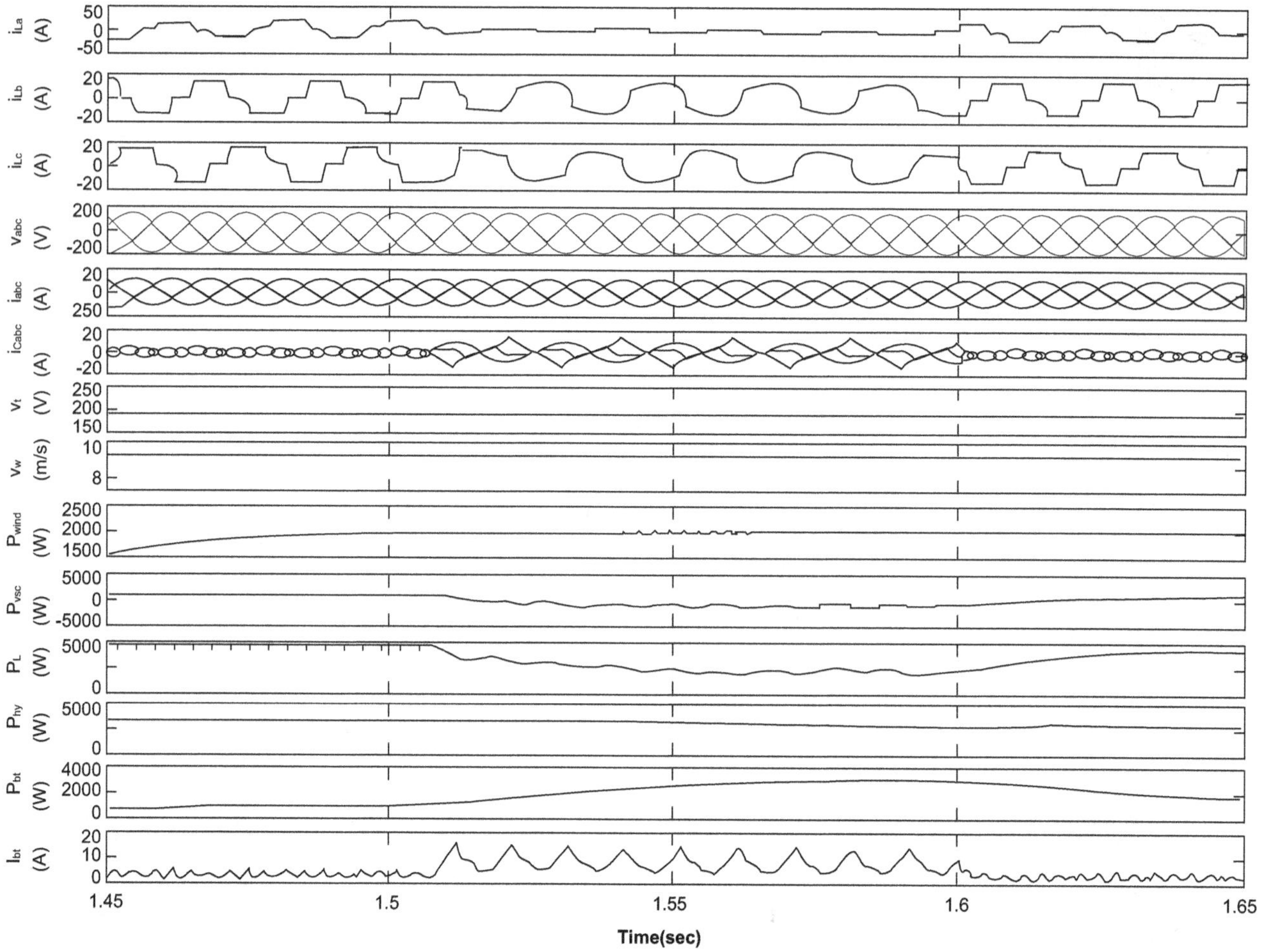

FIGURE 5.9 Dynamic performance of DGS for load imbalance under nonlinear loads.

Calculate the power production and the annual energy production. Air density ρ = 1.225 kg/m3, turbine efficiency η = 0.42 and C_p = 0.48.

Solution

Diameter of the turbine = 14 m
Wind speed (v) = 10 m/s
Air density (ρ) = 1.225 kg/m³
Turbine efficiency (η) = 0.42
Hours in one year = 365 × 24h = (efficiency)(P) (8760 hrs)

$P_t = 0.5\rho\pi R^2 C_p v^3$

$P_t = 0.5 \times \pi \times 1.2225 \times (7)^2 \times 0.48 \times (10)^3$

P_t = 45.165 KW

Therefore:

Power production = η × P_t = 18.969kW
Annual energy production = η x P_t × 365 × 24 = 0.4 × 45.165 × 8760) = 166,172kWh

2. A three-bladed wind-turbine has 50m diameter. At 12m/s wind speed, it produces 750KW power. The air density is given 1.225kg/m³. Calculate the rotor rotational speed in rpm with the tip-speed ratio of 5.0. Betz limit is given as C_p = 16/27.

Solution

Tip-speed ratio (λ) = 5
Diameter of the turbine − 12m
Turbine Power P_t = 750kW
Air density (ρ) = 1.225kg/m³

$P_t = 0.5\rho\pi R^2 C_p v^3$

Wind speed (v) = $\sqrt[3]{\dfrac{750 \times 1000}{0.5 \times 1.2225 \times \pi \times (25)^2 \times {}^{16}\!/_{27}}}$

= 10 m

$\lambda = \dfrac{\omega R}{v}$

Rotational speed (rpm) = 5 × (10/25) × 60/2π = 19.1rpm.

3. A boost converter has input voltage of 15V and the output of 40V. The load is a resistance of 30Ω. Switching frequency is 20kHz. Design for continuous inductor current. Assume ideal components for this design.

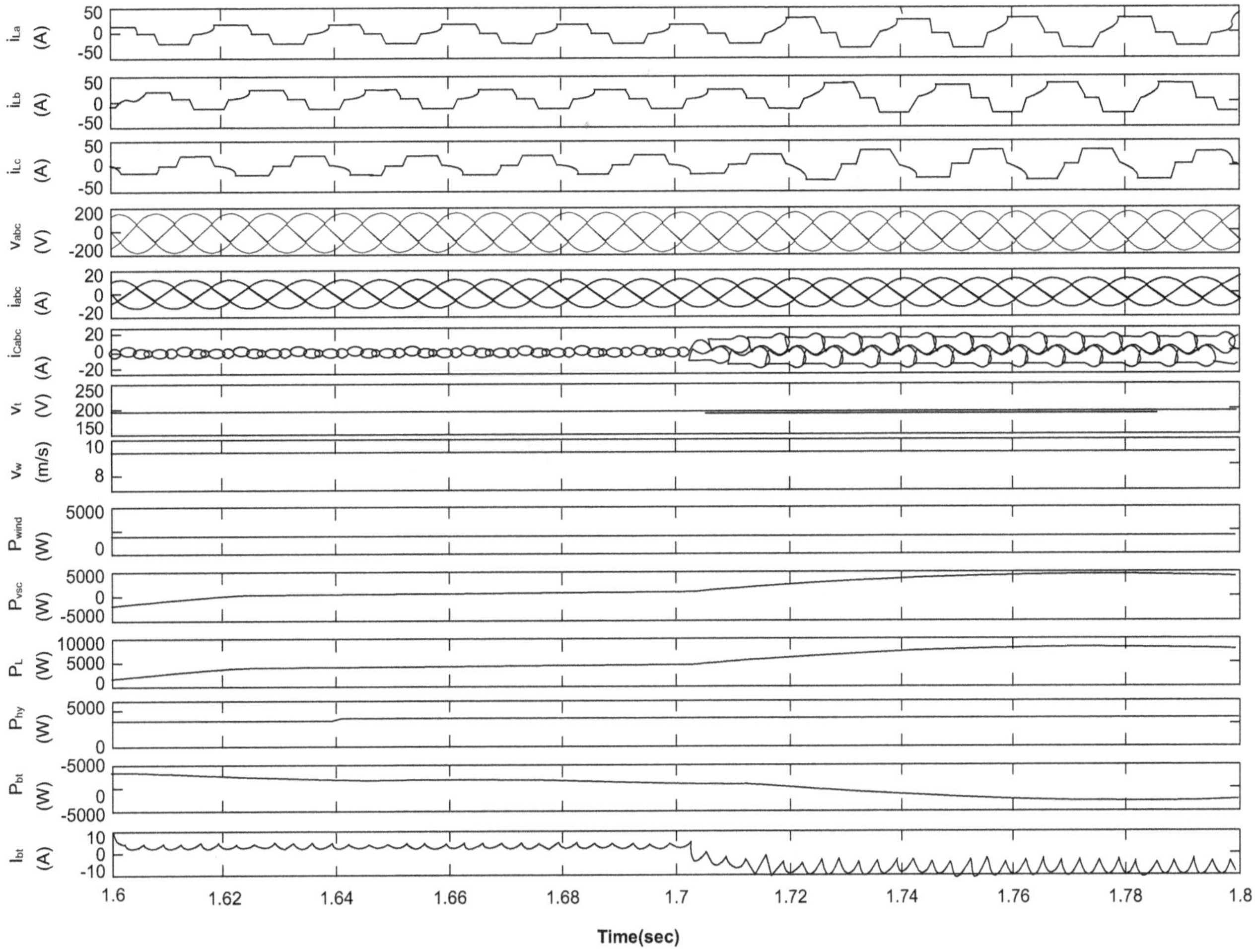

FIGURE 5.10 Dynamic performance of DGS for increased load under nonlinear loads.

Solution

Duty ratio for the boost converter (D) =

$$D = 1 - \frac{V_{in}}{V_{out}} = 1 - \frac{15}{40} = 0.625$$

Switching frequency f_{ss} = 20,000 Hz

Minimum inductance of continuous current mode:

$$L_{\min} = \frac{D(1-D)^2 R}{2f_{ss}} = \frac{0.625 \times (1-0.625)^2 \times 30}{2 \times 20{,}000} = .066mH$$

It will be considered 0.1mH

Continuous load current =

$$I_L = \frac{V_{in}}{(1-D)^2 R_L} = \frac{15}{(1-0.625)^2 \times 30} = 3.55A$$

Load current ripple =

$$\Delta I_L = \frac{V_{in} D T_s}{L} = \frac{15 \times 0.625}{1 \times 10^{-4} \times 20{,}000} = 4.6875A$$

Load current limits =

$$I_{\max} = I_L + \frac{\Delta I_L}{2} = 3.55 + 2.34 = 5.894A$$

$$I_{\min} = I_L - \frac{\Delta I_L}{2} = 1.2125A$$

4. The boost converter has parameter input voltage Vs = 20V, duty ratio $D = 0.62$, load resistance $R_L = 15\Omega$, input inductance of boost converter L_o = 120μH, C = 30μF, and the switching frequency is 20kHz. Calculate: (a) output voltage; (b) average, maximum, and minimum inductor currents; and (c) output voltage ripple. Assume ideal components.

Solution

1. Output voltage of boost converter,

$$V_{out} = \frac{V_{in}}{(1-D)} = \frac{20}{1-0.62} = 52.63V$$

Switching frequency f_s = 20,000 Hz

2. Average inductor current =

$$I = \frac{V_{in}}{(1-D)^2 R_L} = \frac{20}{(1-0.62)^2 \times 15} = 9.22A$$

Load current ripple =

$$\Delta I = \frac{V_{in} D T_s}{L} = \frac{20 \times 0.62}{120 \times 10^{-6} \times 20{,}000} = 5.16A$$

Load current limits =

$$I_{\max} = I + \frac{\Delta I}{2} = 9.22 + 2.58 = 11.8A$$

$$I_{\min} = I - \frac{\Delta I}{2} = 6.64A$$

3. Output voltage ripple:

$$\frac{\Delta V_{out}}{V_{out}} = \frac{D}{R_L \times C \times f_s} = \frac{0.62}{15 \times 30 \times 10^{-6} \times 20{,}000} = 0.138$$

$\Delta V_{out} = 0.138 \times 52.63 = 7.25$
Voltage ripple is 7.25%

5. A three-phase, PWM voltage source converter feeds 25kW to 230V (line) rms, 50Hz, three-phase AC mains from a constant voltage DC bus. The switching frequency is 20kHz; DC bus voltage is 400V. The power factor is corrected close to unity and PWM modulation index is 0.8. Determine: (a) fundamental rms AC current; (b) phase shift in fundamental component of PWM voltage and supply voltage; (c) value of the interfacing inductor; and (d) ripple in AC current is as percentage of the rated rms current.

Solution

AC voltage line to line $V_L = 220V$
Switching frequency $f_s = 20$ KHz
Modulation Index, $m_a = 0.85$
Power factor = 1;
Output power P = 25kW,
DC bus voltage = 400V

1. Fundamental AC rms Current $i_1 = P/\sqrt{3}V_L$ $=25000/\sqrt{3}*230 = 62.75A$
2. Fundamental component per phase of converter $v_{conv1} = \sqrt{(v_s^2+(i_1*x_s)^2)}$

   $V_{conv1} = m_a V_{dc}/2\sqrt{2}$
   $= 232.7V$
   Reactance per phase, $x_s = 0.325\Omega$
   $\delta = \sin^{-1}(P \times x_s)/(v_L \times v_{conv1}) = 8.74^0$
3. $L_s = x_s/\omega = 1.03$ mH
4. AC inductance of the converter, $L_s = \dfrac{\sqrt{3} m_a V_o}{12 a f_s \Delta i}$

   Therefore, $\Delta i = 2.21A$
   Ripple in AC current is 2.21%

REFERENCES

[1] K.R. Smith, H. Frumkin, and K. Balakrishnan, "Energy and Human Health," *Annual Review of Public Health*, vol. 34, no. 1, Jan. 2013.

[2] A. Haines, K. Klingmuller, A. Pozzer, and R.T. Burnett, "Effects of Fossil Fuel and Total Anthropogenic Emission Removal on Public Health and Climate," *Proc. of National Academy of Sciences*, vol. 116, no. 15, 2019.

[3] David Pimentel, *Biofuels, Solar and Wind as Renewable Energy System-Benefits and Risks*, Springer, CO, Dordrecht, 2008.

[4] A. Reinders, P. Verlinden, W.V. Sark, and A. Freundlich, *Photovoltaic Solar Energy: From Fundamentals to Applications*, John Wiley & Sons Ltd., Chichester, West Sussex, United Kingdom, 2017.

[5] Geeta Pathak, Bhim Singh, and B. K. Panigrahi, "Logarithmic Cost Based Adaptive Control for Wind-Diesel Microgrid," In *Proceedings of Annual IEEE India Conference (INDICON)*, New Delhi, 2015, pp. 1–6.

[6] Rajiv Gandhi Grameen Vaidyutikaran Yojana [Online]. Available: https://data.gov.in/keywords/rajiv-gandhi-grameen-vidyutikaran-yojana

[7] I. Bostan, A.V. Gheorghe, V. Dulgheru, I. Sobor, V. Bostan, and A. Sochirean, *Resilient Energy Systems: Renewables: Wind, Solar, Hydro*, vol. 19, Springer Science & Business Media, Dordrecht, Heidelberg, New York, London, 2012.

[8] V. Quaschning, *Renewable Energy and Climate Change*, John Wiley & Sons, New York, 2019.

[9] S. Srivastava, S. Jain, J. Lee, F. Zhao, F. Jayasurya, A. Lathigara, and A. Pek, *India Wind Outlook Towards 2022, Looking Beyond Headwinds*, Globle Wind Energy Council and MEC+, May 2020.

[10] I. Bolden, and S.A. Nasar, *The Induction Machine Handbook*, CRC Press, New York, 2002.

[11] Loi Lei Lai, and Tze Fun Chan, *Distributed Generation, Induction and Permanent Magnet Generators*, John Wiley & Sons, Inc., Chichester, West Sussex, United Kingdom, 2007.

[12] Shailendra Sharma, and Bhim Singh, "Performance Evaluation of Fixed-speed and Variable-speed Stand-alone Wind Energy Conversion Systems," *Electric Power Components and Systems*, 2016, pp. 1–9.

[13] A. Schaffarczyk, *Understanding Wind Power Technology: Theory, Development and Optimization*, Wiley, Chichester, West Sussex, United Kingdom, 2014.

[14] M.K. Mishra, N. Khare, and A.B. Agrawal, "Small Hydro Power in India: Current Status and Future Prospectives," *Renewable and Sustainable Energy Reviews*, vol. 51, Nov. 2015, pp. 101–115.

[15] A. Zidonis, David S. Benzon, and G.A. Aggidis, "Development of Hydro Impulse Turbines and new Opportunities," *Renewable and Sustainable Energy Reviews*, vol. 51, Nov. 2015, pp. 1624–1635.

[16] E.E. Gaona, C.L. Trujillo, and J.A. Guacaneme, "Rural Microgrid and Its Potential Application in Columbia," *Renewable and Sustainable Energy Reviews*, vol. 51, Nov. 2015, pp. 125–137.

[17] K.V.N.S. Pavan Kumar, E. Praveena, and P.V. Kishore, "Isolated Pico-Hydropower Generation Using Asynchronous Generator for Power Quality Improvement," *International Journal of Scientific & Engineering Research*, vol. 2, no. 12, Dec. 2011.

[18] U. Kalla, B. Singh, and S. Murthy, "Modified Electronic Load Controller for Constant Frequency Operation with Voltage Regulation of Small Hydro Driven Single-Phase SEIG," *IEEE Transactions on Industry Applications*, vol. 52, no. 4, July–Aug. 2016, pp. 2789–2800.

[19] Y. Liu, C. Jiang, J. Shen, and J. Hu, "Coordination of Hydro Units with Wind Power Generation Using Interval Optimization," *IEEE Trans. Sustainable Energy*, vol. 6, no. 2, Apr. 2015, pp. 443–453.

[20] Y. Liu, S. Tan, and C. Jiang, "Interval Optimal Scheduling of Hydro-PV-Wind Hybrid System Considering Firm Generation Coordination," *IET Renewable Power Generation*, vol. 11, no. 1, 2017, pp. 63–72.

[21] V.S. Lopes, and C.L.T. Borges, "Impact of the Combined Integration of Wind Generation and Small Hydropower Plants on the System Reliability," *IEEE Trans. Sustainable Energy*, vol. 6, no. 3, July 2015, pp. 1169–1177.

[22] T.K. Malelu, and M. Muteba, "Dynamic Analysis of a Wind Turbine Driven Synchronous Reluctance Generator with Three-Phase Auxiliary Stator Winding," In *Proceedings of IEEE 29th International Symposium on Industrial Electronics (ISIE)*, Delft, Netherlands, 2020, pp. 289–294.

[23] F. Rebahi, A. Bentounsi, H. Khelifa, O. Boulkhrachef, and D. Meherhera, "Comparative Study of a Self-Excited Induction and Synchronous Reluctance Generators Capabilities," In *Proceedings of International Conference on Advanced Electrical Engineering (ICAEE)*, Algiers, Algeria, 2019, pp. 1–5.

[24] Z. Dalala, Z. U. Zahid, and J. Lai, "New Overall Control Strategy for Wind Energy Conversion Systems in MPPT and Stall Regions," In *Proc. of IEEE Energy Conversion Congress and Exposition*, Denver, CO, 2013, pp. 2412–2419.

[25] H. Shin, and J. Hur, "Optimal Energy Storage Sizing with Battery Augmentation for Renewable-Plus-Storage Power Plants," *IEEE Access*, vol. 8, 2020, pp. 187730–187743.

[26] O.M. Akeyo, V. Rallabandi, N. Jewell, A. Patrick, and D.M. Ionel, "Parameter Identification for Cells, Modules, Racks, and Battery for Utility-Scale Energy Storage Systems," *IEEE Access*, vol. 8, 2020, pp. 215817–215826.

[27] S. Haykin, *Adaptive Filter Theory*, Prentice-Hall, Englewood Cliffs, NJ, 1986.

[28] G. Pathak, B. Singh, and B.K. Panigrahi, "Logarithmic Cost Based Adaptive Control for Wind-Diesel Microgrid," In *Proceedings of Annual IEEE India Conference (INDICON)*, New Delhi, India, 2015, pp. 1–6.

[29] G. Pathak, B. Singh, and B.K. Panigrahi, "Wind-Hydro Microgrid and Its Control for Rural Energy System," *7th India International Conference on Power Electronics (IICPE)*, Patiala, India, 2016, pp. 1–5.

6 Modelling of Energy Sources and Loads for Distributed Power Generation

Pagidipala Sravanthi and Vuddanti Sandeep

CONTENTS

NOMENCLATURE

ρ	Friction coefficient
A_s	Area swept by wind speed and rotor blades
ρ	Air density
C_p	Power coefficient
R	Blade radius
ω	Speed ratio of the blade
φ	Pitch angle of the blade
P_w	Power output from a wind turbine
v_{cin}	Cut-in wind speed
v_r	Rated wind speed
v_{cout}	Cut-out wind speed
WEG	Wind energy generator
P_w^r	Rated power from wind energy generator (WEG).
μ	Mean of Weibull distribution
CDF	Cumulative distribution function
d_i	Direct cost coefficient of WEG
k_{pi}	Penalty/underestimation cost coefficient
w	Scheduled wind power
k_{ri}	Reserve/overestimation cost
G	Solar irradiation
V_b	Battery voltage
η_b	Battery efficiency
G_b	Beam solar irradiation
G_d	Diffused solar irradiation
η_{PC}	Power condition efficiency
η_r	Reference module efficiency
T_C	Cell temperature
T_A	Ambient temperature
T_{NC}	Nominal operating temperature of cell
R_{sh}	Shunt resistance
R_s	Series resistance
SOC	State of charge
SOC_{avg}	Average value of SOC
δ	Self-discharge rate of storage
η_d	Discharging efficiency
t	Charging time of an EV
E_B	Capacity of battery in kWh
DOD	Depth of discharge
A_D	Autonomy days
η_{inv}	Inverter efficiency
P_H^r	Rated capacity of PHS plant
h	Hydraulic head
Q_p	Flow rate during the pumping mode
η_p	Pumping efficiency
h_{eff}	Effective available height
K_{he}	Heat coefficient of the cooler

DOI: 10.1201/9781003229124-6

C_{gp}	Price of the gas
L	Net thermal value of gas
C_{ng}	Price of natural gas
CV_{Bio}	Caloric value of biomass

6.1 INTRODUCTION

Nowadays, green energy technologies have gained acceptance to meet increasing demand, to offset instability of fossil fuel prices, to enhance the quality and efficiency of power supplies, and to reduce emissions. Restructured power markets provide competitive incentives to the distribution companies to maintain reliable, economic, and secure electric services. Due to rapid expansion of power networks, the increase in energy demand has made the operation and control of radial distribution system an exigent and complex task for the power utilities. All the countries are working towards increasing their shares of renewable energy and have the target of meeting global power generation by renewable of 45% by the end of year 2030 [1]. However, the large penetration of renewable power generation has brought various new challenges to power and energy systems.

The share of wind energy is growing fast, and the output of wind turbines depends on several factors such as wind velocity and its direction, geographical location, etc. The growing concern about the technological, environmental advances in renewable energy, changes in electric utility infrastructures, and integration of renewable distributed generations (DGs) are considered to be a smart alternative energy solution. A number of methods have been developed in literature to derive the benefits of technical, economical, and environmental of renewable DGs [2]. However, the effect of load variability and the intermittency of non-dispatchable renewable DG power are not considered in these research works. With the emerging deregulated environment of power sectors, the DGs are considering alternative solutions for the improvement of efficiency of power distribution systems. The DGs could be either renewable such as solar photovoltaic, wind power, biomass, solar thermal system, small scale hydro power, or non-renewable energy sources [3].

A stochastic day-ahead optimal operation approach for a distributed integrated energy systems with total operating cost of the system as the decision-making objective has been demonstrated in reference [4]. A combined control approach for a standalone DG approach using squirrel-cage induction generators (SCIGs) and permanent magnet synchronous generators (PMSGs) is proposed in reference [5]. A step-by-step optimization methodology for the optimal allocation of real and reactive powers for both schedulable and non-schedulable DG units with droop characteristics in AC microgrid (MG) has been proposed in [6]. The uncertainties related to wind, PV powers, and load in MG presents various challenges to safety and economy of MG operation and control. An optimization model of MG is proposed in [7] by considering uncertainty which takes into the consideration of robustness and economy of MG operation.

The remainder of this work is organized as follows. Modelling and uncertainty handling of wind power and solar PV power are presented in Section 6.2 and Section 6.3, respectively. The mathematical models of batter energy management and electric vehicles are illustrated in Section 6.4 and Section 6.5, respectively. The modelling of pumped hydro storage and micro turbines are discussed in Section 6.6 and Section 6.7, respectively. The modelling of fuel cell, biomass, thermal generators, and diesel generators and load are presented in Section 6.8, Section 6.9, Section 6.10, and Section 6.11, respectively. Solved and practice problems are described in Section 6.12. Finally, conclusions and findings are presented in Section 6.13.

6.2 MATHEMATICAL MODEL OF WIND TURBINE SYSTEMS

The amount of wind speed depends on the height of the wind turbine (WT), w hich can be modelled as:

$$\frac{v_2}{v_1} = \left(\frac{h_2}{h_1}\right)^{\rho} \tag{6.1}$$

Where ρ is friction coefficient and it depends on wind speed, roughness of terrain, height, temperature, hour of day, and time of year. According to the IEC standards, the value of ρ is 0.20 for normal operating conditions of wind and 0.11 for extreme operating conditions of wind [8]. The amount of wind power produced for a particular wind speed can be expressed as:

$$P_w = \frac{1}{2}\rho A_s v^3 C_p \tag{6.2}$$

Where A_s is area swept by wind speed (in m^2) and rotor blades, and ρ is air density which takes the value of 1.2kg/m^3 at 20°C. C_p is power coefficient, which is the ratio of P_w to P_w^r.

The efficiency of a WT is a function of how fast its rotor runs. The tip-speed ratio (TSR) is the ratio of speed of outer tip of the blade to actual speed of the wind. It plays a major role in the blade design. Generally the slow running wind turbines operate with the TSR of 1:4, whereas fast-running wind turbines operate between 5:1 and 7:1. Power density of a WT is the available wind power per unit area. It can be expressed as:

$$Power\ density = \frac{P}{A} = \frac{\frac{1}{2}\rho A v^3 C_p}{A} = \frac{1}{2}\rho v^3 C_p \tag{6.3}$$

The power output from WEG (P_w) can also be evaluated by using [9]:

$$P_w = \frac{1}{2}\rho\pi R v^3 C_p(\omega,\varphi)\eta_g\eta_{gb} \tag{6.4}$$

Where R is blade radius and its value is between 40m and 60m; φ and ω are pitch angle of blade and speed ratio.

Expected amount of power output from a WT (P_w) depends on the wind speed, and it is represented as the linear relationship. The linear characteristics of wind power output, and wind speed is depicted in Figure 6.1. This power output (P_w) can be represented by [10],

$$P_w = \begin{cases} 0 & v < v_{cin}, v > v_{cout} \\ av + bP_w^r & v_{cin} \le v < v_r \\ P_w^r & v_r \le v < v_{cout} \end{cases} \quad (6.5)$$

where $a = \frac{P_w^r}{v_r - v_{cin}}$, $b = \frac{v_{cin}}{v_r - v_{cin}}$, and P_w^r is rated power from WEG

The power output from WEG as cubic relation can also be expressed as:

$$P_w = \begin{cases} 0 & v < v_{cin}, v > v_{cout} \\ \left(\frac{P_w^r}{v_r^3 - v_{cin}^3}\right)v^3 - \left(\frac{v_{cin}^3}{v_r^3 - v_{cin}^3}\right)P_w^r & v_{cin} \le v < v_r \\ P_w^r & v_r \le v < v_{cout} \end{cases} \quad (6.6)$$

Another way to represent the power output from WEG (P_w) can be represented by [11]:

$$P_w = \begin{cases} 0 & v \le v_{cin},\ v_{cout} \le v \\ P_w^r\left(A + Bv + Cv^2\right) & v_{cin} \le v \le v_r \\ P_w^r & v_r \le v < v_{cout} \end{cases} \quad (6.7)$$

where A, B, C are the coefficients related to P_w and they can be expressed as:

$$A = \frac{1}{\left(v_{ci} - v_r\right)^2}\left[v_{ci}\left(v_{ci} + v_r\right) - 4v_{ci}v_r\left(\frac{v_{ci} + v_r}{v_r}\right)^3\right] \quad (6.8)$$

$$B = \frac{1}{\left(v_{ci} - v_r\right)^2}\left[-\left(3v_{ci} + v_r\right) + 4v_{ci}v_r\left(\frac{v_{ci} + v_r}{2v_r}\right)^3\right] \quad (6.9)$$

$$C = \frac{1}{\left(v_{ci} - v_r\right)^2}\left[2 - 4\left(\frac{v_{ci} + v_r}{2v_r}\right)^3\right] \quad (6.10)$$

6.2.1 Modelling of Wind Speed and Wind Power Distribution

Probability-based approaches are suitable for the modelling the uncertainty of renewable energy sources (RERs) due to the availability of historical data. A Monte Carlo simulation–based (MCS-based) approach is also suitable for modelling the uncertainty of RERs. In this approach, different scenarios are created and the optimization problem is solved for each scenario. The weighted average of all scenarios gives the final solution. The probability distribution of wind speed (that effects the wind power output) is represented by the two-parameter based Weibull distribution, as it is considered as more accurate and appropriate [12], and it can be represented by:

$$f(v) = \left(\frac{k}{c}\right)\left(\frac{v}{c}\right)^{(k-1)} e^{-\left(\frac{v}{c}\right)^k} \quad \text{for } 0 < v < \infty \quad (6.11)$$

Mean of Weibull distribution (μ) is represented by:

$$\mu = c\Gamma\left(1 + k^{-1}\right) \quad (6.12)$$

where the gamma function $\Gamma(x)$ is represented by:

$$\Gamma(x) = \int_0^\infty e^{-t}t^{(x-1)}\, dt \quad (6.13)$$

Cumulative distribution function (CDF) for wind speed (v) is represented by:

$$F(v) = 1 - e^{-\left(\frac{v}{c}\right)^k} \quad (6.14)$$

where k and c are shape and scale parameters of Weibull PDF, and they are derived from the historical data; $k = 1$ for exponential function and $k = 2$ for Reyleigh distribution

The wind speed data can be generated by creating a random number (R) between -1 and 1, and it can be expressed as:

$$v = c\left[-log_e\left(1 - R\right)\right]^{\frac{1}{k}} \quad (6.15)$$

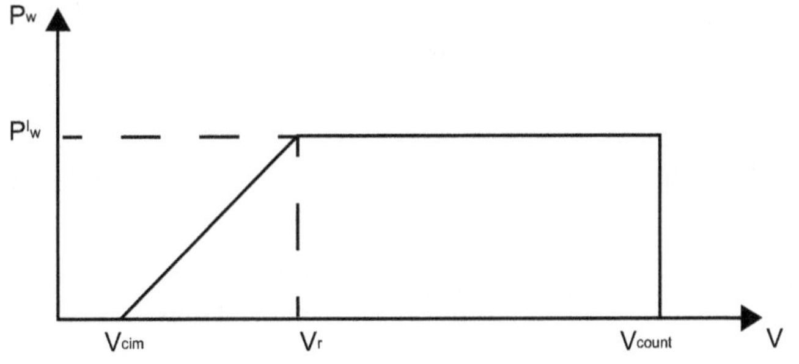

FIGURE 6.1 Characteristics of wind power output and wind speed.

From equations (6.6) and (6.7), it is clear that the wind power distribution is continuous in $v_{cin} \le v < v_r$ and discrete in ($v\langle v_{cin}, v\rangle v_{cout}$) and ($v_r \le v < v_{cout}$). The continuous probability in $\left(v_{cin} \le v < v_r\right)$ range can be defined as [11]:

$$f_w(P_w) = \frac{khv_{cin}}{cP_w^r}\left[\frac{\left(1+\frac{hP_w}{P_w^r}\right)v_{cin}}{c}\right]^{(k-1)} e^{-\left[\frac{\left(1+\frac{hP_w}{P_w^r}\right)v_{cin}}{c}\right]^k} \tag{6.16}$$

where $h = \left(\frac{v_r - v_{cin}}{v_{cin}}\right)$

As mentioned earlier, there are two discrete probabilities. The first discrete probability in ($v\langle v_{cin}, v\rangle v_{cout}$) range can be modelled as [12]:

$$\begin{aligned} \Pr(P_w = 0) &= \Pr(v < v_{cin}) + \Pr(v \ge v_{cout}) \\ &= 1 - e^{-\left(\frac{v_{cin}}{c}\right)^k} + e^{-\left(\frac{v_{cout}}{c}\right)^k} \end{aligned} \tag{6.17}$$

The second discrete probability in ($v_r \le v < v_{cout}$) range can be modelled as:

$$\Pr(P_w = P_w^r) = \Pr(v_r \le v < v_{cout}) = e^{-\left(\frac{v_r}{c}\right)^k} - e^{-\left(\frac{v_{cout}}{c}\right)^k} \tag{6.18}$$

6.2.2 Cost of Wind Power Generation

Cost of power output from ith WEG (C_{wi}) can be expressed as [12]:

$$C_{wi} = d_i P_{wi} \tag{6.19}$$

where d_i is direct cost related to wind power output

When the wind power produced is more than the required value, then it involves the penalty cost, and it can be expressed as:

$$C_{wi,p} = k_{pi}\left(P_{wi,av} - P_{wi}\right) = k_{pi}\int_{P_{wi}}^{P_w^r}\left(w - P_{wi}\right)f_w(w)\,dw \tag{6.20}$$

where k_{pi} is penalty/underestimation cost coefficient, w is scheduled wind power

When the wind power produced is less than the required power, then it involves the reserve cost, and it can be expressed as [13]:

$$C_{wi,r} = k_{ri}\left(P_{wi} - P_{wi,av}\right) = k_{ri}\int_0^{P_{wi}}\left(P_{wi} - w\right)f_w(w)\,dw \tag{6.21}$$

Where k_{ri} is reserve/overestimation cost. The power output from WEGs is between 0 and its rated power, and it can be expressed as:

$$0 \le P_w \le P_w^r \tag{6.22}$$

6.3 MATHEMATICAL MODEL OF A SOLAR PV SYSTEM

Solar PV cells are joined in series or parallel to form the solar PV arrays/panels. The amount of power output from a solar PV panel is directly related to solar irradiation (G) and temperature (T), and it can be calculated by using [14]:

$$P_{pv} = P_{pv}^r \frac{G}{G_{std}}\left[1 + k_t\left(T + 0.0256G - T_{STC}\right)\right] \tag{6.23}$$

Where G is solar irradiation (W/m^2), G_{std} is standard solar irradiation (1,000W/m^2), T is temperature of solar PV cell (°C), T_{STC} is temperature at standard test condition (25°C) and k_t is a constant (–3.7 × 10^{-3} 1/°C)

By neglecting the temperature, power output from solar PV panel (P_{pv}) can be represented by [15]:

$$P_{pv} = \begin{cases} P_{pv}^r\left(\frac{G^2}{G_{std}R_c}\right) & for\ 0 < G < R_c \\ P_{pv}^r\left(\frac{G}{G_{std}}\right) & for\ G \ge R_c \end{cases} \tag{6.24}$$

where R_c is certain solar irradiance (150 W/m^2)

Figure 6.2 represents the grid-connected solar PV energy module, along with the battery energy storage.

Amount of solar PV power injected at a bus (P_s) can be expressed as:

$$P_S = P_{pv} + P_b - P_u \tag{6.25}$$

Where P_u is spillage power. P_b is battery power and it can be calculated by using:

$$P_b = \frac{\left(C_{init} - C_{end}\right)V_b}{\Delta t\eta_b} \tag{6.26}$$

where C_{init} and C_{end} are the state of charge (SOC) of battery at the initial and at the end of operating time interval, V_b is battery voltage, and η_b is battery efficiency

Generally, the solar PV panels are titled to extract maximum amount of solar irradiation on solar PV arrays. Typically, tilt angle is equal to latitude angle for a particular region. Amount of power generated from solar PV panels depends on beam irradiation (G_b) and diffused irradiation (G_d) [16]. Total irradiation (G_T) on a titles surface can be determined by using:

$$G_T = G_b £_b + G_d £_d + \left(G_b + G_d\right)£_r \tag{6.27}$$

where $£_b$, $£_d$, and $£_r$ are the tilt factors for the beam and diffused and reflected solar irradiation

The power output from the solar PV plant can also be represented by:

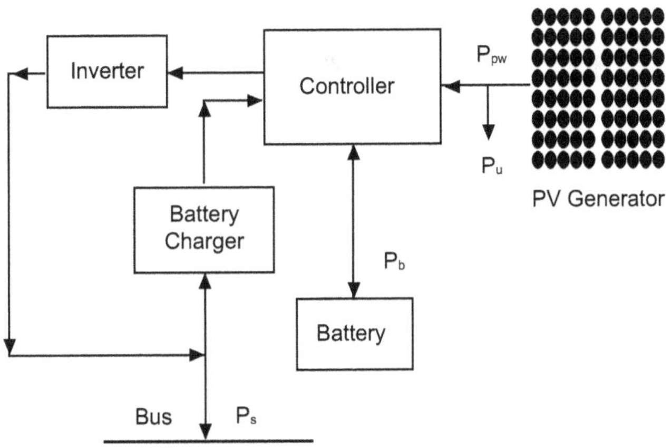

FIGURE 6.2 Solar PV energy system connected with battery energy storage.

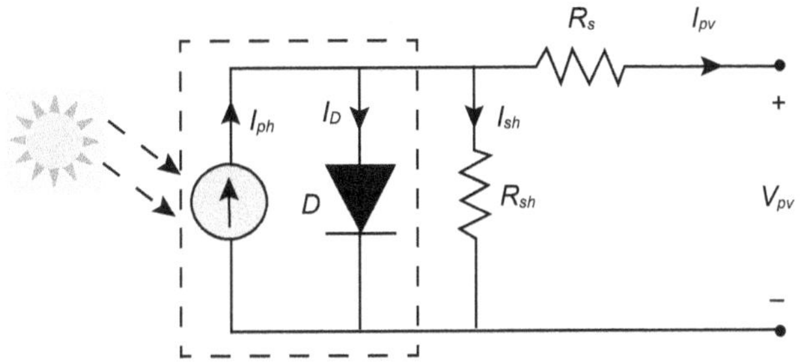

FIGURE 6.3 Equivalent circuit model of a solar PV cell with single diode.

$$P_{pv} = \eta_{pv} G_T A_{pv} \tag{6.28}$$

where η_{pv} is conversion efficiency of solar PV panel, A_{pv} is surface area of solar PV panel

The solar PV panel efficiency (η_{pv}) can be expressed as:

$$\eta_{pv} = \eta_{PC}\eta_r \left[1 - k_t \left(T_C - T_{STC}\right)\right] \tag{6.29}$$

where η_{PC} is power condition efficiency, η_r is reference module efficiency

T_c is cell temperature and it can be expressed as:

$$T_C = T_A + \left(\frac{T_{NC} - 20}{800}\right) G \tag{6.30}$$

where T_A is ambient temperature and T_{NC} is nominal operating temperature of the cell

The identification of solar PV cell can be done by using one-diode and two-diode models. This model is a combination of diode, photo-generated controlled current source, shunt resistance (R_{sh}), and series resistance (R_s) [17]. The equivalent circuit model of solar PV cell with single-diode is depicted in Figure 6.3.

Equations derived from the diode model by using the Kirchhoff's laws are:

$$I_{pv} = I_{ph} - I_D - I_{sh} \tag{6.31}$$

where

$$I_{sh} = \frac{V_{pv} + \left(I_{pv} R_s\right)}{R_{sh}} \tag{6.32}$$

The V-I characteristics of the solar PV cell by single exponential Shockley equation can be expressed as:

$$I_{pv} = I_{ph} - I_{sd}\left[e^{\left(q \cdot \frac{V_{pv} + I_{pv} R_s}{\sigma K T_c}\right)} - 1\right] - \left(\frac{V_{pv} + I_{pv} R_s}{R_{sh}}\right) \tag{6.33}$$

where I_{pv} is output current of cell, I_{ph} is source/photon current generated from the sun rays, q is electron charge, V_{pv} is output voltage of solar PV cell, R_s is series resistance that considers the voltage drop, I_{sd} is reserve saturation current of diode, R_{sh} is shunt resistance that considers the current loss, T_c is cell temperature, K is Boltzmann's constant, and σ is ideality factor of p-n diode

There are five unknown parameters in Eq. (6.33), i.e., I_{ph}, I_{sd}, σ, R_s, and R_{sh}. The two-diode model of solar PV cell can be expressed as:

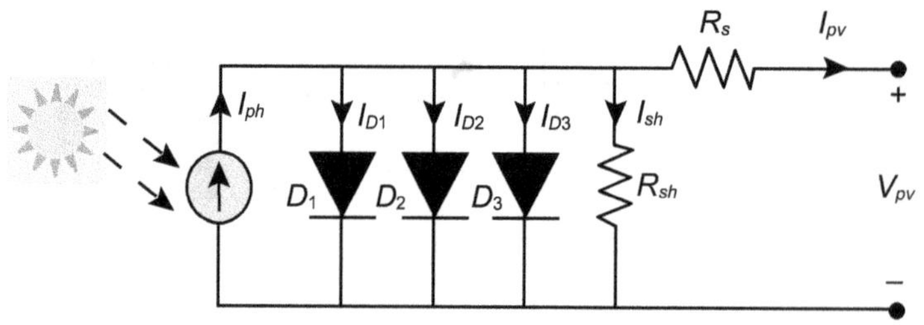

FIGURE 6.4 Equivalent circuit model of a triple-diode solar PV cell.

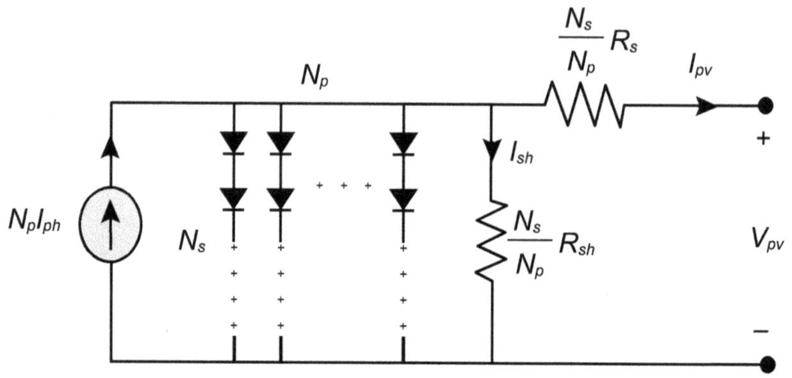

FIGURE 6.5 Equivalent circuit of a solar PV module.

$$I_{pv} = I_{ph} - I_{s1}\left[e^{\left(q.\frac{V_{pv}+I_{pv}R_s}{\sigma_1 KT_c}\right)} - 1\right] - I_{s2}\left[e^{\left(q.\frac{V_{pv}+I_{pv}R_s}{\sigma_2 KT_c}\right)} - 1\right] - \left(\frac{V_{pv}+I_{pv}R_s}{R_{sh}}\right) \quad (6.34)$$

Figure 6.4 depicts the equivalent circuit of triple-diode for a solar PV cell. The output current of cell (I_{pv}) by using the source/photon current (I_{ph}), diode currents (I_{D1}, I_{D2}, I_{D3}), and shunt current (I_{sh}) can be expressed by [18]:

$$I_{pv} = I_{ph} - \sum_{i=1}^{3} I_{Di} - I_{sh} \quad (6.34)$$

where the diode currents (I_{Di}) can be calculated by using:

$$I_{Di} = I_{sdi}\left(e^{\left(\frac{V+I_{pv}R_s}{\sigma_i V_t}\right)} - 1\right) \; for \; i = 1,2,3 \quad (6.34)$$

where $V_t = \frac{KT_c}{q}$ and $I_{sh} = \frac{V_{pv}+I_{pv}R_s}{R_{sh}}$

A typical solar PV module includes several solar PV cells, and its equivalent circuit is depicted in Figure 6.5.

The mathematical representation of solar PV module can be expressed by:

$$I_{pv} = N_p I_{ph} - N_p I_s\left[e^{\left(q.\frac{\frac{V_{pv}}{N_s}+\frac{R_s}{N_p}I_{pv}}{\sigma KT_c}\right)} - 1\right] - \left(\frac{V_{pv}\frac{N_p}{N_s}+I_{pv}R_s}{R_{sh}}\right) \quad (37)$$

where N_s and N_p are number of solar PV cells connected in series and parallel

6.3.1 Uncertainty Modelling of Solar PV Units

The historical data of solar PV irradiation is used to determine the probability distribution of solar PV units. The distribution function of solar PV system follows the bimodal distribution, and it can be modelled by using the Weibull PDF. This Weibull function can be expressed as:

$$f_G = \omega\left(\frac{k_1}{c_1}\right)\left(\frac{G}{c_1}\right)^{(k_1-1)} exp\left[-\left(\frac{G}{c_1}\right)^{k_1}\right] + (1-\omega)\left(\frac{k_2}{c_2}\right)\left(\frac{G}{c_2}\right)^{(k_2-1)} exp\left[-\left(\frac{G}{c_2}\right)^{k_2}\right] \quad (38)$$

where ω is weight factor. c_1 and c_2 are scale factors. k_1 and k_2 are shape factors.

Beta PDF ($f(G)$) can also be used in each unimodal function, and it can be represented by [19]:

$$f_G = \frac{\Gamma(\alpha+\beta)}{\Gamma(\alpha)\Gamma(\beta)} G^{(\alpha-1)}(1-G)^{(\beta-1)} \tag{6.39}$$

where α, β are the parameters of Beta PDF, and they are determined by using the mean (μ) and standard deviation (σ)

They are expressed as:

$$\beta = (1-\mu)\left[\frac{\mu(1+\mu)}{\sigma^2} - 1\right] \tag{6.40}$$

$$\alpha = \frac{\mu\beta}{1-\mu} \tag{6.41}$$

The Beta PDF for the solar irradiation can be expressed as:

$$f_G = \frac{G^{\alpha-1}(1-G)^{\beta-1}}{B(\alpha,\beta)} \tag{6.42}$$

where $B(\alpha,\beta) = \frac{\Gamma(\alpha)\Gamma(\beta)}{\Gamma(\alpha+\beta)}$

The PDF of solar irradiance (G) can be considered to follow lognormal function. The PDF of Lognormal PDF with μ and σ can be represented by:

$$f_G = \frac{1}{\sigma\sqrt{2\pi}G} e^{\frac{-(\ln G-\mu)^2}{2\sigma^2}} \tag{6.43}$$

6.3.2 Cost of Solar PV Power Generation

The cost function of solar PV unit $\left(C\left(P_{pv}\right)\right)$ can be expressed by [20]:

$$C\left(P_{pv}\right) = \Psi_{pv}\int_{t_0}^{t}\left(P_{pv}^{f} + \Delta P_{pv}\right)dt \tag{6.44}$$

where Ψ_{pv} is cost coefficient of solar PV power, P_{pv}^{f} is forecasted solar PV power, and ΔP_{pv} is the difference between actual and forecasted solar PV powers

Due to the uncertainty related to solar PV power generation, there exists a penalty cost $\left(C_p\left(P_{pv}\right)\right)$, and it can be expressed by:

$$C_p\left(P_{pv}\right) = \Psi_{pv}^{p}\int_{t_0}^{t}\left(G_a - G_f\right)dt \tag{6.45}$$

where Ψ_{pv}^{p} is penalty cost coefficient related to solar PV power. G_a and G_f are the actual and forecasted solar irradiation, respectively

The solar PV power restricted between 0 and its rated power, and it can be expressed as:

$$0 \le P_{pv} \le P_{pv}^{r} \tag{6.46}$$

6.4 MATHEMATICAL MODEL OF BATTERY ENERGY STORAGE MANAGEMENT

In recent years, energy storage plays an increasingly vital role in integrating large penetration of non-dispatchable and intermittent RERs into power systems. Battery energy storage units and pumped hydro storage plants are used as important storage technologies for large-scale integration of RERs. The state of charge (SOC) of a battery is calculated by using the ratio of residual capacity/energy to its rated capacity/energy. Determination of SOC is required for economic dispatch of the system and for controlling charging/discharging process. SOC is percentage of energy remaining in a battery. The SOC is calculated as the ratio of charge stored at a particular time t (i.e., $C(t)$) to rated capacity of charge (C_{rated}). It can be expressed as:

$$SOC(t) = \frac{C(t)}{C_{rated}} \times 100 \tag{6.47}$$

SOC is 0% for an empty battery, and 100% for a full battery. For example, a battery has its full capacity of 200Ah and the remaining capacity is 150Ah, then the SOC is 75% ([150Ah/200Ah] × 100).

Depth of discharge (DOD) of a battery pack finds the fraction of power that can be withdrawn from the battery. Suppose, if a 2000Ah battery has a DOD of 25% then this battery can provide only 500Ah (i.e., 2000Ah × 0.25). To extend the life of a battery, minimum SOC is limited to 20%. The maximum value of SOC (SOC_{max}) is 1, and minimum value of SOC (SOC_{min}) can be calculated by using the maximum DOD, which can be expressed by:

$$SOC_{min} = (1-DOD)SOC_{max} \tag{6.48}$$

The charging formula of battery can be expressed as:

$$SOC_t = (1-\delta)SOC_{t-1} - \left(\frac{P_c\eta_c\Delta t}{E_c}\right) \tag{6.49}$$

where δ is self-discharge rate of storage; P_c is negative and it is the charging power; SOC_t and SOC_{t-1} are the SOC of battery storage in time periods t and t–1, respectively; E_c is total capacity of battery storage; and η_c is charging efficiency

The discharging formula of battery can be expressed by,

$$SOC_t = (1-\delta)SOC_{t-1} - \left(\frac{P_d\Delta t}{E_c\eta_d}\right) \tag{6.50}$$

where η_d is discharging efficiency; P_d is positive and it is the discharging power.

The SOC of a battery is limited by:

$$SOC_{min} \leq SOC_t \leq SOC_{max} \quad (6.51)$$

The SOC of a battery energy storage at time t can be represented by:

$$SOC_t = SOC_{init} + \left(\eta_c \sum_{t=1}^{T} P_c^t\right) - \left(\eta_d \sum_{t=1}^{T} P_d^t\right) \quad (6.52)$$

The charging and discharging power limits of a battery pack is restricted by:

$$-P_B^{max} \leq P_B \leq P_B^{max} \quad (6.53)$$

where P_B^{max} is the maximum charging and discharging power of battery pack

SOC at the end of scheduling period (Δt) can be expressed by:

$$SOC_{end} = SOC_{start} + \sum_{t=0}^{T-1} P_B \Delta t = SOC_{start} \quad (6.54)$$

where T is scheduling period

6.5 MATHEMATICAL MODEL OF ELECTRIC VEHICLES

Electric vehicles (EVs) will not release CO_2 and they are drawing attention due to the enforcement of electric vehicle charging stations (EVCSs). In general, charging is considered as grid-to-vehicle (G2V) and discharging is considered as vehicle-to-grid (V2G). Charging ($CH_{EV,i}^t$) and discharging ($DCH_{EV,i}^t$) of EVs cannot be occurred simultaneously, and it can be expressed by:

$$CH_{EV,i}^t + DCH_{EV,i}^t \leq 1 \quad (6.55)$$

$CH_{EV,i}^t$ and $DCH_{EV,i}^t$ are binary variables between 0 and 1. They depict charging and discharging status of ith EV at time t. Energy stored in the battery of EV ($E_{EV,i}^t$) can be calculated by determining the remaining energy from previous period ($E_{EV,i}^{t-1}$), and charge/discharge during the time period t:

$$E_{EV,i}^t = E_{EV,i}^{t-1} + \eta_c P_{ch,EV}^t \Delta t - \frac{P_{Dch,EV}^t \Delta t}{\eta_d} \quad (6.56)$$

where η_c and η_d are charging and discharging efficiencies of EVs

The SOC of ith EV can be expressed by [21]:

$$SOC_i^{min} \leq SOC_i^t \leq SOC_i^{max} \quad (6.57)$$

where SOC_i^{min} and SOC_i^{max} are minimum and maximum SOC limits of ith EV, respectively

Generally, the value of SOC_i^{min} is kept at 20% of battery capacity for allowing the reserve energy to the vehicle in the case of distance travelled requirement being higher than the average distance. The value of SOC_i^{max} is fixed at 85% of battery capacity to account for changes in the aggregated capacity limits of the battery. The battery charging constraint of EV can be expressed as [22]:

$$\left(P_{ch,EV}^t + R_{dn,EV}^t\right)\eta_c \Delta t \leq \left(E_{EV,i}^{max} - E_{EV,i}^t\right) \quad (6.58)$$

The battery discharge of EV can be expressed as:

$$\frac{\left(P_{Dch,EV}^t + R_{up,EV}^t\right)\Delta t}{\eta_d} \leq E_{EV,i}^t \quad (6.59)$$

The uncertainties of EVs charging always pose an impact on the distribution systems. Based on the availability of daily driving data of EVs, the PDF of driving distance can be expressed as:

$$f(d) = \frac{1}{d\sqrt{2\pi\sigma_d}} exp\left[\frac{-(\ln d - \mu_d)^2}{2\sigma_d^2}\right] \quad (6.60)$$

where d is the driving distance. μ_d and σ_d are mean and standard deviation of driving distance

Probability distribution of EV charging state at the end of the day after the last trip, and it follows the normal distribution. This can be expressed as [23]:

$$f(t) = \begin{cases} \frac{1}{\sqrt{2\pi\sigma_t}} exp\left[-\frac{(t-\mu_t)^2}{2\sigma_t^2}\right] & (\mu_t - 12) < t < 24 \\ \frac{1}{\sqrt{2\pi\sigma_t}} exp\left[-\frac{(t+24-\mu_t)^2}{2\sigma_t^2}\right] & 0 < t \leq (\mu_t - 12) \end{cases} \quad (61)$$

Where t is the charging time of an EV for driving distance d

6.5.1 EV Battery Degradation

The degradation of an EV battery depends on the rate of charging, temperature, SOC and DOD, etc. The levels of SOC and DOD must be within their lower and upper limits. Full discharging (i.e., high DOD) may lead to fading of capacity due to the rise in the battery resistance. Capacity fading can also occur by not using the battery for long time with high SOC. The degradation cost by including SOC (i.e., C_d^{SOC}) accounts for capacity fade attributable to SOC, and it can be expressed as [24]:

$$C_d^{SOC} = \left(\frac{mSOC_{avg} - d}{CF_{max} \times 15 \times 8760}\right) C_B \quad (6.62)$$

Where SOC_{avg} is average value of SOC, and CF_{max} is capacity fade at the end of life of a battery. m and d are linear fit parameters. The degradation cost by including DOD (C_d^{DOD}) can be expressed as [25]:

$$C_d^{DOD} = \left(\frac{C_B E_B + C_L}{L_C E_B DOD}\right) E_{dis} \quad (6.63)$$

where E_B is capacity of battery in kWh, DOD is depth of discharge, C_B is battery cost per kWh, L_c is lifecycle of battery at a determined DOD, C_L is battery replacement cost for labour, and E_{dis} is discharge energy

The capacity of battery (E_B) can be determined by using:

$$E_B = \frac{P_L A_D}{\eta_{inv}\eta_B DOD} \tag{6.64}$$

where P_L is the active power load, A_D is autonomy days, η_{inv} is inverter efficiency, and η_B is battery efficiency

6.6 MATHEMATICAL MODEL OF PUMPED HYDRO STORAGE/SMALL HYDRO POWER

A pumped hydro storage (PHS) plant helps in integrating the intermittent RERs and to balance the power generation and load demand in the system, i.e., load balancing. The PHS plant stores the energy in a high-level water reservoir as a hydraulic potential energy (PE) by pumping the water from a low-level water reservoir. During the peak demand period, water flows into low-level water reservoir through the turbines and drives the power generators. However, during low demand periods, water is pumped back to a high-level water reservoir [26]. The amount of power generated from the PHS plant during the pumping/generating mode (P_H^g) can be expressed as:

$$P_H^g = hP_H^r Q_t \eta_g \tag{6.65}$$

where P_H^r is the rated capacity of PHS plant, h is hydraulic head (in metres), Q_t is flow rate through turbine (in m^3/s), and η_g is power generation efficiency

The amount of power used during the pumping mode (P_H^p) to pump the water from lower reservoir to upper reservoir can be expressed as:

$$P_H^p = \frac{hP_H^r Q_p}{\eta_p} \tag{6.66}$$

where Q_p is flow rate during the pumping mode (in m^3/s) and η_p is pumping efficiency during the pumping mode

The volume of water in high-level reservoir of PHS unit during the pumping mode at time (t+1) (i.e., V_H^{t+1}) can be expressed as:

$$V_H^{t+1} = V_H^t + Q_{Ht}^p\left(P_{Ht}^p\right) \tag{6.67}$$

where $Q_{Ht}^p\left(P_{Ht}^p\right)$ is the discharge/generating rate of PHS unit at time t

Water volume in the lower reservoir of a PHS unit during the generating mode at time (t+1) (i.e.,V_H^{t+1}) can be expressed as:

$$V_H^{t+1} = V_H^t - Q_{Ht}^g\left(P_{Ht}^g\right) \tag{6.68}$$

where $Q_{Ht}^g\left(P_{Ht}^g\right)$ is pumping rate of a PHS unit at time t

During the changeover time:

$$V_H^{t+1} = V_H^t \tag{6.69}$$

Upper-reservoir storage of a PHS unit is limited by:

$$V_H^{min} \le V_H^t \le V_H^{max} \tag{6.70}$$

Power generation limits of a PHS unit during the generating mode is expressed by:

$$P_H^{g,min} \le P_{Ht}^g \le P_H^{g,max} \tag{6.71}$$

Power generation limits of a PHS unit during the pumping mode is expressed by:

$$P_H^{p,min} \le P_{Ht}^p \le P_H^{p,max} \tag{6.72}$$

In small hydro power sources, the river flow rate (Q) follows the Gumbel PDF through the location/shape constant (k) and scale constant (c), and it can be expressed as [27]:

$$f_Q = \frac{1}{c}exp\left(\frac{Q-k}{c}\right)exp\left[-exp\left(\frac{Q-k}{c}\right)\right] \tag{6.73}$$

6.6.1 HYDROELECTRIC POWER GENERATION

The amount of hydroelectric power output from hydro plant (P_h) depends on the storage volume of the reservoir (V_h), and water discharge rate (Q_h) is determined by using:

$$P_h = C_1V_h^2 + C_2Q_h^2 + C_3V_hQ_h + C_4V_h + C_5Q_h + C_6 \tag{6.74}$$

where C_1, C_2, C_3, C_4, C_5 and C_6 are power coefficients of jth hydro unit

The constraints on reservoir storage volumes (V_h) and discharge rates (Q_h) can be expressed by:

$$V_h^{min} \le V_h \le V_h^{max} \tag{6.75}$$

$$Q_h^{min} \le Q_h \le Q_h^{max} \tag{6.76}$$

where Q_h^{min} and Q_h^{max} are minimum and maximum water discharge rates; V_h^{min} and V_h^{max} are minimum and maximum storage volumes of reservoir

The continuity equation of hydro power network is given by [28]:

$$V_{hn}^t = V_{hn}^{t-1} + I_{hn}^t - Q_{hn}^t - S_{hn}^t + \sum_{l=1}^{R_n}\left[Q_{hl}^{(t-D_{ln})} + S_{hl}^{(t-D_{ln})}\right] \tag{6.77}$$

where R_n is number of upstream hydro generation units that are connected directly above nth hydro reservoir, I_{hn}^t is inflow rate, S_{hn}^t is rate of spillage discharge of nth reservoir, and D_{ln} is water transport delay from lth hydro reservoir to nth hydro reservoir.

6.6.2 Micro Hydro Power Generation

This generator allows the operation with high volumetric flow rate with low head and also high head with low volumetric flow rate. The amount of power output ($P_{\mu H}$) from this micro hydro power generator can be expressed as:

$$P_{\mu H} = \eta_{\mu H} g \rho_w h_{eff} Q \quad (6.78)$$

where $\eta_{\mu H}$ is the efficiency of micro hydro generator, h_{eff} is the effective available height, g is acceleration due to gravity, ρ_w is water density, and Q is water flow rate in m^3/s

6.7 MODELLING OF A MICRO TURBINE

The amount of emissions released from CO_2 and NO_x of MT are much lower than the conventional fossil fuel based centralized power plants [29]. The residual heat of the exhaust (Q_{MT}) can be expressed as:

$$Q_{MT} = \frac{P_G^{MT}\left(1-\eta_e-\eta_l\right)}{\eta_e} \quad (6.79)$$

where P_G^{MT} is the power output from the MT during the scheduling period (Δt), η_l is heat loss coefficient, and η_e is power generation efficiency of MT

Heat produced by the MT (Q_{he}) as a function of heat coefficient of cooler (K_{he}) can be expressed as:

$$Q_{he} = K_{he} Q_{MT} \quad (6.80)$$

The gas consumption cost of MT (C_{MT}) can be expressed by [29]:

$$C_{MT} = \left(\frac{\sum P_G^{MT}\Delta t}{\eta_e L}\right) C_{gp} \quad (6.81)$$

where L is net thermal value of gas, and C_{gp} is price of the gas

6.8 MODELLING OF A FUEL CELL

A fuel cells (FC) is environmentally friendlier, quiet and exhibits high power efficiencies under variable load conditions. Fuel cells generate a certain amount of heat energy, and if this energy is not used, it will produce a lot of waste. Both heat and electrical energies have a certain conversion efficiency. The voltage of FC (V_{FC}) can be represented by:

$$V_{FC} = N_c\left[E + \frac{K_g T_s}{2K_F} ln\left(\frac{P_H P_O^{0.5}}{P_W}\right)\right] - I_{FC} R_i \quad (6.82)$$

where K_F is Faraday's constant, N_C is number of cells in the stack, R_i is internal resistance of stack, E is the fuel cell energy, T_s is stack temperature, K_g is universal gas constant, I_{FC} is current in the FC system. P_H, P_O and P_W are the partial pressure of hydrogen, oxygen and water, respectively

Gas consumption cost of FC (i.e., C_{FC}) can be expressed as:

$$C_{FC} = \left(\frac{\sum P_G^{FC}\Delta t}{\eta_{FC} L}\right) \quad (6.83)$$

where η_{FC} is electrical efficiency of FC, P_G^{FC} is power output from FC

However, in the case of proton-exchange membrane (PEM) FCs, the fuel cell cost (C_{FC}) can be represented as a linear function of power generation from fuel cell (P_G^{FC}), power generated from the hydrogen (P_H), and the power consumed by auxiliary device (P_{ax}), and it can be expressed as [30]:

$$C_{FC} = \left(\frac{P_G^{FC} + P_H + P_{ax}}{\eta_{FC}}\right) \times C_{ng} \quad (6.84)$$

where C_{ng} is the price of natural gas

6.9 MODELLING OF BIOMASS POWER GENERATION

Generally, biogas is produced from household waste, agriculture, and forest residues. Power output from biomass plant can be calculated by using:

$$P_{Bio} = \eta_{Bio} F_A CV_{Bio} \quad (6.85)$$

where CV_{Bio} is caloric value of biomass, η_{Bio} is efficiency of biomass plant, and F_A is fuel availability

6.10 MODELLING OF THERMAL GENERATORS

During high-demand periods, thermal generators need to open more valves; hence, the hot steam increases by creating the valve point loading (VPL) effect. In this case, the input and output relation becomes non-convex. The non-convex cost of thermal generators with VPL effect can be expressed as:

$$C\left(P_{Gi}\right) = a_i + b_i P_{Gi} + c_i P_{Gi}^2 + \left|d_i \times sin\left[e_i\left(P_{Gi}^{min} - P_{Gi}\right)\right]\right| \quad (6.86)$$

The amount of emission released from the thermal generating units can be represented by:

$$E\left(P_{Gi}\right) = \alpha_i + \beta_i P_{Gi} + \gamma_i P_{Gi}^2 + \tau_i exp\left(\delta_i P_{Gi}\right) \quad (6.87)$$

6.11 MODELLING OF DIESEL GENERATORS AND LOADS

6.11.1 Modelling of Diesel Generators

A diesel generator (DG) has high maintenance and fuel costs, and it is not environmentally friendly. However, it is considered a secondary source of power and plays a major role in maintaining stable operation of the system. The fuel consumption of a DG ($F_c(t)$) can be expressed as:

$$F_c(t) = aP(t) + bP_{DG}^r \quad (6.88)$$

where P_{DG}^r is rated power, $P(t)$ is generated power, and a and b are the fuel consumption coefficients

The fuel cost of ith DG (FC_i) can be represented by:

$$FC_i = a_i + b_i P_{DG,i} + c_i P_{DG,i}^2 \quad (6.89)$$

where a_i, b_i and c_i are cost coefficients of the DG. $P_{DG,i}$ is power produced by DG

Overall efficiency of DG ($\eta_{overall}$) can be calculated by using:

$$\eta_{overall} = \eta_{brake\ thermal} \times \eta_{generator} \quad (6.90)$$

where $\eta_{brake\ thermal}$ is the brake thermal efficiency

6.11.2 Load Modelling

Load modelling represents the mathematical formulation that relates the voltage and power in a load bus. Generally, the loads are classified into static and dynamic load models.

6.11.2.1 Static Load Modelling

In this load model, real and reactive powers can be expressed in bus voltage magnitudes and frequencies. These models can be classified into ZIP model, exponential model, and frequency dependent model.

6.11.2.1.1 *ZIP Load Model*

Generally, in a constant power model, loads are considered as independent of system voltage. The real and reactive powers using constant power (P), constant impedance (Z) and constant current (I)—i.e., ZIP load model—can be expressed as:

$$P_{Di} = P_{Di}^0 \left[a_P + b_I \left(\frac{V_i}{V_i^0} \right) + c_Z \left(\frac{V_i}{V_i^0} \right)^2 \right] \quad (6.91)$$

$$Q_{Di} = Q_{Di}^0 \left[a_P' + b_I' \left(\frac{V_i}{V_i^0} \right) + c_Z' \left(\frac{V_i}{V_i^0} \right)^2 \right] \quad (6.92)$$

where V_i and V_i^0 are the actual and nominal bus voltages at bus i. a_p and a'_p are the constants for P load, b_I and b'_I are the constants for I load, and c_Z and c'_Z are the constants for Z load

6.11.2.1.2 *Exponential Load Model*

In this load model, real and reactive power demands in terms of bus voltages can be expressed as:

$$P_{Di} = P_{Di}^0 \left(\frac{V_i}{V_i^0} \right)^{np} \quad (6.93)$$

$$Q_{Di} = Q_{Di}^0 \left(\frac{V_i}{V_i^0} \right)^{nq} \quad (6.94)$$

where np and nq are the voltage exponents for the exponential load model

6.11.2.1.3 *Frequency Dependent Load Model*

In this load model, a factor is calculated by using the following equation, and it is multiplied to the ZOP or exponential load models:

$$\text{Factor} = \left[1 + a_f \left(f - f_0\right)\right] \quad (6.95)$$

where f and f_0 are the actual and nominal frequency, and a_f is the frequency sensitivity factor

6.11.2.2 Dynamic Load Modelling

Voltage stability studies use dynamic load models as they respect accurate models. Here, real and reactive powers are expressed as a function of voltage, as well as time. The most commonly used models to represent dynamic loads are the induction motor model, composite load model, and exponential recovery load model.

6.12 PROBLEMS

6.12.1 Solved Problems

Problem 1

A 9V battery with a capacity of 1000Ah has the state of charge of 60%. Calculate the charge stored in the battery and energy delivered to load.

Solution

Given, battery voltage (V_B) = 9V, rated capacity (C_{rated}) =1000Ah, $SOC(t)$=80%

$$SOC(t) = \frac{C(t)}{C_{rated}} \times 100 \Rightarrow C(t) = \frac{SOC(t) \times C_{rated}}{100}$$

$$SOC(t) = \frac{80 \times 1000}{100} = 800Ah$$

Therefore, the charge stored at 80% of SOC is 800Ah.
Power delivered to load = 800Ah x 9V = 7200Wh

Problem 2

An initially full charged battery with a voltage of 15V and having 400Ah capacity is connected to a load of 2Ω. Determine the SOC of this battery after four hours.

Solution

As this battery is fully charged, SOC at hour (t–1) is 100%, i.e., SOC(t–1)=100%
Δt = 4 hours
The battery current = Load current = i(t) = (voltage of battery/load resistance) = 15V/2Ω = 7.5A

The SOC of a battery at a time t can be calculated by using:

$$SOC(t) = SOC(t-1) + \left(\frac{i(t)\times\Delta t}{C_{rated}}\times 100\right)$$
$$= 100 + \left(\frac{-7.5\times 4}{400}\times 100\right)$$

$$SOC(t) = 100 - \frac{3000}{400} = 100 - 7.5 = 92.5\%$$

Problem 3

Determine the power output from a wind turbine which is having a blade length of 50m, wind speed of 10m/s, air density of 1.225kg/m^3, and the power coefficient of 0.4.

Solution

Given, blade length = 50m, wind speed (v) = 10m/s, air density (ρ) = 1.225kg/m^3, power coefficient (C_p) = 0.5
Then the swept area is calculated by using:

$$A = \pi r^2 = \pi \times 50^2 = 7850\ m^2$$

Power output of wind turbine is calculated by:

$$P = \frac{1}{2}\rho A v^3 C_p = \frac{1}{2}\times 1.225\times 7850\times 10^3\times 0.5$$
$$= 2.404\ MW$$

Problem 4

A battery has a voltage of 12V, capacity of 1000Ah, and depth of charge (DOD) of 25%. Determine the charge and energy delivered to the load.

Solution

Given, DOD=25%, Battery voltage (V_B) = 12V, C_{rated}=1000Ah

Charge delivered to load = $C(t)$=$C_{rated}\times\left(\frac{DOD}{100}\right)$

$$= (1000Ah)\times\left(\frac{25}{100}\right) = 250Ah$$

Energy delivered to load = $C(t)\times V_B$=250Ah × 12V = 1300Wh

Problem 5

A wind turbine has the following parameters: swept area is 5m^2, wind speed is 6m/s, power output is 100W, air density is 1.23 kg/m^3, power coefficient is 0.6. Calculate the efficiency of wind turbine.

Solution

The following data is given:

Swept area (A) = 5m^2, wind velocity (v)=6m/s, power output (P_{output}) = 100W, air density (ρ) = 1.23 kg/m^3, power coefficient (C_p) = 0.6.

$$\%\ Efficiency = \frac{P_{output}}{P_{input}}\times 100 = \frac{100}{\frac{1}{2}\rho A v^3 C_p}\times 100$$

$$\%\ Efficiency = \frac{100}{\frac{1}{2}\times 1.23\times 5\times 6^3\times 0.6}\times 100$$
$$= \frac{100}{398.52}\times 100 = 25.09\%$$

6.12.2 Practice Problems

1. A 9V battery having a capacity of 600Ah is supplying a load demand of 0.8Ω. Determine the state of charge of battery after four hours of operation. Assume that initially the battery is charged fully.
2. The rated capacity of a wind turbine is 4MW and wind speed is 14m/s. Calculate the length of the blades. Use power coefficient as 0.5 and their air density is 1.225 kg/m^3.
3. A wind turbine has 6.5m long blades and wind speed is 6m/s. Determine the maximum amount of power that can be produced from the wind turbine. Use air density as 1.225 kg/m^3.
4. Determine the power obtained from a wind turbine having 100m long blades, while wind speed is 12m/s with an air density of 1.225 kg/m^3.
5. A particular wind turbine has a blade length of 8m, wind speed of 15m/s, rotor efficiency of 60%, gear box efficiency of 80%, and the generator efficiency of 90%. Find the power output from the wind turbine.

6.13 CONCLUSIONS

This chapter has presented the mathematical models of various RESs such as wind turbine systems, solar PV systems, energy storage systems, electric vehicles (EVs), EV battery degradation, pumped hydro storage systems, micro grids, fuel cells, and biomass. This work also models the uncertainty of wind speed, wind power, solar PV power, EVs driving distance, and EV charging state.

REFERENCES

[1] Available. [online]: www.irena.org/statistics

[2] J. Leithon, S. Werner, V. Koivunen, "Cost-aware renewable energy management: Centralized vs. distributed generation", *Renewable Energy*, vol. 147, no. 1, pp. 1164–1179, 2020.

[3] S. Ullah, A.M.A. Haidar, P. Hoole, H. Zen, T. Ahfock, "The current state of distributed renewable generation, challenges of interconnection and opportunities for energy conversion based DC microgrids", *Journal of Cleaner Production*, vol. 273, 2020.

[4] F. Mei, J. Zhang, J. Lu, J. Lu, Y. Jiang, J. Gu, K. Yu, L. Gan, "Stochastic optimal operation model for a distributed integrated energy system based on multiple-scenario simulations", *Energy*, vol. 219, 2021.
[5] R. Mishra, T.K. Saha, "Modelling and analysis of distributed power generation schemes supplying unbalanced and nonlinear load", *International Journal of Electrical Power & Energy Systems*, vol. 119, 2020.
[6] N.B. Roy, D. Das, "Optimal allocation of active and reactive power of dispatchable distributed generators in a droop controlled islanded microgrid considering renewable generation and load demand uncertainties", *Sustainable Energy, Grids and Networks*, vol. 27, 2021.
[7] J. Yang, C. Su, "Robust optimization of microgrid based on renewable distributed power generation and load demand uncertainty", *Energy*, vol. 223, 2021.
[8] M.A.M. Ramli, H.R.E.H. Bouchekara, A.S. Alghamdi, "Optimal sizing of PV/wind/diesel hybrid microgrid system using multi-objective self-adaptive differential evolution algorithm", *Renewable Energy*, vol. 121, pp. 400–411, 2018.
[9] L.C. Chien, C. Sun, Y. Yeh, "Modeling of wind farm participation in AGC", *IEEE Transactions on Power Systems*, vol. 29, no. 3, pp. 1204–1211, May 2014.
[10] J. Wu, Z. Wu, F. Wu, X. Mao, "A power balancing method of distributed generation and electric vehicle charging for minimizing operation cost of distribution systems with uncertainties", *Energy Science and Engineering*, vol. 5, no. 3, pp. 167–179, Jun. 2017.
[11] S.S. Reddy, P.R. Bijwe, A.R. Abhyankar, "Joint energy and spinning reserve market clearing incorporating wind power and load forecast uncertainties", *IEEE Systems Journal*, vol. 9, no. 1, pp. 152–164, Mar. 2015.
[12] S.S. Reddy, P.R. Bijwe, A.R. Abhyankar, "Optimal posturing in day-ahead market clearing for uncertainties considering anticipated real-time adjustment costs", *IEEE Systems Journal*, vol. 9, no. 1, pp. 177–190, Mar. 2015.
[13] X. Liu, W. Xu, "Minimum emission dispatch constrained by stochastic wind power availability and cost", *IEEE Transactions on Power Systems*, vol. 25, no. 3, pp. 1705–1713, Aug. 2010.
[14] S.S. Reddy, J.A. Momoh, "Realistic and transparent optimum scheduling strategy for hybrid power system", *IEEE Transactions on Smart Grid*, vol. 6, no. 6, pp. 3114–3125, Nov. 2015.
[15] S.S. Reddy, P.R. Bijwe, A.R. Abhyankar, "Real-time economic dispatch considering renewable power generation variability and uncertainty over scheduling period", *IEEE Systems Journal*, vol. 9, no. 4, pp. 1440–1451, Dec. 2015.
[16] A. Chauhan, R.P. Saini, "A review on Integrated Renewable Energy System based power generation for stand-alone applications: Configurations, storage options, sizing methodologies and control", *Renewable and Sustainable Energy Reviews*, vol. 38, pp. 99–120, 2014.
[17] A.K. Bhattacharjee, N. Kutkut, I. Batarseh, "Review of multiport converters for solar and energy storage integration", *IEEE Transactions on Power Electronics*, vol. 34, no. 2, pp. 1431–1445, Feb. 2019.
[18] M.A. Basset, R. Mohamed, A.E. Fergany, M. Abouhawwash, S.S. Askar, "Parameters identification of PV triple-diode model using innovative metaheuristic methodologies: Comparative study", *Mathematics*, vol. 14, pp. 1–24, 2021.
[19] A. Zakariazadeh, S. Jadid, P. Siano, "Smart microgrid energy and reserve scheduling with demand response using stochastic optimization", *International Journal of Electrical Power & Energy Systems*, vol. 63, pp. 523–533, 2014.
[20] T. Niknam, R.A. Abarghooee, M.R. Narimani, "An efficient scenario-based stochastic programming framework for multi-objective optimal micro-grid operation", *Applied Energy*, vol. 99, pp. 455–470, 2012.
[21] M. Meiqin, S. Shujuan, L. Chang, "Economic analysis of the microgrid with multi-energy and electric vehicles", *8th International Conference on Power Electronics—ECCE Asia*, IEEE, 2011, pp. 2067–2072. doi: 10.1109/ICPE.2011.5944505.
[22] M. Honarmand, A. Zakariazadeh, S. Jadid, "Self-scheduling of electric vehicles in an intelligent parking lot using stochastic optimization", *Journal of the Franklin Institute*, vol. 352, no. 2, pp. 449–467, 2015.
[23] S.S. Reddy, J.Y. Park, C.M. Jung, "Optimal operation of microgrid using hybrid differential evolution and harmony search algorithm", *Frontiers in Energy*, vol. 10, no. 3, pp. 355–362, Sept. 2016.
[24] J. Tan, L. Wang, "Integration of plug-in hybrid electric vehicles into residential distribution grid based on two-layer intelligent optimization", *IEEE Transactions on Smart Grid*, vol. 5, no. 4, pp. 1774–1784, July 2014.
[25] A. Hoke, A. Brissette, K. Smith, A. Pratt, D. Maksimovic, "Accounting for lithium-ion battery degradation in electric vehicle charging optimization", *IEEE Journal of Emerging and Selected Topics in Power Electronics*, vol. 2, no. 3, pp. 691–700, Sept. 2014.
[26] P. Punys, R. Baublys, E. Kasiulis, A. Vaisvila, B. Pelikan, J. Steller, "Assessment of renewable electricity generation by pumped storage power plants in EU Member States", *Renewable and Sustainable Energy Reviews*, vol. 26, pp. 190–200, 2013.
[27] P.P. Biswas, P.N. Suganthan, B.Y. Qu, G.A.J. Amaratunga, "Multiobjective economic-environmental power dispatch with stochastic wind-solar-small hydro power", *Energy*, vol. 150, pp. 1039–1057, 2018.
[28] K. Dasgupta, P.K. Roy, V. Mukherjee, "Power flow based hydro-thermal-wind scheduling of hybrid power system using sine cosine algorithm", *Electric Power Systems Research*, vol. 178, 2020.
[29] Hongbin Wu, Xingyue Liu, Ming Ding, "Dynamic economic dispatch of a microgrid: Mathematical models and solution algorithm", *International Journal of Electrical Power & Energy Systems*, vol. 63, pp. 336–346, 2014.
[30] M.Y. El-Sharkh, M. Tanrioven, A. Rahman, M.S. Alam, "Cost related sensitivity analysis for optimal operation of a grid-parallel PEM fuel cell power plant", *Journal of Power Sources*, vol. 161, no. 2, pp. 1198–1207, 2006.

7 Hardware Development of Three-to-Three Phase Indirect Matrix Converters

Payal Patel and Mahmadasraf A. Mulla

CONTENTS

7.1 INTRODUCTION TO MATRIX CONVERTERS

AC/AC power converters are integral parts of various applications such as uninterruptible power supplies, electrical drives, wind energy conversion systems, power quality improvement, etc. The AC/AC converter is utilized to convert AC power with a change in frequency. The preferable features of AC/AC converters are the ability to generate the required voltage at the required frequency, unity power factor, sinusoidal waveforms on supply and load sides, and unity power factor on any load condition, in addition to the compact and simple power circuit. AC/AC converters with energy storing elements in the DC-link, known as the AC/DC/AC converters, are popularly used in the industry. The presence of the energy storing element in the DC-link of the AC/DC/AC converter decouples control of the rectifier and inverter stages of the converter. However, at the same time, the physical volume of the converter becomes large due to the energy storing element which also reduces the converter life. Improvements in power semiconductor technology over the years helped to develop the AC/AC converters without need for energy storing elements which are known as direct frequency converters. The direct AC/AC converters are formed by an array of switching devices which connect all the input phases to each output phase. The first direct frequency converter developed using Thyristor devices was known as a naturally commutated converter. Afterwards, the direct frequency converter developed using power transistors, becoming known as the forced commutated cycloconverter. There are various names by which the AC/AC converter has been recognized in the literature like unrestricted frequency changer or direct frequency changer, generalized transformer, force commutated cycloconverter, direct AC/AC converter, Venturini converter, matrix converter, etc. However, the name "Matrix converter" is the most common term used for these converters.

The matrix converter (MC) is an "all semiconductor" device developed for the direct AC/AC conversion, and it is potentially equivalent to AC/DC/AC converters. The interesting features of the MC are four-quadrant operation, compact design, sinusoidal supply and load waveforms, and the adjustable input power factor.

7.1.1 Matrix Converter Topologies

Depending on the number of phases connected at the input and output, the MCs are classified as single phase, single- to two-phase, three-to-three (3 × 3) phase, three-to-five phase and three-to-seven phase MCs. In this section, the various 3 × 3 phase MC topologies used for low- and high-power applications are reviewed. The conventional 3 × 3 phase MC consists of nine bidirectional switches for connecting input phases to the output phases in the form of matrix as shown in Figure 7.1(a). This conventional MC topology is recognized as direct matrix converter (DMC). The DMC topology consists of 18 insulated gate bipolar transistors (IGBTs) and diodes. An alternative structure of the DMC is known as indirect matrix converter (IMC), which contains the input bidirectional current source rectifier (BCSR) and output voltage source inverter (VSI) connected back-to-back, as shown in Figure 7.1(b). The performance of the IMC is equivalent to the DMC topology.

The structure of the IMC with the lower number of switches which is known as sparse MC (SMC) consists of 15 IGBTs and 18 diodes. With this topology, the bidirectional power flow is possible. The very sparse MC (VSMC) topology further reduces the switching devices to12 IGBTs and 30 diodes. The switching devices of the VSMC topology are further reduced to form the ultra sparse MC (USMC). The topology of USMC consists of nine IGBTs and allows power flow in only one direction. The multi-step commutation and simplified two-step commutation are required for all these MC topologies. In the IMC, SMC, USMC and VSMC topologies, the output DC-to-AC stage is equivalent to the conventional VSI.

DOI: 10.1201/9781003229124-7

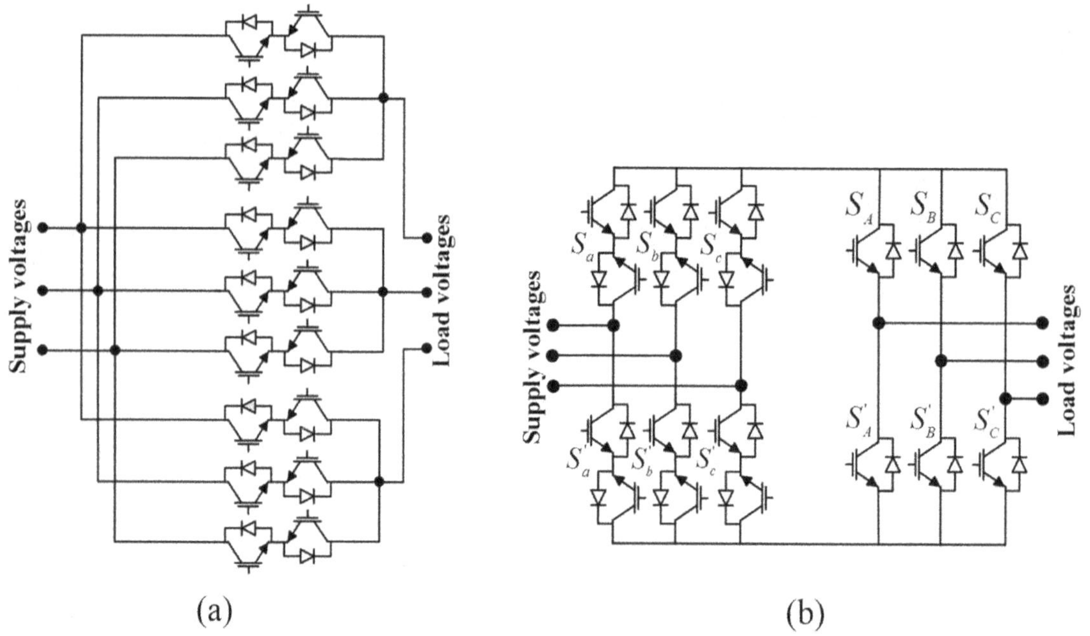

FIGURE 7.1 Matrix converter topology: (a) direct matrix converter; and (b) indirect matrix converter.

The crucial problem of low voltage transfer ratio (VTR) in the linear modulation range for the MCs is overcome by including Z-source and quasi–Z-source networks with the conventional DMC, known as Z-source DMC and quasi–Z-source DMC, respectively. The IMC topology integrated with switched inductor and quasi–Z-source networks is known as switched inductor Z-source IMC and quasi–Z-source IMC, respectively. In order to use the MC topologies into the higher power applications, several new MC topologies have been developed based on the concepts of conventional multi-level converters. Combining the key features of multi-level converters and MC, many multi-level MC topologies such as diode clamp MC, capacitor clamp MC and multi-modular MC have been introduced.

7.1.2 Matrix Converter Switches

The conventional MC consists of the nine bidirectional switches which conduct the currents and block the voltages in the both directions. However, the single switching device with such capabilities is currently unavailable commercially. By appropriately connecting the available unidirectional switching devices, the bidirectional switches are formed. Basically, four configurations of the bidirectional switches are commonly used which are diode bridge configuration, common emitter configuration, common collector configuration and reverse block IGBTs configuration. The simplest form of bidirectional switch is diode bridge configuration, which consists of single IGBT and four diodes, as shown in Figure 7.2(a). Another form of bidirectional switch which consists of anti-parallel connection of the two IGBTs is shown in Figure 7.2(b). For the realization of this bidirectional switch, the ultra-thin wafer technology with deep boron diffusion technique is used. The use of this configuration eliminates the use of diodes to assume the reverse blocking capability. This results in the reduced conduction losses. The most common forms of the bidirectional switches are common emitter configuration and common collector configuration which consist of two diodes and two IGBTs as shown in Figure 7.2(c) and Figure 7.2(d), respectively. The IGBTs control the direction of the current, whereas diodes give the reverse blocking capability.

The power devices other than IGBTs like gate turn off (GTO) Thyristors, metal-oxide-silicon (MOS) turn-off Thyristors (MTOs) and pure junction-gate field effect transistors (JFETs), with the reverse blocking capability, are also used in MCs. The properties of the switching devices like voltage blocking capacity, current and voltage ratings, switching frequency, etc., are highly influenced by the advancement of the semiconductor technology.

To decrease the volume of the filter elements, the switching devices which can handle high switching frequencies are preferable. The newly invented SiC (silicon carbide) and GaN (gallium nitride) switching devices can highly improve the converter performances.

7.1.3 Matrix Converter Modulation Strategies

Developing the appropriate modulation strategies for the MC is quite a difficult task, owing to its complex structure and large number of switching devices. Many modulation strategies have been developed for the 3 × 3 phase DMC and IMC topologies which are reviewed in this section.

In 1980, Alesina-Venturini proposed the first algorithm called "Venturini modulation method" to synthesize the required output voltages from the three-phase balanced power supply [1]. This is known as the "direct transfer function" method of the modulation, with a maximum VTR of 0.5 and limited control of input power factor. The modified Alesina-Venturini modulation strategies, presented in

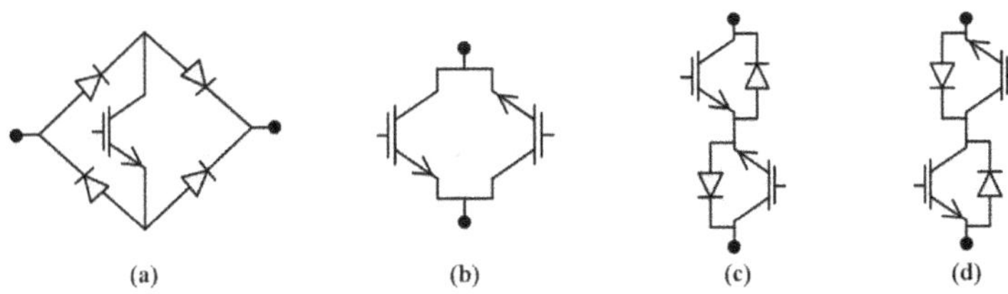

FIGURE 7.2 Configurations of bidirectional switches: (a) diode bridge configuration; (b) reverse blocking IGBTs; (c) common emitter configuration; and (d) common collector configuration.

[2], increase the limit of the maximum VTR to 0.866 by introducing the third harmonic into the fundamental signal. In 1983, Rodriguez presented the new control approach for the MC which was based on the "fictitious DC link" [3]. By using the "indirect transfer function" approach of modulation, it is possible to achieve VTR up to 1.053 for the MC [4]. A new "scalar" modulation strategy was introduced by Roy and April in 1987 which used the active and zero voltage states of the MC switches [5]. The modified scalar method with the improved VTR is presented in [6].

The principle of space vector pulse width modulation (SVPWM) for the MC was first introduced by Huber et. al. [7]. To obtain the required performance of the IMC in the presence of abnormal supply voltage conditions, the modifications are introduced in the SVPWM strategy as presented in [8–10]. The SVPWM strategy presented in [11, 12] decreases the switching losses of the MC. However, due to the involvement of the complex duty cycle computation and look-up tables for determining the switching patterns, the implementation of SVPWM strategy is a highly complicated job.

The conventional voltage and current source converter topologies used in the industry are popularly known for their mature and efficient carrier based pulse width modulation (CBPWM) strategies. Inspired by this, several CBPWM strategies for the DMC topologies have been investigated [13–16]. In the literature, control techniques like predictive control and sliding mode control are also presented for the DMC. The predictive torque control for the DMC is described in [17]. The sliding mode control technique for the DMC topology is presented in [18].

Due to the decoupled inversion and rectification stages of the IMC topology and simple commutation process, the control of the IMC has been the topic of interest for the researchers. The SVPWM approach for the IMC topology was first presented in [19]. The indirect vector modulation technique for the IMC was developed in [20] using the theory of space vectors. The performance of the SVPWM modulated IMC is compared and analyzed in the presence of harmonics in the supply voltages for the four simple methods in [21]. The SVPWM techniques presented in [22, 23] are developed to decrease the common mode voltage.

Various CBPWM strategies for the IMC have been described in [24–27]. The CBPWM schemes presented in [27, 28] minimize the switching losses of the IMC. All these CBPWM strategies resort to the variable slope carrier for coordinating the switching events of the two IMC stages. The concept of CBPWM strategy with a single triangular carrier for modulating the two stages of IMC is presented in [29]. The predictive control technique for the IMC is presented in [30].

7.2 INDIRECT MATRIX CONVERTER FUNDAMENTALS

In this section, the fundamental working principles and space vector modulation approach of the IMC are explained. The detailed explanation of 3 × 3 phase IMC topology is presented with the basic operation principles of the two converter stages. The principle of space vector approach to control the two stages of the IMC is explained.

7.2.1 Indirect Matrix Converter Topology

Figure 7.1(b) shows the 3 × 3 phase IMC topology consisting back-to-back connection of BCSR and VSI stages. The upper and lower group of switches S_a, S_b, S_c and S_a', S_b', S_c' of the BCSR stage are shown in Figure 7.1(b). The controlling of six bidirectional switches on the BCSR stage synthesize the sinusoidal currents on input side and positive DC-link voltage V_{dc} on the DC-link. The upper and lower group of switches S_A, S_B, S_C and S_A', S_B', S_C' of the VSI stage are shown in Figure 7.1(b). Control of the six unidirectional switches on the VSI stage generates the desired output voltages. Figure 7.1(b) shows the LC filter connected on the input side of the IMC. The three IMC output terminals are connected to the three-phase inductive load.

To ensure the safe operation of the switches on the BCSR stage of the IMC, the conditions represented by Eq. 7.1 should be satisfied while developing the switching functions.

$$\begin{cases} S_a + S_b + S_c = 1 \\ S_A + S_B + S_C = 1 \end{cases} \tag{7.1}$$

To ensure the safe operation of the switches on the VSI stage of the IMC, the conditions represented by Eq. 7.2 should be satisfied while developing the switching functions.

$$\begin{cases} S_A + S_A' = 1 \\ S_B + S_B' = 1 \\ S_C + S_C' = 1 \end{cases} \tag{7.2}$$

The output and the input voltages, $\left[V_A, V_B, V_C\right]^T$ and $\left[V_a, V_b, V_c\right]^T$, of the IMC are associated with each other by Eq. 7.3.

$$\begin{bmatrix} V_A \\ V_B \\ V_C \end{bmatrix} = \begin{bmatrix} S_A\, S_A' \\ S_B\, S_B' \\ S_C\, S_C' \end{bmatrix} \begin{bmatrix} S_a & S_b & S_c \\ S_a' & S_b' & S_c' \end{bmatrix} \begin{bmatrix} S_a \\ S_b \\ S_c \end{bmatrix} \tag{7.3}$$

where the switch S_x is considered ON when $S_x = 1$ and OFF when $S_x = 0$

Appropriate modulation strategies are required to derive the switching functions for the input and output stages of the IMC. In order to maintain the balance between input and output operations, it is required to synchronize the modulations of both converter stages of the IMC.

7.2.1.1 Space Vector Modulation (SVPWM)

In this section, the SVPWM scheme for the IMC is explained in detail. The SVPWM scheme of the IMC is derived by applying the SVPWM approaches of the conventional CSR and the VSI on the BCSR and VSI stages of the IMC, respectively.

7.2.1.1.1 Modulation of BCSR Stage

Clarke's transformation is used to convert the three input reference current signals I_{ra}, I_{rb} and I_{rc} from *abc* to $\alpha\beta$ as represented in Eq. 7.4.

$$\begin{bmatrix} i_\alpha \\ i_\beta \end{bmatrix} = \frac{2}{3} \begin{bmatrix} 1 & -\frac{1}{2} & -\frac{1}{2} \\ 0 & -\frac{\sqrt{3}}{2} & \frac{\sqrt{3}}{2} \end{bmatrix} \begin{bmatrix} I_{ra} \\ I_{rb} \\ I_{rc} \end{bmatrix} \tag{7.4}$$

where i_α and i_β are the $\alpha\beta$ components of the input reference currents

The input modulation vector I_{ref} and the angle θ_i are calculated as given in Eqs. 5 and 6, respectively.

$$I_{ref} = \sqrt{i_\alpha^{\,2} + i_\beta^{\,2}} \tag{7.5}$$

$$\theta_i = tan^{-1}\frac{i_\beta}{i_\alpha} \tag{7.6}$$

There are six active vectors ($I_1 - I_6$) and three zero vectors (I_0) to synthesize the input vector I_{ref}. The six active vectors $I_1 - I_6$—in the form of hexagon—are shown in Figure 7.3(a).

The input angle θ_i determines the position of I_{ref} within the six sectors of the hexagon. Figure 7.3(b) shows the vector I_{ref} lying in Sector-1. The input vector I_{ref} is shown to be synthesized by the two adjacent active vectors I_1 and I_6 based on the duty cycles as expressed by Eq. 7.7. The duty cycles d_α and d_β are calculated by Eqs. 8 and 9, respectively.

$$I_{ref} = d_\alpha \cdot I_1 + d_\beta \cdot I_6 \tag{7.7}$$

$$d_\alpha = m_c \cdot sin(\theta_{in}) \tag{7.8}$$

$$d_\beta = m_c \cdot sin(60 - \theta_{in}) \tag{7.9}$$

where θ_{in} represents the position of I_{ref} in the given sector, as shown in Figure 7.3(b)

The variable m_c represents the modulation index of the BCSR stage, which is defined as the ratio of the input current to the DC-link current. In the IMC operation, generally the value of m_c is kept at unity. In order to eliminate the zero switching vectors from the BCSR operation, to produce the maximum DC voltage, the duty cycles are normalized as given in Eqs. 10 and 11.

$$d_\alpha' = \frac{d_\alpha}{d_\alpha + d_\beta} \tag{7.10}$$

$$d_\beta' = \frac{d_\beta}{d_\alpha + d_\beta} \tag{7.11}$$

Figure 7.4 depicts the modulation of the BCSR stage for the period T_s. The comparison of the duty cycles d_α' and $d_\alpha' + d_\beta'$ with the triangular carrier produces the switching durations as shown. It is noted that the intersection of d_α' with the triangular carrier produces the switching durations $t_1/2$ and $t_2/2$, as indicated in Figure 7.4. These switching durations,

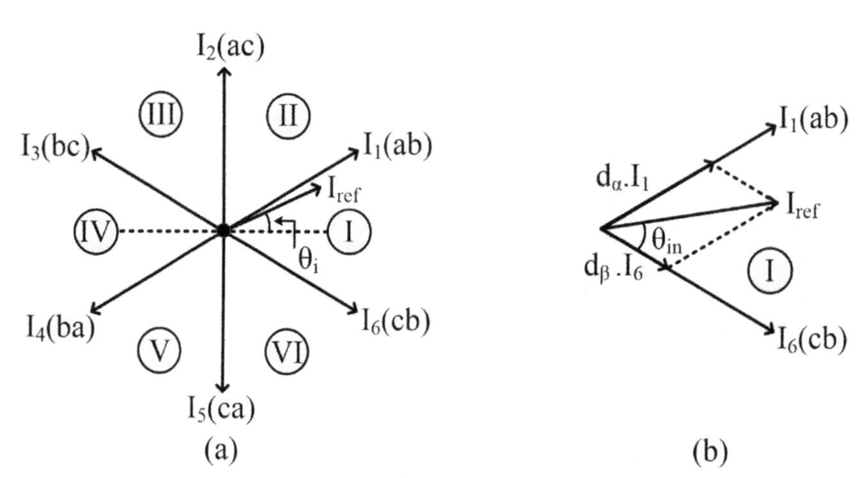

FIGURE 7.3 BCSR stage space vectors: (a) six input sectors; and (b) input vector I_{ref} in Sector-1.

together with the sector information, determine the gating signals for the BCSR stage switching devices. By considering the position of I_{ref} in Sector-1, the switching signals S_a', S_b' and S_c' are determined, as shown in Figure 7.4. Now, in order to synchronize the modulations of the two stages of the IMC, the variable slope carrier is required for modulation of the VSI stage. This variable slope carrier is designed based on the switching instants created on the BCSR stage. The intersection of d_α' with the carrier wave creates the switching instants on the BCSR stage, as shown in Figure 7.4. This intersection produces the durations $t_1/2$ and $t_2/2$ in both halves of the symmetrical triangular carrier of the BCSR stage. These durations $t_1/2$ and $t_2/2$ are used to derive the slopes of the carrier required for the LSC stage, as shown in Figure 7.4. The rising and falling slopes of the carrier, dS_1, dS_2, dS_3 and dS_4, are calculated by Eqs. 12–15, respectively.

$$\delta S_1 = \frac{2}{t_1} = \frac{2}{T_s \cdot d_\alpha'} \tag{7.12}$$

$$\delta S_2 = 1 - \frac{2}{t_1} = 1 - \frac{2}{T_s \cdot (1 - d_\alpha')} \tag{7.13}$$

$$\delta S_3 = \frac{2}{t_2} = \frac{2}{T_s \cdot (1 - d_\alpha')} \tag{7.14}$$

$$\delta S_4 = 1 - \frac{2}{t_1} = 1 - \frac{2}{T_s \cdot d_\alpha'} \tag{7.15}$$

7.2.1.1.2 *Modulation of VSI Stage*

The reference output signals V_{rA}, V_{rB} and V_{rC} are converted to $\alpha\beta$ coordinates using Clarke's transformation, as given in Eq. 7.16.

$$\begin{bmatrix} v_\gamma \\ v_\delta \end{bmatrix} = \frac{2}{3} \begin{bmatrix} 1 & -\frac{1}{2} & -\frac{1}{2} \\ 0 & -\frac{\sqrt{3}}{2} & \frac{\sqrt{3}}{2} \end{bmatrix} \begin{bmatrix} V_{rA} \\ V_{rB} \\ V_{rC} \end{bmatrix} \tag{7.16}$$

where v_γ and v_δ are the $\alpha\beta$ components of the output references

The output modulation vector V_{ref} and the output angle θ_o are calculated as given in Eqs. 17 and 18, respectively.

$$V_{ref} = \sqrt{{v_\gamma}^2 + {v_\delta}^2} \tag{7.17}$$

$$\theta_o = tan^{-1} \frac{v_\delta}{v_\gamma} \tag{7.18}$$

Figure 7.5(a) shows the six active vectors $(V_1 - V_6)$ and two zero vectors (V_0, V_7) available to modulate the VSI stage.

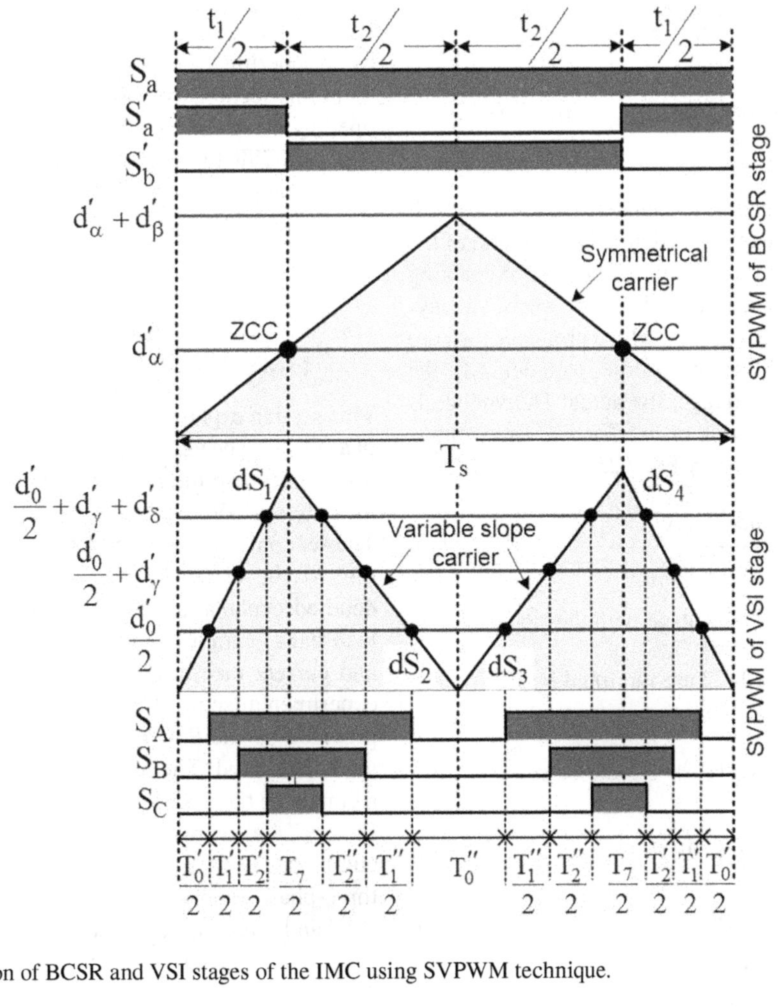

FIGURE 7.4 Modulation of BCSR and VSI stages of the IMC using SVPWM technique.

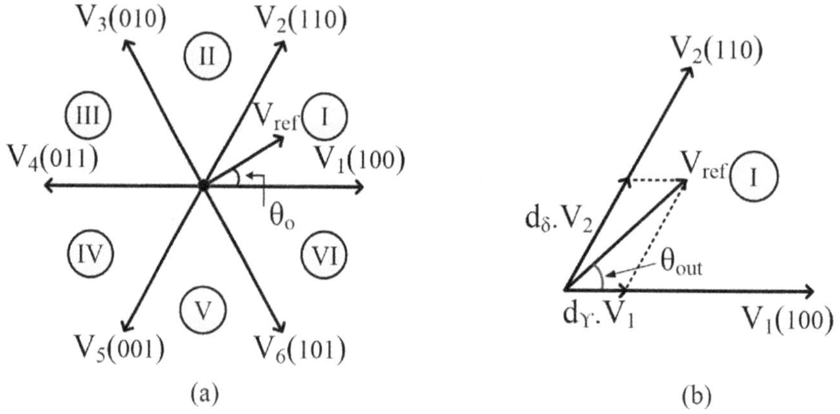

FIGURE 7.5 VSI stage space vectors: (a) six output sectors; and (b) output vector V_{ref} in Sector-1.

Figure 7.5(b) shows the vector V_{ref} lying in the Sector-1 of the hexagon. The two adjacent active vectors V_1 and V_2 along with the zero vector V_0 synthesize the output vector V_{ref} based on the duty cycles as expressed by Eq. 7.19. The duty cycles d_γ, d_δ and d_0 are calculated by Eqs. 20–22, respectively.

$$V_{ref} = d_\gamma \cdot V_1 + d_\delta \cdot V_6 + d_0 \cdot V_0 \tag{7.19}$$

$$d_\gamma = m_v \cdot sin(\theta_{out}) \tag{7.20}$$

$$d_\delta = m_v \cdot sin(60 - \theta_{out}) \tag{7.21}$$

$$d_0 = 1 - d_\gamma - d_\delta \tag{7.22}$$

where θ_{out} represents the position of V_{ref} in the given sector as depicted in Figure 7.5(b)

The variable m_v represents the modulation index of the VSI stage. The modulation index m_v is defined as the ratio of the output reference vector to the DC-link voltage. For enabling the ZCC on the BCSR stage, the value of m_v should always be selected less than unity ($m_v < 1$), which produces the zero voltage states on the VSI stage. In order to indemnify the ripple present in the DC voltage, the actual DC voltage is incorporated in the calculation of m_v. The DC-link voltage with the ripple is represented by Eq. 7.23.

$$V_{rip} = \frac{max(|V_{ab}, V_{bc}, V_{ca}|)}{\sqrt{3} \cdot V_m} \tag{7.23}$$

where V_{rip} represents the DC voltage with the ripple

The duty cycles of Eqs. 20–22 are modified by incorporating V_{rip} as given in Eqs. 24–26.

$$d_\gamma' = \frac{m_v}{V_{rip}} \cdot sin(\theta_o) \tag{7.24}$$

$$d_\delta' = \frac{m_v}{V_{rip}} \cdot sin(60 - \theta_{out}) \tag{7.25}$$

$$d_0' = 1 - d_\gamma' - d_\delta' \tag{7.26}$$

Figure 7.4 depicts the modulation of the VSI stage of the IMC. The duty cycles and $d_0'/2$, $\left(d_0'/2\right) + d_\gamma'$ and $\left(d_0'/2\right) + d_\gamma' + d_\delta'$ are compared with the variable slope carrier which produce the various switching durations $T_0'/2$, $T_1'/2$, $T_2'/2$, $T_7/2$, $T_1''/2$ and $T_2''/2$, as shown. Based on these durations, together with the sector information, the gating pulses of the VSI stage switching devices are determined. Figure 7.4 shows the determination of switching signals S_A, S_B and S_C by considering the position of V_{ref} in Sector-1 of the hexagon.

In this section, the SVPWM strategy of the IMC is developed by using the SVPWM approaches of the conventional VSI and CSR topologies to control the two stages of IMC with the appropriate modifications. The concept of synchronizing the two stages of IMC by using the variable slope carrier is also explained.

7.3 INDIRECT MATRIX CONVERTER HARDWARE DEVELOPMENT

This section explains the hardware development of the 3 × 3 phase IMC. The various power and control circuit equipment used to develop the experimental rig, as well as the implementation of SVPWM algorithm for the IMC, are described. The section starts with the introduction of the overall structure of the experimental setup. This is followed by the detailed explanation of the input filter, switching devices, load bank, clamp circuit, DSP and FPGA boards, voltage and current measurements and driver circuits used in the experimental setup. The automatic code generation facility provided by the PSIM software for the DSP hardware target is presented. The implementation of the SVPWM algorithm using the DSP and FPGA control devices is presented. Figure 7.6(a) represents the various power and control circuit blocks used in the experimental setup of the IMC. The three-phase supply, input filter, two converter stages of the IMC and three-phase R-L load are the main elements of the

TABLE 7.1
Different Parameters of Hardware Setup

Parameter	Values
Supply voltage	80 V
Supply frequency	50 Hz
Filter inductance	2.7 mH
Filter resistance	68 Ω
Filter capacitance	15 μF
Load resistance	11 Ω
Load inductance	10 mH
Carrier frequency	5 kHz

power circuit. The digital controllers DSP and FPGA, driver and sensor circuits are the main elements of the control circuit. Figure 7.6(b) shows the photograph of the laboratory prototype of the 3 × 3 phase IMC. The various power circuit parameters selected for the experimental testing are given in Table 7.1.

7.3.1 Power Circuit Implementation

The various power circuit elements used in the experimental setup are explained in this section.

7.3.1.1 Switching Devices

Figure 7.6(a) represents the power circuit of the IMC including the BCSR and VSI stages. The IGBT FGA25 N120ANTD by Fairchild semiconductor is used for the development of the power circuit of the IMC. The six bidirectional switches and six unidirectional switches are required to form the BCSR and VSI stages of the IMC, respectively. The common emitter configuration of the bidirectional switch is used to form the BCSR stage. The voltage and current ratings of these IGBTs are 1200 V and 25 A, respectively.

7.3.1.2 Input Filter

To minimize the switching frequency components from the supply currents, the input filter is connected between the three-phase source and IMC topology as shown in Figure 7.6(a). The second-order low-pass filter circuit with inductance L_f, capacitance C_f and resistance R_f used in this work is shown in Figure 7.7(a). The resistance R_f connected in the parallel with the inductance L_f provides the appropriate damping for the filtering action. The roll-off rate of the second order filter is −40 dB/decade. The relation between cut-off frequency f_c, filter inductance L_f and capacitance C_f is expressed by Eq. 7.27. The cut-off frequency f_c is 1000 Hz for the chosen value of L_f = 1 mH, the value obtained for the filter capacitor C_f = 25 mF.

$$f_c = \frac{1}{2\pi\sqrt{L_f C_f}} \tag{7.27}$$

7.3.1.3 Clamp Circuit

A clamp circuit in the IMC is used to prevent the destruction of the switching devices due to overvoltage spikes. The clamp circuit required for the IMC topology is very simple. The IMC requires a single diode and a capacitor to form the clamp circuit, as shown in Figure 7.7(b). In the IMC, the clamp circuit ensures the safe operation of the inverter stage during the shutdown. The dimensions of the capacitor in the clamp circuit depend on the peak supply voltage and maximum energy stored in the inductive load. The diode MUR420 and a capacitor with 1 mF, 1000 V rating are used to build the clamp circuit of the experimental setup.

7.3.1.4 Load Bank

Three-phase R-L load is used to carry out the experimental testing, as shown in Figure 7.7(c). The load consists of three rheostats with 12 Ω range and 10 A current capacity in series with the three inductors of 10 mH range and 10 A current capacity. The R-L load is connected in the star without neutral connection.

7.3.2 Control Circuit Realization Using DSP TMS320F28335 and FPGA Spartan-6 Development Boards

The implementation of the IMC modulation schemes requires a huge number of mathematical operations to be performed in every switching period. Therefore, it is required to use the control platforms capable of performing all these calculations in a short period of time. The DSP TMS320F28335 by Texas Instruments and FPGA by Spartan-6 were selected as the control platforms for implementing the control scheme of the IMC.

7.3.2.1 DSP TMS320F28335 Development Board

Figure 7.8(a) shows the photograph of the TMS320F28335 development board used in the present work. The DSP is used for calculating the duty cycles/modulation signals, identifying the sectors and performing PWM.

The DSP board encompasses the high speed 16-bit architecture and 32-bit CPU with 88 programmable and multiplexed GPIO pins with input filtering. The board integrates the UART, LEDs, LCD, PWM motor control interface, SPI DAC, 12-bit A/D converter, CAN, SPI EEPROM on-board XDS100 USB V1 emulator, enhanced control peripherals like PWM outputs, HRPWM outputs, event capture inputs, quadrature encoder interfaces, 32-bit/16-bit timers and three 32-bit CPU timers. The TI component devices and several peripheral options in the code composer studio (CCS) are supported by the DSP board. The crystal frequency of this DSP board is 20 MHz (clock frequency 150 MHz after PLL). The code development is carried out through an application suite called CCS. It is made by Texas Instruments. It supports programming in both C and Assembler.

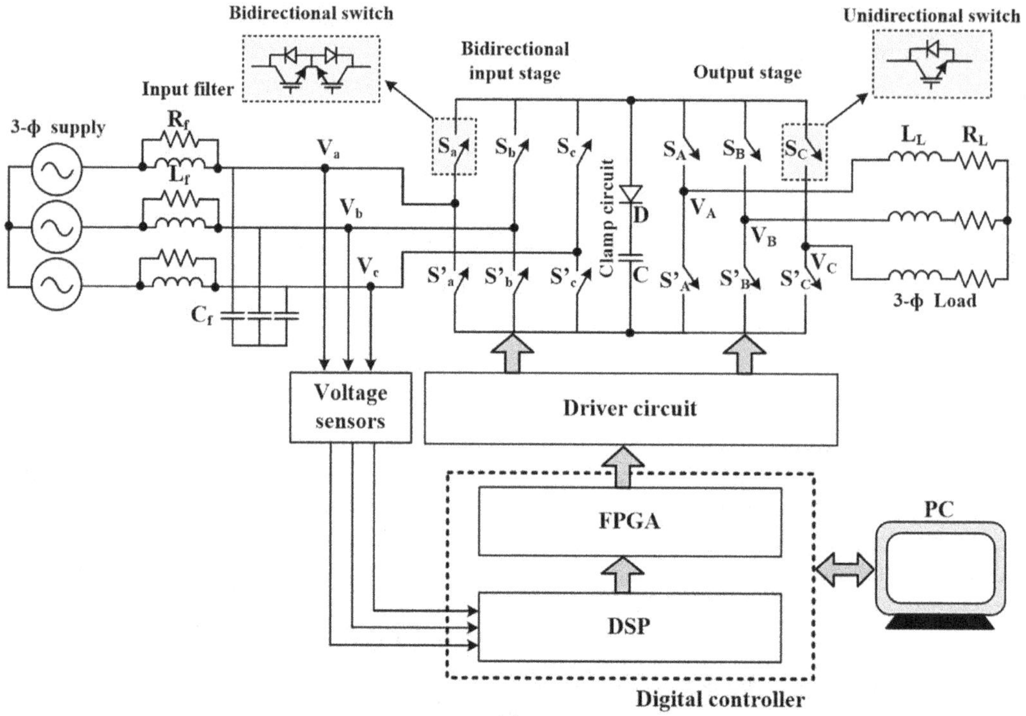

(a)

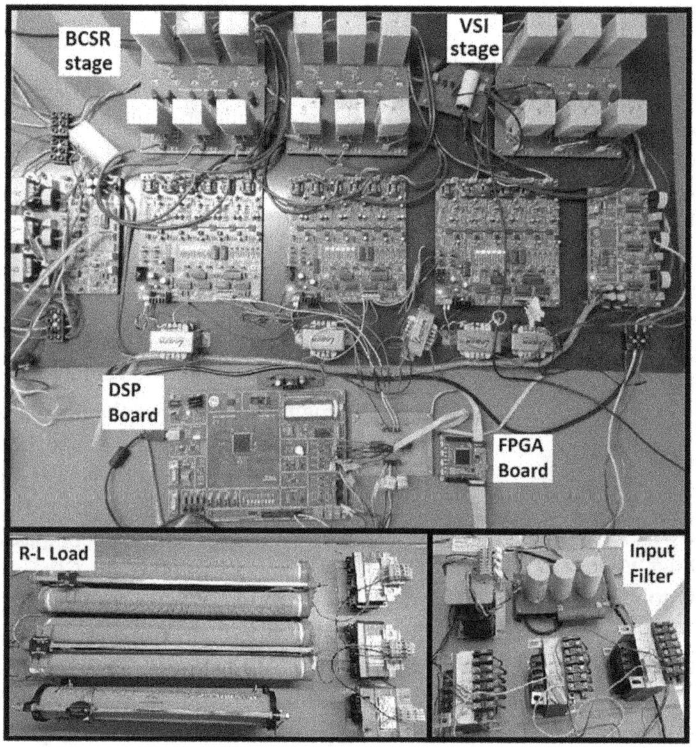

(b)

FIGURE 7.6 Laboratory prototype of IMC: (a) block diagram of control and power circuit; and (b) photograph of setup.

7.3.2.2 FPGA Spartan-6 Development Board

The FPGA is used to perform the logical operations and to produce the delay on the rising edges of the switching signals. The FPGA development board designed by Mimas features Xilinx XC6SLX9 TQG144 FPGA with 70 user IO pins is shown in Figure 7.8(b). The FPGA board encompasses 16 MB SPI flash memory, 100 MHz CMOS oscillator, USB 2.0 interface, FPGA configuration via JTAG and USB, eight LEDs and four switches for users and on-board voltage regulators.

The part of the IMC control scheme implemented by the FPGA uses the Xilinx ISE (Integrated Synthesis

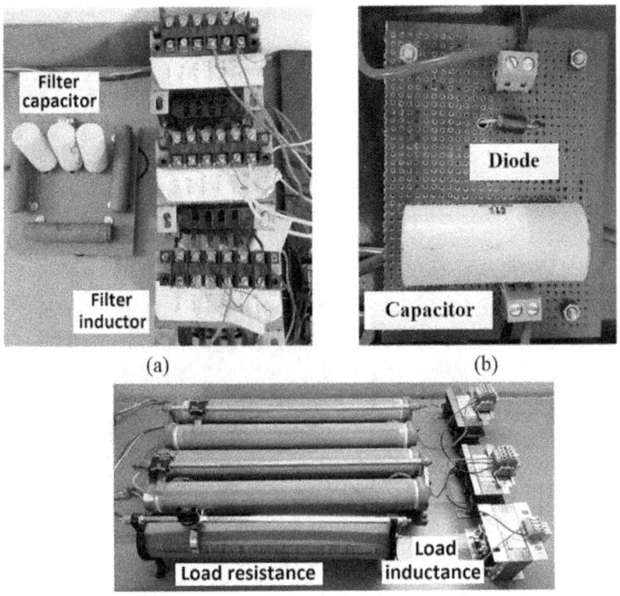

FIGURE 7.7 Components of setup: (a) input filter; (b) clamping circuit; and (c) three-phase R-L load.

Environment) design suite. Xilinx ISE is a software tool from Xilinx which primarily targets the development of embedded firmware for Xilinx FPGA integrated circuit (IC) product families. ISE enables the developer to compile their design codes and configure the target device with the programmer.

7.3.3 Driver Circuit, Voltage Measurement and Current Measurement

The main function of the driver circuit is to provide the interfacing between the control circuit and the power circuit in the experimental setup. The gate drive circuit converts the lower power signals of the controller to the voltage and current levels required to switch ON/OFF the IGBT module and to provide the isolation between power and control circuits. The driver card used in this work provides six optically isolated gate signals for the IGBTs. The driver card consists of six isolated power supplies for the driver ICs. A total of three driver cards are used for the 18 IGBTs of the IMC. Figure 7.9(a) represents the internal driver circuit designed using IED020I12-F2. The sensor circuit is used in this work to measure the three input phase voltages of the IMC to produce the modulation signals for the BCSR stage. Figure 7.9(b) shows the internal circuit of the sensor card. This sensor card is capable of sensing three AC voltage and current signals. The signal conditioning circuit is provided to add the offset voltage for interfacing sensed signal with unipolar ADC, in addition with the amplitude calibration.

7.3.4 Automatic Code Generation Using PSIM Software

The simcoder module of the PSIM software provides the facility to generate the C-code automatically from the simulation model, which is ready to run for the specific hardware target. This facility provided by the PSIM greatly simplifies the implementation process of the hardware development. The simulation of the IMC is developed in the PSIM software using the simcoder module to generate the C-code for the TI F28335 (DSP TMS320F28335) hardware device. It is required to ensure that all the elements used in the simulation support the code generation. All the elements included in the simcoder menu support the code generation. In the standard PSIM library, some elements with the symbol "C" displayed next to them also support code generation. The various control blocks used for the code generation are: A/D converter, digital output, PWM, simplified C-block, hardware board configuration, DSP configuration, SPI configuration and simulation control. The functions of these blocks are as follows.

1. **A/D Block:** The simcoder provides the 12-bit A/D converter block with 16 channels for the TI F28335 hardware target. The 16 channels are divided into group-A and group-B. The A/D converter block is used to select the A/D converter mode, channel mode and channel gain, according to the requirement of the hardware device.
2. **Digital Output Block:** The TI F28335 provides the 88 GPIO ports which can be configured either as the digital inputs or digital outputs. An eight-channel digital output block provided by the simcoder is used to select the digital output pins.
3. **Two-phase PWM generators:** Six two-phase PWM generators, PWM1–PWM6, are provided by the simcoder for the TI F28335 hardware target. Each of these PWM generators may be configured in one of the six operating modes. These six

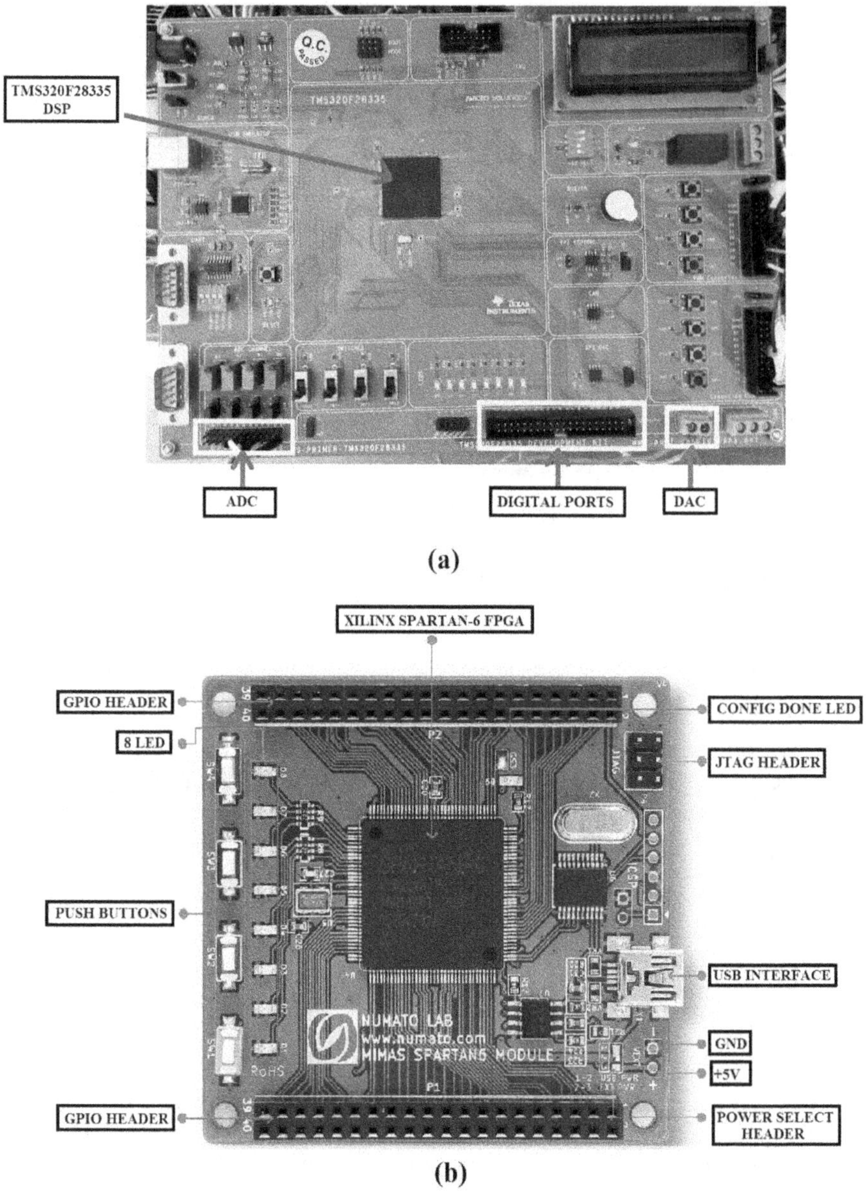

FIGURE 7.8 Development boards: (a) DSP TMS320F28335; and (b) FPGA Spartan-6

operation modes differ from each other based on the types of carrier and output signals of the PWM. The PWM block enables the selection of the mode, carrier frequency and peak amplitude of the PWM operation according to the requirement.

4. **Simplified C-block:** The simplified C-block is used to enter the C-code directly without compiling. The outputs of this block are written in terms of the inputs and independent variables. This block is used to perform the various mathematical operations associated with the modulation scheme.
5. **Hardware Board Configuration Block:** The TI F28335 provides 88 GPIO pins (GPIO0–GPIO87). Each pin is configured for multiple functions. By using the hardware board configuration block, the simcoder can be configured for a particular port pins of the DSP board.
6. **DSP Configuration and ZOH Blocks:** The DSP configuration block is used to define the external clock frequency and speed of the DSP. The zero order hold (ZOH) block is used to discretize the simulation. The input signal to this block is sampled and hold until the next clock cycle. This block exhibits a very important role in the code generation. If the sampling rate is set at a very high value, the calculations cannot be completed in the given sampling interval, and consequently, poor quality of waveforms is obtained. If this value is too low, the resolution of the waveforms gets affected. Therefore, the care should be taken while setting the value of sampling rate in the ZOH block.
7. **SPI Blocks:** With the use of SPI blocks in the control schematic, communication with the external SPI devices like A/D and D/A converters are

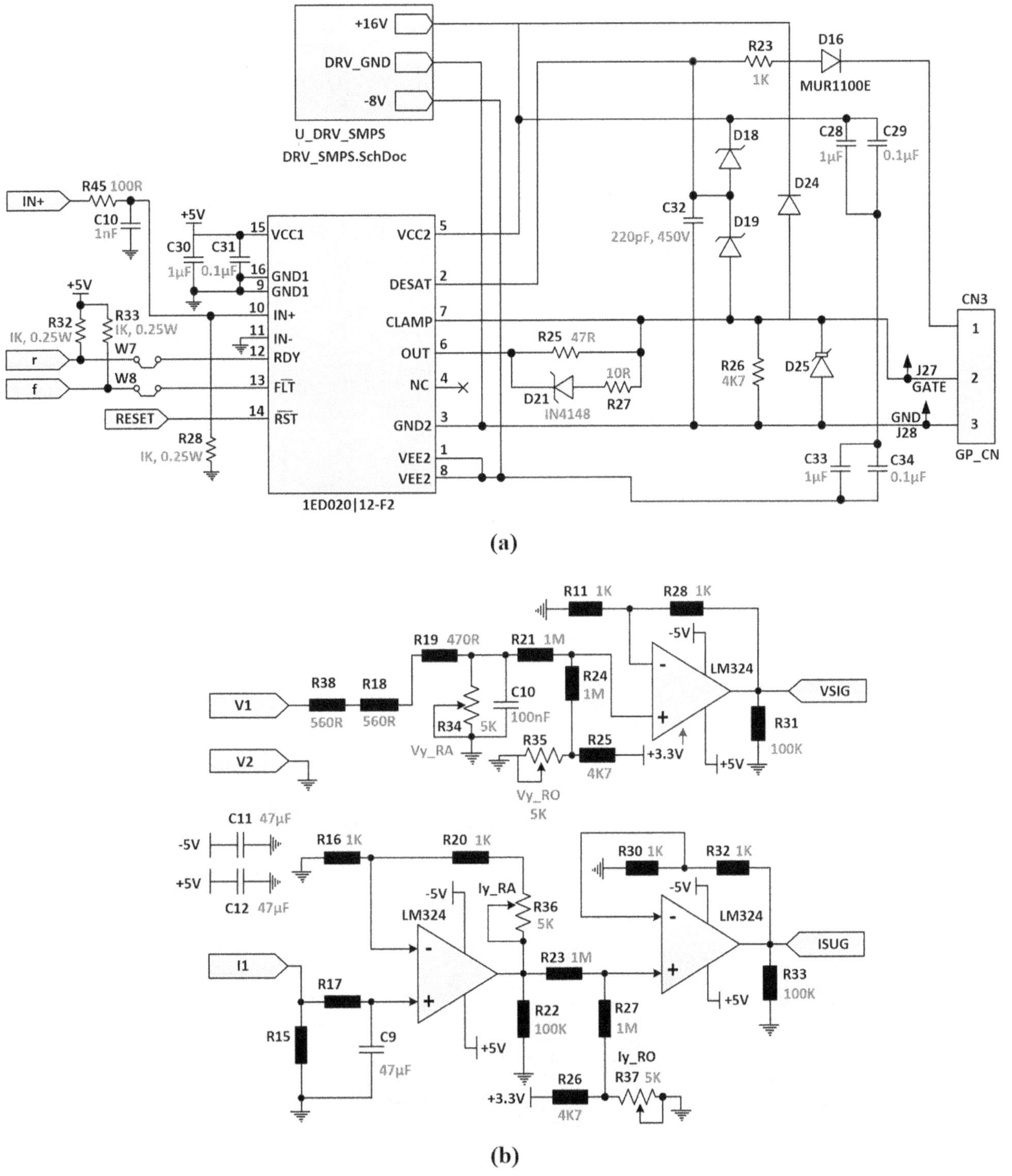

FIGURE 7.9 Circuit schematic: (a) IGBT driving circuit; (b) sensing circuit.

conveniently and easily established. The important purpose of using SPI is to visualize the control signals calculated by the DSP on the DAC. For this, the three SPI function blocks—SPI configuration block, SPI device block and SPI input block—are required in the main control schematic. The information of the SPI hardware device is defined by SPI device block. The SPI configuration block is used to define the SPI port, the chip selection pins and the SPI buffer size. The input channel properties of the SPI communication are defined by using the SPI input block.

8. **Simulation Control Block:** The simulation control block is used in the main control schematic to control the various timings related to the simulation. When this block is used in the control schematic developed to generate the code for TI F28335 target, the "hardware target" is set to "TI F28335".

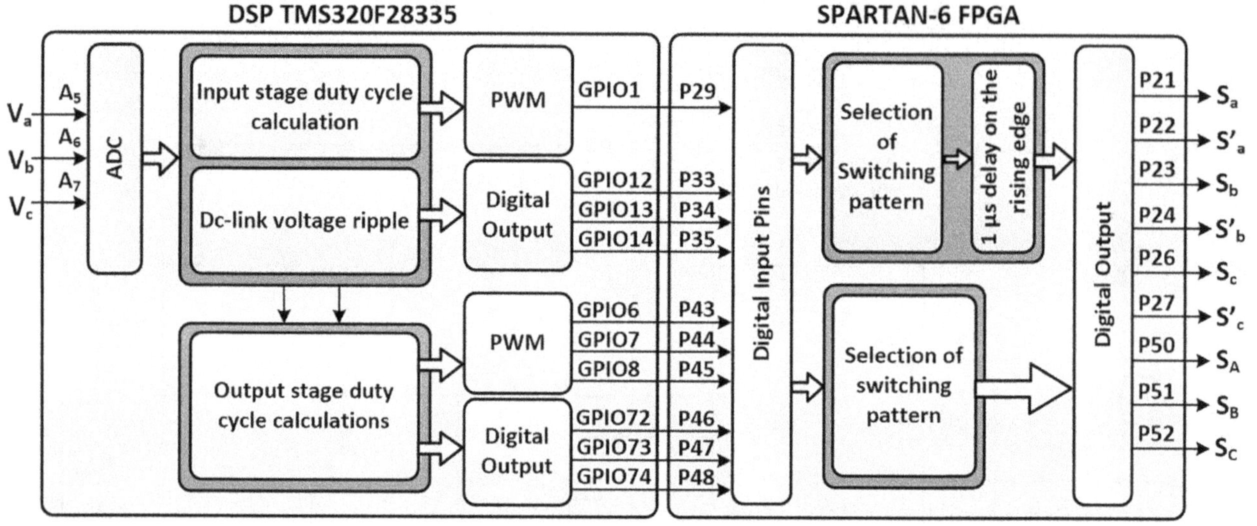

FIGURE 7.10 Implementation of SVPWM strategy using DSP TMS320F28335 and FPGA Spartan-6.

The simulation is run to validate the performance of the system developed in the control schematic. Then, the code is generated for the given hardware target by selecting the option "Generate code" in the "Simulate" menu in the main control schematic. The generated code is ready to compile in the CCS.

7.3.4.1 Implementation of SVPWM Control Strategy

This section demonstrates the real time implementation of SVPWM strategy of the IMC. The control devices DSP TMS320F28335 and FPGA Spartan-6 are programmed to produce the switching signals for the two stages of IMC by SVPWM strategy. For the purpose of implementation, the SVPWM strategy is divided into two parts: (1) the computations of sector wise duty cycles, DC voltage ripple and comparison of the duty cycles with the carrier; and (2) determination of the switching patterns based on the sector information and generating delay on the rising edge of BCSR stage switching signals. Part 1 of the SVPWM strategy is implemented by DSP TMS320F28335 and Part 2 is implemented by FPGA Spartan-6.

Figure 7.10 shows the implementation of SVPWM strategy using the control devices DSP and FPGA. Figure 7.10 also indicates the input and output port pins selected for the both control devices. On the DSP TMS320F28335 board by Pantech, the port pins A5–A7 of the A/D converter are used as the input pins for the analog signals. The port pin GPIO1 is configured as the PWM output and GPIO12–GPIO14 are configured as the digital output pins for the BCSR stage modulation of the IMC. Similarly, GPIO6–GPIO8 are configured as the PWM output pins and GPIO72–GPIO74 are configured as the digital output pins for the VSI stage modulation of the IMC. The port pins P29–P32, P33–P35 and P43–P48 on the FPGA board are configured as the digital input pins. The port pins P21–P24, P26–P27 and P50–P52 are configured as the digital output pins. The look-up tables implemented in the FPGA determine the gating signals for the input and output stages of the IMC based on the information provided by the DSP output port pins.

The actual supply voltages V_a, V_b and V_c are converted into the digital form using the A/D converter to calculate the input reference signals as shown in Figure 7.10. The Clarke's transformation, input angle calculation, sector identification and duty cycle calculations are performed for the input stage of the IMC, then the Clarke's transformation, output angle calculation, output sector identification and duty cycle calculations for the output stage are performed. The duty cycles calculated for the input and output stages are compared with the carrier to determine the pulse durations. This part of the SVPWM algorithm to be implemented by DSP is developed and compiled in CCS. The PWM signals, in addition to the sector information, are brought to the output port pins of the DSP board as shown in Figure 7.10. The look-up tables for the input and output stages are implemented in FPGA to determine the gate signals. The gate signals are brought out by the output port pins, as shown in Figure 7.10.

This part of the SVPWM algorithm to be worked out by FPGA is developed and compiled through Xilinx ISE.

7.4 EXPERIMENTAL RESULTS AND DISCUSSION

In this section, the hardware test results of the IMC using SVPWM techniques are presented. The experimental waveforms of the various signals generated by the SVPWM technique are represented in Figures 7.11(a)–(b). The switching pulses obtained for the upper switches of the BCSR stage are shown in Figure 7.11(a). Figure 7.11(b) depicts the switching operation of the BCSR switch S_a with ZCC. The experimental results of the supply voltage V_a, supply current I_a, DC-link voltage V_{dc}, output current I_A and output phase and line voltages V_A and V_{AB} the for the SVPWM technique are shown in Figure 7.12. Figure 7.12(a) shows the supply current I_a and supply voltage V_a in phase resulting into the unity power factor. The sinusoidal nature

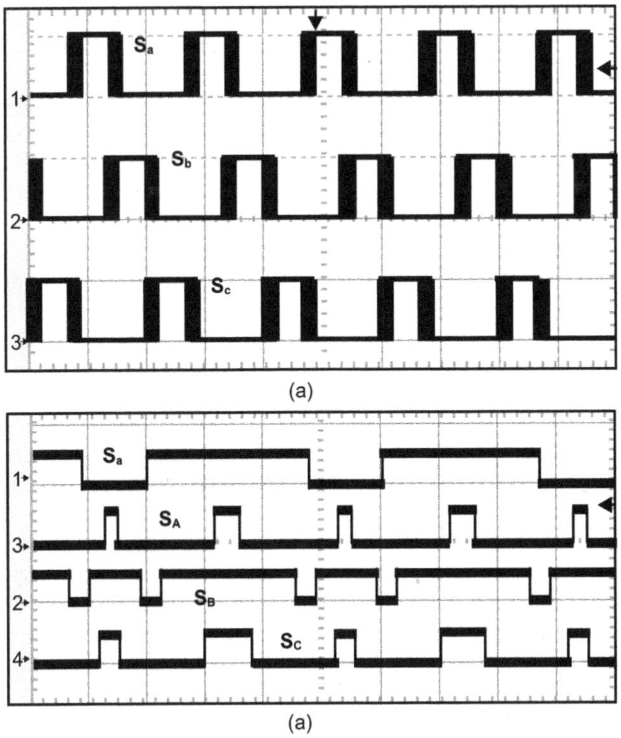

FIGURE 7.11 Experimental results of SVPWM technique: (a) BCSR stage gate signals; and (b) gate signals S_a, S_A, S_B and S_C.

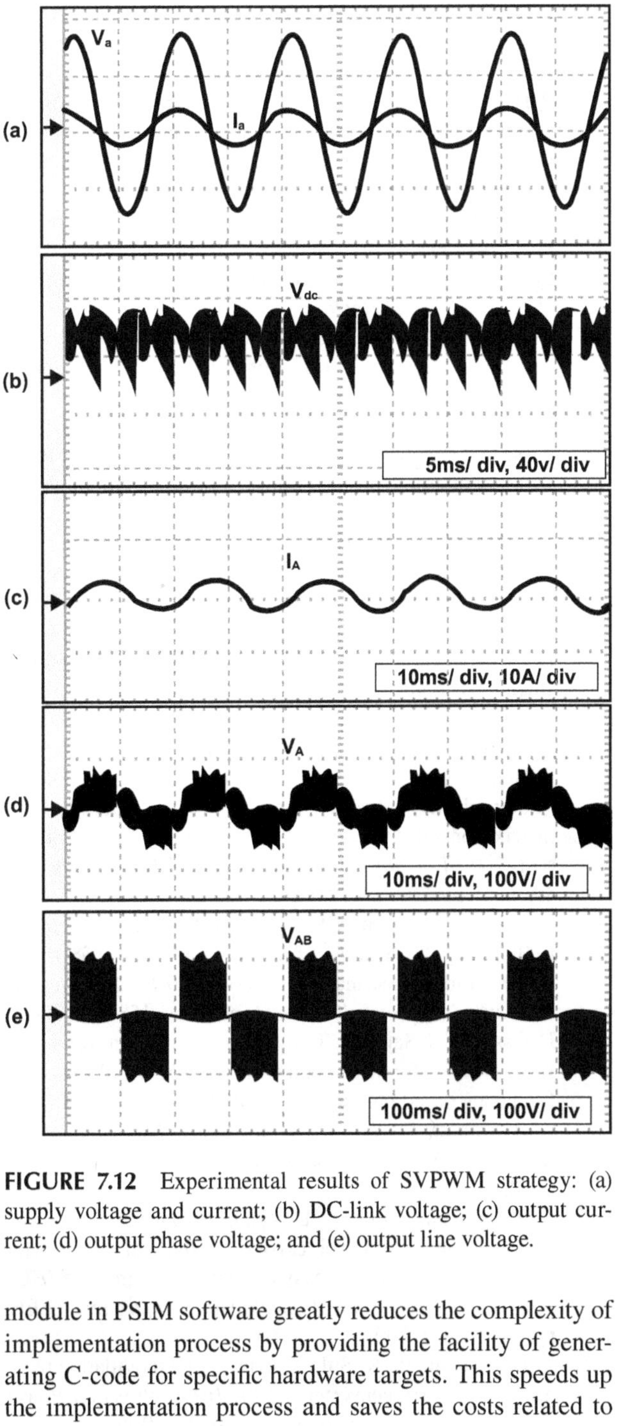

FIGURE 7.12 Experimental results of SVPWM strategy: (a) supply voltage and current; (b) DC-link voltage; (c) output current; (d) output phase voltage; and (e) output line voltage.

of the supply current is clearly observed. The DC-link voltage V_{dc} shown in Figure 7.12(b) does not contain any zero voltage states, as per the requirement. The output current I_A shown in Figure 7.12(c) is sinusoidal in nature. Figure 7.12(d)–(e) represent the phase and line voltages which contain the DC voltage ripple profile due to the absence of the DC-link storage element.

7.5 SUMMARY AND CONCLUSIONS

The chapter provides the introduction to the various matrix converter topologies, switches and modulation strategies. The fundamental principles of the IMC and space vector modulation technique to control the IMC are also presented, as is the hardware development of the IMC. The power and control circuit components used in the hardware development of the IMC are described in detail. The important features of the DSP TMS320F28335 and FPGA Spartan-6 development boards used in the experimental setup are presented. The descriptions of the driver and sensor cards used in the experimental rig are included. The implementation of the SVPWM algorithm for the IMC using the control devices DSP and FPGA is discussed. The automatic code generation facility provided by the simcoder module of the PSIM software is explained. The hardware test results are presented to verify the performance of the IMC.

This chapter provides the complete information of working principle and hardware development of the IMC. A simple approach of realizing the control algorithms for the IMC is presented using DSP TMS320F28335. The simcoder module in PSIM software greatly reduces the complexity of implementation process by providing the facility of generating C-code for specific hardware targets. This speeds up the implementation process and saves the costs related to the hardware development.

REFERENCES

[1] M. Venturini and A. Alesina, "The generalised transformer: A new bidirectional, sinusoidal waveform frequency converter with continuously adjustable input power factor," in *1980 IEEE Power Electronics Specialists Conference*. IEEE, pp. 242–252, 1980.

[2] A. Alesina and M. Venturini, "Intrinsic amplitude limits and optimum design of 9-switches direct pwm ac-ac converters," in *PESC'88 Record., 19th Annual IEEE Power Electronics Specialists Conference*. IEEE, pp. 1284–1291, 1988.

[3] J. Rodriguez, "A new control technique for ac-ac converters," in *Control in Power Electronics and Electrical Drives 1983*. Elsevier, 1984, pp. 203–208.

[4] P. D. Ziogas, S. I. Khan, and M. H. Rashid, "Analysis and design of forced commutated cycloconverter structures with improved transfer characteristics," *IEEE Transactions on Industrial Electronics*, no. 3, pp. 271–280, 1986.

[5] G. Roy, L. Duguay, S. Manias, and G.-E. April, "Asynchronous operation of cycloconverter with improved voltage gain by employing a scalar control algorithm," in *Proceedings of IEEE-IAS Annual Meeting*, pp. 889–898, 1987.

[6] G. Roy and G.-E. April, "Direct frequency changer operation under a new scalar control algorithm," *IEEE Transactions on Power Electronics*, vol. 6, no. 1, pp. 100–107, 1991.

[7] L. Huber and D. Borojevic, "Space vector modulated three-phase to three-phase matrix converter with input power factor correction," *IEEE Transactions on Industry Applications*, vol. 31, no. 6, pp. 1234–1246, 1995.

[8] X. Wang, H. Lin, H. She, and B. Feng, "A research on space vector modulation strategy for matrix converter under abnormal input-voltage conditions," *IEEE Transactions on Industrial Electronics*, vol. 59, no. 1, pp. 93–104, 2012.

[9] D. Casadei, G. Serra, and A. Tani, "Reduction of the input current harmonic content in matrix converters under input/output unbalance," *IEEE Transactions on Industrial Electronics*, vol. 45, no. 3, pp. 401–411, 1998.

[10] F. Blaabjerg, D. Casadei, C. Klumpner, and M. Matteini, "Comparison of two current modulation strategies for matrix converters under unbalanced input voltage conditions," *IEEE Transactions on Industrial Electronics*, vol. 49, no. 2, pp. 289–296, 2002.

[11] L. Helle, K. B. Larsen, A. H. Jorgensen, S. Munk-Nielsen, and F. Blaabjerg, "Evaluation of modulation schemes for three-phase to three-phase matrix converters," *IEEE Transactions on Industrial Electronics*, vol. 51, no. 1, pp. 158–171, 2004.

[12] D. Casadei, G. Serra, A. Tani, and L. Zarri, "A novel modulation strategy for matrix converters with reduced switching frequency based on output current sensing," in *2004 IEEE 35th Annual Power Electronics Specialists Conference (IEEE Cat. No. 04CH37551)*, vol. 3. IEEE, pp. 2373–2379, 2004.

[13] K. Mohapatra, P. Jose, A. Drolia, G. Aggarwal, N. Mohan, and S. Thuta, "A novel carrier-based pwm scheme for matrix converters that is easy to implement," in *2005 IEEE 36th Power Electronics Specialists Conference*. IEEE, pp. 2410–2414, 2005.

[14] Y.-D. Yoon and S.-K. Sul, "Carrier-based modulation technique for matrix converter," *IEEE Transactions on Power Electronics*, vol. 21, no. 6, pp. 1691–1703, 2006.

[15] S. Thuta, K. Mohapatra, and N. Mohan, "Matrix converter over-modulation using carrier-based control: Maximizing the voltage transfer ratio," in *2008 IEEE Power Electronics Specialists Conference*. IEEE, pp. 1727–1733, 2008.

[16] S. Kim, Y.-D. Yoon, and S.-K. Sul, "Pulsewidth modulation method of matrix converter for reducing output current ripple," *IEEE Transactions on Power Electronics*, vol. 25, no. 10, pp. 2620–2629, 2010.

[17] R. Vargas, U. Ammann, B. Hudoffsky, J. Rodriguez, and P. Wheeler, "Predictive torque control of an induction machine fed by a matrix converter with reactive input power control," *IEEE Transactions on Power Electronics*, vol. 25, no. 6, pp. 1426–1438, 2010.

[18] F. Pinto and F. Silva, "Sliding mode control of space vector modulated matrix converter with sinusoidal input/output waveforms and near unity input power factor," in *Proceedings of European Conference on Power Electronics and Applications, EPE*, vol. 99, pp. 1–9, 1999.

[19] L. Wei and T. A. Lipo, "A novel matrix converter topology with simple commutation," in *Conference Record of the 2001 IEEE Industry Applications Conference 36th IAS Annual Meeting (Cat. No. 01CH37248)*, vol. 3. IEEE, pp. 1749–1754, 2001.

[20] M. Jussila, M. Salo, and H. Tuusa, "Realization of a three-phase indirect matrix converter with an indirect vector modulation method," in *IEEE 34th Annual Conference on Power Electronics Specialist, 2003. PESC'03.*, vol. 2. IEEE, pp. 689–694, 2003.

[21] M. Jussila and H. Tuusa, "Comparison of simple control strategies of space-vector modulated indirect matrix converter under distorted supply voltage," *IEEE Transactions on Power Electronics*, vol. 22, no. 1, pp. 139–148, 2007.

[22] T. D. Nguyen and H.-H. Lee, "A new svm method for an indirect matrix converter with common-mode voltage reduction," *IEEE Transactions on Industrial Informatics*, vol. 10, no. 1, pp. 61–72, 2013.

[23] K. Rahman, N. Al-Emadi, A. Iqbal, and S. Rahman, "Common mode voltage reduction technique in a three-to-three phase indirect matrix converter," *IET Electric Power Applications*, vol. 12, no. 2, pp. 254–263, 2018.

[24] B. Wang and G. Venkataramanan, "A carrier based pwm algorithm for indirect matrix converters," in *2006 37th IEEE Power Electronics Specialists Conference*. IEEE, pp. 1–8, 2006.

[25] P. C. Loh, R. Rong, F. Blaabjerg, and P. Wang, "Digital carrier modulation and sampling issues of matrix converters," *IEEE Transactions on Power Electronics*, vol. 24, no. 7, pp. 1690–1700, 2009.

[26] G. T. Chiang and J.-I. Itoh, "Comparison of two overmodulation strategies in an indirect matrix converter," *IEEE Transactions on Industrial Electronics*, vol. 60, no. 1, pp. 43–53, 2012.

[27] Q.-H. Tran and H.-H. Lee, "An effective carrier-based modulation strategy to reduce the switching losses for indirect matrix converters," *Journal of Power Electronics*, vol. 15, no. 3, pp. 702–711, 2015.

[28] J.-I. Itoh, T. Hinata, K. Kato, and D. Ichimura, "A novel control method to reduce an inverter stage loss in an indirect matrix converter," in *2009 35th Annual Conference of IEEE Industrial Electronics*. IEEE, pp. 4475–4480, 2009.

[29] D.-T. Nguyen, H.-H. Lee, and T.-W. Chun, "A carrier-based pulse width modulation method for indirect matrix converters," *Journal of Power Electronics*, vol. 12, no. 3, pp. 448–457, 2012.

[30] P. Correa, J. Rodriguez, M. Rivera, J. R. Espinoza, and J. W. Kolar, "Predictive control of an indirect matrix converter," *IEEE Transactions on Industrial Electronics*, vol. 56, no. 6, pp. 1847–1853, 2009.

APPENDIX: SOME QUESTIONS

1. Design the appropriate input filter parameters R_f, L_f and C_f for the IMC which operates at 20kHz switching frequency.
2. What is the function of the damping resistor in the input filter? How do the low and high values of the damping resistors affect the performance of the filter?
3. How do the size of the filter elements affect the performance of the system?
4. How is the simulation developed in the PSIM for code generation discretized? How does sampling rate affect the performance of the system?
5. Explain the various library elements of the PSIM software which are used for automatic code generation for the hardware target TI F28335.

8 Grid-Interactive Solar Energy Conversion Systems

Shailendra Kumar and More Raju

CONTENTS

8.1 GENERAL

The role of renewable energy is becoming significant to resolve the issues of energy crisis [1–4]. The growth of solar photovoltaic (SPV)-based energy sources resolves the issues of global warming, carbon dioxide pollution, and low installation cost [5, 6]. Moreover, in case of non-renewable energy–based plants, the energy required by the consumer loads is supplied through long transmission lines, which increases the transmission and distribution losses. However, by installing rooftop PV energy systems, there is significant reduction in these losses near to load end. This chapter proposes a grid-interactive solar energy conversion systems (SECS), while the initial stage is a DC/DC stage and another stage is an utility interactive voltage source converter (VSC) [7, 8]. The main control philosophy of SECS is to inject balanced sinusoidal currents to the utility at power factor corrections [9]. However, The variation in voltages is realized in the far radial ends of the distribution transformer. Harvesting maximum power from the renewable energy source improves the efficiency, compactness, and cost of a solar energy system [10–14].

The main objective of this chapter is to address both energy security and power quality improvement, the challenges being faced by the distribution system. The research is focused toward the design, analysis, and implementation of grid-interfaced solar PV energy conversion systems. A grid-interfaced solar PV system can integrate renewable power of solar PV with the utility grid and simultaneously enhance the power quality at their point of common coupling (PCC) [15–17]. As compared to multiple device performing different operations, the proposed solar PV system can save a great amount of capital investment and system space. This also helps in increasing the effective utilization of solar PV energy conversion systems, and the cost recovery is expected to be faster. Simulated results of the presented topology and control for the grid-interfaced solar energy conversion system have been demonstrated, and their performances are also reported through simulation and test results. Topology of grid-interfaced PV energy conversion has been studied and its success fully implemented in the laboratory. Therefore, in this chapter, the performance of SPV generating systems have been simulated and tested under different operating conditions.

8.2 CLASSIFICATION OF SPV ENERGY CONVERSION SYSTEMS

Figure 8.1 shows the major classification of SPV generating systems on the basis of type of connection, converter topology, and number of stages. The basic types of SECS are grid-interactive and standalone. In standalone, energy storage is needed to store the energy, then to supply to the loads.

8.2.1 Classification of the Basis of Number of Stages

On the basis of number of stages, SPV systems are classified in three stages: single-stage systems, two-stage systems, and multi-stage systems. Stages are categorized on the basis of performance, efficiency, complexity, reliability, etc.

8.2.1.1 Single-Stage SPV Systems

Figure 8.2 and Figure 8.3 show the single-stage grid-interfaced SPV systems. Basically, there are two single-stage

DOI: 10.1201/9781003229124-8

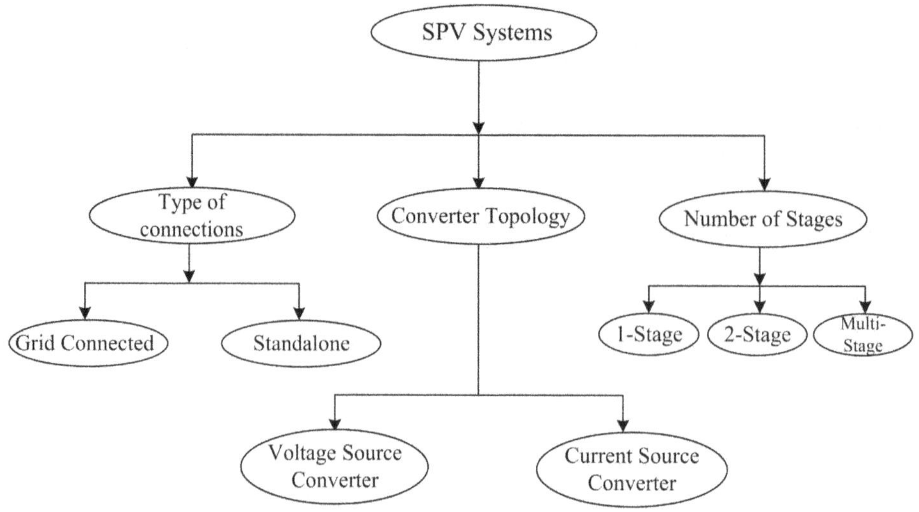

FIGURE 8.1 Classification of solar PV generating systems.

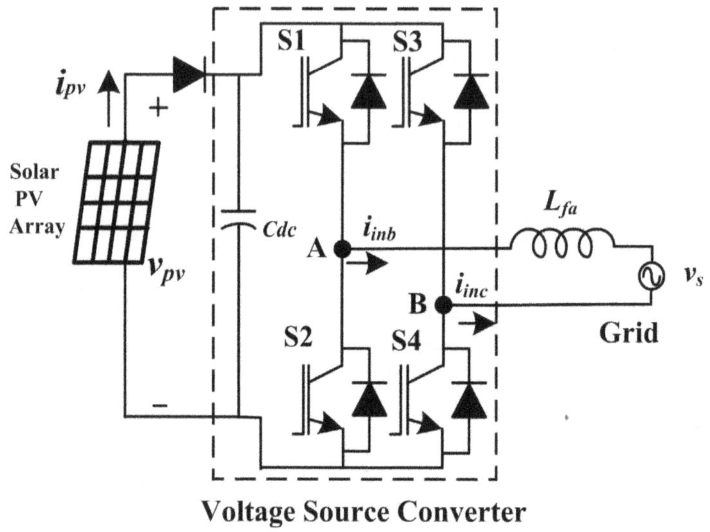

FIGURE 8.2 Single-stage single-phase grid-interfaced voltage source converter configurations.

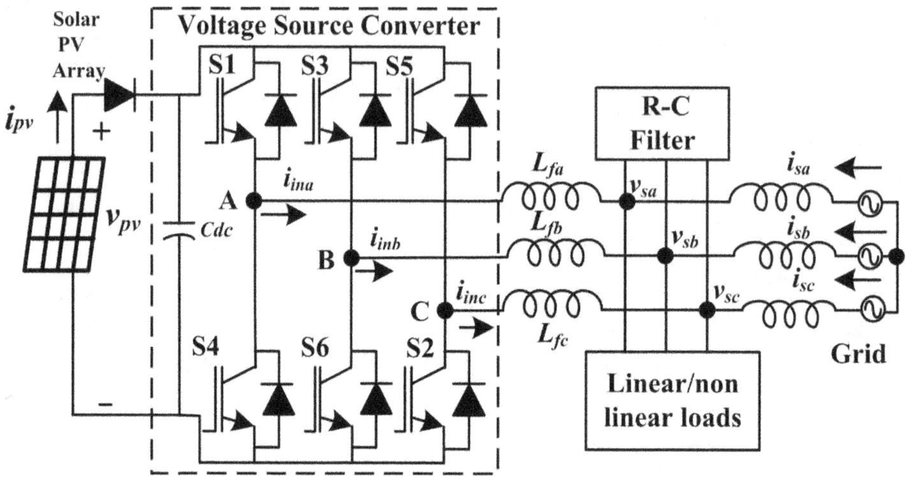

FIGURE 8.3 Single-stage three-phase grid-connected VSC configurations.

systems: one is DC to AC (grid connected) and other is DC to DC, which is used for battery charging applications.

In a single-stage DC to AC system, solar generator is fed to the capacitor of voltage source converter, and the number of series and parallel modules determines the DC link voltage power rating of the VSC. This DC link voltage is used for the MPPT purpose.

8.2.1.2 Two-Stage SPV Systems

Figures 8.4 and Figure 8.5 show the two-stage system. In two stages, the first stage is used for extracting maximum power from the solar PV array by using an MPPT controller, so this is the DC to DC conversion; in the second stage, a VSC is used to transfer AC power to the grid. So as the name suggests, the first stage is DC to DC conversion and the second stage is DC to AC conversion.

8.2.1.3 Multi-Stage SPV Systems

The multi-stage solar PV energy conversion system may be used for decrement in current and voltage ratings of the system. Many researchers have proposed lots of configurations for multi-stage systems. For example, in a three-stage system, a boost converter is followed by the Ćuk converter, followed by the VSC. This three-stage system has a very low power rating.

8.2.2 Classification on the Basis of Connections

On the basis of connections, a solar PV energy conversion system is divided into grid-connected systems and standalone systems.

8.2.2.1 Grid-Connected SPV Energy Systems

In grid-connected systems, maximum power is extracted from the solar array and feeds the power to the AC grid. There are various issues while using utility-tied SECS like efficiency, grid tied several techniques which are also presented in the many literatures. The utility-tied SECS can be three-phase type or single-phase type, depending on the ratings. Single-phase SECS performs the MPPT and grid synchronization.

8.2.2.2 Standalone SPV Systems

Figure 8.6 shows the standalone SPV system, in which SPV power extracted from the solar array is stored in the battery, flywheel, or some storage device. This energy can be used as requirement by the loads. Standalone solar PV systems are independent of the grid.

8.2.3 Classification on the Basis of Types of Converter

On the basis of converters, SPV systems are classified as voltage-source or current-source converters.

8.2.3.1 Voltage-Source Converters

Topologies based on voltage-source converters are very popular for SPV systems. The VSC technology is chosen as the basis for several various applications due to its simplicity, controllability, compact modular design, simple integration, and small environmental effect. It can be classified as follows.

- Three-phase three-wire single-stage SECS
- Three-phase four-wire two-stage SECS
- Single-phase single-stage SECS
- Single-phase two-stage SECS

8.2.3.2 Current-Source Converters

The primarily objective in a SECS is the accurate technique of real and reactive powers flow to sustain the stability of the SPV system. This is realized using the power converters and their capability of changing electrical power from DC to AC, or vice versa. In a current-source converter (CSC), the input current is constant with a low harmonics through a

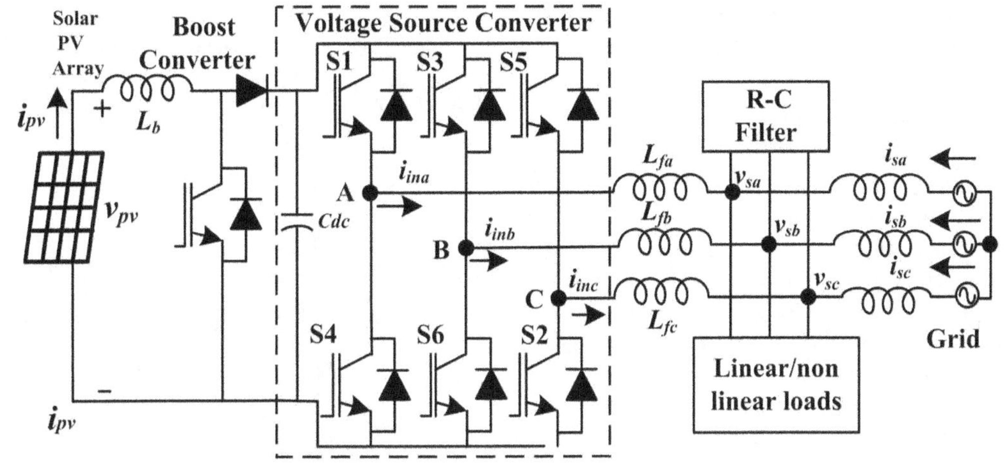

FIGURE 8.4 Two-stage three-phase SECS.

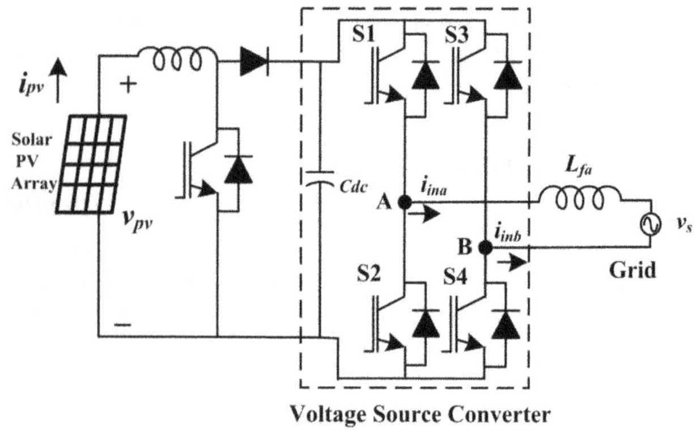

FIGURE 8.5 Two-stage single-phase SECS.

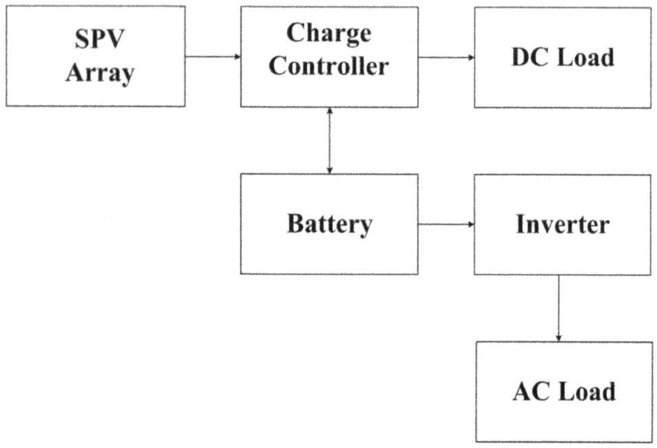

FIGURE 8.6 Standalone SPV system with battery storage system.

sizable smoothing reactor. Therefore, this creates a current source at the input side. Therefore, CSC is also becoming popular day by day; however, cost and size of DC inductors are still troublesome.

8.2.4 Classification on the Basis of Phases

On the basis of number of phases, SPV systems can be categorized into single-phase and three-phase types.

8.2.4.1 Single-Phase SPV Systems

Figure 8.7 shows a single-phase SPV system. Commercially available photovoltaic modules are low power. In a single-phase DC/AC system, a series parallel combination of such low power modules is made out to attain string of the required power rating. The number of series modules is decided by output voltage required, and the number of parallel modules is then calculated according to power rating of the system. In single-phase single-stage grid connected VSC (voltage source converter)-based topologies, the string is directly connected to DC link of VSC and a floating DC link is used for MPPT, while in the event of single-phase double stage, the boost converter is utilized.

8.2.4.2 Three-Phase SPV Systems

Three-phase grid-interfaced SPV systems can be further classified into interfaced SPV systems.

8.2.4.2.1 Three-Phase Three-Wire SPV Systems

Figure 8.8 shows the three-phase three-wire SPV system. The three-phase three-wire SECS consists of six IGBTs to form a three-phase VSC. On the input side, a electrolyte capacitor is used to maintain the DC voltage, and on the AC side, a smoothing inductor is used to interface with the grid.

8.3 PV ARRAY MODELING TECHNIQUES

Normally, numerous PV cells are connected in series to form a PV module for obtaining a higher voltage rating. Similarly, in parallel, they increase the current rating. The modeling has been explained with the reference to one PV module. A review of the literature reveals that a number of modeling techniques of PV array have been reported. Mathematical modeling of PV cells is described in following sub-section.

8.3.1 Ideal Single-Diode Models

An ideal single diode model is the basic of all mathematical modeling, in which only three parameters—i.e. open circuit current, open circuit voltage, and ideality factor—are involved. MPPT performance of this model is poor, as there

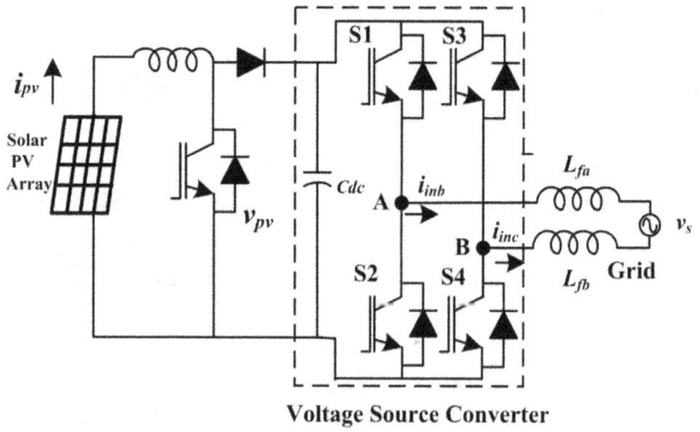

FIGURE 8.7 Single-phase grid-interfaced SPV system.

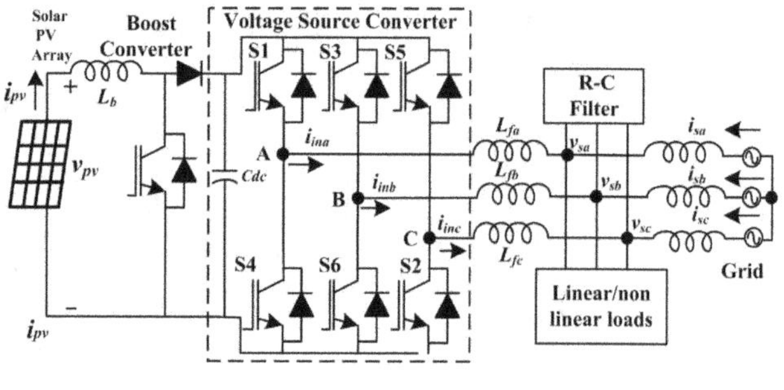

FIGURE 8.8 Three-phase three-wire grid-interfaced SPV system.

is no parallel and series resistance included, so, I-V characteristic is not very accurate. Figure 8.9 shows the ideal single-diode model.

8.3.2 Simplified Single-Diode Models

In this case, series resistance is included, but the parallel resistance is still not considered. Figure 8.10 shows the simplified diode model. The main disadvantage is poor accuracy when subjected to large temperature. The following equation defines the simplified diode model:

$$I = I_{pv} - I_0\left[e^{\left[\frac{q(V_{PV}+IR_S)}{N_s KTA}\right]} - 1\right] \tag{8.1}$$

The proposed MATLAB/Simulink model of the PV array is shown in Figure 8.11 using single diode equivalent for sake of simplicity.

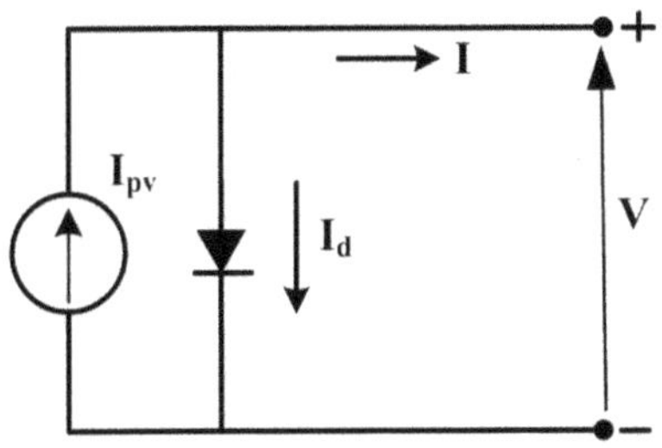

FIGURE 8.9 Single-diode model for PV cell.

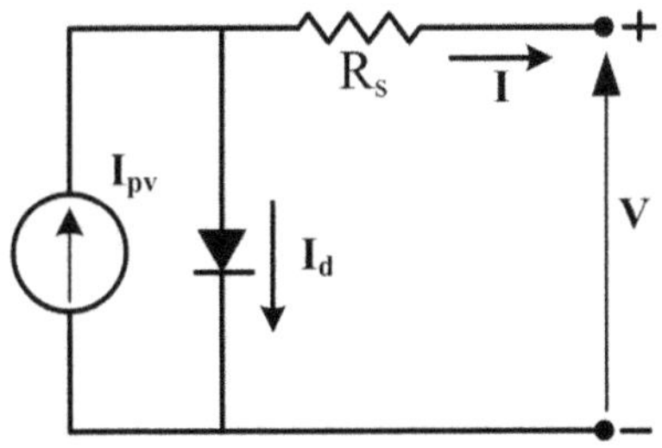

FIGURE 8.10 Simplified single-diode model for a PV cell.

The I-V and P-V characteristics of PV arrays at different insolation and different temperatures are shown in Figures 8.12(a)–(b) and Figures 8.13(a)–(b).

8.4 CIRCUIT CONFIGURATIONS OF GRID-INTERACTIVE SOLAR ENERGY CONVERSION SYSTEMS

The circuit topology for the SECS is presented in Figure 8.14. A two-stage SECS is investigated for utility-interfaced solar energy conversion systems. The initial stage is a boost converter for maximum power, and the second stage is the grid-side converter. The solar generator is tied to the DC/DC converter and it offers constant impedance such that solar generator produces the peak energy. The R-C and inductive filters control the ripple in the PCC voltages and currents.

8.5 DESIGN AND SELECTION OF GRID-INTERACTIVE SOLAR ENERGY CONVERSION SYSTEMS COMPONENTS

This section describes the design of each stage of the proposed 50kW SPV power generating system.

8.5.1 Design of a PV Generator String

A PV string is developed for 50kW maximum power rating to be connected to 415V, 50Hz utility. The peak power is derived as:

$$P_{mpp} = \left(N_s * V_{mpp}\right) * \left(N_p * I_{mpp}\right) = 50\ kW \tag{8.2}$$

where n_s and n_p are the numbers of series and parallel PV modules, open circuit PV voltage and short circuit PV current are selected 21V and 3.8A; moreover, the V_{mp} and I_{mp} are the MPP voltage and current of a PV generator

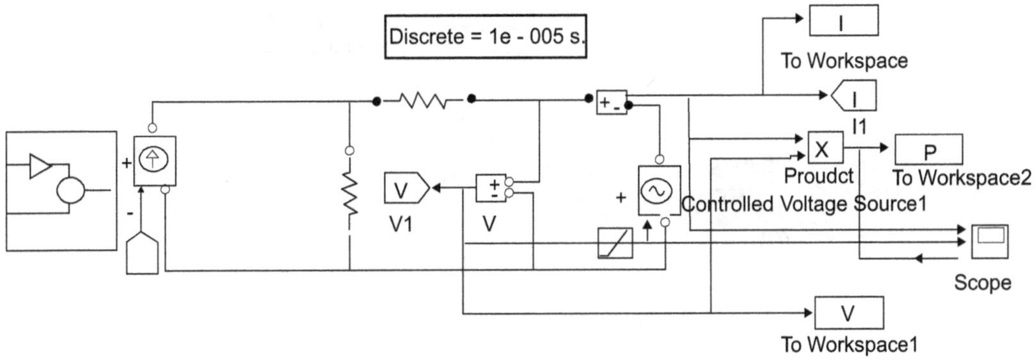

FIGURE 8.11 Simulink model of a PV array.

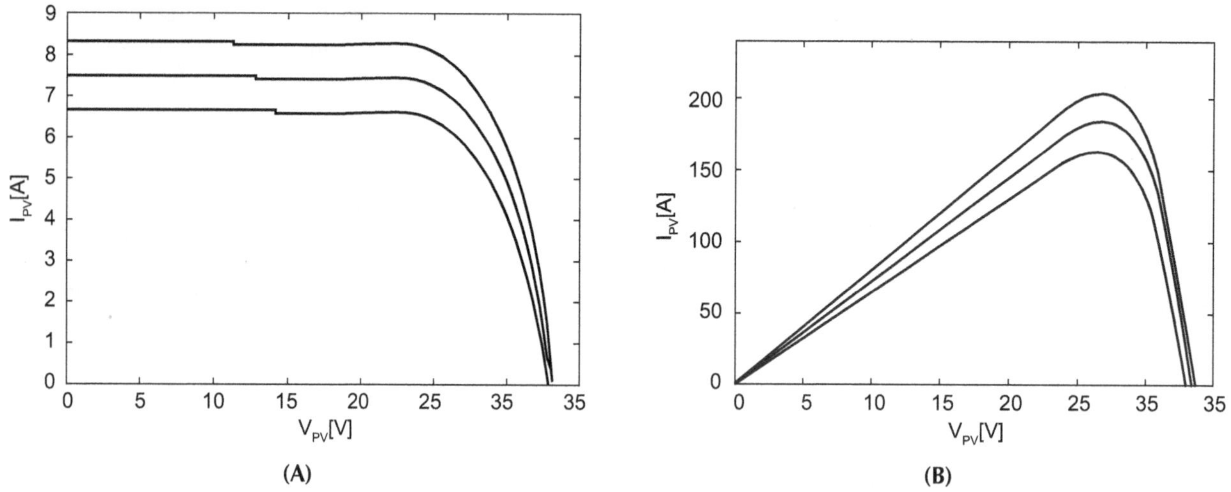

FIGURE 8.12 (a) I-V; (b) P-V characteristics of PV array at different insolation.

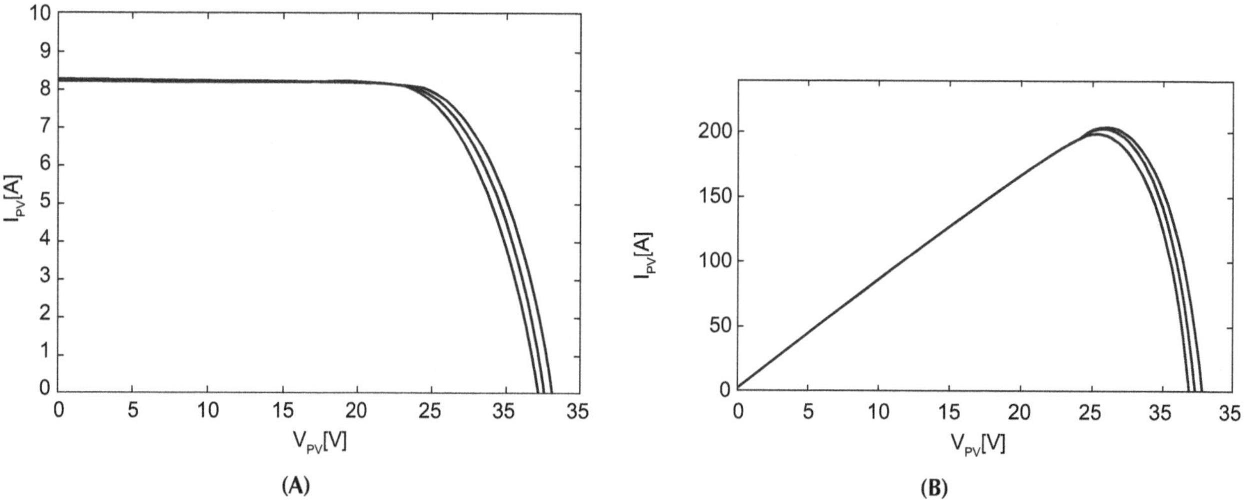

FIGURE 8.13 (a) I-V; (b) P-V characteristics of PV array at different temperatures.

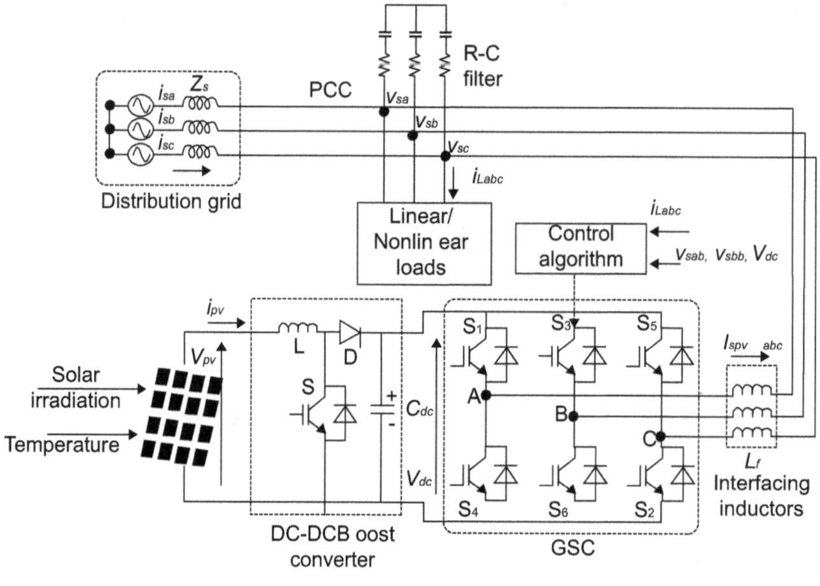

FIGURE 8.14 Solar energy conversion system.

The P_{mp} is calculated as:

$$P_{mpp} = \left(N_s * 85\% \ of \ V_{ocn} * N_p * 85 \ \% \ of \ I_{sc}\right) = 50 \ kW \quad (8.3)$$

The OC voltage V_{OCT} is selected as 700V equal to DC bus voltage of VSC interfaced to AC mains. The number panels required for series connected are calculated as:

$$V_{OCT} = n_s * V_{ocn} \rightarrow N_s = V_{OCT} / V_{OCn} = 700V / 21V = 33 \ Modules \quad (8.4)$$

The peak MPP current is estimated as:

$$I_{mp} = peak \ power / \left(0.85 * \text{O.C. voltage}\right) = 84 \ A \quad (8.5)$$

The parallel solar panels are given as,

$$I_{mp} = n_p * I_{sc} \rightarrow n_p = I_{mp} / I_{sc} = 25 \text{ A} \quad (8.6)$$

Therefore, the PV string of 50kW rating is developed with 25 x 33 panels in parallel and series.

8.5.2 Estimation of VSC Input DC Voltage

A proper selection of input VSC DC voltage is very necessary. For proper selection it is calculated as:

$$V_{dc} = \frac{\left(2\sqrt{2}V_{LL}\right)}{\sqrt{3}m} = \frac{\left(2\sqrt{2} * 415\right)}{\sqrt{3} * 0.95} = 713.27 \approx 700 \text{ V} \quad (8.7)$$

where V_{LL} is the PCC voltage

8.5.3 Design of DC/DC Boost Converter

The DC/DC boost converter is used for MPPT so that the PV panel can be utilized at its maximum efficiency. For the boost converter, the input voltage is MPP voltage of the PV array (V_{string}), which is 85% of open circuit voltage i.e. 595 V (700 x 0.85). The duty ratio considering V_{string} as input voltage and V_{DC} (700V) as output voltage is 0.15. Boost inductor value is calculated as:

$$L_b = \frac{V_{dc}D(1-D)}{(2\Delta i_1 f_{sw})} = \frac{700 * 0.15(1-0.15)}{(2 * 5.04 * 10000)} = 0.88 \ mH \approx 1.00 \ mH \quad (8.8)$$

where Δi_1 is ripple current in PV array, which is selected as 6% of inductor current i_1 (= P_{pv}/V_{string}) = 84A. Thus, the Δi_1 is 5.04A; the estimated value of L_b is derived as 0.88 mH and is chosen as 1.00mH

8.5.4 Design of AC Interfacing Inductor

The design of coupling inductor is calculated as,

$$L_f = \frac{\sqrt{3}mV_{dc}}{12hf_s\Delta i} = \frac{\sqrt{3} * 1 * 700}{12 * 1.2 * 10000 * (0.05 * 69.55)} = 2.42 \, mH \approx 2.50 \, mH \quad (8.9)$$

Selecting Δi, = 5% of AC line current, switching frequency = 10kHz. The L_f is calculated as 2.42mH. The chosen value is 2.50mH.

8.5.5 Estimation of VSC Capacitor

The design and selection of VSC capacitor is presented as:

$$C_{dc} = \frac{I_d}{\left(2 * \omega * v_{dcrip}\right)} = \frac{(P_{dc} / V_{dc})}{\left(2 * 314 * 0.025 * 700\right)} = \frac{(50000 / 700)}{\left(2 * 314 * 0.025 * 700\right)} = 4513.5 \mu F \approx 4700 \mu F \quad (8.10)$$

8.5.6 Design of Ripple Filter

The R-C value with the following equation, is estimated as:

$$f_s = 1 / \left(2 * \pi * R_f * C_f\right) \quad (8.11)$$

After examination, $R_f = 5\Omega$, a $C_f = 5\mu F$ are chosen for R-C filter.

Therefore, the design of various components of 50kW, 415V AC three-phase, four-wire solar PV array is summarized in Table 8.1.

8.6 CONTROL APPROACH FOR GRID-INTERACTIVE SOLAR ENERGY CONVERSION SYSTEMS

This chapter shows the SECS controlled through the quaternion-based adaptive method. For harvesting the peak

TABLE 8.1
Components and Their Design Parameters

S. No.	Name of Component	Design Value
1	SPV array	25 modules in parallel and 33 modules in series, with PV array of 25 x 33 modules
2	DC/DC boost converter	1.75mH
3	DC capacitor voltage	700V
4	AC inductor	2.5mH
5	DC link capacitor	4700μF
6	Ripple filter	5μF, 5Ω

energy from PV generator, an adaptive perturb and observe (P&O) technique is utilized. To provide the pulses to VSC, a quaternion-based adaptive method is utilized, along with harmonics mitigation.

8.6.1 MPP Tracking Using the Boost Converter

In the last few years, various MPPT control methods have been developed to derive peak energy from a solar generator. Moreover, the merit of the P&O MPPT method is its simplicity. In this method, the duty cycle of the DC/DC converter is to regularly perturb to achieve the peak point in the MPP curve. The variable step is generally sizable when far from MPP, and minimizes as the point achieves MPP. The expression of perturb and observe technique are given as:

$$\delta(k) = \delta(k-1) - K(\frac{dp}{dpv}) \tag{8.12}$$

The duty cycle of boost converter is represented by δ(n) and K is a step size. The varying step size process is observed that the (dV_{pv}/dI_{pv}) must be vanish at peak point as:

$$\frac{dP_{pv}}{dI_{pv}} = \frac{d(I_{pv}V_{pv})}{dI_{pv}} = V_{pv} + I_{pv}\frac{dV_{pv}}{dI_{pv}} \tag{8.13}$$

The V_{pv}/I_{pv} and dV_{pv}/dI_{pv} are subtracted to achieve the minimum diversion point, and MPP is obtained by minimizing the error to zero.

8.6.2 Quaternion-Based Adaptive Method for Control of VSC

A quaternion-based adaptive method is a well-known algorithm for noise and error minimization, which is proposed for other application by many researchers. The main disadvantage of the existing methods is poor dynamic response. This issue is suppressed by using the quaternion-based adaptive method. Hence, the quaternion-based adaptive method is superior as compared to other algorithms. Moreover, this method enhances the transient behavior of SECS.

The grid voltages help in the reference currents generation, which are further compared with the sensed grid current. The terminal voltage at point of common coupling (PCC) and unit vectors are computed as:

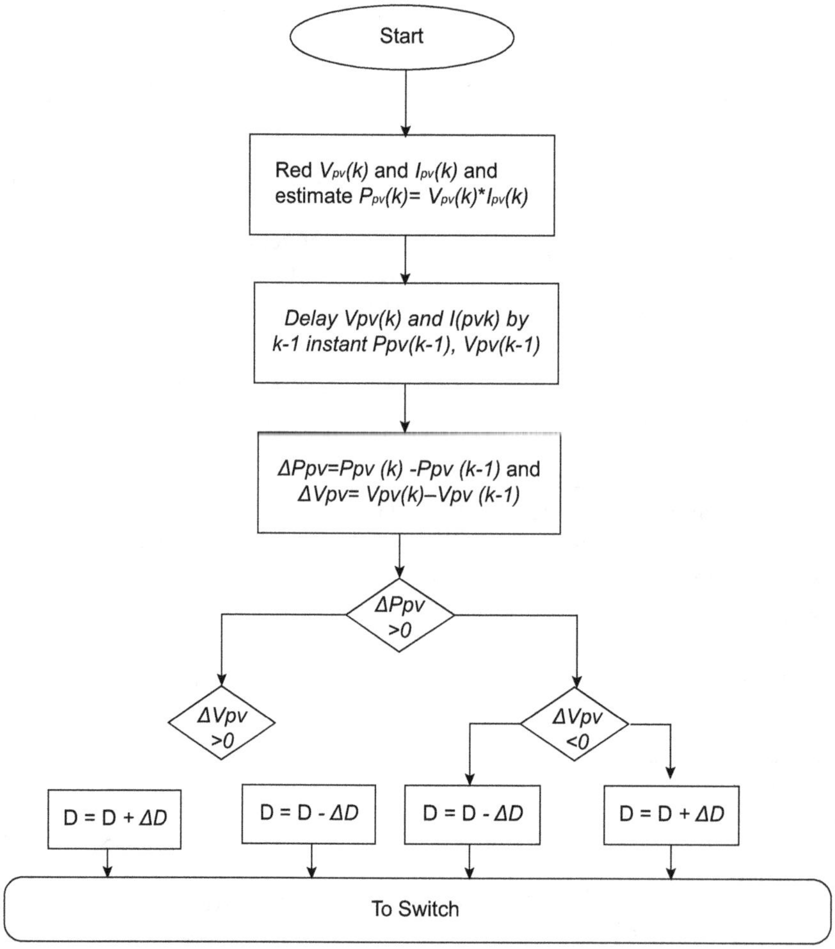

FIGURE 8.15 Perturb and observe flowchart.

$$v_{sa} = \frac{2v_{sab} + v_{sbc}}{3},\ v_{sb} = \frac{-v_{sab} + v_{sbc}}{3},\ v_{sc} = \frac{-v_{sab} - 2v_{sbc}}{3} \quad (8.14)$$

$$V_t = \sqrt{(2/3)(v_{sa}^2 + v_{sb}^2 + v_{sc}^2)} \quad (8.15)$$

The in-phase unit templates *(S_{pa}, S_{pb}, S_{pc})* are computed as:

$$S_{pa} = \frac{v_{sa}}{V_t},\ S_{pb} = \frac{v_{sb}}{V_t},\ S_{pc} = \frac{v_{sc}}{V_t} \quad (8.16)$$

The development of quaternion-based adaptive method for estimation of weight signal and mean square error is illustrated in Figure 8.16(a)–(b). The estimation of weight signals *(Y_{La}, Y_{Lb}, Y_{Lc})* with the help of load currents *(i_{La}, i_{Lb}, i_{Lc})* and the in-phase synchronizing signal *(S_{pa}, S_{pb}, S_{pc})* are shown in Figure 8.16(a), there are shown as:

$$Y_{La}(n+1) = f_{SQVM}\{i_{La}(n), S_{pa}(n)\} \quad (8.17)$$

$$Y_{Lb}(n+1) = f_{SQVM}\{i_{Lb}(n), S_{pb}(n)\} \quad (8.18)$$

$$Y_{Lc}(n+1) = f_{SQVM}\{i_{Lc}(n), S_{pc}(n)\} \quad (8.19)$$

The recursive updated weight governing equation for the quaternion-based adaptive method is given as:

$$Y_{La}(n+1) = Y_{La}(n) + \frac{1}{2}(\mu_a - 4\sigma_a)[e_a(n)S_{pa}(n)] + \mu_a \tilde{N}_w J_0(n) - \frac{1}{4}\mu_a\delta_a.sign(Y_{La}(n)) \quad (8.20)$$

$$Y_{Lb}(n+1) = Y_{Lb}(n) + \frac{1}{2}(\mu_b - 4\sigma_b)[e_b(n)S_{pb}(n)] + \mu_b \tilde{N}_w J_0(n) - \frac{1}{4}\mu_b\delta_b.sign(Y_{Lb}(n)) \quad (8.21)$$

$$Y_{Lc}(n+1) = Y_{Lc}(n) + \frac{1}{2}(\mu_c - 4\sigma_c)[e_c(n)S_{pc}(n)] + \mu_c \tilde{N}_w J_0(n) - \frac{1}{4}\mu_c\delta_c.sign(Y_{Lc}(n)) \quad (8.22)$$

The cost function *(J_0)* of quaternion-based adaptive method is estimated as:

$$J_0(n) = [(1-\delta_a)\times e_a(n)e_a^*(n)] + \delta_a\|Y_{La}(n+1)\| \quad (8.23)$$

The condition of active weight of phase with sign function is given as:

$$sign[(Y_{La}(n)] = \begin{cases} \frac{Y_{La}(n)}{|Y_{La}(n)|}, & Y_{La}(n) \neq 0 \\ 0, & Y_{La}(n) = 0 \end{cases} \quad (8.24)$$

where δ_a is a step size and the value is less than 1

The MSE is calculated as:

$$e_a(n) = i_{La}(n) - [Y_{La}(n) \times S_{pa}(n)]$$

The cost function in terms of MSE and conjugate of MSE is estimated as:

$$J_0(n) = [e(n)e^*(n)] \quad (8.25)$$

where e(n) and e × (n) represent the mean square errors

Here, there is PV array acting as a renewable energy source, the contribution for PV array is given as:

$$Y_{pvp} = 2P_{pv}/(3\times V_t) \quad (8.26)$$

The DC link voltage is regulated through the PI regulator and Y_{loss} represents the current component which is required to regulated the DC voltage as:

$$Y_{loss}(n) = Y_{loss}(n-1) + k_p\{V_{dce}(n) - V_{dce}(n-1)\} + k_i V_{dce}(n) \quad (8.27)$$

The reference active weight is calculated as:

$$Y_{pLnet} = Y_{pavg} + Y_{loss} - Y_{pvp} \quad (8.28)$$

The reference current for the switching pulse generation for the six IGBT switches, two in each phase of VSC, is achieved as:

$$i_{sa}^* = Y_{pLnet} * u_{pa},\ i_{sb}^* = Y_{pLnet} * u_{pb},\ i_{sc}^* = Y_{pLnet} * u_{pc} \quad (8.29)$$

The switching signals are produced by comparing the *(i_{sa}*, i_{sb}*, i_{sc}*)* and sensed *(i_{sa}, i_{sb}, i_{sc})*. The extracted signals are supplied to the controller to turn on the VSC.

8.7 MATLAB SIMULINK MODELING OF SECS

A 50kW SECS is designed and modeled using MATLAB Simulink software, as shown in Figure 8.17 and Figure 8.18. Figure 8.17 shows the MATLAB modeling of a solar PV array and a boost converter. After a solar PV array, a boost converter is used. To regulate the duty cycle of DC/DC boost converter, a P&O-based MPPT algorithm is used. Moreover, Figure 8.18 shows the MATLAB

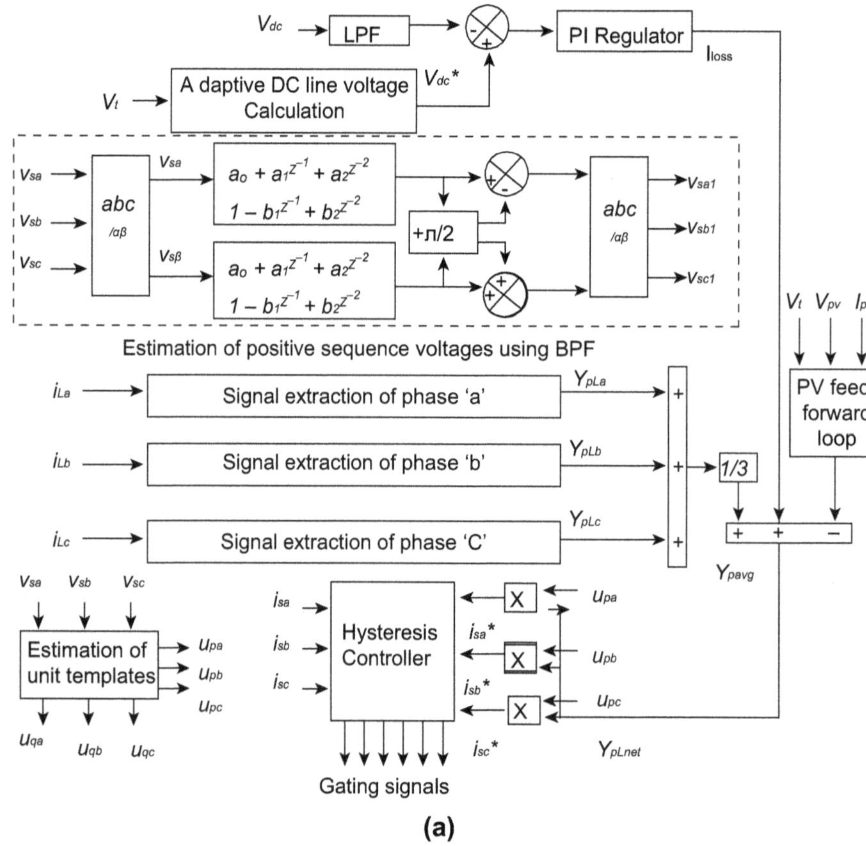

FIGURE 8.16 (a) Quaternion-based adaptive method for VSC

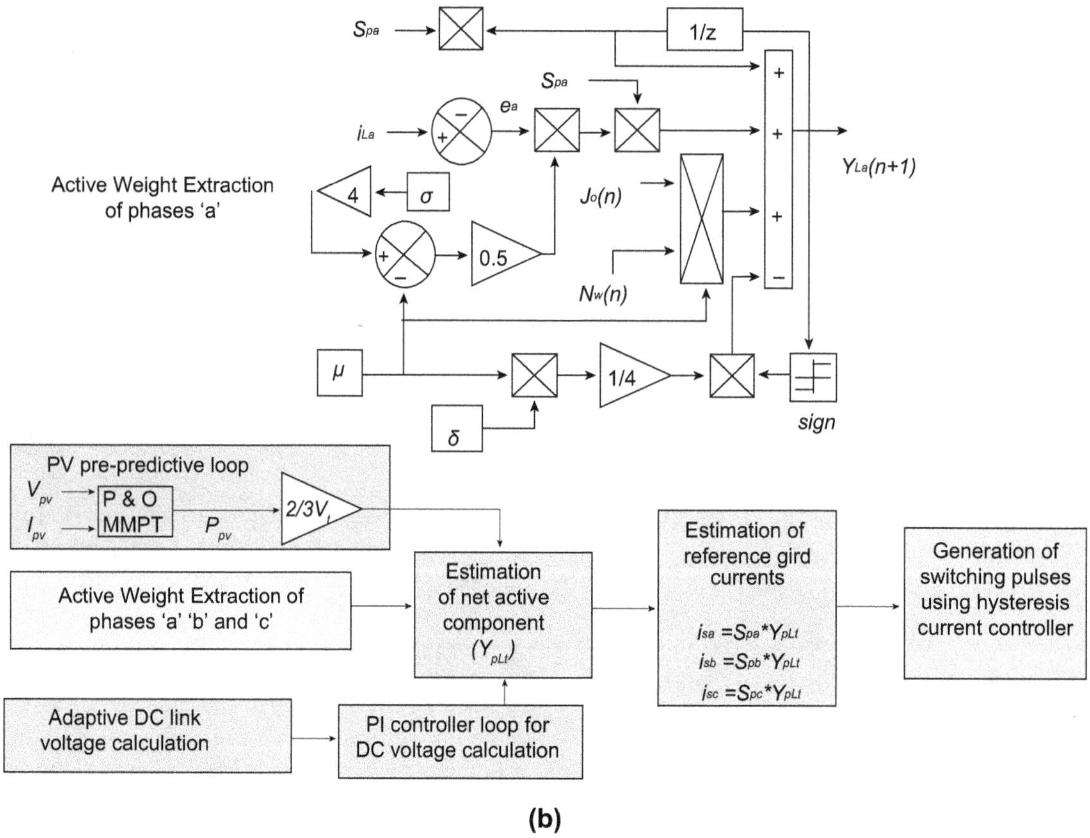

FIGURE 8.16 (b) Extraction of active weight signal.

Simulink model of grid-interfaced solar PV system. It consists 415V, 50Hz AC grid, ripple filter, linear and nonlinear loads, interfacing inductors, VSC as a shunt compensator, and DC link capacitor, which is connected to the solar PV array. The MATLAB modeling of quaternion-based adaptive method is shown, and load currents have been sensed. For nonlinear loads, load currents are square wave, and after rectification, it is modified into sine wave. To convert this signal into absolute value, an absolute block is used.

8.8 RESULTS AND DISCUSSION

In this section, simulated performances of the proposed SECS system under different operating conditions with linear and nonlinear loads are discussed.

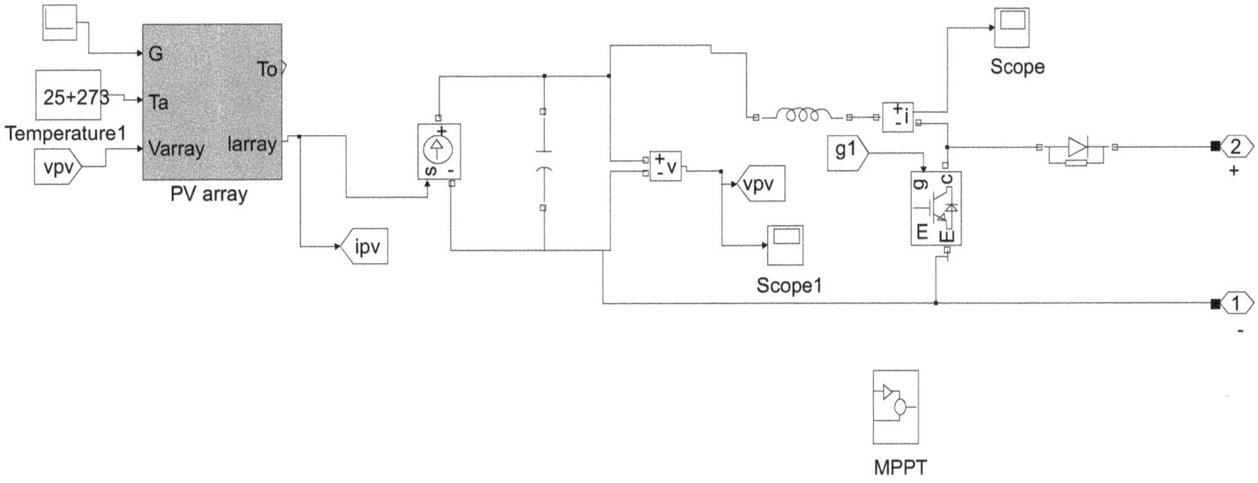

FIGURE 8.17 Model of a PV generator and boost converter.

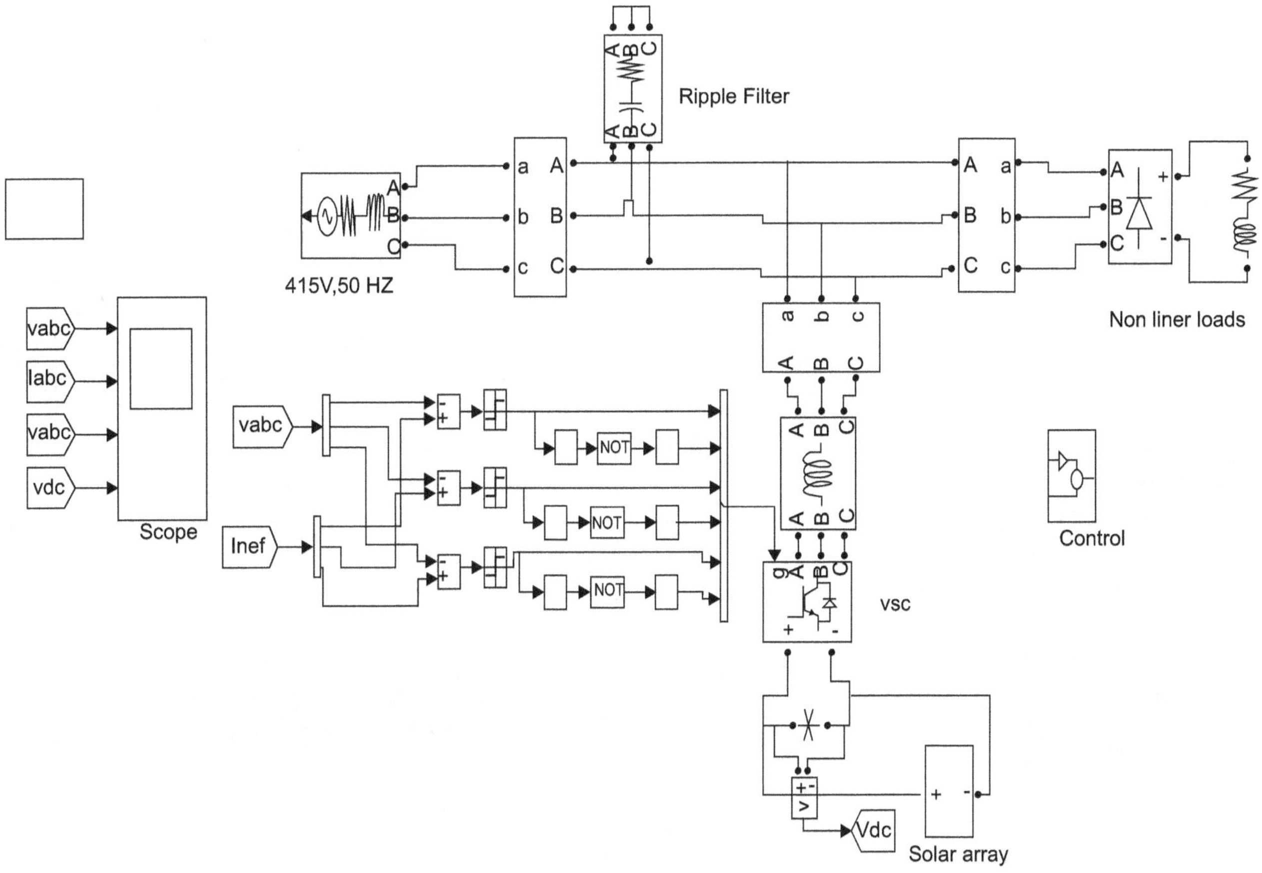

FIGURE 8.18 MATLAB Simulink model of a system configuration.

8.8.1 Simulated Response at Linear Loads

Figure 8.19 and Figure 8.20 present the analysis of the proposed SECS at linear loads. In linear loads, a household load with resistive and inductive circuit is considered. The power factor of the grid side is sustained to unity even at 0.8PF lagging load. Figure 8.20 illustrate the dynamic response, which is realized from 0.4s to 0.5s. Moreover, the dynamic load is realized by removing the one phase of line. The DC bus voltage is regulated at its reference value. The load imbalance is sustained by the quaternion-based adaptive method. The VSC currents are unbalanced to sustain the supply currents balanced and sinusoidal.

8.8.2 Simulated Response at Nonlinear Loads

In the case of nonlinear loads, the power factor at the grid is maintained. Figure 8.21 illustrates the waveform of SECS with nonlinear loads. However, the disconnection of phase a is observed from 0.4s to 0.5s, as presented in Figure 8.22. It is observed that the i_s are balanced, and sinusoidal and DC bus voltage is maintained to actual value. The simulated response has presented the acceptable behavior of SECS at transient load scenarios. Figure 8.23 shows the harmonic spectra of v_{sa}, i_{sa}, and i_{La}, which are 1.72%, 3.54%, and 22.44%, respectively, and it can be observed that grid current THDs are within the IEEE-519 standard.

8.8.3 Simulated Performance at Change in Insolations

The performance of SECS under change in irradiance is presented Figure 8.24. It is observed that under the 500W/m^2 irradiance, the magnitude of utility current is small. However, under increase in PV insolations, the large amount of power is fed to the grid as the extracted energy

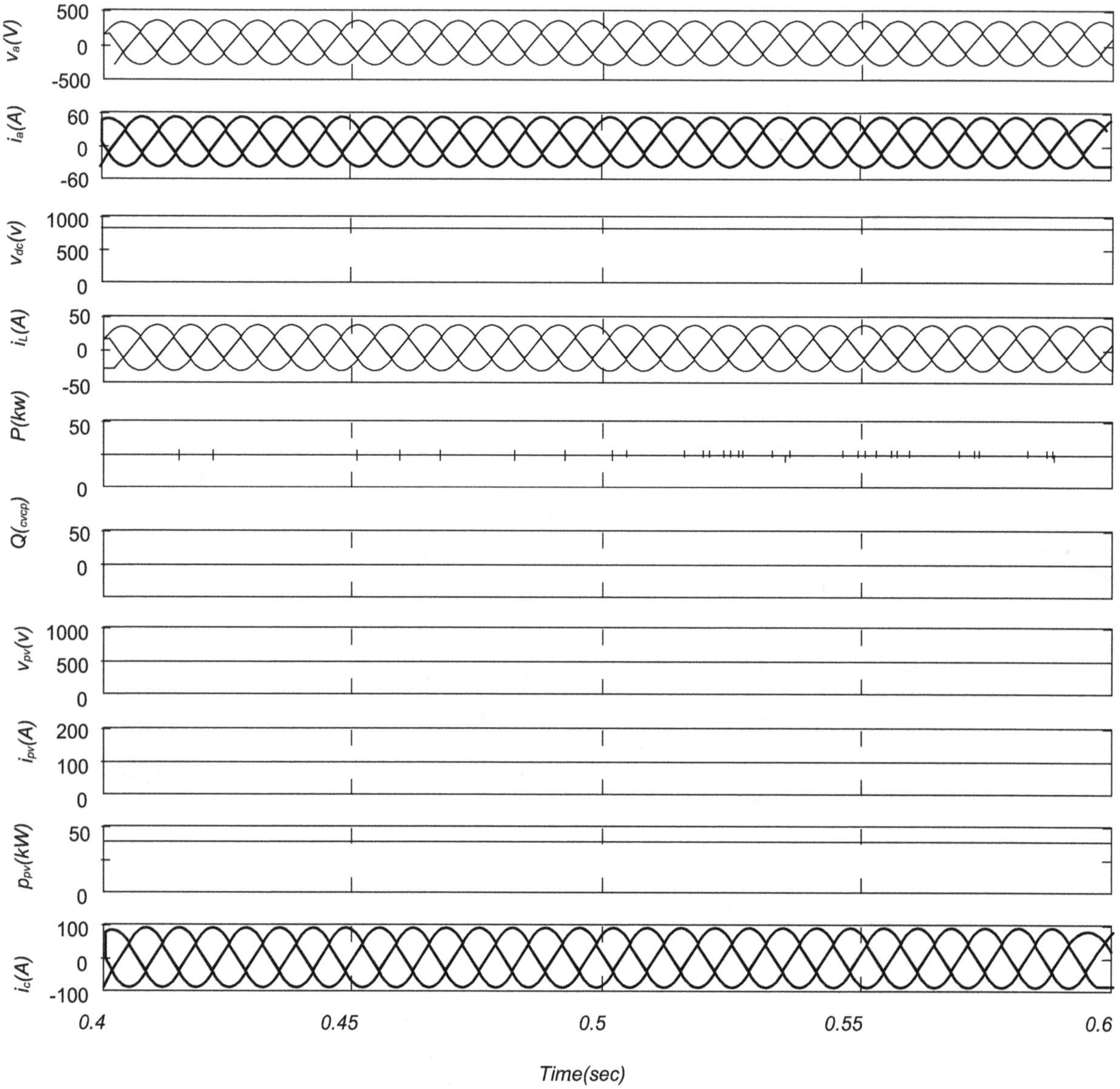

FIGURE 8.19 Simulated steady behavior at linear load.

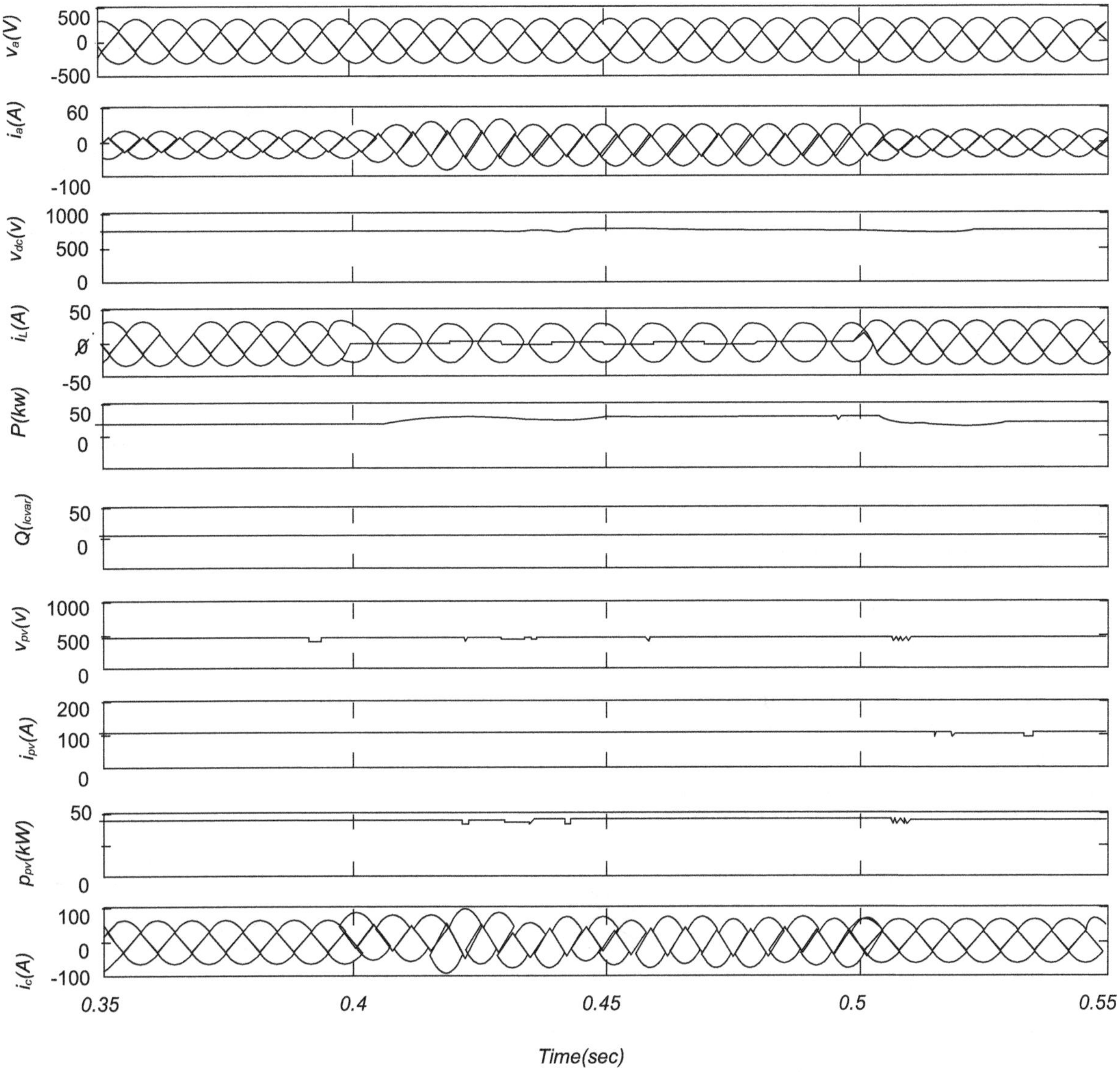

FIGURE 8.20 Dynamic behavior of SECS with linear load.

from the PV array is increased, which is observed in 0.5s to 0.6s. The behavior of the SECS is good and satisfactory at various dynamic scenarios.

8.9 CONCLUSIONS

On the basis of two-stage configuration system and different control algorithms implemented in the laboratory using DSP during the thesis work, the following conclusions may be reached.

1. Grid-interfaced SPV two-stage three-phase systems have been successfully designed, controlled, and implemented in the laboratory.
2. It can be observed during the research work that solar PV systems have been the better option for feeding power to the grid as compared to conventional systems feeding only real power.
3. Performance of "incremental conductance MPPT technique" has been found better than the conventional "perturb and observe MPPT technique" in terms of MPPT tracking efficiency, and it loses the track when step sizes are large.
4. The EMI problems due to high switching operation of the DC/DC boost converter at high-voltage and high-current levels are reduced by the use of a filter capacitor.
5. As closed-loop controls have been used here, it has been observed that this loop prevents the DC link voltage from going beyond its reference value.

8.10 SUGGESTION FOR FURTHER WORK

The objective of this chapter is achieved successfully. However, there is a further scope and new challenges on

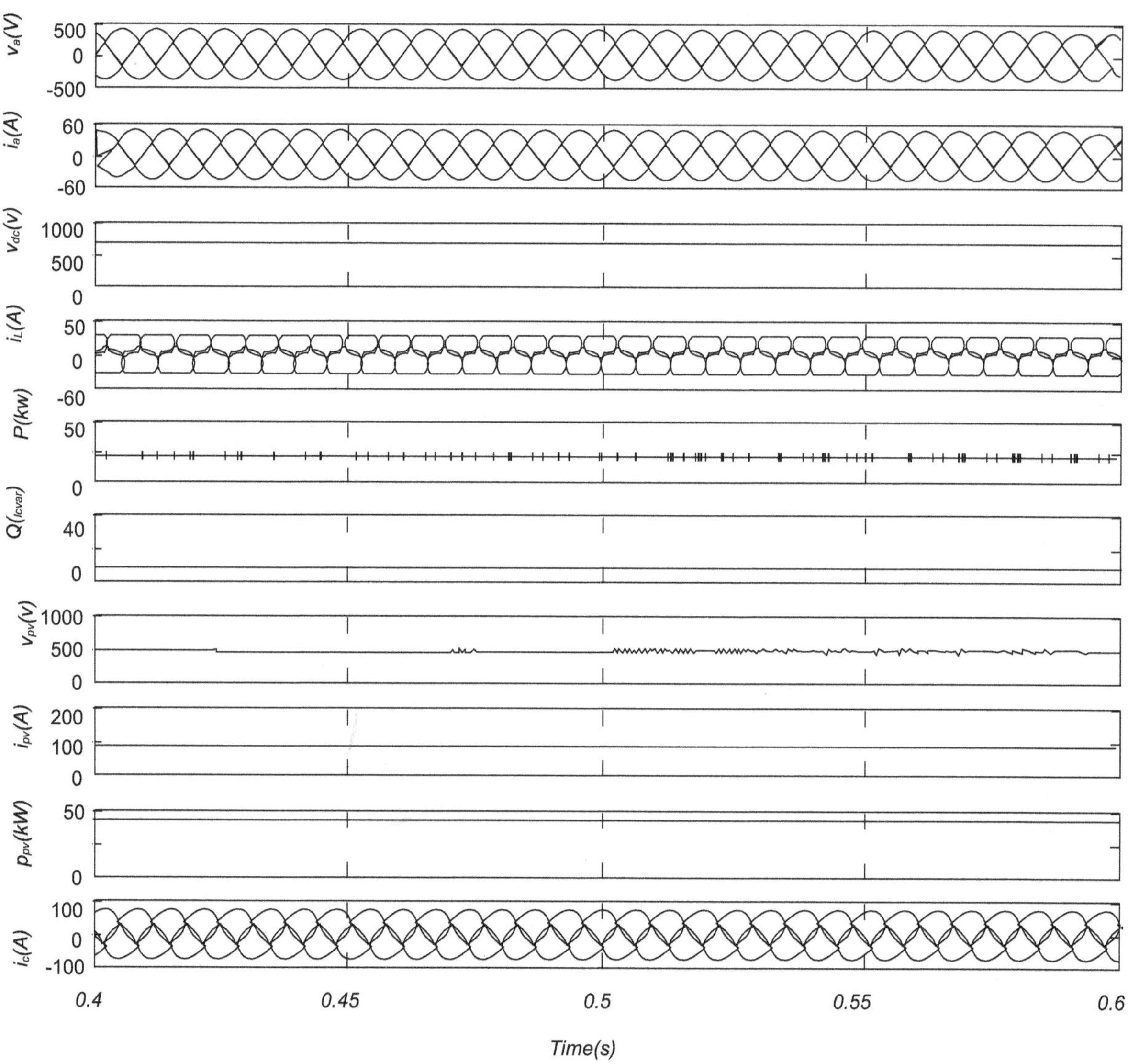

FIGURE 8.21 Behavior of SECS at nonlinear load.

high power solar PV technology. The following topics may be investigate for further study.

1. Use of multi-level converters for solar PV energy conversion applications.
2. Study of EMI problems caused due to operation of DC/DC converters and their mitigation.
3. Study and analysis of anti-islanding techniques for solar PV energy conversion systems.
4. Development of new MPPT techniques for PV array under partial-shading conditions.
5. Use of current-source converters for solar PV applications.
6. Use of space vector modulation for grid-interfaced solar PV energy conversion systems.

8.11 SOME NUMERICAL EXAMPLES

1. In a grid integrated solar PV system, a single-phase diode rectifier based load is drawing AC current at 0.95DPF, THD of AC current is 0.95 and at crest factor of 3. It is drawing 1000W at 230V, 50Hz grid, under zero insolations from the PV array. Estimate: (a) power factor; (b) rms current; and (c) peak current of grid.

Solution

DPF = 0.95, THD = 0.95, P = 1000W, V = 230 V, CF= 3

$DF = 1/\sqrt{(THD^2+1)} = 0.7249$

a. Power factor (PF) = DPF × DF = 0.6887
b. RMS current = P/(V × PF) = 6.3126 A
c. Peak current of AC mains = CF × I_s = 18.937A

2. A single-stage grid connected three-phase SPV system is supplying a linear load of 2kW and 0.5kVAR. The grid is operating at 415V and 50Hz. The SPV operating point is observed to be 450V and 10A at its maximum power point at 1000W/

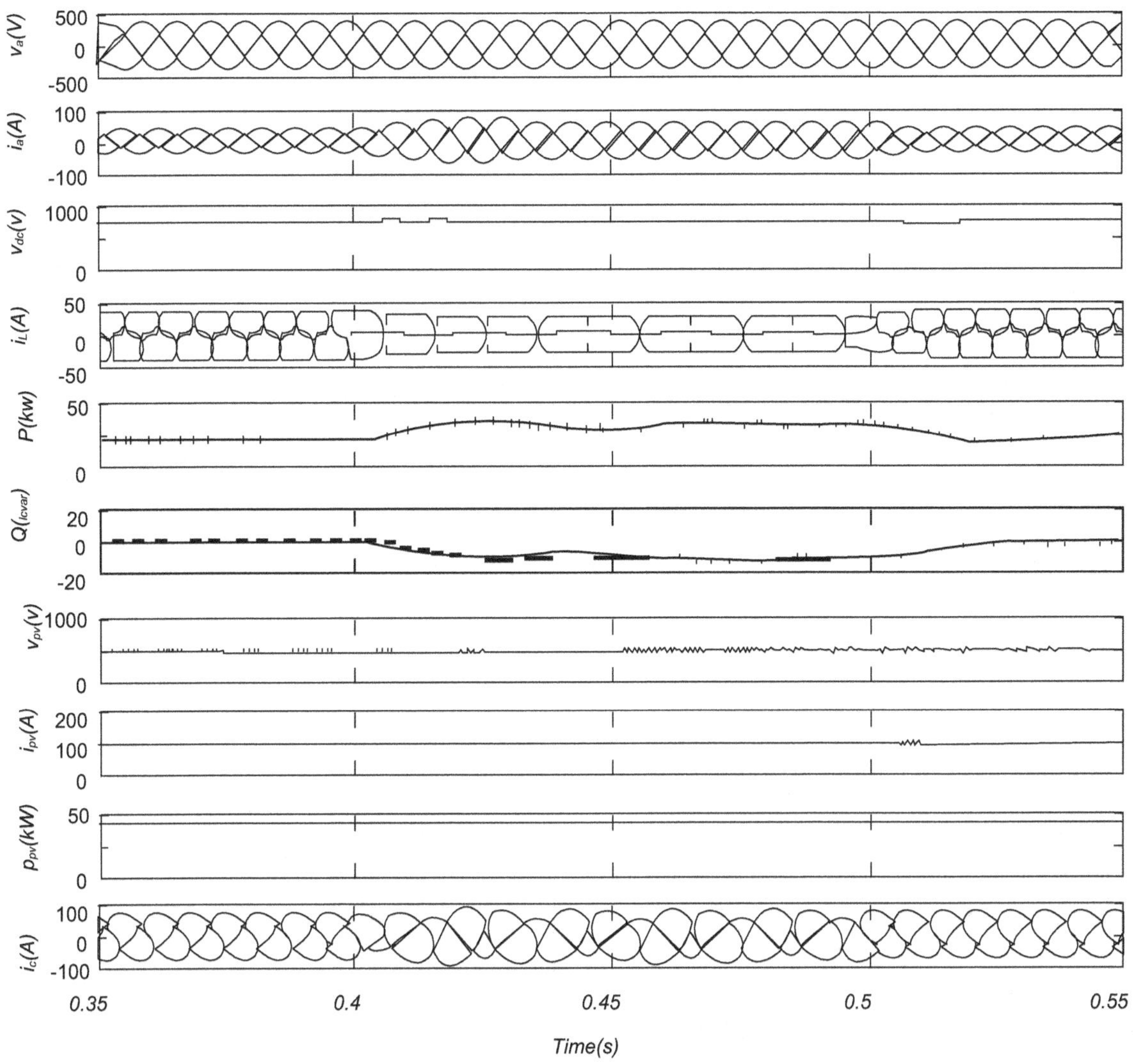

FIGURE 8.22 Behavior of SECS under dynamic conditions at nonlinear load.

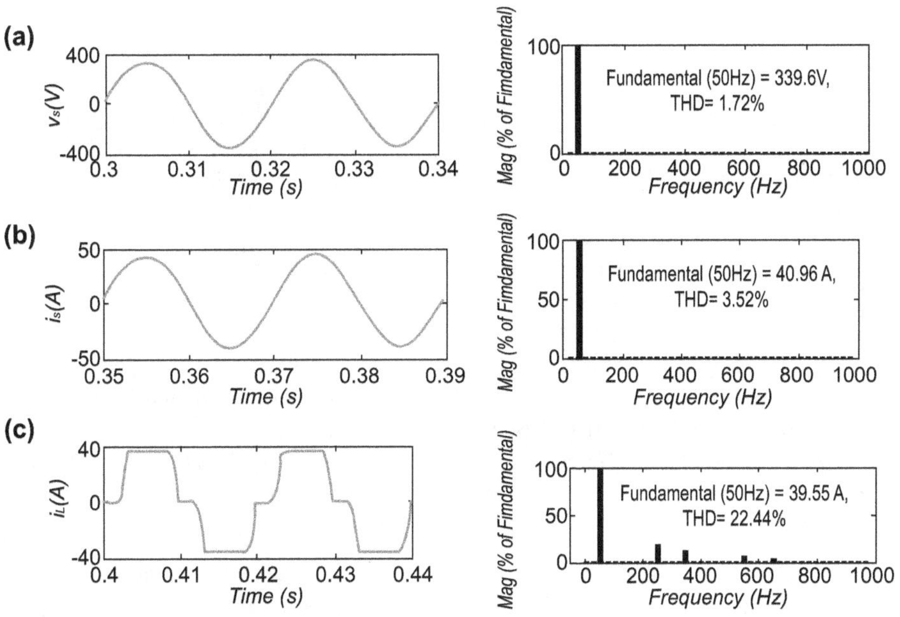

FIGURE 8.23 Total harmonics distortion analysis: (a) v_{sa}; (b) i_{sa}; (c) i_{La}.

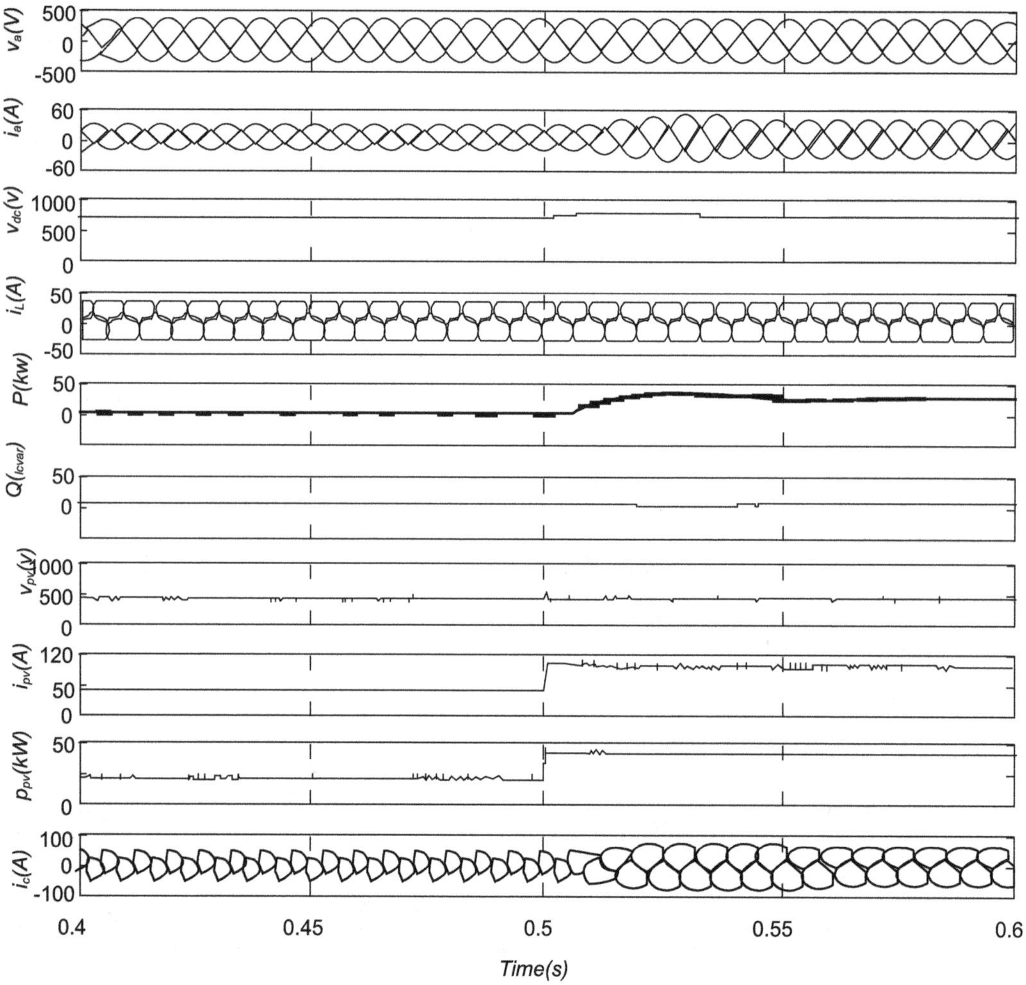

FIGURE 8.24 Simulated response at varying insolations.

m². The MPPT algorithm efficiency of PV is observed to reduce cording to the equation:

$$\eta_x = \frac{1}{1 + c\sqrt{[(1000/x)^4 - 1]}}$$

where c is the efficiency gain parameter holding a value of 0.04

The VSC control scheme ensures that the grid always operates at UPF. The VSC losses can be assumed to be 5% of its power. Estimate: (a) the load current and the grid current under the current MPPT operation (1000W/m²) of the PV array; and (b) the grid power factor and phase shift in VSC phase currents with respect to phase voltages.

Solution

(a) Solar PV is operating at 450V and 10A. Thus the net power supplied by the PV is 4500 or 4.5 Kw.

Considering the inverter losses at 5%, net inverter active power supplied = 0.95 × 4.5 = 4.275kW.

Load active power = 2kW.

Thus, the net power fed to the grid = 2.275kW.

Since the grid voltage is 415V (line to line, RMS), the line current flowing in the load can be computed as:

$$I_{Load,line} = \frac{S(3\phi)}{\sqrt{3}V_{line-line}} = \frac{\sqrt{2^2 + 0.5^2} \times 10^3}{\sqrt{3} \times 415} \approx 2.87A$$

Since the grid voltage is 415 V (line to line, RMS), the line current in the grid can be computed as:

$$I_{Grid,line} = \frac{P(3\phi)}{\sqrt{3}V_{line-line}} = \frac{2.275\times \sim 10^3}{\sqrt{3} \times 415} \approx 3.16A$$

It is noteworthy that only the three-phase active power (P) is considered for calculating the grid line current. This is because the VSC is ensuring the UPF operation always. Otherwise, net apparent power (S) needs to be considered.

(b) Since the VSC control scheme ensures that the grid always operates at UPF, the grid line currents are always in phase with the corresponding voltages.
However, in this case, the energy injected to the grid and power factor is negative. Thus, grid power factor = −1.
Owing to UPF operation of grid, the VSC supplies total load reactive power factor of 0.5kVAR. Since the active power fed through VSC is 4.275kW, the phase shift in VSC phase currents with respect to phase voltages can be computed as:

$$\varphi = \tan^{-1}(Q/P) = \tan^{-1}(0.5/4.275) = 6.67^0$$

3. Compute the load current and the grid current in Question-2 for the solar insolation dip to 600W/m^2 and load increase by 1.5 kW and 1 kVAR. The MPPT algorithm efficiency of PV is observed to reduce with insolation increase according to the equation

$$\eta_x = \frac{1}{1 + c\sqrt{[(1000/x)^4 - 1]}}$$

where c is the efficiency gain parameter holding a value of 0.03

After change in insolation, estimate grid power factor and phase shift in VSC phase currents with respect to phase voltages. The PV array is operating at 450V with solar insolation of 600W/m^2

Solution

Here, solar PV is operating at 450V but not at 10A. Ideally, since the power supplied by the PV array is directly proportional to solar insoltion:

$$P_{600W/m^2}(Ideal) = (600/1000)P_{1000W/m^2} = 0.6 \times 4.5\ kW = 2.7\ kW$$

However, since the efficiency of MPPT algorithm varies with insolation, the net power extracted from the PV array is computed as:

$$\eta_{600W/m^2} = \frac{1}{1 + 0.03\sqrt{((1000/x)^4 - 1)}} \approx\sim 0.93$$

$$P_{600W/m^2}(Net) = (600/1000)P_{1000W/m^2} = 0.93 \times 2.7\ kW = 2.5\ kW$$

Considering the inverter losses as 5%, net inverter active power supplied = 0.95 × 2.5 = 2.375 kW.

The load active power = 2kW + 1.5kW = 3.5 kW.

Thus, the net power fed to the grid is P_g = 2.375–3.5kW = −1.125kW; i.e. active power is being drawn from the grid.

Since the grid voltage is 415V (line to line, RMS), the line current flowing in the load can be computed in this case as:

$$I_{Load,line} = \frac{S(3\phi)}{\sqrt{3}V_{line-line}} = \frac{\sqrt{2^2 + 3.5^2} \times 10^3}{\sqrt{3} \times 415} \approx 5.6A$$

Since the grid voltage is 415V (line to line, RMS), the line current in the grid can be computed as:

$$I_{Load,line} = \frac{S(3\phi)}{\sqrt{3}V_{line-line}} = \frac{\sqrt{2^2 + 3.5^2} \times 10^3}{\sqrt{3} \times 415} \approx 5.6A$$

It is worth noting that only the tree phase active power (P) is considered for calculating the grid line current. This is because the VSC is ensuring the UPF operation always. The whole three-phase apparent power (S) needs to be considered otherwise:

$$I_{Load,line} = \frac{P(3\phi)}{\sqrt{3}V_{line-line}} = \frac{1.125 \times 10^3}{\sqrt{3} \times 415} \approx 1.565A$$

Since the VSC control scheme ensures that the grid always operates at UPF, the grid line currents are always in phase with the corresponding voltages. In this case, the power is being given by the grid, the grid power factor is positive. Thus, grid power factor = +1.

Owing to UPF operation of grid, the VSC supplies total load reactive power of 1.5kVAR. Since the active power fed through VSC is 2.375kW, the phase shift in VSC phase currents with respect to phase voltages can be computed as:

$$\varphi = \tan^{-1}(Q/P) = \tan^{-1}(1.5/2.375) = 32.27^0$$

4. In Q2, if VSC is used for power factor correction to supply a maximum reactive power 0.3kVAR, the solar irradiance is now reduced to 800W/m^2. Estimate the grid current under the MPPT operation (1000/m^2) of the PV array. Also, find the power factor of the grid and phase shift in VSC phase currents with respect to phase voltages. The MPPT algorithm efficiency of PV is observed to

be reduce with insolation increase according to the equation:

$$\eta_x = \frac{1}{1 + c\sqrt{[(1000/x)^4 - 1]}}$$

where c is the efficiency gain parameter holding a value of 0.03

Solution

Here, solar PV is operating at 450V. Since the power supplied by the PV array is directly proportional to solar insolation:

$$P_{600W/m^2}(Ideal) = (800/1000)P_{1000W/m^2} = 0.8 \times 4.5\ kW$$
$$= 3.6\ kW$$

However, since the efficiency of MPPT algorithm varies with insolation, the net power extracted from the PV array is computed as:

$$\eta_{600W/m^2} = \frac{1}{1 + 0.03\sqrt{((1000/800)^4 - 1)}} \approx\sim 0.965$$

$$P_{600W/m^2}(Net) = 0.965 \times 3.6kW \approx 3.47kW$$

Considering the inverter losses as 5%, net inverter active power supplied = 0.95 x 3.47=3.295 kW.

The load active power = 2 kW. Thus, the net power fed to the grid = 3.295–2 kW = 1.295kW.

Since the grid voltage is 415V (line to line, RMS), the line current flowing in the load can be computed as:

$$I_{Load,line} = \frac{S(3\phi)}{\sqrt{3}V_{line-line}} = \frac{\sqrt{2^2 + 1^2} \times 10^3}{\sqrt{3} \times 415} \approx 3.11A$$

The reactive power limitation imposed is 0.3 kVAR; thus, out of 0.5 kVAR load, only 0.3kVAR is supplied by the VSC and the rest of the reactive power is supplied by the grid. In this case, the grid power is not the active power alone, but the reactive power of 0.2kVAR.

Thus, apparent power supplied by the grid = 1.72.

The line current in the grid can thus be computed as:

$$I_{Grid,line} = \frac{S(3\phi)}{\sqrt{3}V_{line-line}} = \frac{1.72 \times 10^3}{\sqrt{3} \times 415} \approx 2.39A$$

The grid is operating at a power factor of:

$$\varphi_{grid} = \tan^{-1}(Q/P) = \tan^{-1}(0.2/1.295) = 8.78^0$$

For VSC:

$$\varphi_{VSC} = \tan^{-1}(Q/P) = \tan^{-1}(0.3/3.295) = 5.2^0$$

Thus, the VSC here operates in power factor correction mode rather than fully nullifying the grid power factor.

8.12 UNSOLVED QUESTIONS

1. Calculate the number of series and parallel strings of a PV array for 5kW, 600V load connected at the output of PV array. The open circuit voltage (Voc) and short circuit current (I_{sc}) of one PV string are given as 32.8V and 8.3A, respectively. Assume a factor of 0.85 of open and short circuit parameters to determine the maximum possible PV array voltage and current at MPPT.
2. What is the necessity of solar energy estimation for solar power generation?
3. What is the need of maximum power point tracker? Present a comparative analysis of conventional and intelligent MPPT methods for solar energy extraction.
4. Discuss the reasons for the low efficiency of solar cells. What methods are used to improve solar cell efficiency?
5. What are the major issues and challenges in the grid integration of PV systems?
6. Write the procedure to implement MPPT technique on two PV panels connected in series with partial shading.
7. How is stable voltage acquired in boost converters for varying insolations and temperatures with load changes?
8. How do the insolations and temperature affect PV array characteristics? Brief about the effect of shadowing on the performance of solar cell and how it can be solved?
9. In Q8, also design the inductor of a boost converter, if the duty ratio is considered as 20% at MPPT with 10kHz switching frequency and 10% input current ripple.
10. Design a 100kW three-phase two-stage solar PV grid-interfaced power generating system.

REFERENCES

[1] F. A. Farret and M. G. Simoes, *Integration of Alternative Sources of Energy*, Hoboken, NJ: John Wiley & Sons, Inc., 2005.

[2] A. Keyhani, M. N. Marwali and M. Dai, *Integration of Green and Renewable Energy in Electric Power Systems*, Hoboken, NJ: John Wiley & Sons, Inc., 2010.

[3] Fang Lin Luo and Hong Ye, *Advanced DC/AC Inverters: Applications in Renewable Energy*, Boca Raton, FL: CRC Press, Taylor & Francis Group, 2013.

[4] U. K. Kalla, B. Singh and S. S. Murthy, "Modified Electronic Load Controller for Constant Frequency Operation with Voltage Regulation of Small Hydro-Driven Single-Phase SEIG," *IEEE Transactions on Industry Applications*, vol. 52, no. 4, pp. 2789–2800, July–Aug. 2016.

[5] R. Carbone, "Grid-Connected Photovoltaic Systems with Energy Storage," in *Proc. of IEEE International Conference on Clean Electrical Power, Capri, Italy,* pp. 760–767, Aug. 2009.

[6] U. K. Kalla, H. Kaushik, B. Singh and S. Kumar, "Adaptive Control of Voltage Source Converter Based Scheme for Power Quality Improved Grid-Interactive Solar PV—Battery System," *IEEE Transactions on Industry Applications*, vol. 56, no. 1, pp. 787–799, Jan.–Feb. 2020.

[7] Shailendra Kumar and Bhim Singh, "A Multipurpose PV System Integrated to Three-Phase Distribution System Using LWDF Based Control Approach," *IEEE Transactions on Power Electron*, vol. 33, no. 1, pp. 739–748, Jan. 2018.

[8] Q. C. Zhong and T. Hornik, *Control of Power Inverters in Renewable Energy and Smart Grid Integration*, Hoboken, NJ: John Wiley & Sons. Inc., 2013.

[9] A. Ghosh and G. Ledwich, *Power Quality Enhancement Using Custom Power Devices*, Delhi: Springer International Edition, 2009.

[10] D. Sera, L. Mathe, T. Kerekes, S. Spataru and R. Teodorescu, "Perturb-and-Observe and Incremental Conductance MPPT Methods for PV Systems," *IEEE Transactions on Photovoltaic*, vol. 3, no. 3, pp. 1070–1078, July 2013.

[11] B. Subudhi and R. Pradhan, "A Comparative Study on Maximum Power Point Tracking Techniques for Photovoltaic Power Systems," *IEEE Transactions on Sustainable Energy*, vol. 4, no. 1, pp. 89, 98, Jan. 2013.

[12] T. Esram and P. L. Chapman, "Comparison of Photovoltaic Array Maximum Power Point Tracking Techniques," *IEEE Transactions on Energy Conversion*, vol. 22, no. 2, pp. 439–449, June 2007.

[13] M. Elgendy, B. Zahawi and D. J. Atkinson, "Assessment of the Incremental Conductance Maximum Power Point Tracking Algorithm," *IEEE Transactions on Sustainable Energy*, vol. 4, no. 1, pp. 108–117, Jan. 2013.

[14] A. Safari and S. Mekhilef, "Simulation and Hardware Implementation of Incremental Conductance MPPT with Direct Control Method Using Cuk Converter," *IEEE Transactions on Industrial Electronics*, vol. 58, no. 8, pp. 1154–1161, Apr. 2011.

[15] R. Teodorescu, M. Liserre and P. Rodríguez, *Grid Converters for Photovoltaic and Wind Power Systems*, Hoboken, NJ: John Wiley & Sons. Inc., 2011.

[16] Bhim Singh, Shailendra Kumar and C. Jain, "Damped-SOGI-Based Control Algorithm for Solar PV Power Generating System," *IEEE Transactions on Industry Applications*, vol. 53, no. 3, pp. 1780–1788, May–June 2017.

[17] Shailendra Kumar and Bhim Singh, "Implementation of High Precision Quadrature Control for Single Stage SECS," *IEEE Transactions on Industrial Informatics*, vol. 13, no. 5, pp. 2726–2734, Oct. 2017.

LIST OF SYMBOLS

I	Output current
I_{ph}	Module photo-current
I_o	Module saturation current
q	Electron charge
V_{pv}	Voltage across PV module
I_{pv}	Current across PV module
R_s	Series resistance of PV cell model
R_p	Parallel resistance of PV cell
Ns	Number of cells in series
N_p	Number of cells in parallel
R_{sh}	Shunt resistance for PV cell
K	Boltzman constant
T	PV module operating temperature in kelvin
a	Ideality factor
Isc	Short-circuit current of PV module
I_{d1}	Current through diode d1
I_{d2}	Current through diode d2
I_{mpp}	PV current at maximum power point
V_{OCT}	Open-circuit PV voltage
V_{mpp}	PV voltage at maximum power point
C_{dc}	DC-link capacitor capacitance
V_{dc}	DC-link voltage
k_v	Voltage temperature coefficient of PV cell
k_i	Current temperature coefficient of PV cell
V_{LL}	AC line voltage L_f inductance of AC interfacing inductor
V_t	Terminal voltage peak value
S_{pa}, S_{pb}, S_{pc}	Unit templates in-phase with the PCC voltages
S_{qa}, S_{qb}, S_{qc}	Unit templates in quadrature with the PCC voltages
Y_{pavg}	Average magnitude of active component
i_{sa}, i_{sb}, i_{sc}	Three-phase grid currents
i_{La}, i_{Lb}, i_{Lc}	Instantaneous three-phase load currents
K_{ploss}	Proportional gain of DC-link voltage PI controller
K_{iloss}	Integral gain of DC-link voltage PI controller
i_{sa}^*, i_{sb}^*, i_{sc}^*	Instantaneous reference grid currents
i_{VSCa}, i_{VSCb}, i_{VSCc}	Instantaneous VSC currents
v_{sa}, v_{sb}, v_{sc}	Instantaneous PCC phase voltages
v_{sab}, v_{sbc}, v_{sca}	Instantaneous PCC line voltages
Y_{loss}	Output of DC-link voltage PI controller
M	Modulation index
ρ	Damping factor
D	Duty cycle of boost-converter
L_b	Boost inductance
f_s	Switching frequency of DC/DC boost converter
ΔI_1	Input current ripple
L_f	Interfacing inductor
H	Overloading factor
Ppv	Active power from the solar PV array
R_f	Ripple filter resistance
C_f	Ripple filter capacitance
v_{dcrip}	Percentage of ripple voltage in DC link voltage
I_k	Sampling instant of current
V_k	Sampling instant of voltage

9 Multifunction Grid-Interactive Solar Photovoltaic Water Pumping System for Water Pumping and Distributed Generation Application

Vamja Rajan Vinodray and Mahmadasraf A. Mulla

CONTENTS

9.1 INTRODUCTION

In developing countries, household electricity and water are two essential needs at remote locations. These requirements are currently satisfied utilizing diesel generators and pumps [1, 2]. Diesel pumps become inoperable due to higher fuel prices. In these times, solar photovoltaic water pumping systems (SPVWPS) become favourable for water pumping at remote regions over conventional diesel pump systems due to advantages like being fuel free, maintenance free and eco-friendly [3, 4].

In reported studies, the standalone SPVWPS is categorized in two configurations based on the power electronics converter (PEC) stages, viz., single-stage standalone SPWPS [5] and two-stage standalone SPWPS [6]. The block diagram of these standalone SPVWPS is as shown in Figure 9.1. In these topologies, PEC 1 is utilized for motor drive. The voltage source inverter (VSI) is used for induction motor (IM), permanent magnet synchronous motor (PMSM) [7], brushless direct current motor (BLDC) [8], and synchronous reluctance motor (SyRM) [9]; whereas a mid-point converter is used for switched reluctance motor (SRM) [10]. The PEC 2 is a DC/DC converter used to control and track the maximum SPV power. In a two-stage system, a step-up DC/DC converter is used for maximum power point tracking (MPPT), and VSI is used to control DC-bus voltage. In a single-stage system, VSI is utilized to serve both the purposes, i.e., DC-link voltage control and MPPT control. In SPWPS, the IM, PMSM, and SyRM are operated by Volt/Hz controls, whereas the SRM and the BLDC motors are controlled with the PWM duty cycle control method.

The Volt/Hz control has the disadvantage that the flux may drift, and as a result, the torque will vary. Therefore, the scalar-controlled SPWPS shows the primitive performance. The advanced speed control schemes like a field-oriented control (FOC) [11], a direct torque control (DTC) [12], sliding mode control (SMC)-based FOC [13], SMC-based DTC [14] and predictive current control [15] are introduced to get better precision and efficiency for SPVWPS. The FOC or DTC implementation adds to control complexity and computational burden. With this, a realization of the system becomes difficult and costly. However, the implementation of these high-end controls improves the system performance.

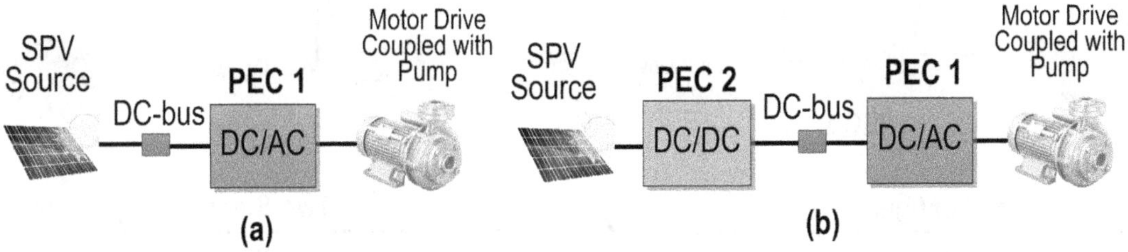

FIGURE 9.1 Block diagram of standalone SPVWPS topologies: (a) single-stage; (b) two-stage.

DOI: 10.1201/9781003229124-9

A field application of standalone SPVWPS is limited due to higher initial cost, intermittent nature of SPV power generation, and significantly high idle periods. The SPVWPS infrastructure remains non-functional or idle for the non-pumping period. In the past few years, the electrical grid supply is interfaced with standalone SPVWPS utilizing extra PEC to alleviate the water delivery fluctuation. Maximum water delivery is possible with these systems. This hybrid pumping system is categorized as grid-assisted SPVWPS, where SPV source and grid both delivers the power to the motor. The next section discusses the details of grid-assisted SPVWPS, along with the limitations of the system.

9.2 PREVIOUS WORK ON GRID-ASSISTED SPVWPS AND MOTIVATION

In recent studies, the grid-assisted SPVWPS are further divided into two categories based on the source interfacing method, viz., AC-bus interfaced SPVWPS [16] and DC-bus interfaced SPVWPS [17, 18]. In AC-bus interfaced SPVWPS, the grid is interfaced directly at AC-bus terminals of the motor drive converter (PEC 1), as shown in Figure 9.2(a). In DC-bus interfaced SPVWPS, the grid supply is interfaced utilizing additional grid synchronize inverter (PEC 3) at the DC-bus terminals of the motor drive converter (PEC 1), as shown in Figure 9.2(b). Both the topologies are capable to operate in standalone and grid-connected mode. The power and the water delivery are variable in the standalone operation. Motor power and water delivery are variable in the standalone mode, and are maintained at rated condition in the grid-connected mode by delivering deficit power from the grid.

Figure 9.3(a) shows the single-stage and the two-stage arrangements for the AC-bus interfaced SPVWPS. In these systems, VSI is utilized as a PV-inverter in the grid-connected mode and as a motor drive in the standalone mode. In DC-bus interfaced SPVWPS topologies, a single-phase or a three-phase grid supply [19] is connected utilizing an additional grid-interfacing PEC (PEC 3), as shown in Figure 9.3(b). The boost-PFC converter [20] or single-phase voltage source converter (VSC) is utilized to interface single-phase grid supply. The Vienna rectifier [21] or three-phase VSC [22] is utilized to interface a three-phase grid supply.

In a two-stage configuration (Figure 9.3[b][i]), SPV source is connected to the DC-bus utilizing the step-up DC/DC converter, which also ensures MPPT of the SPV source. The motor is driven at rated speed using a three-phase VSI. In a single-stage configuration (Figure 9.3[b][ii]), the SPV source is directly connected to DC-bus. The grid-interface converter, i.e., VSC or VR, is utilized for both MPPT control and grid current control [23]. The sharing of the motor power between SPV source and grid supply is the key requirement of the control system in the grid-assisted SPVWPS. Many smart power sharing control algorithms have been reported in the last decade for AC or DC-bus interfaced SPVWPS. These grid-assisted SPVWPS are feasible at the place where grid supply is available.

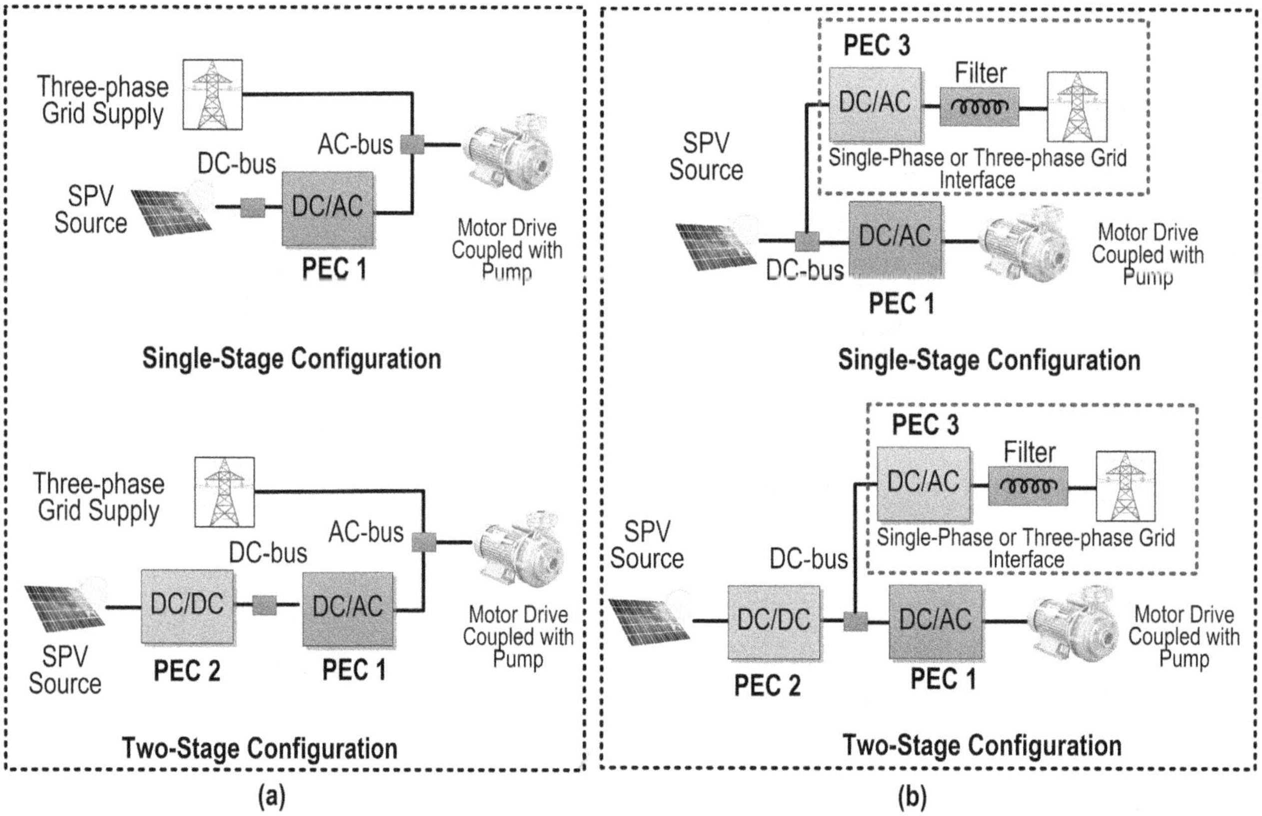

FIGURE 9.2 Block diagram of grid-assisted SPVWPS topologies: (a) AC-bus interfaced system; (b) DC-bus interfaced system.

The comparative features of a diesel fuel-based pumping system, a grid supply-based pumping system, a standalone SPVWPS, and a grid-assisted SPVWPS are summarized in Table 9.1. From the comparison, the following points are focused.

- The standalone SPVWPS is a suitable technology for a remote location, but a higher initial cost; variable water delivery, and an underutilization of the system components, make it unreliable.
- The grid-assisted SPVWPS is a solution for reducing water delivery fluctuation. The addition of PEC for grid-interface further adds to the system cost and control system complexity.
- These SPVWPS topologies remain idle for a long non-pumping period, which underutilizes the capacity of pumping infrastructure like SPV source, PECs, motor drive, etc.

To address the issue of component underutilization, a new topology of grid-interactive SPVWPS is presented by Rajan et al. [24], where two-stage SPVWPS is manifested as a distributed generator. In this system, the two-stage SPVWPS is directly connected with a grid supply. The motor drive inverter is controlled as a single-phase grid synchronous converter (GSC) and the motor winding inductance is utilized as a current filter. The system exhibits multifunctionality by operating in two modes, i.e., standalone SPVWPS and distributed generator. In this system, an extra grid-interface PEC and inductors are not needed for realizing grid-interactive distribution generation operation, and the utilization of motor drive components increases considerably. In the next section, the system configuration of this multifunction grid-interactive SPVWPS is discussed, along with the details of operating modes.

9.3 SYSTEM CONFIGURATION AND OPERATING MODES

The system configuration of the multifunction SPVWPS is shown in Figure 9.3. The two-stage standalone SPVWPS is connected with the single-phase grid supply to construct an innovative distributed generator without the addition of extra power electronics components. The system is with a two PEC, i.e., three-phase VSI, a boost converter, and a star-connected induction motor. Using the contactor, the utility grid is connected between the star terminal of an IM winding and the mid-point of the DC-bus. The SPV source is interfaced at DC-bus using a boost converter.

The IM is operated as pumping drive using the three-phase VSI in a standalone mode. For the non-pumping period. The IM and three-phase VSI are utilized as the grid-interface inductor and the grid synchronize inverter, respectively. The PECs are operated with a suitable control system to establish the required multifunctionality. Five sensor outputs are feedback to the control system and the required switching pulses are generated, as shown in Figure 9.3. Depending on the SPV power availability and pumping requirements, the system operates in two modes: as a standalone SPVWPS and as a grid-interactive distributed generator.

9.3.1 Mode I: Standalone SPVWPS

The circuit connections in standalone SPVWPS mode are shown in Figure 9.4. The contactor is open and disconnects the single-phase grid from the system. An induction motor is driven by modulating the three-phase AC voltages using VSI, where it forces three-phase balanced currents in the induction motor windings. The winding currents are shifted by 120° to each other. It is given by:

$$i_x = \sqrt{2}I\sin\left(wt - \left(\frac{y\pi}{3}\right)\right) \tag{9.1}$$

where, x = {*a,b,c*} and y = {*0,2,4*}, *I* is the magnitude of the motor winding current, and *w* is the angular frequency of the winding current

TABLE 9.1
Comparison of Different Pumping Technologies

	Parameter	Diesel Pump	Electric Pump	Standalone SPVWPS	Grid-Assisted SPVWPS
1.	Initial cost	Medium	Low	High	High
2.	Maintenance cost	High	Medium	Low	Low
3.	Time dependency forpumping operation	No	Yes	No	No
4.	Water delivery	Constant	Constant	Variable	Constant
5.	Eco-friendly	No	Yes	Yes	Yes
6.	Feasibility at remotelocations	Yes	No	Yes	No
7.	Idle non-pumpingperiod	Yes	Yes	Yes	Yes
8.	Underutilization ofsystem components	Yes	Yes	Yes	Yes
9.	Payback period	-	-	High	Low

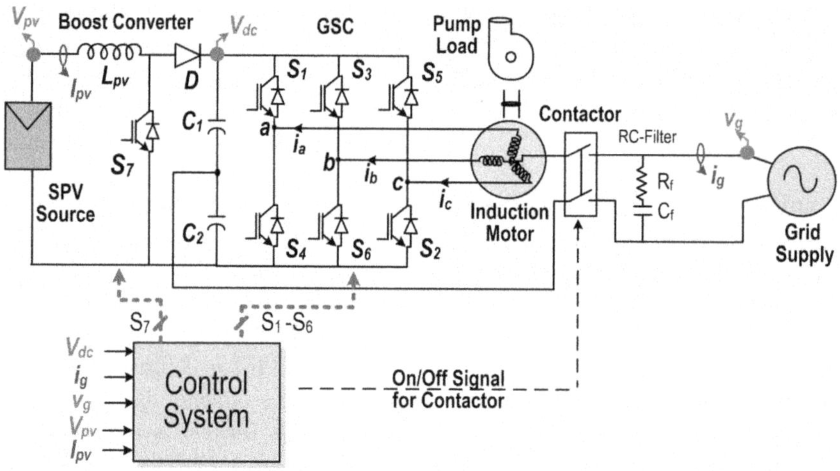

FIGURE 9.3 Block diagram of grid-interactive SPVWPS.

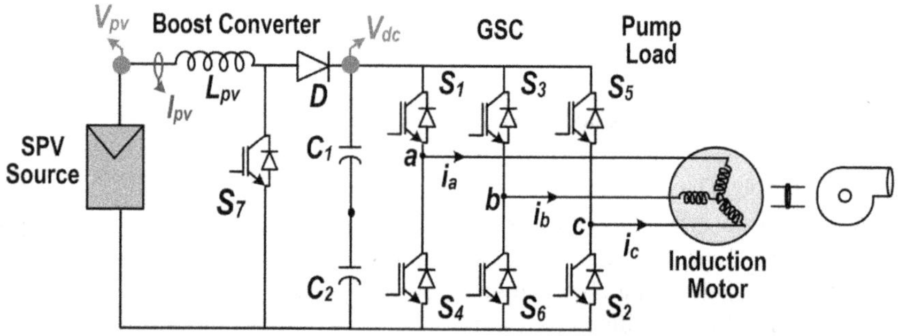

FIGURE 9.4 System connections under standalone SPVWPS operation (Mode I).

These three-phase balanced winding currents produces a rotating magnetic field (RMF) in the induction motor, which is expressed as:

$$\phi_r(t) = \phi_{rm} e^{jwt} \tag{9.2}$$

where, ϕ_{rm} is the peak value of rotating magnetic fluxThe rotating magnetic flux interacts with the rotor circuit and effective torque produces in the machine. It results in electromechanical energy conversion and rotor rotates. It drives the pump coupled with the motor. In this standalone SPVWPS operation, the available peak SPV energy is extracted using a boost converter and fed to DC-bus. The VSI is operated in variable frequency mode, where voltage and frequency of motor are varied to regulate the motor power. With this control, motor power is always matched with SPV generation to achieve energy balance in the system. Hence, motor speed, power and water delivery are variable in this condition. The details of this mode are summarized in Table 9.2.

TABLE 9.2
Operating Modes of the Multifunction Grid-Interactive SPVWPS

Parameters	Mode I	Mode II	
System operates as	Standalone SPVWPS	Distributed generation	
SPV power availability	✓	✓	✗
Irrigation requirement	✓	✗	✗
Status of contactor	OFF	ON	
Inverter functions as	VSI	GSC	
Motor functions as	Pump drive	Filter	
Resultant magnetic flux	$\phi_r(t) = \phi_{rm} e^{jwt}$	Zero	
Net torque	Finite value	Zero	

9.3.2 Mode II: Grid-Interactive Distributed Generator

The system is utilized as a grid-interactive distributed generator in this mode of operation. The single-phase utility grid is connected directly with a system using a contactor. Single-phase GSC operation is achieved by operating the inverter switches using synchronous sine pulse width modulations (SySPWM). In SySPWM, all the upper switches are operated simultaneously with a sinusoidal varying duty ratio, whereas the lower switches are operated in a complement. With this SySPWM switch control, a VSI equivalently operates as a single-phase GSC, and the overall system functions as a distributed generator, as shown in

Figure 9.5. Each motor winding carries the cophasal currents, which is expressed as:

$$i_x = \sqrt{2}I\sin(wt - \delta) \quad (9.3)$$

where, x={a,b,c}, and δ is the power factor angle

The cophasal winding currents are the zero-sequence currents as shown in Eq. (9.3). The resultant alternating magnetic field is produced due to these zero-sequence currents in the induction motor. There is no electromechanical energy transfer from the stator to the rotor, and net torque in a machine is zero. As a result, the rotor remains standstill during distributed generation operation. The motor winding and the three-phase VSI operate equivalently as filter inductor and single-phase mid-point inverter, respectively. The details of distributed generator mode are summarized in Table 9.2. The system design equations and calculations are presented in the next section by considering the exemplary data.

9.4 SYSTEM DESIGN

The stepwise system design procedure is given in this section by considering the data of 10 HP pump. Also, the described procedure is applicable for other rating system. The value of different system parameters are derived as follows:

Step 1: *Motor Selection*

For an exemplary system design, an induction motor with power 10 HP, voltage 400 V, full load speed 1430 rpm, and four poles is considered to drive a centrifugal pump of compatible rating.

Step 2: *Pump Load Design*

The power of pump (P_{pump}) is in cubic relation with speed [25]. It is given by:

$$P_{pump} = K_{pump} \times w_r^3 \quad (9.4)$$

where, K_{pump} is the pump constant, P_{pump} is the power of coupled system, and w_r is the full load angular speed of the pump drive

The pump constant is calculated as:

$$K_{pump} = \frac{P_{pump}}{w_r^3} = \frac{7500}{(149.67)^3} = 2.23 \times 10^{-3} W.s^3 / rad^3 \quad (9.5)$$

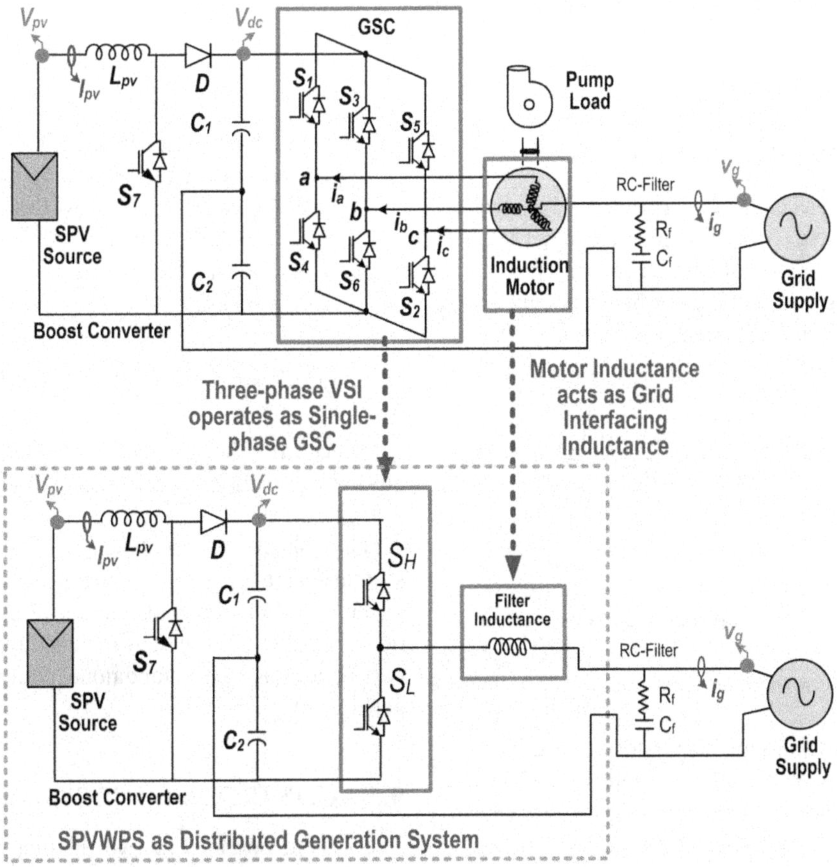

FIGURE 9.5 System connections under multifunction grid-interactive distributed generation operation (Mode II).

The derived value of pump constant is used to evaluate the feed forward reference frequency for the control system in standalone operation.

Step 3: ***SPV Source***

An SPV source with a peak power of 7.68 kW is designed to supply power to a 10 HP pump drive. A solar photovoltaic module produced by Waarre Energies Ltd. is selected with a maximum power of 160 W_p. The 160 W_p module is with 4.57 A and 35 V of peak current peak voltage at 1000 W/m^2 solar radiation and 25°. These 160 W_p modules are connected in series and parallel combinations to get the power matching to the pump drive. A combination of 12 modules in series and such four parallel circuits are formed to provide a peak power of 7.68 kW. The equivalent peak voltage is 420 V (V_{mpp}) and the peak current is 18.28 A (I_{mpp}) at standard test condition, 1,000 W/m^2 solar radiation and 25°.

Step 4: ***DC-Bus Voltage***

In the motoring mode, the VSI switches are controlled using SPWM control. With this reference, the required DC-bus voltage is calculated as [26]:

$$V_{dc} = \frac{2\sqrt{2}}{\sqrt{3}} \times V_{LL} = \frac{2\sqrt{2}}{\sqrt{3}} \times (400) = 653V \quad (9.6)$$

where, V_{LL} is the motor line voltage

Step 5: ***Selection of DC-Bus Capacitor***

The value of DC-bus capacitors are selected by considering the energy storage capability under transient conditions. The DC-bus capacitors provide sufficient energy when the disturbance occurs due to a change in sun radiation or pump load. It is calculated as [20]:

$$C_{dc} = \frac{6\alpha_0 V_{ph} I_{ph} \Delta t}{V_{dc}^2 - V_{dc1}^2} = \frac{6 \times 1.2 \times 230 \times 13.7 \times 0.004}{[650^2 - 620^2]} = 2380uF \quad (9.7)$$

where, α_o is an overloading factor, V_{dc1} denotes lower critical DC-bus voltage under transient variations, I_{ph} is the motor phase current, and Δt is the time duration for reaching DC-bus voltage to lower critical value

From the standard available capacitors, two capacitor of 4,700 uF value are selected for DC-bus, which have the series equivalent capacitance value (2,350 uF) is near to the derived in Eq. (9.7).

Step 6: ***Boost Converter Design***

The boost converter duty cycle at V_{mpp} is obtained as:

$$D = \frac{V_{dc} - V_{mpp}}{V_{dc}} = \frac{650 - 420}{650} = 0.35 \quad (9.8)$$

The inductor value is calculated as,

$$L_{pv} = \frac{V_{mpp} D}{\Delta I \times F_{sw}} = \frac{420 \times 0.35}{1 \times 25000} = 5.94mH \quad (9.9)$$

where, ΔI denotes the ripple current, and F_{sw} denotes the switching frequency

Step 7: *Design of Ripple Filter*

A first-order RC circuit is utilized to filter out the higher-order voltage harmonics, which is tuned at the half of F_{sw}. It is expressed as:

$$R_f \times C_f <= \frac{1}{2 \times f_{sw}} \quad (9.10)$$

R_f of 5W is selected, and F_{sw} is considered as 10 kHz. Then, C_f is calculated using Eq. (9.10) as:

$$5 \times C_f <= \frac{1}{2 \times 10000} \Rightarrow C_f \approx 10uF \quad (9.11)$$

Fundamental inverter voltage is generated as 230 V and the impedance offered by the RC filter circuit at 50 Hz is 637W. Due to this, the fundamental current of 0.361 A is drained from the supply. The power loss in the RC filter due to this current is calculated as:

$$P_{RC} = I_{RC}^2 \times R_f = 0.361^2 \times 5 \Rightarrow P_{RC} = 0.65W \quad (9.12)$$

Hence, the filter register is selected as 5 W, 5W, which has higher power dissipating capacity than calculated in Eq. (9.12).

The solved and unsolved practice questions related to system design are given in the appendix. The control system of the multifunction grid-interactive SPVWPS is discussed in the next section, which includes the control system for two operations, i.e., standalone pumping and distributed generation.

9.5 CONTROL OF THE SYSTEM

The system operates in two distinct operations as standalone SPVWPS and distributed generation source. The control systems required for these operations are also different, which is explained next in two parts.

9.5.1 Control Scheme for Standalone SPVWPS Mode

In standalone operation, the SPV power is variable due to the atmospheric conditions; hence, the motor speed control is requisite to adjust the motor power equal to SPV power generation.

The control scheme for standalone water pumping operation is as shown in Figure 9.6(a). To achieve speed control in this condition, a sinusoidal modulation technique is considered which allows the frequency variation and voltage magnitude variation simultaneously. The reference frequency of modulating scheme (f^*) consists of two components, viz., a DC-bus PI regulator output (f_{dc}^*) and an estimated feed forward frequency (f_{pv}^*). f_{pv}^*, and is derived as:

$$f_{pv}^*[n] = \left[\frac{P_{pv}[n]}{K_{pump}}\right]^{\frac{1}{3}} \frac{p}{4\pi} = KP_{pv}[n]^{\frac{1}{3}} \tag{9.13}$$

where, p is the number of pole, K_{pump} denotes the pump constant, and K is the proportionality constant derived in Eq. (9.13)

This feed forward frequency (f_{pv}^*) is added to the voltage regulator output. The overall scheme is given by:

$$V_{dcer}[n] = (V_{dc}^*[n] - V_{dc}[n]) \tag{9.14}$$

$$f_{dc}^*[n] = f_{dc}^*[n-1] + k_p(V_{dcer}[n] - V_{dcer}[n-1]) + k_i T_s V_{dcer}[n] \tag{9.15}$$

$$f^*[n] = f_{pv}^*[n] - f_{dc}^*[n] \tag{9.16}$$

where, V_{dc} is DC-bus voltage, V_{dcer} is the error of DC-bus voltages, V_{dc}^* is reference DC-bus voltage, k_p and k_i are the voltage regulator gains and T_s denotes sample time of DC-bus control loop.

The two inputs of the modulation technique, modulation index and reference frequency, are calculated using f^* value derived from Eq. (9.16). The modulation index is the ratio of fundamental component of VSI output line voltage (V_{LL1}) to input DC-bus voltage (V_{dc}). For SPWM technique, it is expressed as:

$$M_a = \frac{f^*}{f_{rated}} = \frac{2\sqrt{2}V_{LL1}}{\sqrt{3}V_{dc}}, 0 < M_a < 1 \tag{9.17}$$

where, f_{rated} is the rated frequency

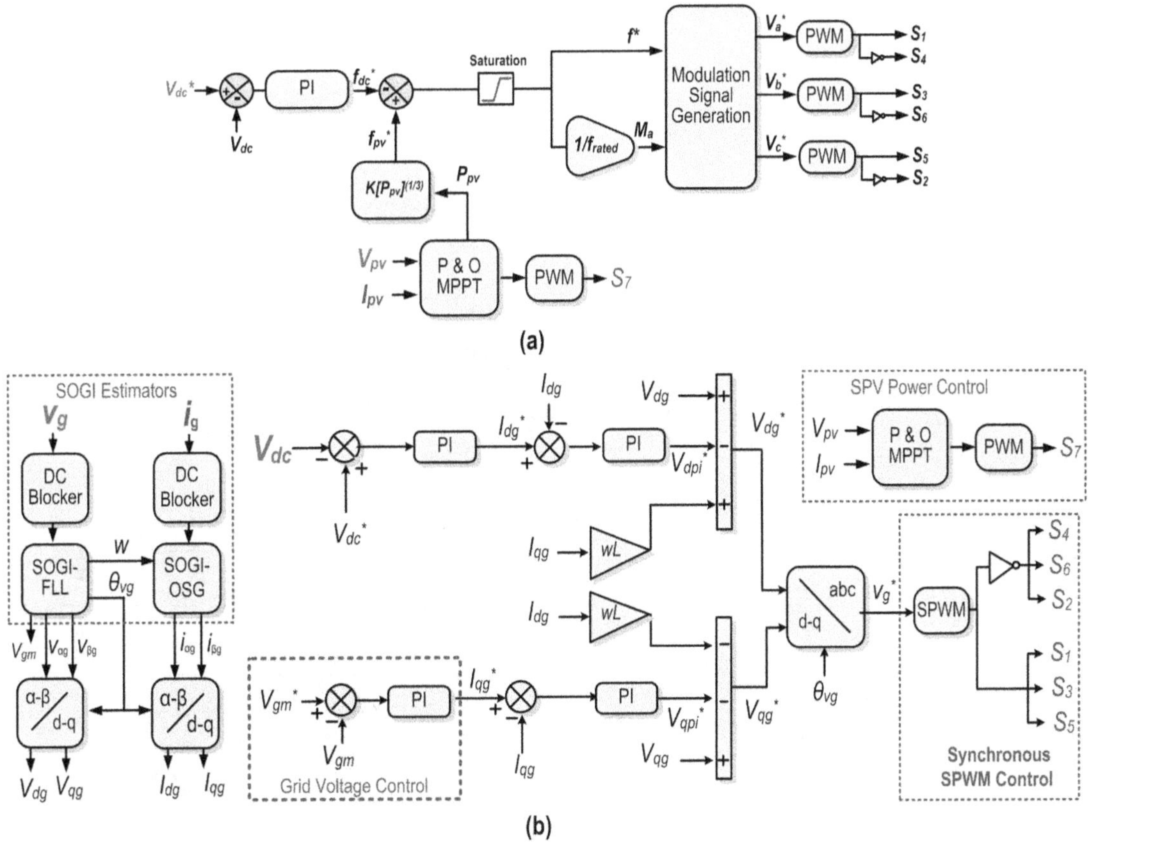

FIGURE 9.6 Control scheme of multifunction grid-interactive SPVPWS: (a) water pumping mode; (b) distributed generator mode.

The time angle for modulating signal is derived using a reference frequency (f^*) as:

$$\theta_m = \int 2\pi f^* dt = \int w^* dt = w^* t \tag{9.18}$$

Using Eqs. (9.17) and (9.18), the modulating signals are expressed as:

$$v_a^* = M_a \sin(\theta_m) \tag{9.19}$$

$$v_b^* = M_a \sin(\theta_m - \frac{2\pi}{3}) \tag{9.20}$$

$$v_c^* = M_a \sin(\theta_m - \frac{4\pi}{3}) \tag{9.21}$$

The PWM pulses are generated by comparing these sinusoidal reference signals, v_a^*,v_b^*, and v_c^*, with a high-frequency carrier signal. The VSI switches are driven with derived PWM pulses. To achieve soft starting, f^* and M_a are increased gradually at starting, which increases motor voltages and current gradually. In the steady-state, motor currents are equal in magnitude and phase-shifted by 120° with each other. Such currents produce the RMF in the induction motor. This establishes the net torque in the induction motor. The motor speed and power are updated according to solar power generation, and at the same time, DC-bus voltage is maintained to the desired fixed value.

9.5.2 Control Scheme for Distribution Generation Mode

Figure 9.6(b) shows the overall control scheme for distributed generator operation. Voltage-oriented control (VOC) is utilized to obtain the decoupled power control. The modified second-order generalized integrator (SOGI) is used to estimate grid angle. The optimum utilization of the SPV source is ensured by the implementation of MPPT control. The MPPT scheme, SOGI estimators, and VOC are explained next.

9.5.2.1 Maximum Power Point Tracking Scheme

The perturb and observed (P&O) and incremental conductance (INC) MPPT algorithms are very popular for photovoltaic systems. P&O is simple in implementation and also tracks the peak SPV power effectively. In P&O scheme, the value of duty cycle (D) is derived based on the sign of the slope $\Delta P_{pv}/\Delta V_{pv}$, as shown in Figure 9.7(a). In the flowchart, ΔP_{pv} shows SPV power change and ΔV_{pv} shows the SPV voltage change. Depending on the sign of slop $\Delta P_{pv}/\Delta V_{pv}$, the value of duty cycle is altered by δD, which is expressed as:

$$D^* = \begin{cases} D_{prev}^* + \delta D & \text{For } \Delta P_{pv}/\Delta V_{pv} > 0 \\ D_{prev}^* - \delta D & \text{For } \Delta P_{pv}/\Delta V_{pv} < 0 \end{cases} \tag{9.22}$$

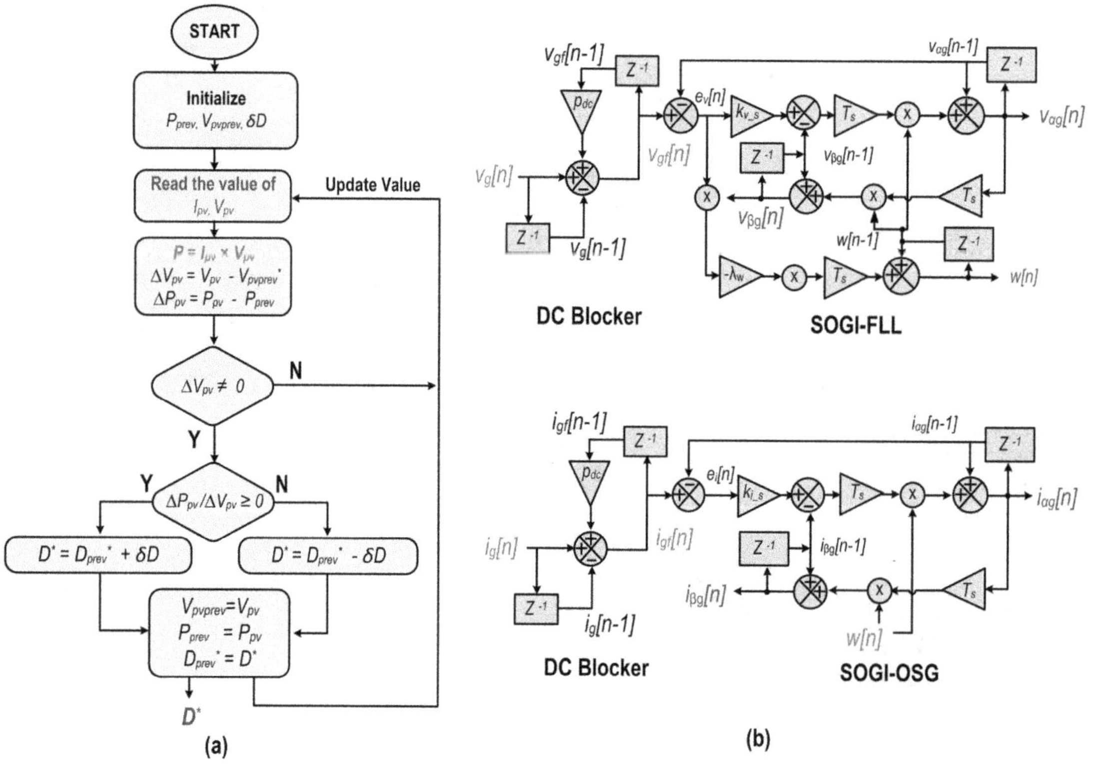

FIGURE 9.7 Flow diagrams: (a) perturb and observe peak power tracking scheme; (b) second-order generalized integrators.

where, the D_{prev}^* is the previous value of duty, and δD is an incremental value.

The PWM pulse is generated by comparing D^* with the high-frequency carrier signal, which drives the boost converter switch.

9.5.2.2 Modified SOGI Estimator

The conventional SOGI algorithm cannot process the AC signals containing DC offset [27]. To overcome this limitation, the modified SOGI is given by cascading digital DC blocker and conventional SOGI. The digital DC blocker is a cascaded form of the digital differentiator and integrator. The discrete realization of this filter is given by [28]:

$$x_{gf}[n] = x_{gf}[n-1]p_{dc} + x_g[n] - x_g[n-1] \quad (9.23)$$

where, $x = \{v, i\}$, x_g and x_{gf} indicate the sensed variable and the filtered variable, respectively, and p_{dc} is the filter constant

In Eq. (9.23), a zero is added to remove the effect of DC offset from the input signal. The pole is to cancel the differentiation effect. The location of the pole ($z = p_{dc}$) allocates a trade-off between bandwidth and the transient response. The value of p_{dc} is selected as 0.998 to get minimum magnitude attenuation and better DC-filtering in of applied AC signal [28].

The block diagram of the SOGI estimators is shown in Figure 9.7 (b). An orthogonal components of grid parameters (x_{ag}, x_{bg}) are estimated from x_{gf} as:

$$e_x[n] = x_{gf}[n] - x_{\alpha g}[n] \quad (9.24)$$

$$x_{\alpha g}[n] = x_{\alpha g}[n-1] + (((e_x[n]k_{x\_s}) - x_{\beta g}[n])w[n-1]T_s) \quad (9.25)$$

$$x_{\beta g}[n] = x_{\beta g}[n-1] + (x_{\alpha g}[n]w[n-1]T_s) \quad (9.26)$$

where, $x = \{v, i\}$, e_x shows the error in x_{gf} and x_{ag}, w is the supply frequency, T_s is a sampling time, k_{x_s} is the gain of the SOGI estimators. The choice of k_{x_s} affects the filtering of the signal

For stable response, value of k_{x_s} is selected between 1 to 1.4 [29, 30].

In Eqs. (9.25) and (9.26), estimation of the orthogonal components depends on frequency input (w), which varies with time. Therefore, estimation of input signal frequency is important part to achieve effective orthogonal signal generation. In present implementation, w is derived by implementing the frequency look loop (FLL) in the voltage SOGI structure as shown in Figure 9.7(b). The voltage error (e_v) and estimated v_{bg} are utilized to achieve FLL as follows:

$$w[n] = w[n-1] - (\lambda_w x_{\beta g}[n]e_v[n]T_s) \quad (9.27)$$

where, the parameter l_w is gain of the FLL

The peak value of grid voltage and grid angle (q_{vg}) are calculated using Eqs. (9.25) and (9.26) as:

$$V_{gm}[n] = \sqrt{(v_{\alpha g}[n])^2 + (v_{\beta g}[n])^2} \quad (9.28)$$

$$\theta_{vg}[n] = \tan^{-1}\left(\frac{v_{\beta g}[n]}{v_{\alpha g}[n]}\right) \quad (9.29)$$

The q_{vg} is required for reference frame transformation. V_{gm} is required to monitor the grid voltage at PCC. VOC is discussed in the next section.

9.5.2.3 Voltage-Oriented Control

A block diagram of VOC is as shown in Figure 9.6(b). First, the stationary orthogonal grid voltage components (v_{ag}, v_{bg}) and the grid current components (i_{ag}, i_{bg}) are transformed using a park transformation into (V_{dg}, V_{qg}) and (I_{dg}, I_{qg}), respectively, given by:

$$\begin{bmatrix} X_{dg}[n] \\ X_{qg}[n] \end{bmatrix} = \begin{bmatrix} \sin(\theta_{vg}[n]) & -\cos(\theta_{vg}[n]) \\ \cos(\theta_{vg}[n]) & \sin(\theta_{vg}[n]) \end{bmatrix} = \begin{bmatrix} x_{\alpha g}[n] \\ x_{\beta g}[n] \end{bmatrix} \quad (9.30)$$

where, $x = \{v, i\}$, and $X = \{V, I\}$

As shown in Figure 9.6(b), outer control loop consists of two PI regulators, which are to control the DC-bus voltage and the grid voltage. The reference dq-axis currents, I_{dg}^*, and I_{qg}^*, are estimated from the DC-bus voltage regulator and the grid voltage regulator, respectively. The outer control loop regulators are expressed as:

$$I_{xg}^*[n] = I_{xg}^*[n-1] + k_{po}(V_{xer}[n] - V_{xer}[n-1]) + k_{io}T_{s\_dc}V_{xer}[n] \quad (9.31)$$

where, $x = \{d, q\}$, k_{po}, k_{io} and T_{s_dc} are the proportional gain, integral gain and sampling time of outer control loop regulators, respectively, and V_{xer} represents the voltage error in outer loops

I_{dg}* and I_{qg}* from Eq. (9.31) are compared with I_{dg} and I_{qg}, respectively. These current errors are processed with two PI controllers, respectively. It is given by:

$$V_{xpi}^*[n] = V_{xpi}^*[n-1] + k_{pi}(I_{xger}[n] - I_{xger}[n-1]) + k_{ii}T_sI_{xger}[n] \quad (9.32)$$

where, $x = \{d, q\}$. k_{pi}, k_{ii}, and T_s are the proportional gain and integral gain, and the sampling time of inner control loop regulators, I_{xger} presents the current error in inner control loops

The reference inverter voltages in a dq reference frame (V_{dg}^*, V_{qg}^*) are calculated by considering the internal decoupling of system. It is given by:

$$V_{dg}^*[n] = V_{dg}[n] + V_{dpi}^*[n] + I_{dg}[n]w[n]L \quad (9.33)$$

$$V_{qg}^*[n] = V_{qg}[n] + V_{qpi}^*[n] + I_{qg}[n]w[n]L \quad (9.34)$$

V_{dg}^*, V_{qg}^* are transformed in to a stationary reference frame voltage (v_g^*) as:

$$v_g^*[n] = V_{dg}^*[n]\sin(\theta_{vg}[n]) + V_{qg}^*[n]\cos(\theta_{vg}[n]) \quad (9.35)$$

where, v_g^* is the reference inverter voltage

The synchronous PWM pulses are generated by comparing v_g^* with the high-frequency carrier signal. All the upper switches of the inverter function simultaneously. The lower switches are operated in complement to respective upper switches as shown in Figure 9.6(b). With this synchronous modulation, three-phase VSI functions as equivalent to a single-phase GSC. Each motor winding carries the cophasal currents, which are one-third of grid current (*ig*), i.e., $i_x = ig/3$, $x = \{a,b,c\}$. This achieves the torque-free grid-interactive distributed generator operation of the system.

9.6 EXPERIMENTAL RESULTS AND DISCUSSIONS

The experiment prototype is prepared as shown in Figure 9.8. The system consists of the SPV source of 640 W_p power, the 1 kW boost converter, the general-purpose 10 kVA high power inverter module, induction motor and single-phase grid supply. The transfer switches are utilized to connect the SPV source and grid supply. The one phase of the three-phase autotransformer is utilized as a single-phase grid supply. A hall effect or an inductive effect sensor cards are utilized to measure required AC and DC electrical quantities.

The inverter module saves the major prototype development time because it consists of an on-board combination of many required circuits like a three-phase IGBT power card, an Infineon integrated circuit (IC)-based smart driver circuit, a hall effect sensor to measure the DC quantities, and inductive effect sensors to measure AC side electrical quantities. Similarly, the boost converter module is made up of silicon carbide switches, SiC switch driver cards and hall effect sensor cards.

A four-channel digital signal oscilloscope (DSO) is utilized to record the test results. The photograph of the experimental setup is as shown in Figure 9.8(b). These signals from the sensor cards are feedback to digital to analogue converter (ADC) pins of 32-bit microcontroller (STM32F407VG). The detailed rating of components used in the setup and details of the control system parameters are given in Table 9.3. The system is tested in standalone SPVWPS operation and grid-interactive distributed generator operation, and corresponding results are discussed.

In the standalone pumping operation, the single-phase grid is disconnected using transfer switch-2 and motor power speed is regulated corresponding to peak solar power generation. The system performance for the standalone pumping operation is as shown in Figure 9.9. According to peak power condition, the SPV voltage is 120 V. The SPV current is 2.4 A. The corresponding solar power is 288 W. This power is less than the rated system condition. Hence, the speed of the induction motor is adjusted to 1050 rpm with control action and the DC-bus voltage is regulated at 300 V.

The balanced three-phase winding currents are observed, which are equal in magnitude and 120° phase-shifted with each other. This evidences the creation of RMF in the induction motor, and corresponding torque is generated in the induction motor. It drives a coupled pump at 1050 rpm.

The single-phase grid is connected using transfer switch-2 and the system is operated in grid-interactive distributed generator operation. The performances for implemented schemes—i.e., SOGI estimators, MPPT scheme, decoupled power control are analyzed. Figure 9.10 (a) shows the steady-state performance of the system for grid angle and frequency estimation using the SOGI scheme. The single-phase grid voltage is set at 80 V. Figure 9.10(a) demonstrates the grid voltage (v_g) and estimated α-b voltages. An estimated q_{vg} varies from 0–2p, and the estimated *f* is 50 Hz is depicted.

Figure 9.10(b) shows the system performance during the real power transfer to the grid. The grid current is phase-shifted by 180° to the grid voltage. This shows that the direction of power transfer is from constructed distributed generator to grid. In this condition, the DC-bus voltage is maintained at 300 V by outer loop control action. The motor winding currents and grid voltage angle (q_{vg}), are as shown in Figure 9.10(b). The winding currents are cophasal and the magnitude is one-third of the grid current. The i_g is equally shared between all the motor winding. It confirms the effective synchronous parallel operation of three inverter legs. There is no electromechanical energy conversion in an induction motor. Thus, the motor does not rotate and the motor winding just acts as a grid current filter.

Figure 9.10(c) shows the v_g, i_g, $v_{\alpha g}^*$ and V_{dc} when the leading VAr is exchanged between a grid and constructed distributed generator. The grid current leads to the grid voltage by 90°. In the second figure of Figure 9.10(c), the grid current lags to the grid voltage by 90° for the lagging VAr exchange condition under distributed generation operation. Under both reactive power exchange conditions, the DC-bus voltage is unaltered at 300 V.

Figure 9.11(a) shows the system performance when the SPV power is varied with solar radiation. i_g, I_{qg}, and V_{gm}

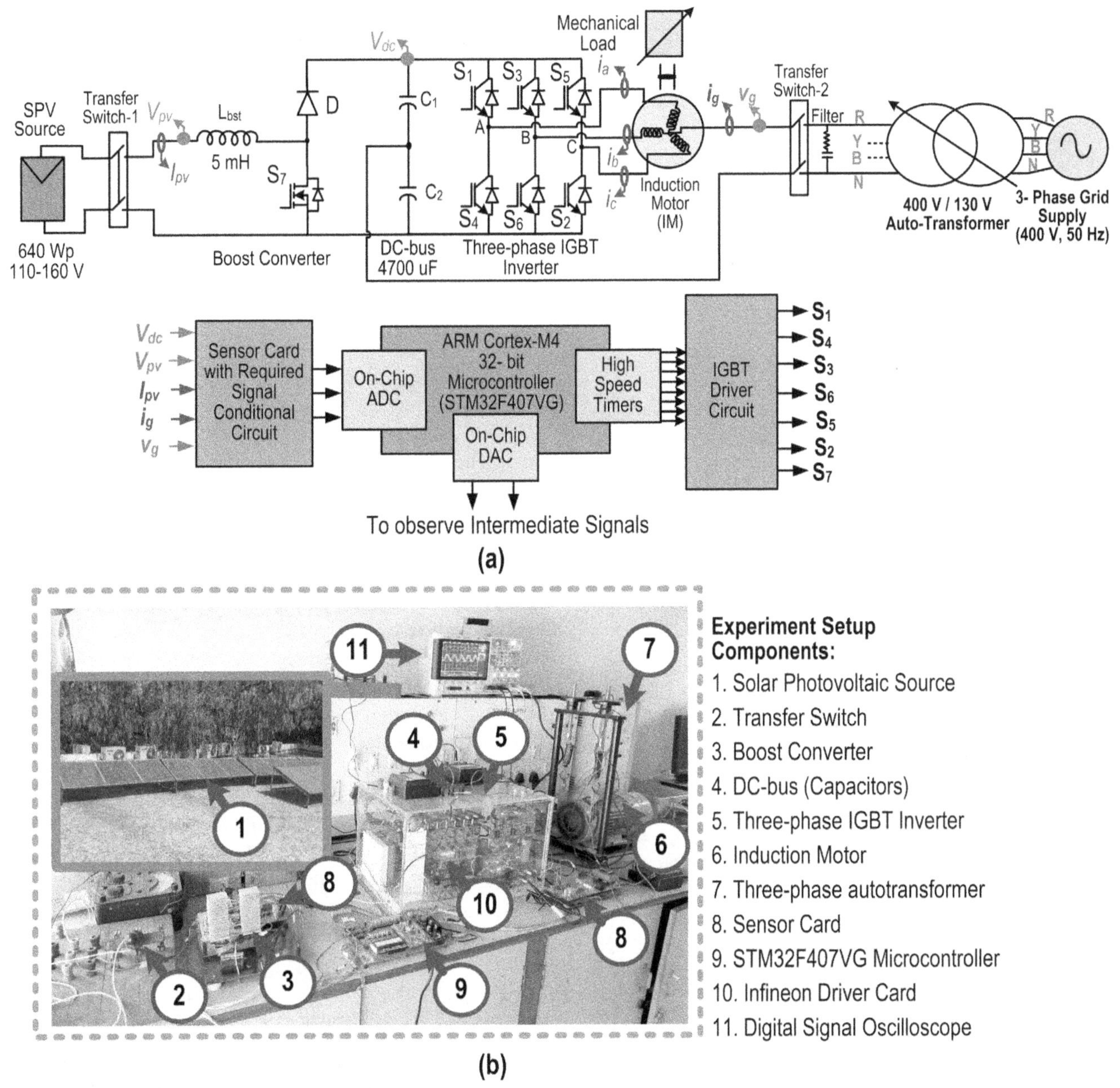

FIGURE 9.8 Experimental prototype: (a) block diagram; (b) picture of the system.

are observed in this experimental examination. When the reactive support is not activated, I_{qg} is zero and grid current is corresponding to active SPV power generation. It is observed that grid current is varying as active energy generated changes with time. V_{gm} is varied significantly from 98–110 V. Figure 9.11(b) shows performance when the reactive support is activated. These steady-stage voltage deviations are alleviated by delivering the reactive power from the constructed distributed generator. I_{qg} is varied adequately and the voltage peak (V_{gm}) is maintained at 99 V (70 Vrms). In the next section, the comparison of multifunction grid-interactive SPVWPS is carried out with grid-assisted SPVWPS.

9.7 COMPARISON OF DIFFERENT GRID-INTERACTIVE SPVPWSS

A brief cost analysis is carried out in Table 9.4 for the grid-interactive systems, whereby a single-phase grid is interfaced with standalone SPVWPS. The cost values presented in Table 9.4 are only for the power electronic converter (PEC) and motor-pump set used.[COMP: Set the below Table as in Original]

The PV panel cost is assumed to be nearly similar since the same power rating is considered for all three compared systems. As exemplary data, all the costs are given for 3 HP solar pump drive applications.

TABLE 9.3
Details of Components and Control System Parameters Considered for the Experimental Study

Parameters	Value
SPV Array:	
Peak power output of module	160 Wp
I_{sc} and V_{oc} of module	4.96 A, 43 V
I_{mp} and V_{mp} of module	4.57 A, 35 V
No. of modules connected in series	4
Maximum power of PV source	0.64 kW
Induction motor	
Power	1 HP
Voltage (V_{LL})	400 V
Full load speed	1430 rpm
Number of poles	4
Frequency	50
DC-Bus and Grid Parameter:	
DC-bus voltage (V_{dc})	300 V
DC-bus capacitor value (C_1, C_2)	4700 uF
Grid voltage (v_g)	80 V
Boost Converter and VSI:	
Switching frequency of boost converter switch	25 kHz
Switching frequency of VSI switches	10 kHz
Inductor value (L_{pv})	5.0 mH
Controller Gains (Proportional Gain, Integral Gain, Sampling Time):	
DC-bus voltage controller in standalone operation	0.1, 0.2, 4000 us
Outer loop grid voltage controller	0.1, 0.1, 4000 us
Outer loop DC-bus voltage controller	0.01, 0.09, 4000 us
q-axis inner loop current controller	6.7, 8.7, 400 us
d-axis inner loop current controller	4.2, 2.6, 400 us

In [17, 18], a single-phase grid is interfaced at DC-bus using extra PEC and inductor; however, in the multifunction grid-interactive SPVWPS, a single-phase grid is interfaced directly with the SPVWPS system without addition of power electronics components. The major cost-affecting factors in these topologies are several semiconductor switches, sensors and motor design. Compared to other systems, the multifunction grid-interactive SPVWPS is low cost, as it does not require an extra grid-interface PEC and bulky grid-interface inductors.

9.8 SUMMARY

The SPVWPS is an effective solution for remote water irrigation compared to a diesel pump. However, the standalone SPVWPS suffers from the issues of variable water delivery, higher initial cost, and system underutilization because of the large idle non-pumping period. The grid-assisted SPVWPS solves the issue of variable water delivery by supporting deficit energy, but it does not solve other aforementioned issues, and it also adds to the cost and complexity of the system. To alleviate the issue of underutilization, the grid-interactive SPVWPS is presented, whereby the standalone SPVWPS infrastructure is utilized as distributed generation source during a large idle non-pumping period. The steps to design passive components of the system are discussed in detail. The control system for standalone SPVWPS operation and multifunction grid-interactive distributed generation operation is presented. For distributed generator operation, the active and reactive power exchange with the grid is established for idle period. The active power transfers correspond to the available SPV energy, and the

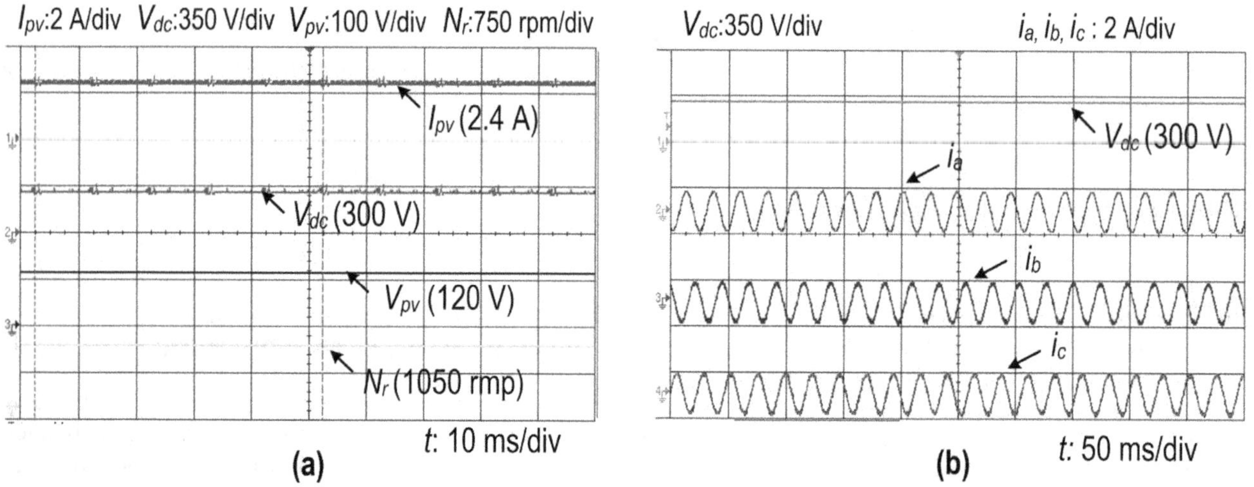

FIGURE 9.9 System performance in standalone SPVWPS mode.

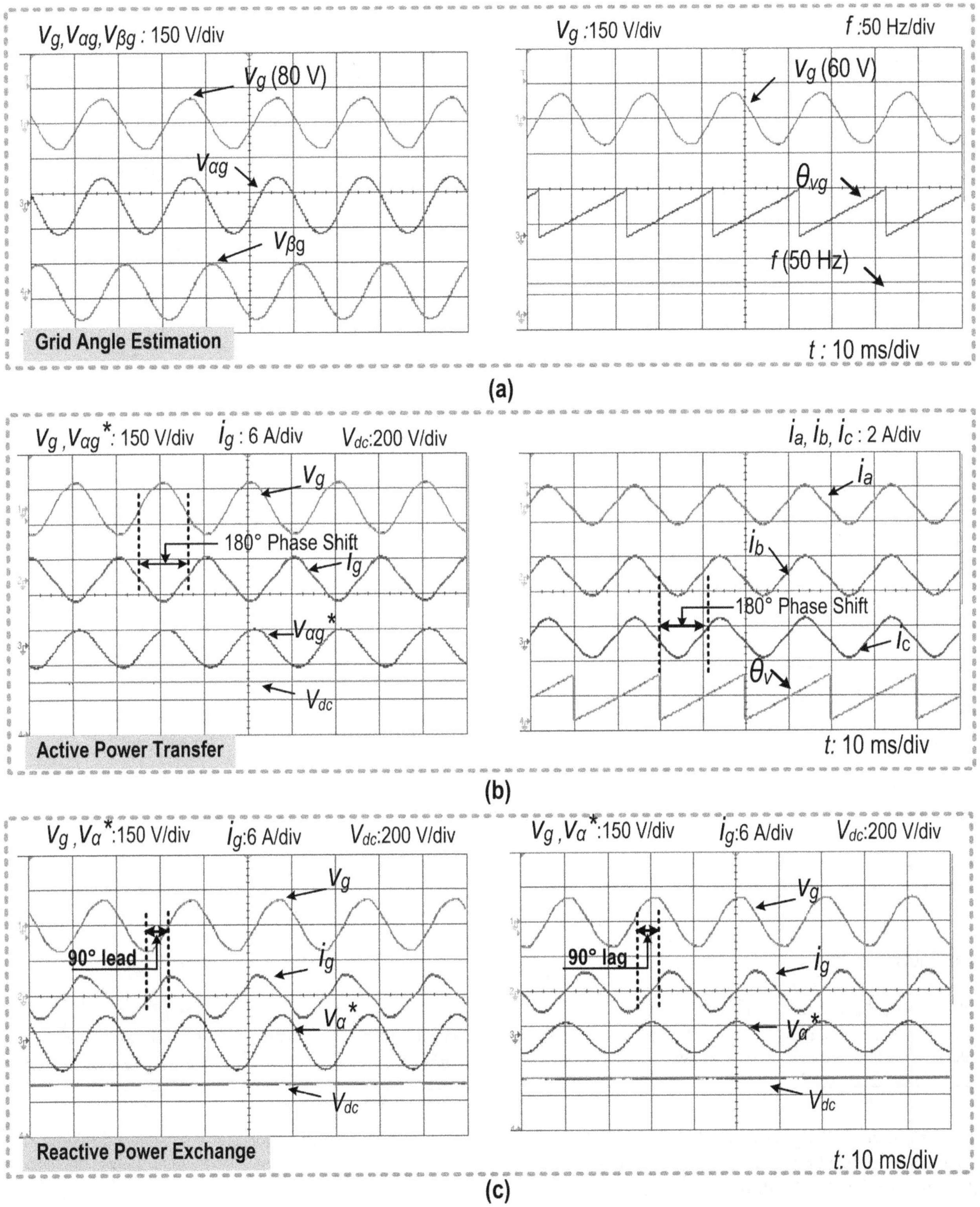

FIGURE 9.10 Steady-state performance in distributed generator mode: (a) SOGI estimator performance; (b) performance during active power transfer; (c) performance during reactive power exchange.

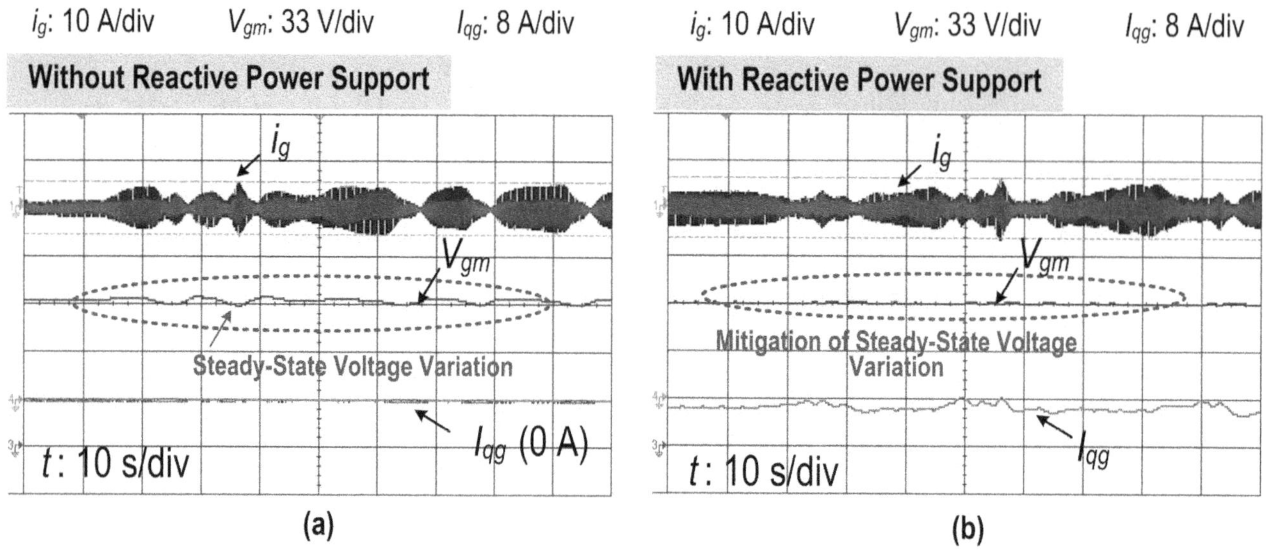

FIGURE 9.11 Grid voltage profile under the variation of SPV power generation: (a) without reactive power support; (b) with reactive power support.

TABLE 9.4
Techno-Economic Comparison of Different Grid-Interactive SPVWPS

	[17]	[18]	Multifunction Grid-Interactive SPVWPS
System Configuration Details			
Grid-interface PEC	*Boost PFC*	*Single-phase VSC*	-
SPV interface PEC	*Boost converter*	-	*Boost converter*
Motor drive PEC	*Three-phase VSI*	*Three-phase VSI*	*Three-phase VSI*
Total no. of PECs	*3*	*2*	*2*
No. of sensors	*5*	*4*	*4*
Cost Analysis			
Voltage sensors (LV 25-P)	*3, [$180]*	*2, [$120]*	*3, [$180]*
Current sensors (LV 55-P)	*2, [$96.50]*	*2, [$96.50]*	*2, [$96.50]*
DC-link capacitors (2700 uF, 400 V)	*1, [$19]*	*1, [$19]*	*2, [$38]*
IGBT switches	*11, [$210](STGW40M120DF3,SKM50GB12V)*	*10, [$200](SKM50GB12V)*	*(STGW25H120F2,A1P25S12M3)*
IGBT driver circuit	*[$165]*	*[$150]*	*[$70]*
Grid-interface inductor (5 mH, 15 A)	*1, [$45]*	*1, [$45]*	-
SPV interface inductor (5 mH, 15 A)	*1, [$45]*	-	*1, [$45]*
Submersible motor-pump set	*230 V, 3 HP[$400](special design)*	*230 V, 3 HP[$400](special design)*	*440 V, 3 HP[$289](conventional)*
Total	**$1,160**	**$1,030**	**$758**

reactive power exchanges to alleviate the grid voltage fluctuation. The realized grid-interactive operation enhances the utilization of SPVWPS infrastructure, and simultaneously solves the problem of grid voltage fluctuations on the irrigation feeder. The presented topology is low cost, as it does not require an extra grid-interface PEC and bulky inductor.

REFERENCES

[1] Chandel, S. S., M. Nagaraju Naik, and Rahul Chandel. "Review of solar photovoltaic water pumping system technology for irrigation and community drinking water supplies." *Renewable and Sustainable Energy Reviews* 49 (2015): 1084–1099.

[2] Bhattacharyya, Subhes C., and Debajit Palit. "Mini-grid based off-grid electrification to enhance electricity access in developing countries: What policies may be required?" *Energy Policy* 94 (2016): 166–178.

[3] Dadhich, Gitika, and Vivek Shrivastava. "Economic comparison of solar PV and diesel water pumping system." In *2017 International Conference on Information, Communication, Instrumentation and Control (ICICIC)*, pp. 1–6. IEEE, 2017.

[4] Pandikumar, M., and R. Ramaprabha. "Financial analysis of diesel and solar photovoltaic water pumping systems." In *Advances in Smart Grid Technology*, pp. 179–188. Springer, Singapore, 2020.

[5] Shukla, Saurabh, and Bhim Singh. "Solar powered sensorless induction motor drive with improved efficiency for water pumping." *IET Power Electronics* 11, no. 3 (2018): 416–426.

[6] Madark, M., A. Ba-razzouk, and M. El Malah. "Linear and nonlinear controllers of a solar photovoltaic water pumping system." *Bulletin of Electrical Engineering and Informatics* 9, no. 5 (2020): 1861–1872.

[7] Prabhakaran, K. K., A. Karthikeyan, S. Varsha, B. Venkatesa Perumal, and Sukumar Mishra. "Standalone single stage PV-Fed reduced switch inverter based PMSM for water pumping application." *IEEE Transactions on Industry Applications* 56, no. 6 (2020): 6526–6535.

[8] Kumar, Rajan, and Bhim Singh. "Single stage solar PV fed brushless DC motor driven water pump." *IEEE Journal of Emerging and Selected Topics in Power Electronics* 5, no. 3 (2017): 1377–1385.

[9] Zaky, Alaa A., Mohamed N. Ibrahim, Hegazy Rezk, Eleftherios Christopoulos, Ragab A. El Sehiemy, Evangelos Hristoforou, Antonios Kladas, Peter Sergeant, and Polycarpos Falaras. "Energy efficiency improvement of water pumping system using synchronous reluctance motor fed by perovskite solar cells." *International Journal of Energy Research* 44, no. 14 (2020): 11629–11642.

[10] Koreboina, Vijay Babu, Narasimharaju BL, and Vinod Kumar DM. "Performance investigation on ANFIS and FFNN assisted direct and indirect PV-fed switched reluctance motor water pumping system." *International Journal of Modelling and Simulation* (2021): 1–13.

[11] Vitorino, Montiê Alves, Maurício Beltrão de Rossiter Corrêa, Cursino Brandão Jacobina, and Antonio Marcus Nogueira Lima. "An effective induction motor control for photovoltaic pumping." *IEEE Transactions on Industrial Electronics* 58, no. 4 (2010): 1162–1170.

[12] Rekioua, Djamila, and Ahmed Mohammedi. "Direct torque control for autonomous photovoltaic system with MPPT control." *Turkish Journal of Electromechanics and Energy* 5, no. 2 (2020).

[13] Rai, Rashmi, Saurabh Shukla, and Bhim Singh. "Sensorless field oriented SMCC based integral sliding mode for solar PV based induction motor drive for water pumping." *IEEE Transactions on Industry Applications* 56, no. 5 (2020): 5056–5064.

[14] Dubey, Menka, Shailendra Kumar Sharma, and Rakesh Saxena. "Solar energy based ZSI fed induction motor drive for water pumping." In *2020 IEEE International Conference on Computing, Power and Communication Technologies (GUCON)*, pp. 135–139. IEEE, 2020.

[15] Riedemann, Javier, Werner Jara, Ruben Pena, Iván Andrade, Ramón Blasco-Gimenez, and Pedro Melín. "Predictive control of a multi-drive solar pumping system." In *2019 21st European Conference on Power Electronics and Applications (EPE'19 ECCE Europe)*, pp. 1–7. IEEE, 2019.

[16] Vamja, Rajan Vinodray, and Mahmadasraf A. Mulla. "Upgradation of grid-connected water pumping system integrating multifunction PV inverter with reduced sensor." *Electrical Engineering* 103, no. 3 (2021): 1629–1646.

[17] Sharma, Utkarsh, and Bhim Singh. "Utility-tied solar water pumping system for domestic and agricultural applications." *Journal of the Institution of Engineers (India): Series B* 101, no. 1 (2020): 79–91.

[18] Shukla, Saurabh, and Bhim Singh. "Single-stage PV-grid interactive induction motor drive with improved flux estimation technique for water pumping with reduced sensors." *IEEE Transactions on Power Electronics* 35, no. 12 (2020): 12988–12999.

[19] Murshid, Shadab, and Bhim Singh. "Analysis and control of weak grid interfaced autonomous solar water pumping system for industrial and commercial applications." *IEEE Transactions on Industry Applications* 55, no. 6 (2019): 7207–7218.

[20] Sharma, Utkarsh, Bhim Singh, and Shailendra Kumar. "Intelligent grid interfaced solar water pumping system." *IET Renewable Power Generation* 11, no. 5 (2017): 614–624.

[21] Shukla, Saurabh, and Bhim Singh. "Improved power quality converter for three-phase grid-interfaced PV array fed reduced current sensor-based induction motor drive for water pumping." *International Transactions on Electrical Energy Systems* 30, no. 4 (2020): e12304.

[22] Murshid, Shadab, and Bhim Singh. "A multiobjective GI-based control for effective operation of PV pumping system under abnormal grid conditions." *IEEE Transactions on Industrial Informatics* 16, no. 11 (2019): 6880–6891.

[23] Sharma, Utkarsh, Chinmay Jain, and Bhim Singh. "Three phase grid interfaced solar water pumping system." In *2017 National Power Electronics Conference (NPEC)*, pp. 1–6. IEEE, 2017.

[24] Vamja, Rajan V., and Mahmadasraf A. Mulla. "Development of grid-interactive inverter utilising induction motor driven photovoltaic water pumping system." *IET Power Electronics* 13, no. 15 (2020): 3373–3383.

[25] Jones, Walter V. "Motor selection made easy: Choosing the right motor for centrifugal pump applications." *IEEE Industry Applications Magazine* 19, no. 6 (2013): 36–45.

[26] Rashid, Muhammad H. *Power Electronics: Circuits, Devices, and Applications*. Pearson Education India, New Delhi, India, 2009.

[27] Możdżyński, Kamil, Krzysztof Rafał, and Małgorzata Bobrowska-Rafał. "Application of the second order generalized integrator in digital control systems." *Archives of Electrical Engineering* 63, no. 3 (2014): 423–437.

[28] Yates, Randy, and Richard Lyons. "DC blocker algorithms [DSP Tips & Tricks]." *IEEE Signal Processing Magazine* 25, no. 2 (2008): 132–134.

[29] Ahmed, Hafiz, and Mohamed Benbouzid. "Simplified second-order generalized integrator-frequency-locked loop." *Advances in Electrical and Electronic Engineering* 17, no. 4 (2019): 405–412.

[30] Rodriguez, Pedro, Alvaro Luna, Raul Santiago Munoz-Aguilar, Ion Etxeberria-Otadui, Remus Teodorescu, and Frede Blaabjerg. "A stationary reference frame grid synchronization system for three-phase grid-connected power converters under adverse grid conditions." *IEEE Transactions on Power Electronics* 27, no. 1 (2011): 99–112.

Appendix

Solved Problems

Example 9.1

Find the value of pump constant and DC-bus voltage value for the 15 HP induction motor–driven SPVWPS. The induction motor rating is 220 V (line voltage), 50 Hz, four-pole, 1430 rpm. Consider the scalar space vector pulse width modulation control for motor drive inverter.

Solution

The induction motor rating is 220 V (line voltage), 50 Hz, four-pole, 1430 rpm. The value of angular frequency is calculated as:

$$w_r = \frac{2\pi N}{60} = \frac{2\times 3.14\times 1430}{60} = 149.67\ rad/s$$

For the given power 15 HP (11.18 kW), the value of pump constant is calculated as:

$$K_{pump} = \frac{P_{pump}}{w_r^3} = \frac{11180}{(149.67)^3} = 3.33\times 10^{-3}\,W.s^3/rad^3$$

The line voltage of motor (V_{LL}) is 220 V and space vector modulation is implemented for inverter switches. The value of DC-bus voltage is calculated as:

$$V_{dc} = \sqrt{2}\times V_{LL} = \sqrt{2}\times(220) = 311\ V$$

Example 9.2

Calculate the value of DC-bus capacitor for 15 HP induction motor–driven SPVWPS. The induction motor rating is 220 V (line voltage), 50 Hz, four-pole, 1430 rpm, 0.85 PF. Consider the overloading factor as 1.2. The motor drive inverter is controlled with space vector modulation control. The DC-bus capacitors should provide sufficient energy for 5 ms when the disturbance is there due to change in sun radiation or pump load. The acceptable voltage dip is 30 V.

Solution

The phase current of the motor is calculate as:

$$I_{ph} = \frac{P}{3V_{ph}(PF)} = \frac{11180}{3\times 127\times 0.85} = 34.52\ A$$

The line voltage of motor (V_{LL}) is 220 V and space vector modulation is implemented for inverter switches. The value of DC-bus voltage is calculated as:

$$V_{dc} = \sqrt{2}\times V_{LL} = \sqrt{2}\times(220) \approx 310\ V$$

For the given acceptable DC-bus deap, V_{dcl} is calculated as 280 V.

Using the given data, the value of capacitor is selected as:

$$C_{dc} = \frac{6\alpha_0 V_{ph} I_{ph}\Delta t}{V_{dc}^2 - V_{dc1}^2} = \frac{6\times 1.2\times 127\times 34.52\times 0.005}{[310^2 - 280^2]} = 8916\ uF$$

Example 9.3

Calculate the value of boost converter duty cycle at MPP and value of inductance required for a 5 HP two-stage induction motor–driven SPVWPS. The induction motor rating is 400 V (line voltage), 50 Hz, four-pole, 1430 rpm. The motor drive inverter is controlled with third harmonic injection pulse width modulation control. The peak SPV voltage at standard test condition is 400 V. Consider the switching frequency as 50 kHz, and acceptable current ripple for the inductor current is 1 A.

Solution

The line voltage of motor (V_{LL}) is 400 V and third harmonic injection pulse width modulation control is implemented for motor drive inverter. The value of DC-bus voltage is calculated as:

$$V_{dc} = \sqrt{2}\times V_{LL} = \sqrt{2}\times(400) = 565\ V$$

The boost converter duty cycle at V_{mpp} (400 V) is obtained as:

$$D = \frac{V_{dc} - V_{mpp}}{V_{dc}} = \frac{565 - 400}{565} = 0.29$$

The inductor value is calculated as:

$$L_{pv} = \frac{V_{mpp} D}{\Delta I\times F_{sw}} = \frac{400\times 0.29}{1\times 50000} = 2.32\ mH$$

UNSOLVED PROBLEMS

1. Find the value of pump constant and DC-bus voltage value for the 5 HP induction motor–driven SPVWPS. The induction motor rating is 220 V (line voltage), 60 Hz, four-pole, 1710 rpm. Consider the scalar sine pulse width modulation for motor drive inverter.

   [Answer: $K_{pump} = 0.65 \times 10^{-3}$ W.s³/rad³, $V_{dc} = 359$ V]

2. Calculate the value of DC-bus capacitor for 15 HP induction motor–driven SPVWPS. The induction motor rating is 400 V (line voltage), 50 Hz, four-pole, 1430 rpm, 0.85 PF. Consider the overloading factor as 1.2. The motor drive inverter is controlled with space vector modulation control. The DC-bus capacitors should provide sufficient energy for 4 ms when the disturbance is there due to change in sun radiation or pump load. The acceptable voltage dip is 50 V.

   [Answer: C_{dc} = 2307 uF]

3. Calculate the value of boost converter duty cycle at MPP and value of inductance required for 15 HP two-stage induction motor–driven SPVWPS. The induction motor rating is 220 V (line voltage), 50 Hz, four-pole, 1430 rpm. The motor drive inverter is controlled with third harmonic injection pulse width modulation control. The peak SPV voltage at standard test condition is 250 V. Consider the switching frequency as 20 kHz and acceptable current ripple for inductor current is 1.5 A.

   [Answer: D = 0.19, L_{pv} =1.5 mH]

10 Fixed-Speed Wind and Solar Power System with Multifunctional Voltage Source Converter

Geeta Pathak, Bhim Singh and B. K. Panigrahi

CONTENTS

10.1 INTRODUCTION

In the present era, the world is facing an energy crisis as conventional fossil fuel–based sources are in a shrinking state due to their exhaustive utilization for the development and comfort of the human race. However, advanced technology is capable of exploiting energy from renewables such as wind, tidal, solar, biogas, etc., and is also able to provide it at reasonable rates [1–4]. In such situations, renewable-based distributed generation can become a hope for energy sector for smooth conduction of electricity-based applications and activities for day-to-day life.

Wind and solar photovoltaics (PV) are the fastest growing renewable energy resources at present. Solar PV is becoming popular because of its everlasting and pollution-free nature. They are almost maintenance free and possess good reliability. However, installation cost and efficiency have always been the challenging factors for this sector. Government schemes and subsidies are taking up the installation cost issues and advance research in PV field for material quality, accurate estimation of parameters; system resistance etc. is helping to improve its efficiency [5, 6].

Wind energy applications have travelled a journey of more than 100 years from windmills to wind farms for electricity generation. Two types of wind power generations are available: 1) fixed speed and 2) variable speed [7, 8]. For fixed-speed generation, wind variation is almost fixed or has a very small range of wind variation; hence, the power will be generated at constant frequency and frequency regulation is not required in this type. The installation costs of such systems are low and control complexity is also reduced. For wind power generation, permanent magnet synchronous generator (PMSG) is used. In PMSG, slip-rings and rotor windings are replaced by rotor permanent magnets; therefore, it becomes maintenance free and has good thermal capacity and reduced field losses. Overall size and weight of the generator become compact and light weighted therefore, the efficiency is improved [9, 10].

Worldwide, wind installation capacity has increased to 651 GW and in India will reach 60 GW by 2022. With commercialization and latest technology, wind energy has become cheaper. Its cost has been reduced from Rs. 2.8 to Rs. 2.2 per kWh [11]. Similarly, the PV energy is also in boom and total installation capacity of solar PV will be 100 GW by 2022 in India. It is expected to acquire 40 GW each through decentralized rooftop projects and utility-scale solar plants, and 20 GW ultra-mega solar parks [12]. The electricity cost of solar PV generation is reaching Rs. 2.3/kWh.

Indian green energy potential is immense and mostly untapped. Solar potential is more than 750 GW and wind potential is 302 GW as per the recent estimations in the Indian scenario. In reality, it could be larger than 1,000 GW. India Energy Security (IES)-2047 predicts wind and solar PV peaks of 410 GW and 479 GW by 2047 [13].

Application of wind/solar distributed generation systems can be useful for electricity supply to local community loads where grid access is not available. In this chapter, a fixed-speed wind and solar system with battery storage is employed to supply power to the AC loads. Battery banks provide reliable operation with such highly unpredictable generating sources, so that the consumers may not face difficulties during peak load demands or with lower generation or load variations. A multifunctional VSC is used to perform various tasks, i.e. harmonics reduction from voltage and currents at the point of common coupling (PCC), power balancing, voltage regulation etc. Various functions are performed by VSC implementing least logarithmic absolute difference (LLAD) control algorithm for appropriate VSI switching. This algorithm has fast rate of conversance and good tracking performance for transient and steady state analysis [14].

DOI: 10.1201/9781003229124-10

The chapter is organized in various sections. Section 10.2 summarizes the configuration of wind/solar distributed generation (DG) systems. Section 10.3 presents the DG system components' design and modelling to make the chapter more relevant and better understandable. Section 10.4 details various control algorithms to make the system functioning smooth and accurate. Section 10.5 highlights the results of the DG system performance under nonlinear loads with various power scenarios. Conclusions are reached in Section 10.6, which is followed in Section 10.7 by some numerical problems based on system components to make the chapter more reader friendly and interesting. Finally, this chapter is concluded with descriptions of system utilization and future scope.

10.2 STRUCTURE OF WIND/SOLAR DISTRIBUTED GENERATION SYSTEMS

A wind/solar distributed system is illustrated in Figure 10.1. The PMSG generates electricity at fixed wind speed and it is directly supplied to AC loads. The PV array generates DC power and delivers it to DC bus after MPPT using a DC/DC converter. At DC bus, a battery bank is connected to store the excess energy during low load demand or high SPV generations. An incremental conductance (INC) approach is used for computation of the boost converter duty cycle. During high load demands, this SPV power is converted into AC power by a voltage source converter (VSC) and fed to the loads. When both the renewables are optimally utilized and demand is still high, the battery discharges to fulfil the load demand.

A 3-Φ VSC is connected at common coupling point (PCC) through L_f (interfacing inductors), and a capacitor bank is connected at the PMSG terminals to deliver excitation currents for voltage buildup in a standalone system. A ripple filter is used to suppress the harmonics of source voltage.

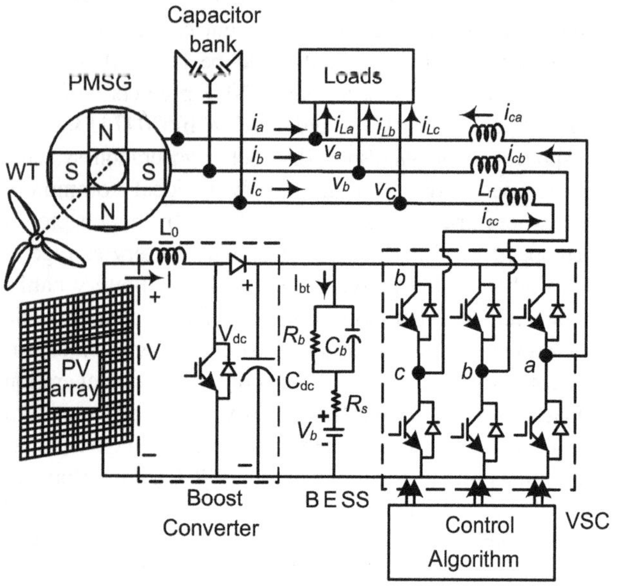

FIGURE 10.1 Wind/solar distributed generation system.

10.3 MODELLING AND DESIGN OF WIND/SOLAR DGS COMPONENTS IN MATLAB ENVIRONMENT

Various components of the wind/solar distributed generation system are modelled in MATLAB Simulink and design parameters also explained in this section.

10.3.1 Solar PV Modelling

Modelling of a solar PV array is important for describing and understanding of the V-I characteristic of the solar system, which can be helpful to study the dynamic behaviour of the DC/DC power converters and for the maximum power point tracking (MPPT) using various techniques [15].

Solar photovoltaic cells are modelled with the equivalent circuit depicted in Figure 10.2. The mathematical equation for V-I characteristic of ideal PV cell is given as:

$$I = I_{SPV,c} - I_{0,c}\left[\exp\left(\frac{qV}{akT}\right) - 1\right] \tag{10.1}$$

where I_{PV} is the generated by the PV cell by the incident light (directly proportional to sun insolation), $I_{0,c}$ is the reverse saturation current of the diode, q is electron charge, k is Boltzmann constant, a is diode identity constant and T is temperature if p-n junction

The solar photovoltaic array is series and parallel arrangements of the PV cells. Therefore, the equation for V-I characteristic will be expressed differently as:

$$I = I_{SPV} - I_0\left[\exp\left(\frac{V + R_{sr}I}{V_t a}\right) - 1\right] - \frac{V + R_{sr}I}{R_{prl}} \tag{10.2}$$

where I_{SPV} and I_0 are the photovoltaic and saturation currents of the array; V_t is thermal voltage, and cells connected

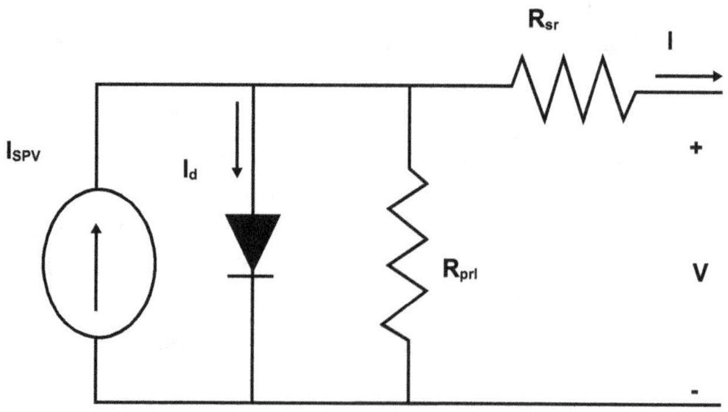

FIGURE 10.2 SPV equivalent circuit.

in series (N_{sr}) and parallel (N_{prl}) provide increased voltage or current, respectively; R_{sr} and R_{prl} are series and parallel resistances of the array respectively

$$V_t = \frac{N_{sr}kT}{q} \quad (10.3)$$

Expressions for the photovoltaic and saturation currents arecalculated as:

$$\begin{aligned} I_{SPV} &= I_{PV,c}N_{prl} \\ I_0 &= I_0 N_{prl} \end{aligned} \quad (10.4)$$

In case of SPV array having series and parallel PV modules ($N_{sr} \times N_{prl}$), the output current for is given as [16]:

$$I = I_{SPV}N_{prl}\left(\exp\left(\frac{V + IR_{sr}(N_{sr}/N_{prl})}{aV_t N_{sr}}\right) - 1\right) - \frac{V + IR_{sr}(N_{sr}/N_{prl})}{R_{prl}\left(N_{sr}/N_{prl}\right)} \quad (10.5)$$

The I-V characteristic of solar photovoltaic (SPV) array for 1,000 W/m^2 insolation is illustrated in Figure 10.3. The P-V characteristic of solar PV is also depicted in Figure 10.4, which shows the maximum power point of the system.

10.3.2 Boost Converters

In this chapter, a boost converter is employed for second stage solar power transfer to the VSC after maximum power tracking using INC (incremental conductance) technique. Calculations for input inductor of boost converter to limit ripple current (ΔI_L) with switching frequency f_{ss} is as [17]:

$$L_o = \frac{V_{dc}}{4 \times f_{ss} \times \Delta I_L} \quad (10.6)$$

where V_{dc} is DC bus voltage

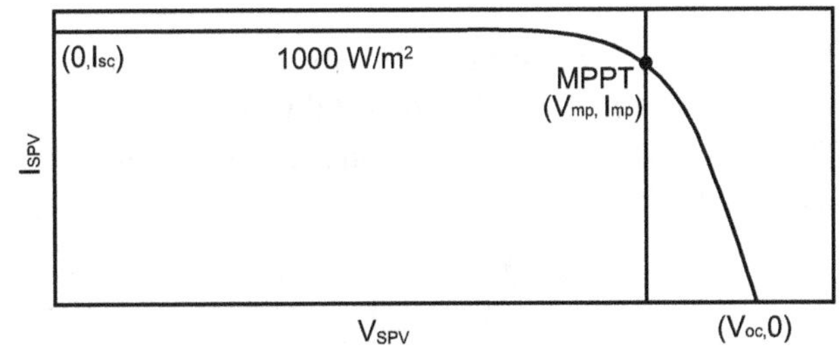

FIGURE 10.3 V-I characteristics of a solar PV system.

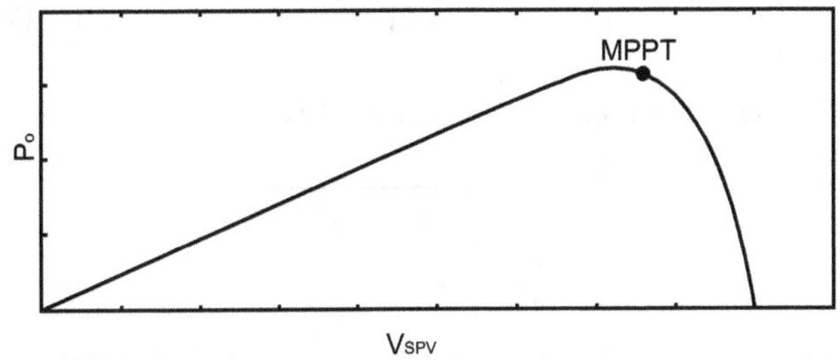

FIGURE 10.4 PV curve of a solar PV system.

10.3.3 Battery Storage Systems and VSC

The voltage at the DC capacitor of the VSC is derived as:

$$V_{dc} = \frac{2\sqrt{2}}{\sqrt{3}m}V_L \quad (10.7)$$

where m is the modulation index, V_L is the line rms voltage and V_{dc} is the DC bus voltage

A Thevénin's theorem for battery storage system is described, whereby C_b and R_b are parallelly connected capacitance and resistances, in series with an ideal voltage source (V_{oc}) having internal resistance (R_s). Equivalent capacitance C_b is computed as:

$$C_b = \frac{(kW.h*3600*1000)}{0.5(V_{oc\max}^2 - V_{oc\min}^2)} \quad (10.8)$$

where V_{ocmax} and V_{ocmin} are the maximum and minimum open circuit voltage values under fully charged and discharged states of battery (±10% V_{oc}).

10.3.4 Wind Turbines and PMSG Generators

At present, though, variable speed wind generation is very popular. Fixed-speed wind generation is also considered in many applications due to the following benefits.

1. The installation cost and maintenance of such systems is lower.
2. Necessity for gearboxes or variable speed drives is reduced.
3. There is no requirement of maximum power point tracking, which reduces the converter costs and also the control complexity.
4. They deliver power at a constant frequency, so frequency regulation is therefore not required.

With such features, constant speed wind generation is included in this chapter with permanent magnet synchronous generation.

Wind turbines convert kinetic energy of wind into useful electrical energy. This power is characterized by coefficient of performance (C_p), which is the function of tip speed ratio (λ) and is defined as:

$$\lambda = \frac{r\omega_t}{v} \tag{10.9}$$

where, r is the radius of the wind turbine blade in metres, ω_t is turbine shaft speed in rad/sec, v is the wind velocity in m/sec

The C_p can be calculated in terms of λ and blade pitch angle (β) as:

$$C_p(\lambda) = C_1\left(\frac{C_2}{\lambda_i} - C_3\beta - C_4\right)e^{\frac{-C_5}{\lambda_i}} + C_6\lambda \tag{10.10}$$

where λ_i is defined as:

$$\frac{1}{\lambda_i} = \frac{1}{\lambda + 0.08\beta} - \frac{0.035}{1+\beta^3} \tag{10.11}$$

The mechanical power (P_t) and torque (T_t) made available by the wind turbine can be calculated as:

$$P_t = 0.5\rho\pi R^2 C_p v^3 \tag{10.12}$$

$$T_t = \frac{P_t}{\omega_t} \tag{10.13}$$

PMSG generator has very high efficiency due to absence of rotor windings and its compact size. It has good thermal capacity and does not require maintenance.

The mathematical model of PMSG can be described in d-q reference systems as:

$$\begin{aligned}\frac{di_d}{dt} &= (-R_s i_d + L_q p\omega_r i_q + v_d)/L_d \\ \frac{di_q}{dt} &= (-R_s i_d + L_d p\omega_r i_d - p\psi f\omega_r + v_q)/L_q\end{aligned} \tag{10.14}$$

$$T_e = 1.5p(\psi_f i_q - (L_d - L_q)i_d i_q) \tag{10.15}$$

$$\omega_e = p\omega_r \tag{10.16}$$

where v_d, v_q and i_d, i_q are the d- and q-axis voltages and currents, respectively; R_s is the stator resistance; L_d and L_q are the d- and q-axis inductances, respectively; ω_r is the mechanical angular speed and ω_e is the electrical rotating speed of generator; T_e is the electromagnetic torque; $\psi_f$is the magnitude of the flux produced by the permanent magnet; and p is the number of pole pairs of the generator

10.3.5 Selection of RC Filter and Interfacing Inductor of VSC

The PCC voltage noise is filtered out with a high-pass ripple filter. This filter filters out the high-frequency voltage ripples by providing a low-impedance path. This offers high impedance to the fundamental frequency values at 10 kHz switching frequency. RC filter capacitance (c) and resistance (R) are considered 10 μF and 5 Ω, respectively:

$$Z = \sqrt{\left(R^2 + \left(\frac{1}{2\pi fc}\right)^2\right)} \tag{10.17}$$

Interfacing inductor is formulated as,

$$L_f = \left(\frac{\sqrt{3}}{2}\right) \times \frac{mV_{dc}}{6af_s I_{rpl}} \tag{10.18}$$

where V_{dc} and L_f are voltage at DC bus capacitance and interfacing inductance, respectively, and $I_{ri.}$ is the ripple current and switching frequency f_s

10.4 CONTROL ALGORITHMS FOR WIND/SOLAR DGS

In this section, two control algorithms are described as current control algorithm for VSC switching and MPPT algorithm for boost converter switching.

10.4.1 Voltage Source Converter Control Technique

For a satisfactory control operation of VSC, the control technique for the VSC switching should be reliable and satisfactory. Least logarithmic absolute difference (LLAD) control is applied for harmonics reduction and voltage regulation under dynamics. Figure 10.5 depicts the complete approach for estimation of various current components for VSC switching.

10.4.1.1 Calculation for In-Phase and Quadrature Templates of Source Voltages

The phase voltages (v_{abc}) are computed from the line voltages as [18, 19]:

$$v_a = \frac{1}{3}(2v_{ab} + v_{bc}) \tag{10.19}$$

$$v_b = \frac{1}{3}(-v_{ab} + v_{bc}) \tag{10.20}$$

$$v_c = \frac{1}{3}(-v_{ab} - 2v_{bc}) \tag{10.21}$$

The terminal voltage V_t is calculated as:

$$V_t = \sqrt{\left(2\left(v_a^2 + v_b^2 + v_c^2\right)/3\right)} \tag{10.22}$$

In-phase templates of all 3-Φ voltages are estimated as:

$$u_{pa} = v_a / V_t, u_{pb} = v_b / V_t, u_{pc} = v_c / V_t \tag{10.23}$$

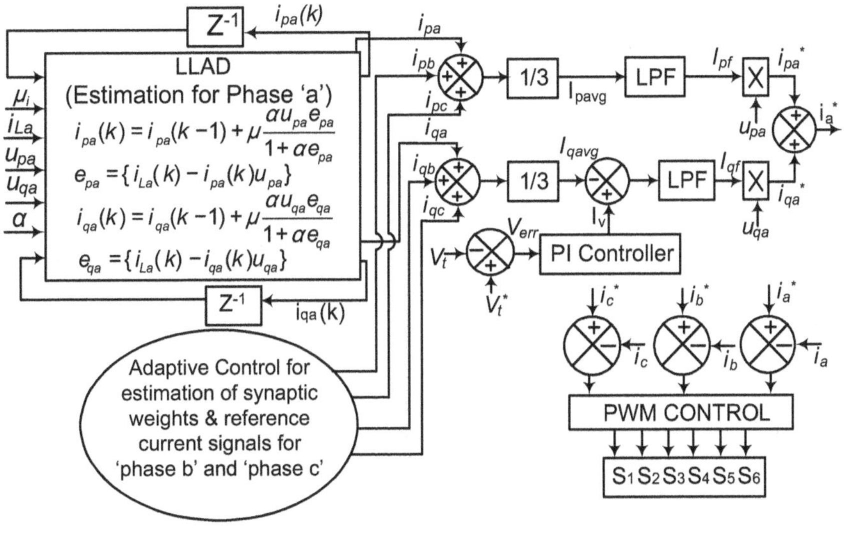

FIGURE 10.5 LLAD control approach.

The quadrature voltage templates of all three phase voltages are calculated as:

$$u_{qa} = (-u_{pa} + u_{pc}) / \sqrt{3} \tag{10.24}$$

$$u_{qb} = (3u_{pa} + u_{pb} - u_{pc}) / \sqrt{3} \tag{10.25}$$

$$u_{qc} = (-3u_{pa} + u_{pb} - u_{pc}) / \sqrt{3} \tag{10.26}$$

10.4.1.2 Extraction of Active and Reactive Components of Load Currents Using Least Logarithmic Absolute Difference (LLAD) Algorithm

Load current's load active component for phase a is extracted by implementing LLAD approach at *k*th sample as:

$$i_{pa}(k) = i_{pa}(k-1) + \mu \frac{\alpha u_{pa} e_{pa}}{1+\alpha |e_{pa}|} \tag{10.27}$$

The error between actual and desired outputs e_{ak} (k) at *k*th sample is calculated as:

$$e_{pa}(k) = \{i_{La}(k) - i_{pa}(k) u_{pa}\} \tag{10.28}$$

Estimation of load active components for phases b–c is made in the manner same as that of phase a using the LLAD control technique.

Load current's load reactive component of for phase a applying the same approach is computed at *k*th sample as:

$$i_{qa}(k) = i_{qa}(k-1) + \mu \frac{\alpha u_{qa} e_{qa}}{1+\alpha |e_{qa}|} \tag{10.29}$$

$$e_{qa}(k) = \{i_{La}(k) - i_{qa}(k) u_{qa}\} \tag{10.30}$$

Estimation of load active components for phases b–c is made in the similar manner that of phase a, applying same technique.

The step size μ is selected as 0.45 and constant α as 0.06.

The equivalent load active component is estimated as:

$$I_{pav} = \frac{i_{pa} + i_{pb} + i_{pc}}{3} \tag{10.31}$$

The equivalent load reactive component is estimated as:

$$I_{qav} = \frac{i_{qa} + i_{qb} + i_{qc}}{3} \tag{10.32}$$

After filtering out through low-pass filter (LPF), this component is called I_{pf}.

A proportional and integral (PI) regulator is tuned with k_p (proportional gain) and k_i (integral gain) to maintain common coupling (PCC) voltage. The error between sensed and reference PCC voltage V_t and V_t* is forwarded to the PI regulator, and the output current becomes:

$$I_v(k) = I_v(k-1) + k_p\{V_{err}(k) - V_{err}(k-1)\} k_i V_{err}(k) \tag{10.33}$$

where V_{err} is the reference and actual voltage error and is shown as:

$$V_{err} = V_t^* - V_t \tag{10.34}$$

Total reactive component of source current is calculated as:

$$I_{qt} = I_v - I_{qav} \tag{10.35}$$

After filtering out through LPF, this component is called I_{qf}.

10.4.1.3 Computation of Fundamental Active and Reactive Components of Supply Currents

The positive sequence active and reactive power components of reference three-phase source currents are defined as:

$$i_{pa}^* = I_{pf}u_{pa} \quad i_{pb}^* = I_{pf}u_{pb} \quad i_{pc}^* = I_{pf}u_{pc} \tag{10.36}$$

$$i_{qa}^* = I_{qf}u_{qa} \quad i_{qb}^* = I_{qf}u_{qb} \quad i_{qc}^* = I_{qf}u_{qc} \tag{10.37}$$

10.4.1.4 Derivation of Three-Phase Reference Supply Currents

The three phases—a, b and c—source reference currents are computed as:

$$i_a^* = i_{pa}^* + i_{qa}^* \quad i_b^* = i_{pb}^* + i_{qb}^* \quad i_c^* = i_{pc}^* + i_{qc}^* \tag{10.38}$$

The three-phase reference source currents (i_a^*, i_b^*, i_c^*) and sensed currents (i_a, i_b, i_c) are compared for generating PWM to provide switching to three legs of VSC.

10.4.2 Maximum Power Point Tracking Control Algorithm for Boost Converters

INC MPPT technique is used to calculate the duty ratio of the boost converter. The flowchart for incremental conductance (INC) method is depicted in Figure 10.6.

Derivative of output power and output voltage for MPPT of solar systems should be zero. It states that instantaneous conductance addition incremental conductance, i.e. $Z = (I_d/V_d + \Delta I_d/\Delta V_d)$ is zero. Due to the variation in any factor, if point shifting towards the right side value of Z becomes negative and to maintain maximum power point, the duty ratio increases [20]. Similarly, if left side point shifting takes place and Z becomes positive, the duty ratio decreases to keep the MPPT. With the variation in duty cycle, the boost converter keeps maintains the point of maximum power tracking. Figure 10.7 depicts the generation of PWM pulse for boost converter switching after duty cycle comparison with triangular wave.

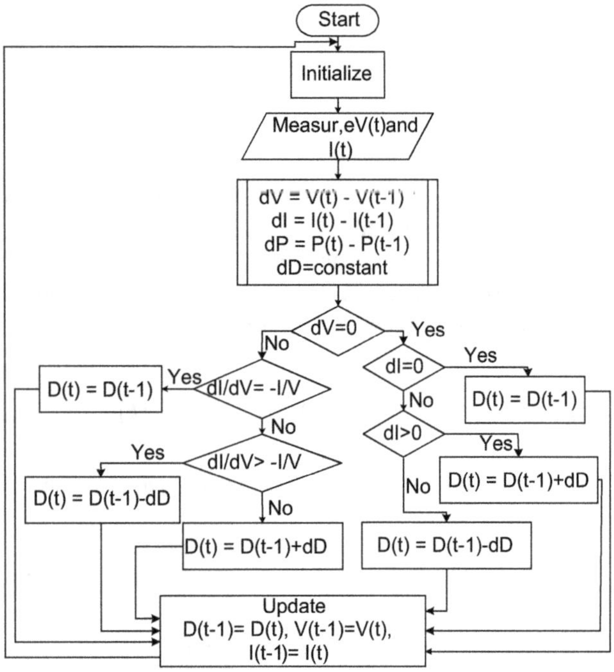

FIGURE 10.6 Flowchart for incremental conductance maximum power tracking.

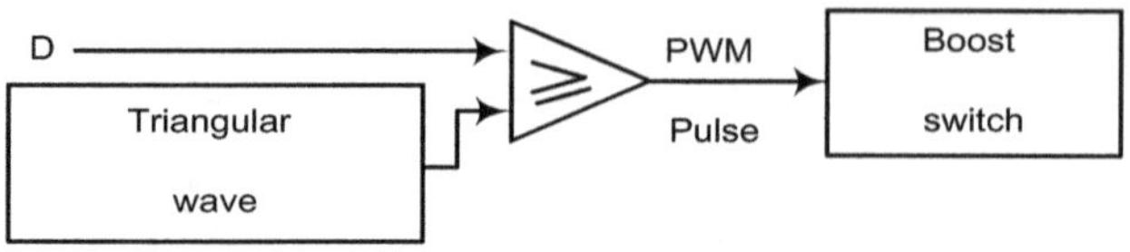

FIGURE 10.7 Boost controller switching pulse generation.

10.5 RESULTS AND DISCUSSION

The results of the wind/solar DGS are shown in this section for steady state and dynamic conditions under nonlinear loads. The simulations are carried out in MATLAB/Simulink.

The behaviour of the proposed system has been observed with steady state and dynamic conditions. The intermediate signals of the control algorithm are shown to see the behaviour of the internal signals during dynamics under nonlinear loads.

10.5.1 Internal Parameter Performance of the Control Algorithm at Nonlinear Loads

The performance of control technique is shown in Figure 10.8 with internal parameters at nonlinear loads. The dynamic condition under unbalanced load is realized by throwing a switch of phase a for 0.2 s; under this duration, the load current of phase a (i_{La}) becomes zero, and load active and load reactive components for phase a (i_{pa} and i_{qa}) therefore reduce to zero value. The step size (μ_i) remains constant throughout. The average load active and average load reactive components (I_{pavg}, I_{qavg}) are reduced with decrease in demand due to disconnection of phase a. The output PI controller current (I_v) provides voltage regulation at point of common coupling. With load variation, reference current active component (I_{pf}) and that of reactive component (I_{qf}) are also changed. Therefore, reduction is observed in reference currents (i^*_{abc}) during unbalanced load. The control technique is capable of settling the response very fast with adequate performance during dynamics. The response is stable in both of the steady states before and after the imbalance. This shows the robustness of the control algorithm in steady state and dynamic conditions.

10.5.2 Steady State Performance

FFT analysis of the wind/solar DGS at steady state at nonlinear load is demonstrated in Figure 10.9. It shows the harmonic content and wave shapes of common coupling point

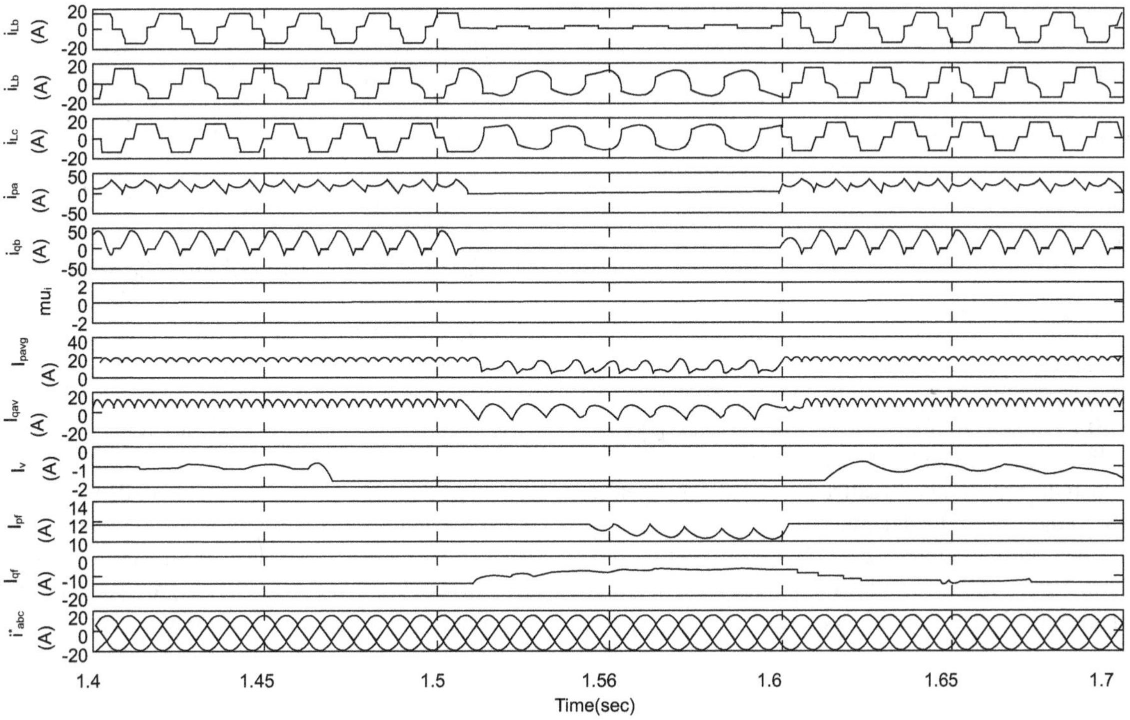

FIGURE 10.8 Intermediate parameters of control algorithm under nonlinear load.

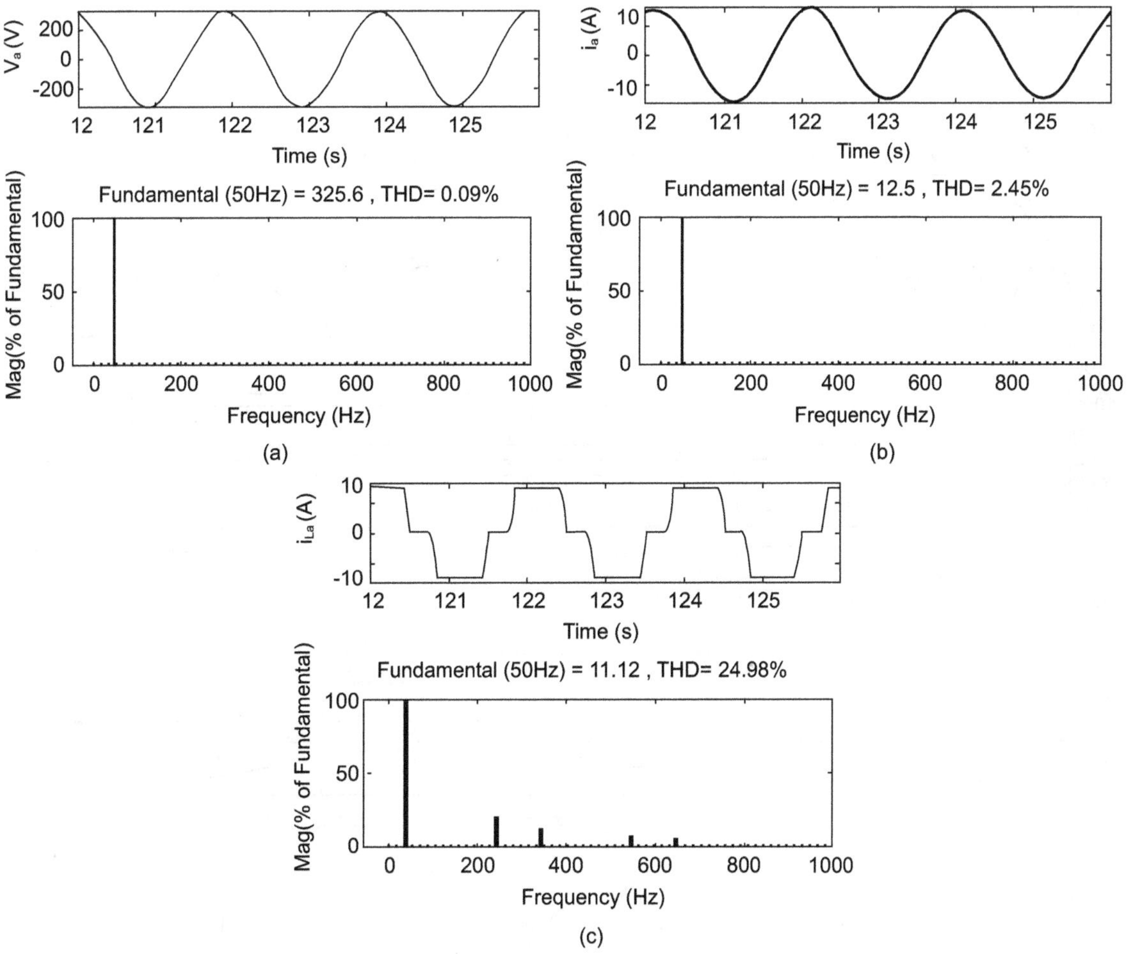

FIGURE 10.9 Steady-state performance of the DGS.

(PCC) voltage (v_a), source current (i_a) and load current (i_{La}) of phase a during normal running conditions. The sinusoidal PCC voltage and source current are obtained with their total harmonic distortion (THD) less than 5% at the nonlinear load THD of 25%. The satisfactory performance of DGS is demonstrated during normal operating conditions, and the IEEE-915 standard for current and voltage are followed.

10.5.3 Dynamic Performance

The dynamic responses of wind/solar distributed generation system are illustrated in Figure 10.10, Figure 10.11 and Figure 10.12 under nonlinear loads. Contingency (dynamics) conditions are created by: 1) increasing the solar insolation; 2) load disconnection is realized at phase a by opening a switch; 3) increase in load demand.

First, the solar radiation rose 800 W/m^2 to 1000 W/m^2 at t = 1.4 s in Figure 10.10. With fixed wind generation (P_{wind}) and constant load demand (P_L), load currents (i_{La}, i_{Lb} and i_{Lc}) and wind source currents (i_{abc}) are maintained constant throughout. The voltage source converter delivers power to fulfil the excess load demand by utilizing the solar and battery energy when P_L is more than P_{wind}. With insolation rise, the solar power generation is increased (P_{pv}) at t = 1.4 s and that increased power is stored in a battery bank, which can be seen in terms of battery power (P_{bt}) and battery current (I_{bt}) increase. For fixed load (P_L), the VSC is delivering constant deficit power utilizing the solar and battery power after complete exhaustion of wind power. Therefore the VSC current (i_{Cabc}) and power (P_{VSC}) are maintained throughout constant; with this change, the terminal voltage (V_t) at PCC is maintained constant.

Second, the imbalance is realized at t = 1.5 s by disconnecting phase a via throwing a switch, as shown in Figure 10.11. Due to this imbalance, total load demand is reduced (P_L) and load currents between phases b and c are realized as single-phase loads (i_{Lb}, i_{Lc}). With this contingency, the VSC current (i_{VSC}) and power (P_{VSC}) are varied to maintain the power balance and terminal voltage.

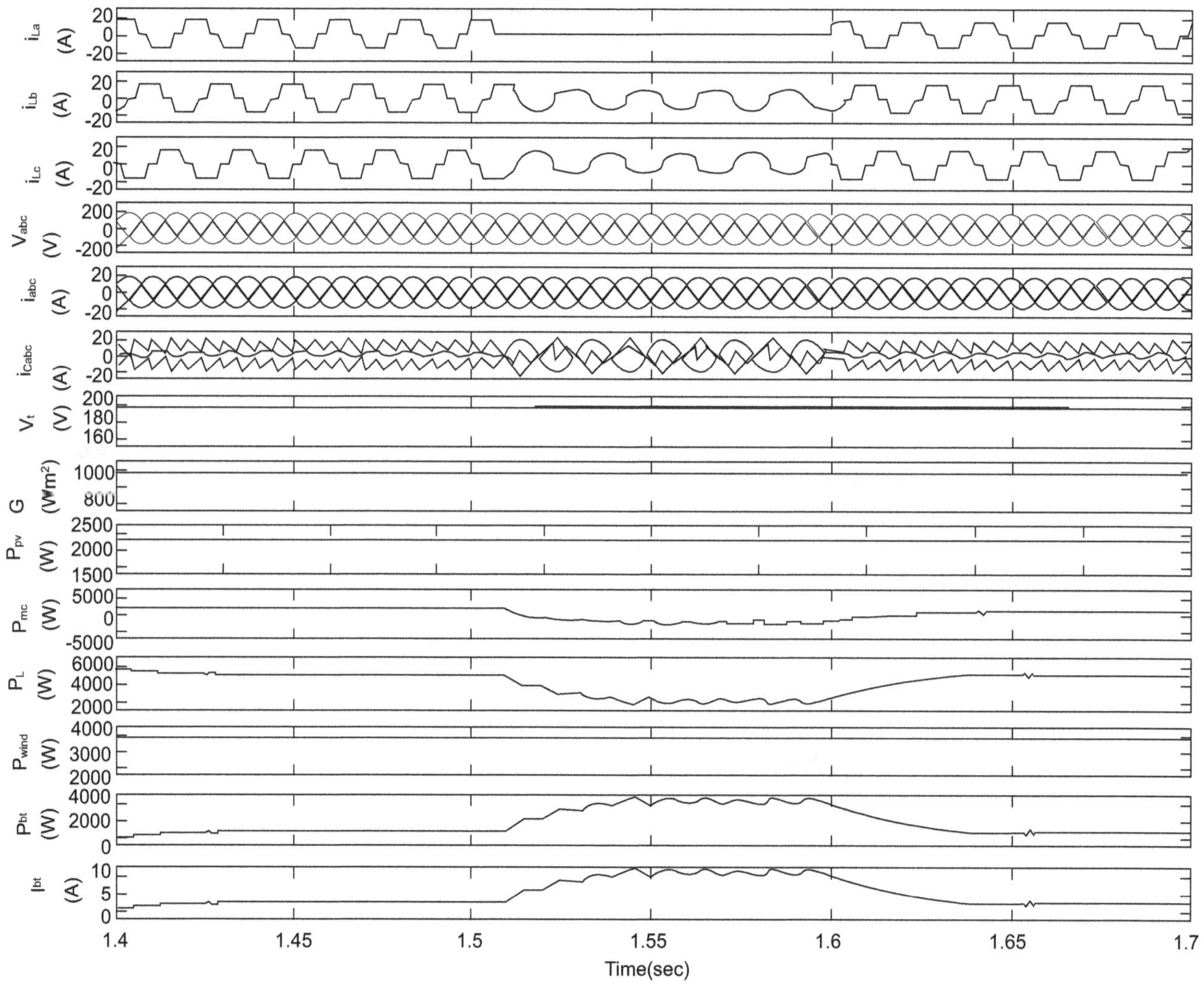

FIGURE 10.10 Insolation change under nonlinear loads.

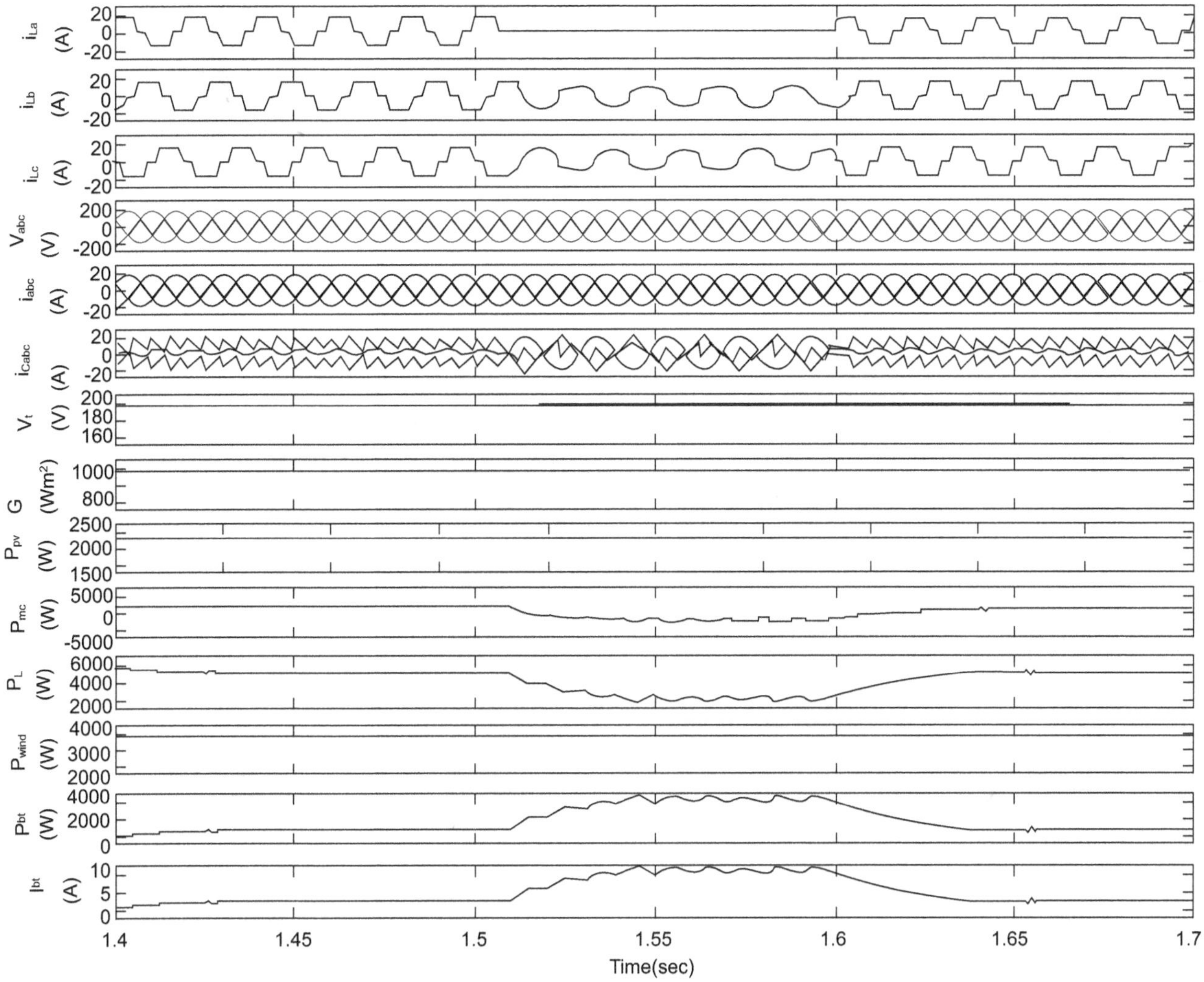

FIGURE 10.11 Load imbalance at phase a.

This undelivered power is stored into the battery; therefore, P_{bt} and I_{bt} are increased during imbalance. Under such dynamic conditions, the PCC voltage (v_{abc}) remains constant and sinusoidal. The source currents (i_{abc}) are maintained balance and sinusoidal. The terminal voltage (V_t) is also constant at its reference value. Terminal voltage (v_t) is constant, and due to fixed insolation (1,000 W/m^2), the solar power generation (P_{PV}) is also constant. With the variation in load demand (P_L), the converter power (P_{VSC}) is also varied but the source power (P_{wind}) remains constant due to fixed wind speed. Though the demand is reduced, the solar and wind generation systems are providing maximum power which can be stored in the batteries. Therefore, the battery power and currents are increased during load imbalance or lower load conditions.

Due to sudden increase in load demand at t = 1.7 s, load currents (i_{La}, i_{Lb}, i_{Lc}) are increased, as depicted in Figure 10.12. P_{wind} is delivered power by the wind generator at its full capacity with constant magnitude and sinusoidal source currents (i_{abc}). The VSC increases its power supply (P_{vsc}, I_{Cabc}) by converting the DC power of PV (P_{PV}) and battery power (P_{bt}) into AC power to supply the increased AC load demand. The battery discharges under this process; therefore, the P_{bt} and I_{bt} become negative. With such dynamics, terminal voltage is maintained and the PCC voltage (v_{abc}) is also regulated and remains sinusoidal. These results demonstrate the satisfactory performance of wind/solar DGS during times of such dynamics.

10.6 CONCLUSIONS

In this chapter, a wind/solar renewable distributed generation system is described for electricity generation to minimize the effects of energy crisis on development. Fixed-speed wind generation and solar power generation is utilized under various power scenarios, i.e. unbalanced load, generation and load demand variation. Brushless generator (PMSG) is implemented due to its cost and maintenance benefits. DG system structure is explained in detail and modelling, and design of wind turbine, PMSG generator, battery storage system and VSC are described in a systematic and detailed manner. The LLAD control technique is implemented on

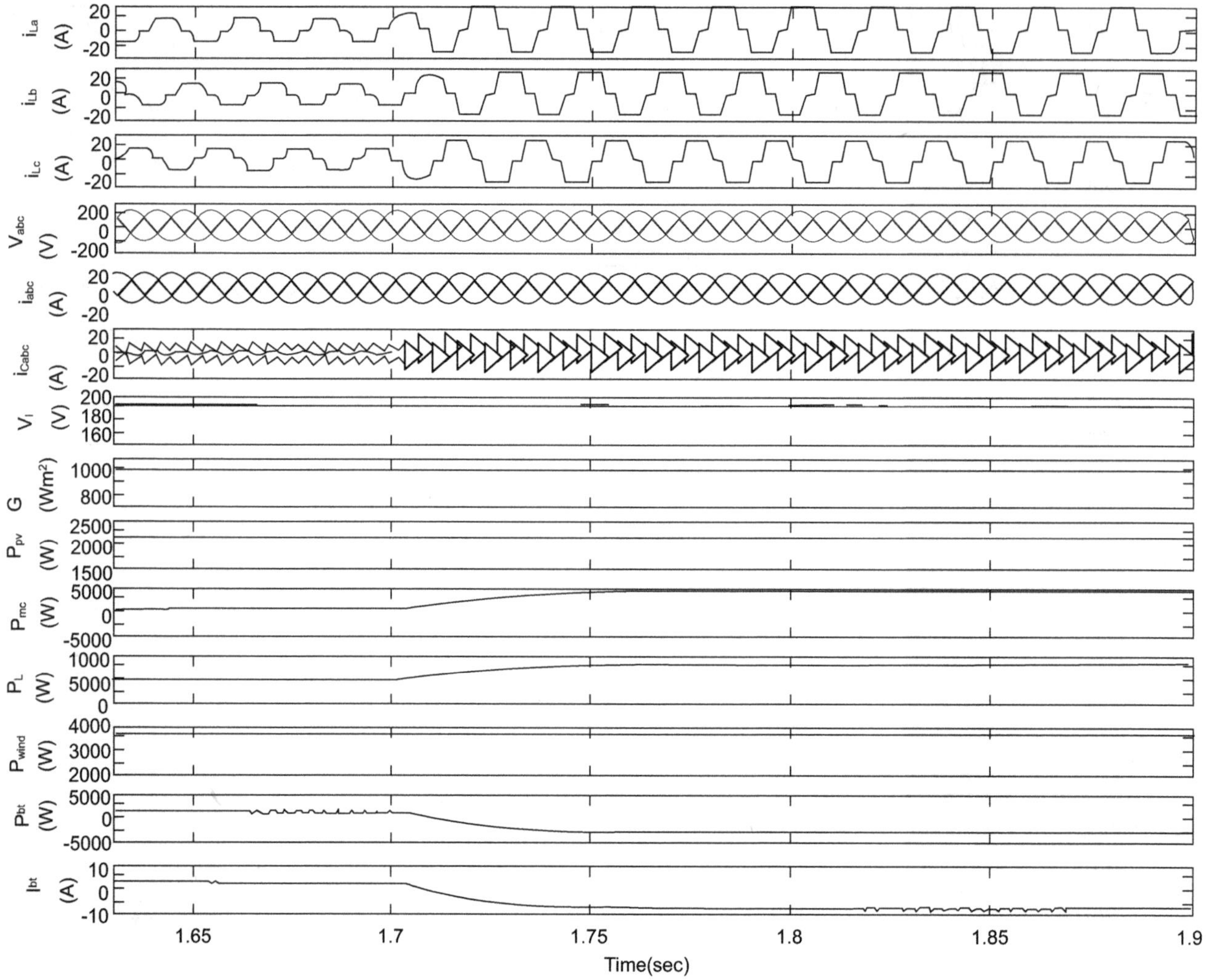

FIGURE 10.12 Sudden load increase.

TABLE 10.1
Wind/Solar DGS Components

S No.	Distributed Generation System Components	Rating
1	Three-phase permanent magnet synchronous generator	3.7 kW, 13 A, 230 V,1500 rpm
2	Solar PV panel:	2.5 kW, short ckt current = 16.5 A, open circuit voltage of PV array = 360 V
3	Battery bank: V_b	360 V
	C_b	1125 F
	R_b	10 kΩ,
	R_s	0.1 Ω
4	PI controller (voltage regulation) gains	
	k_p	0.6
	k_i	0.1
5	Interfacing inductance (L_f):	4 mH
6	RC filter: R	5 Ω
	C	10 μF
7	DC link: V_o	400 V
	C_{dc}	1600 μF

VSC for its multifunctional performance like voltage regulation, harmonics mitigation, load levelling, etc. The waveforms of the source currents and voltages are sinusoidal and their THDs are also under the limits, according to the IEEE standards. The voltage source converter (VSC) control also maintains the power balance among the all energy generation resources as solar, wind, battery storage, etc., under changing power generation and demand scenarios. Such systems can provide energy independence in rural places and can also contribute in reducing the fossil consumption and its bad impact on the environment. To make the chapter more interactive and to build better understanding, some numerical problems on the covered topics are also included.

10.7 NUMERICAL PROBLEMS

1. Output voltage of a boost converter is 12 V and load current of 1.5 A. variation in input voltage is from 4 to 6.2 V. To maintain the output voltage constant duty ratio of the converter is adjusted. Select the switching frequency 20 kHz. Determine the input inductance of the boost converter with the in inductor current variation not more than 40% of the average inductor current.

Solution

Duty ratios for both the input voltages:

$$D_1 = 1 - \frac{V_{in\_1}}{V_{out}} = 1 - \frac{4}{12} = 0.67$$

$$D_2 = 1 - \frac{V_{in\_2}}{V_{out}} = 1 - \frac{6.2}{12} = 0.48$$

Average inductor current due to input voltage variation:

$$I_1 = \frac{V_{out} I_L}{V_{in\_1}} = \frac{12 \times 1.5}{4} = 4.5A$$

$$I_2 = \frac{V_{out} I_L}{V_{in\_2}} = \frac{12 \times 1.5}{6.2} = 2.9A$$

Inductor ripple current due to 40% variation in inductor current:

$$\Delta I_1 = 0.4 \times 4.5 = 1.8A$$

$$\Delta I_2 = 0.4 \times 2.9 = 1.16A$$

Inductance values:

$$L_{o1} = \frac{V_1 D_1}{\Delta I_1 f_s} = \frac{4 \times 0.67}{1.8 \times 20{,}000} = 74.4\mu H$$

$$L_{o2} = \frac{V_2 D_2}{\Delta I_2 f_s} = \frac{6.2 \times 0.48}{1.16 \times 20{,}000} = 128.3\mu H$$

To satisfy the range of input, voltage inductor value must be (L_o)=128.3 μH. Hence, inductor ripple current due to 40% variation in inductor current and input voltage 4 V will be calculated as:

$$\Delta I_1 = \frac{V_1 D_1}{L_o f_s} = \frac{4 \times 0.67}{128.3 \times 10^{-6} \times 20{,}000} = 1.04A$$

Maximum inductor currents are:

$$I_{\max,V_{in\_1}} = I_{o1} + \frac{\Delta I_L}{2} = 4.5 + (1.04/2) = 5.02A$$

$$I_{\max,V_{in\_2}} = I_{o2} + \frac{\Delta I_L}{2} = 2.9 + 0.52 = 3.42A$$

Minimum inductor currents are:

$$I_{\min,V_{in\_1}} = I_{o1} - \frac{\Delta I_L}{2} = 4.5 - 0.52 = 3.98A$$

$$I_{\min,V_{in\_2}} = I_L - \frac{\Delta I_L}{2} = 2.9 - 0.52 = 2.38A$$

2. A three-phase, PWM voltage source converter feeds 50 kW to 415 V (line) rms, 50 Hz, three-phase AC mains from a constant voltage DC bus. The switching frequency is 20 kHz and AC inductor is 2.40 mH. The power factor is corrected close to unity and PWM modulation index is 0.85. Determine an input DC voltage, rms AC current and phase shift in fundamental component of PWM voltage and supply voltage.

Solution

line to line AC voltage = 415 V
Switchng frequency = 20 kHz
Inductor value = 2.40 mH
Modulation index, m = 0.85
Power factor = 1
Active power (P) = 50 kW
Fundamental AC current (I_{s1}) = 50 k/3 × (415/√3) × 1 = 69.56 A
Inductive reactance (X_L) = 314 × 2.4 × 10^{-3} = 0.754 Ω
Converter voltage $V_{con1}= \sqrt{V_s^2 + (I_s \times X_L)^2} = \sqrt{(239.6^2 + (69.56 \times 0.754)^2)}$ = 245.3 V
Modulation index (m_a) = 0.85
Input DC bus voltage (V_{dc})= 2√2 × V_{con1}/m_a = 816.14 V
RMS AC current (I_s) = 69.56 A
Phase shift in fundamental component of PWM voltage and supply voltage (δ) = $\sin^{-1}(P \times X_c/(3 \times V_s \times V_c))$ = 12.35 degree

3. A three-phase, PWM VSI operates from DC at 6250 ± 5 V. It supplies a motor load 50 kW, 460 V (line), 60 Hz at 1800 rpm. It is desired to operate the motor at 1400 rpm (speed is linear function of frequency). What is the modulation index under these conditions with nominal DC bus voltage? What are the modulation indexes at the high and low bus values?

Solution

Motor load (P) = 50 KW
Line-to-line voltage V_{LL} = 460 V, 60 Hz
Supply current (I_s) = P/(√3 × 460) = 62.57 A
Supply phase voltage (V_s) = 265.58 V
Four-pole motor runs at 1800 rpm, 60 Hz
For same number of poles at 1400 rpm, operation 46.66 Hz frequency is required.
Taking L_s =1.5 mH
X_s = 0.439 H
V inverter fundamental voltage, $V_{conv1} = \sqrt{\{V^2 + (I_s * X_s)^2\}}$
V_{conv1} = 267 V
Modulation indexes:
$M_a = 2\sqrt{2}\, V_{conv1}/V_{dc}$
= 1.2
At higher V_{dc} = 630 V
M_a = 1.19
At lower V_{dc} = 620 V
M_a = 1.218

REFERENCES

[1] John Love, John A. Bryant, *Biofuels and Bioenergy*, John Wiley & Sons Ltd., 2017.
[2] A. Schaffarczyk, *Understanding Wind Power Technology: Theory, Development and Optimization*, Wiley, 2014.
[3] A. Reinders, P. Verlinden, W. V. Sark, A. Freundlich, *Photovoltaic Solar Energy*, John Wiley & Sons Ltd., 2017.
[4] Siegfried Heier, *Grid Integration of Wind Energy, Onsore and Offshore Conversion Systems*, John Wiley & Sons, Ltd, 2014.
[5] M.A. Green, "Recent Advances in Silicon Solar Cell Performance," In Proceedings of 10th E. C. Photovoltaic Solar Energy Conference, Springer, pp. 250–253, Dordrecht, 1991.
[6] M. B. Hayat, D. Alli, K. C. Monyake, L. Alagha, N. Ahmed, "Solar energy-A look into Power Generation, Challenges, and a Solar—Powered Future," *International Journal of Energy Research*, vol. 43 (3), pp. 1049–1067, Mar. 2019.
[7] Bin Wu, Y. Lang, N. Zargari, S. Kouro, *Power Conversion and Control of Wind Energy Systems*, John Wiley & Sons, Inc., 2011.
[8] Gilburt M. Master, *Renewable and Efficient Electric Power Systems*, John Wiley & Sons, Inc., 2004.
[9] C. N. Wang, W. C. Lin, and X. K. Le, "Modelling of a PMSG Wind Turbine with Autonomous Control," *Mathematical Problems in Engineering*, vol. 2014, Article ID 856173, pp. 9, May 2014.
[10] E. Hossain, J. Hossain, N. Sakib, R. Bayindir, "Modelling and Simulation of Permanent Magnet Synchronous Generator Wind Turbine: A Step to Microgrid Technology," *International Journal of Renewable Energy Research*, vol. 7 (1), 2017.
[11] S. Srivastava, S. Jain, J. Lee, F. Zhao, F. Jayasurya, A. Lathigara, A. Pek, "India Wind Outlook Towards 2022, Looking Beyond Headwinds," *Globle Wind Energy Council and MEC+*, May 2020.
[12] National Institution for Transforming India, Report Of The Expert Group On 175 GW RE By 2022, Dec. 2015. https://www.niti.gov.in/sites/default/files/energy/175-GW-Renewable-Energy.pdf
[13] www.indiaenergy.gov.in/docs/RE_Documentation.pdf.
[14] M.O. Sayin, N.D. Vanli, S.S. Kozat, "A Novel Family of Adaptive Filtering Algorithms Based on the Logarithmic Cost," *IEEE Transactions on Signal Processing*, vol. 62 (17), pp. 4411–4424, Sept. 2014.
[15] M. G. Villalva, J. R. Gazoli, E. Ruppert F., "Modeling and Circuit-Based Simulation of Photovoltaic Arrays," *Brazilian Journal of Power Electronics*, vol. 14 (1), pp. 35–45, 2009.
[16] Shengyi Liu, R. A. Dougal. "Dynamic Multiphysics Model for Solar Array," *IEEE Transactions on Energy Conversion*, 17 (2), pp. 285–294, 2002.
[17] Geeta Pathak, Bhim Singh, B. K. Panigrahi, "Back Propagation Algorithm Based Controller for Autonomous Wind-DG Microgrid," *IEEE Transactions on Industry Applications*, vol. 52 (5), pp. 4408–4415, Sept.–Oct. 2016.
[18] Bhim Singh, B. K. Panigrahi, GeetaPathak, "Control of Wind-Solar Microgrid for Ruraelectrification," *In proc. of IEEE 7th Power India International Conference (PIICON)*, 2016.
[19] G. Pathak, B. Singh, B. K. Panigrahi, "Wind—Hydro Microgrid and Its Control for Rural Energy System," *IEEE Transactions on Industry Applications*, vol. 55 (3), pp. 3037–3045, May–June 2019.
[20] M. A. Khafagy, M. M. Eshak, P. Makeen, S.O. Abdellatif, H. A. Ghali, "Investigating the Utility of Perturbation and Observation, Incremental Conductance and Grey-Wolf MPPT Techniques for On- Grid PV System Under Fluctuated Environmental Conditions," *In Proceedings of IEEE International Conference on Power and Energy (PECon)*, Penang, Malaysia, pp. 338–343, 2020.

11 DROGI-FLL-Based Control Algorithm for Coordinated Operation of 3P4W Wind/Solar Hybrid System

Sombir, Madhusudan Singh, Ashutosh K. Giri, and Sabha Raj Arya

CONTENTS

11.1 INTRODUCTION

An enhanced adaptive filter (EAF) with an incremental conductance maximum power point tracking (INC MPPT) method is implemented to improve power quality and to harness the maximal solar power from the SPV. The EAF output is yields to VSC to enhance the power quality as load imbalance, harmonics eradication, and reactive power [1]. A perturb and observe method is implemented to harness the MPPT from the wind turbine (WT). Control with VSC provides voltage and frequency regulation at common point of interfacing (CPI) and manages the power in isolated mode and diesel mode. The battery energy storage system (BESS) is interfaced at DC link to maintain the power of the micro-grid during variation in load and wind speed [2, 3]. An AVR-based VSC is designed to regulate the voltage and SPV is interfaced at DC link of the VSC to support the reactive power, harmonic mitigation, and DG current balancing, and to enhance the system power quality [4]. A MSOGI-FLL–based DSTATCOM control has been designed with variation in frequency and high harmonic filtering capacity [5]. MSOS-based FLL control is used to accurately evaluate the frequency and phase angle under unbalanced and distorted grid voltages [6]. SOGI-based PLL control technique is implemented to improve the DC offset rejection. An NVLF-based approach is utilized to compensate the power quality problems in off-grid conditions [7–9]. A FOGI-FLL is implemented with better frequency tracking capabilities; the control provides multifunction like reactive support, load leveling, harmonic mitigation, and power factor improvement. A SPV is interfaced at DC link side of the VSC for support of the system frequency by providing real power compensation and enhancing the voltage profile and current profile quality under varying load demand [10]. An ASMC-based control technique is implemented to evaluate the control signals for the single-phase VSC and maintain the micro-grid parameters in addition to harmonic reduction [11]. A super capacitor battery–based HES is implemented to control the power fluctuations due to wind speed, solar insolation, and sudden load variations [12]. A RZALMS-based VSC control algorithm is implemented to generate PWM switching signals and the control effectively maintains the micro-grid (MG) parameters [13]. Different types of FLL control technique are implemented such as DSC-FLL, CBF-FLL, and SOSF-FLL [14]. To improve the system power quality of PMSG, an MS-TOGI based VSC control is used [15]. A LAF-based VSC control algorithm is designed to control the terminal voltage/frequency of the DG system, and the control also provides reactive support, harmonic eradication, neutral current compensation, and load leveling [16]. A modified DSC-PLL removes the double frequency oscillations under unbalanced loading conditions and rejects the harmonic component under nonlinear load [17]. A CC-ROGI-FLL with VSC is designed to evaluate the fundamental load current component under varying wind speeds, solar irradiance, and unbalanced linear/nonlinear load [18]. A VLGMS-based control approach is used to calculate the real and reactive weight components of fundamental load current for the evaluation of reference current [19]. An AVF-based approach maintains the frequency and regulate the voltage profile of single-phase SEIG under different operating conditions [20]. An INC MPPT–based control approach is utilized to extract maximum power under intermittent condition of solar intensity [21], and BES with PV system maintains the stability in voltage profile and current profile under variation in generator power and load [22, 23]. Different types of FLL-based control algorithms are discussed for three-phase systems [24]. PLL-based algorithms (DSOGI-PLL and VCO-less

DOI: 10.1201/9781003229124-11

PLL) are discussed to mitigate the power quality problems under dynamic conditions [25–27].

The main contributions of the author in this study areas follows.

- A DROGI-FLL with enhanced source current imbalance rejection ability is implemented in standalone hybrid system and the control techniques have components of the fundamental frequency–positive (FFP) and fundamental frequency–negative (FFN) sequences.
- The key approach of this technique is making the ROGI-I of the control immune to disturbing the effect of the wind/solar hybrid system load current harmonic component, helping to remove the disturbing harmonic component.
- The DROGI-FLL extracts the fundamental active load component under worst load current conditions and harmonic distortion. The control approach is used to generate reference current for VSC switching.
- The DROGI-FLL with VSC works as a static compensator (STATCOM), which executes various functions like harmonics eradication, reactive power demand, fast convergence, load leveling, and enhancing the power quality of the standalone hybrid system.
- The SPV with battery storage is interfaced at DC link of the VSC to decline the oscillations of a hybrid system current under changing solar insolation/wind speed feeding changing linear/nonlinear load. By providing active/reactive power assistance, the battery system regulates the power between the source and the linear/nonlinear load.

11.2 SYSTEM DESCRIPTION

The implemented 3P4W standalone wind/solar hybrid system is illustrated in Figure 11.1. It comprises a wind turbine, squirrel-cage induction generator (SCIG), lithium-ion battery system, DROGI-FLL control algorithm, SPV, boost converter, interfacing inductors, VSC, and linear/nonlinear loads. The VSC is connected to the source through an interfacing inductance (L_f) at CPI (common point of interfacing), which is utilized to reduce ripple current. An INC MPPT control–based SPV is interfaced at DC link to harness maximal power from a SPV. The battery system is also connected at DC link, which is used to support active/reactive power and maintain the power level between the source and the load under changing load demand in the standalone hybrid system.

11.3 CONTROL STRATEGY

The DROGI-FLL–based control method with an INCMPPT is utilized in a 3P4W standalone wind/solar hybrid system is seen in Figure 11.2. The MPPT is achieved in this hybrid system through a VSC, with the controller assisting in maintaining DC link voltage. The proposed hybrid system's goal is to provide high-quality power to the load while managing the generator's voltage/frequency and the battery system to compensate for the SPVs and intermittent

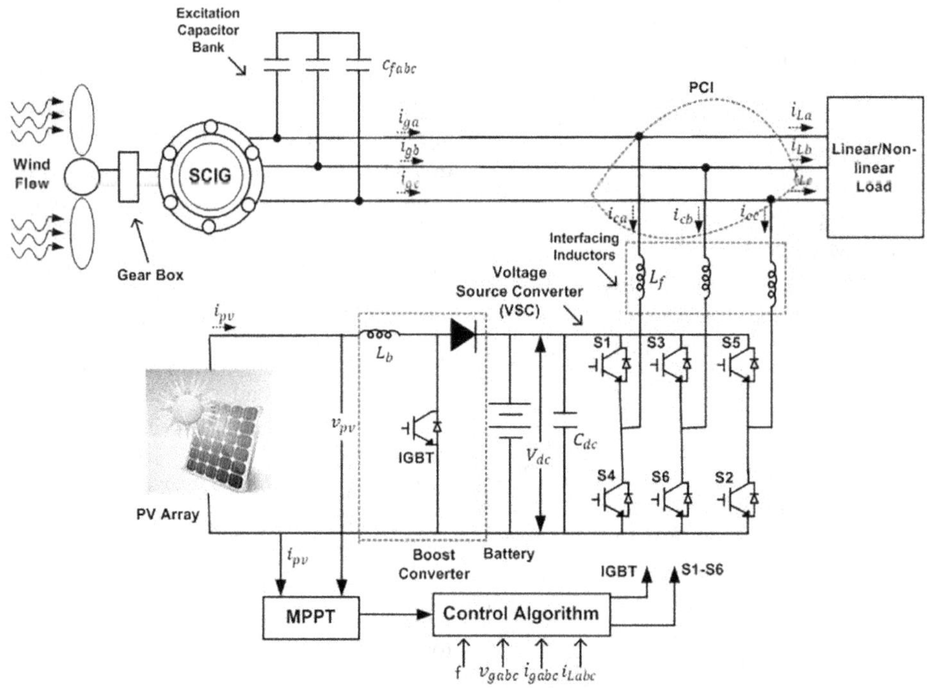

FIGURE 11.1 Schematic diagram of standalone hybrid system.

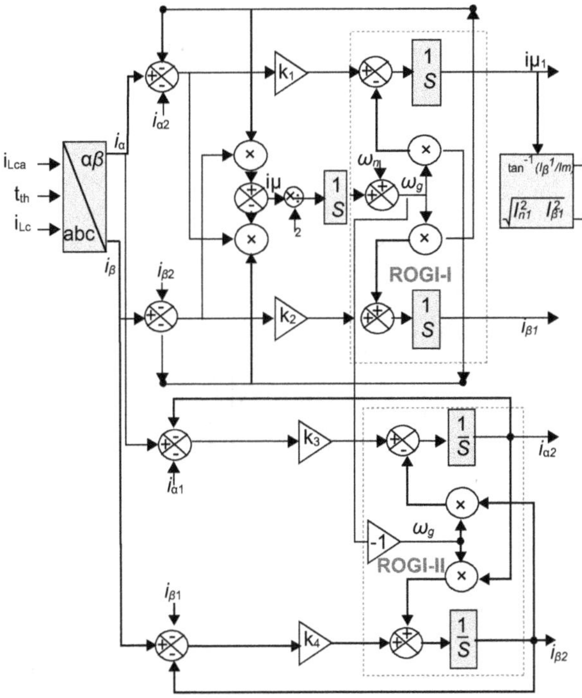

FIGURE 11.2 DROGI-FLL control algorithm.

power of winds. This control approach also maximizes the use of solar power and wind energy for providing uninterruptible power to the loads. The VSC switching pulses are obtained by feeding the hysteresis controller the error generated by the reference generator current and sensed generator current.

11.3.1 Evaluation of Phase Angle, Frequency, and the Amplitude of Fundamental Load Current Component

A DROGI-FLL control algorithm is designed to evaluate the components such as phase angle, frequency, and fundamental load current. Two parallel ROGIs centered at the fundamental positive frequency and a μ-order disturbance frequency, as well as a FLL for adapting their center frequencies to frequency changes, are used in a ROGI-based signal decomposition algorithm. The ROGI is explained in the Laplace domain as:

$$G_{ROGI}(s) = \frac{1}{s - j\omega_g} \tag{11.1}$$

where ω_g denotes the center frequency of the ROGIs

Figure 11.2 shows the DROGI-FLL control algorithm; $k_1, k_2, k_3, k_4$and λ are the gain parameters, and $i_{\alpha 1}$ and $i_{\beta 1}$ ($i_{\alpha 2}$ and $i_{\beta 2}$) are evaluation of the FFPS (μ-order frequency) component of the load current in the αβ frame. $\theta_1$and I_p are the evaluation of the phase angle and the peak value of the FFNS (μ-order frequency) component of the load current. ω_g is an evaluation of the fundamental angular frequency of the load current. μ = −1 represents extracting fundamental negative sequence component of the DROGI-FLL control algorithm [23]:

$$\theta_1 = \tan^{-1}\left(i_{\beta 1} / i_{\alpha 1}\right) \tag{11.2}$$

$$I_p = \sqrt{\left(i_{\alpha 1}\right)^2 + \left(i_{\beta 1}\right)^2} \tag{11.3}$$

11.3.2 Amplitude of Generator Voltage and Unit Template Generation

$$v_{ga} = \frac{1}{3}\left(2v_{ab} + v_{bc}\right) \tag{11.4}$$

$$v_{gb} = \frac{1}{3}\left(-v_{ab} + v_{bc}\right) \tag{11.5}$$

$$v_{gc} = \frac{1}{3}\left(-v_{ab} - 2v_{bc}\right) \tag{11.6}$$

The quadrature and in-phase unit vector are extracted from the peak value of generator voltage and it is calculated as:

$$V_t = \sqrt{\frac{2}{3}} \times \sqrt{v_{ga}^2 + v_{gb}^2 + v_{gc}^2} \tag{11.7}$$

The in-phase unit vector are evaluated as:

$$u_{pa} = \frac{v_{ga}}{V_t},\ u_{pb} = \frac{v_{gb}}{V_t},\ \text{and } u_{pc} = \frac{v_{gc}}{V_t} \tag{11.8}$$

The quadrature unit vectors are evaluated by using the in-phase unit vectors and it is calculated as [24]:

$$u_{qa} = \frac{u_{pa}}{\sqrt{3}} + \frac{u_{pc}}{\sqrt{3}} \tag{11.9}$$

$$u_{qb} = \frac{\sqrt{3}u_{pa}}{2} + \frac{(u_{pb} - u_{pc})}{\sqrt{3}} \tag{11.10}$$

$$u_{qc} = \frac{\sqrt{3}u_{pa}}{2} + \frac{(u_{pb} - u_{pc})}{2\sqrt{3}} \tag{11.11}$$

11.3.3 Switching Pulses Generation for VSC

The DROGI-FLL-based control scheme for the generation of switching pulses for VSC is illustrated in Figure 11.3. The load current input is given to the abc–αβ stationary frame. The DROGI-based FLL estimates the phase angle and peal value of fundamental load current. The peak value of the active/reactive component of the load current can be calculated by using two different sample and hold (S/H) circuits and zero crossing detectors (ZCD). The in-phase/quadrature unit vectors yield input to ZCD; the ZCD generates edge triggering pulses for S/H. The output of S/H is given to the low-pass filter (LPF), and this produces active component of load current. Similarly, a parallel computation path is used for the evaluation of reactive component of load current.

The generator voltage is fed to the three-phase PLL to determine the actual value of the frequency. The error between the actual frequency and reference frequency is fed to the DC voltage PI controller. The summation of the output of PI controller and PV current is subtracted from active load current component, which yields active component for reference generator current:

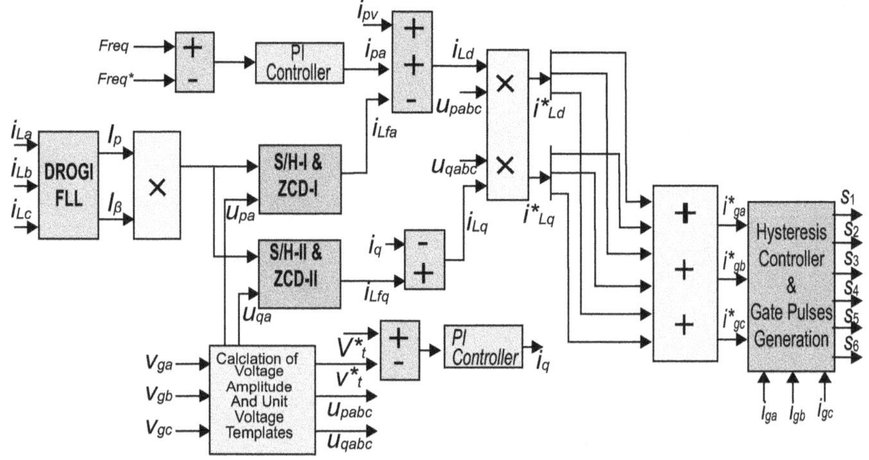

FIGURE 11.3 Control scheme of a standalone hybrid system.

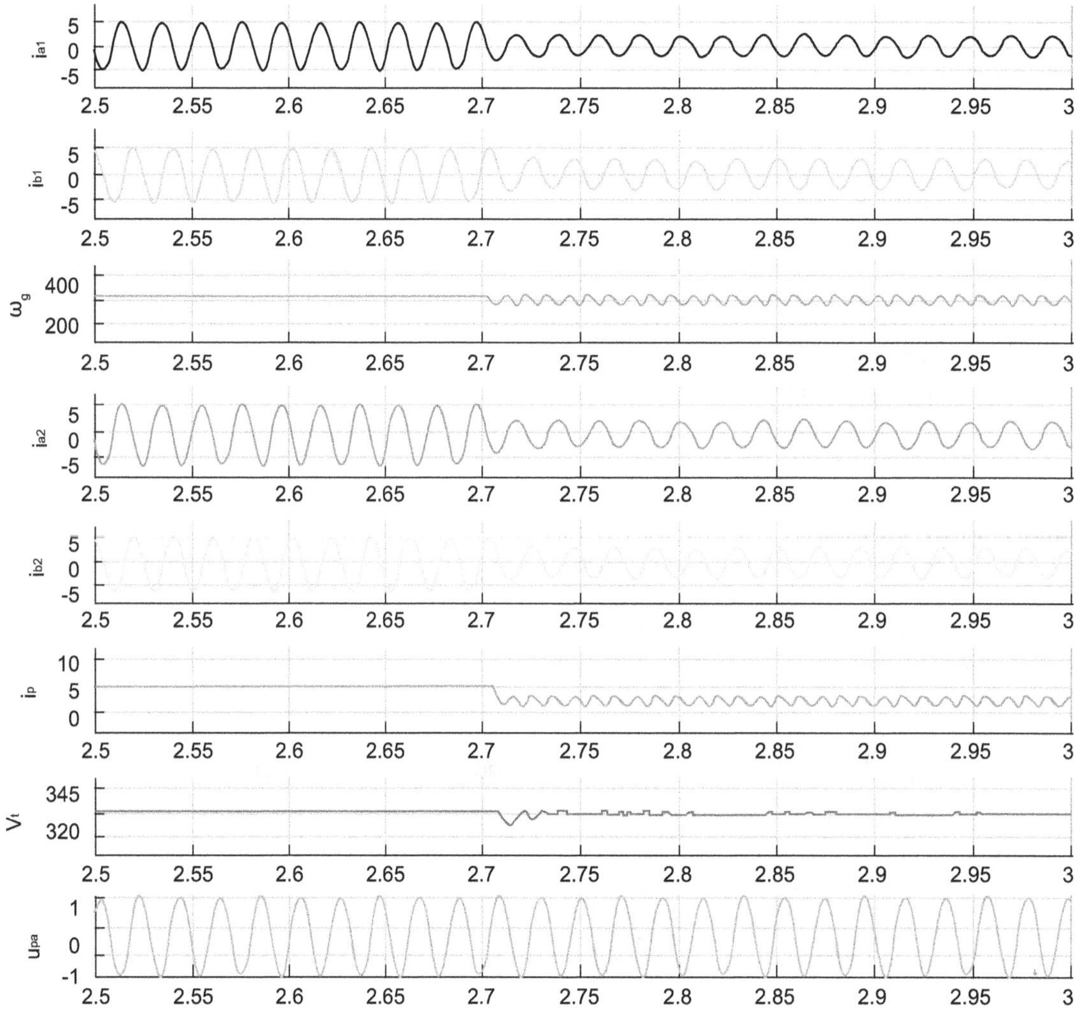

FIGURE 11.4 Internal control signal of DROGI-FLL control algorithm: (a) active/reactive load current component of ROGI-I and II, center frequency, amplitude of PCC voltage; (b) active/reactive load current component, DC PI controller current, PV current; (c) extracted active/reactive component, extracted reference source current.

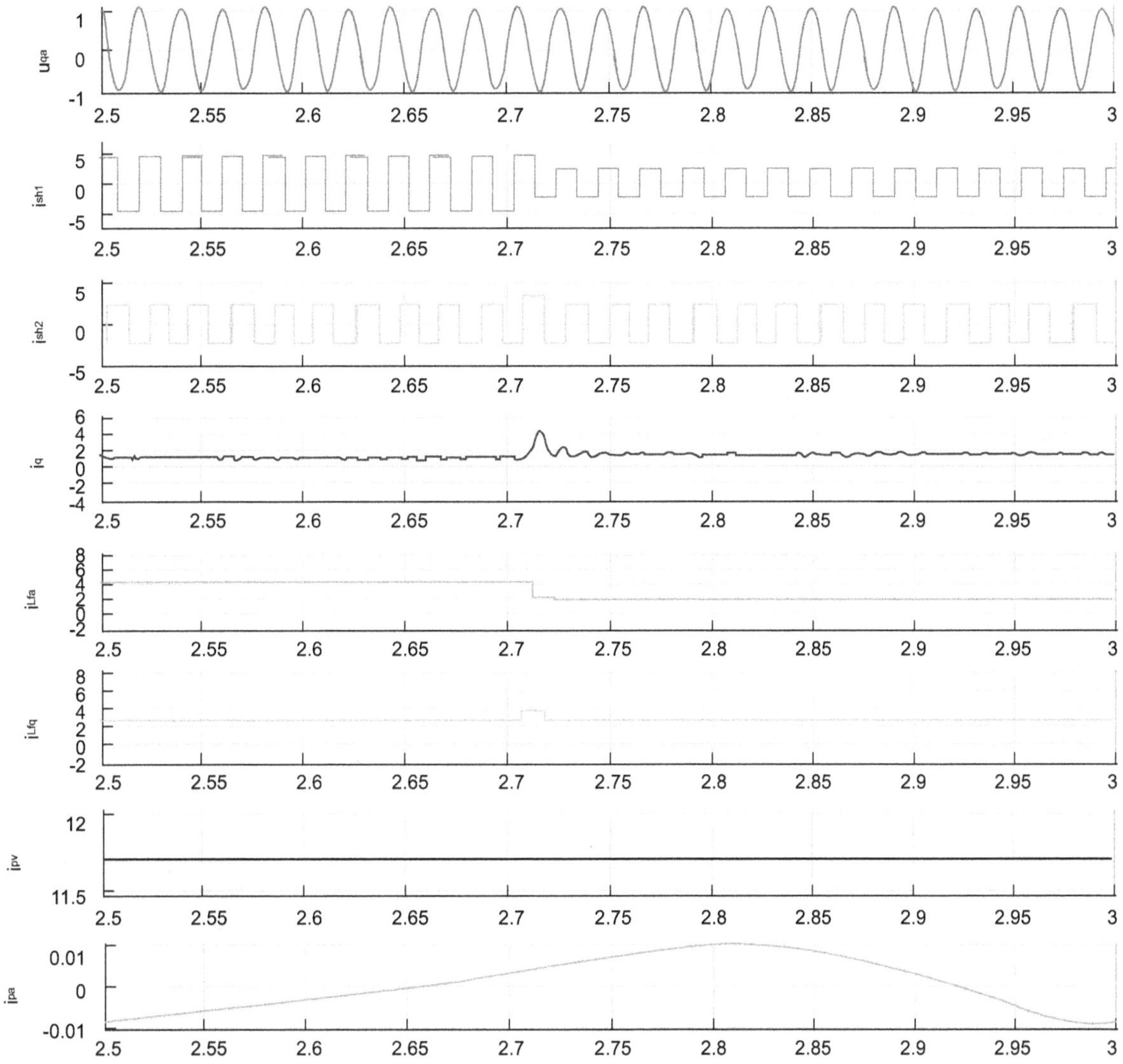

(b)

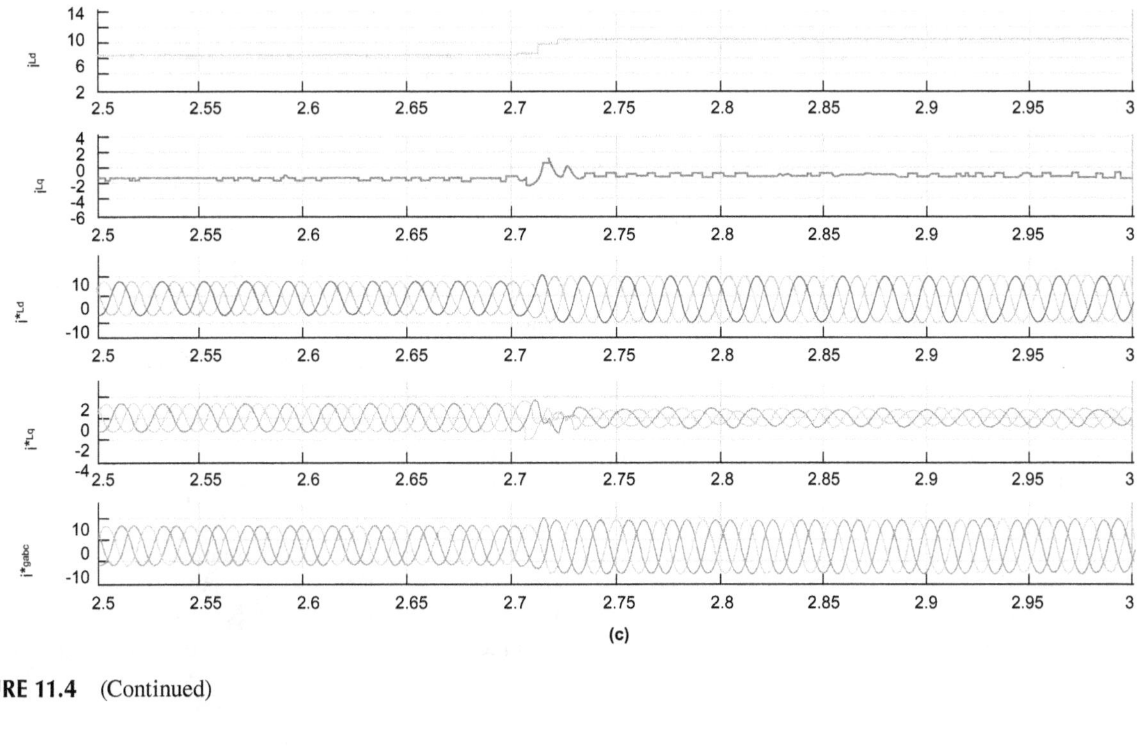

(c)

FIGURE 11.4 (Continued)

$$i_{Ld} = i_{pa} + i_{pv} - i_{Lfa} \tag{11.12}$$

$$i^*_{Ld} = i_{Ld} \times u_{pabc} \tag{11.13}$$

The peak value of reactive component of reference generator current is calculated as:

$$i_{Lq} = i_q - i_{Lfq} \tag{11.14}$$

$$i^*_{Lq} = i_{Lq} \times u_{qabc} \tag{11.15}$$

The summation of Eqs. (11.14) and (11.15) yields the reference generator current:

$$i^*_{gabc} = i^*_{Ld} + i^*_{Lq} \tag{11.16}$$

For the generation of the VSC switching signal, the sensed reference generator current and extracted sinusoidal reference generator current are fed to the hysteresis controller [25, 26].

11.4 RESULTS AND DISCUSSION

The simulation results show the dynamic response of 3P4W hybrid system for isolated locations. The behavior of the proposed standalone hybrid system is studied and analyzed under fixed/varying wind speed/solar intensity/linear/nonlinear loads. To analyze dynamic response of the hybrid system, simulation results accommodate significant signals, like generator voltage (v_{gabc}), generator current (i_{gabc}), load current (i_{Labc}), controller current (i_{cabc}), frequency (f), neutral current (i_{sn}), battery voltage (v_b), battery current (i_b), PV voltage (v_{pv}), PV current (i_{pv}), solar PV power (p_{pv}), and duty cycle. The internal signals of the DROGI-FLL control algorithms that have been implemented are also deeply analyzed.

11.4.1 Internal Signals of DROGI-FLL Algorithm

Figure 11.4 shows the all internal control signals of 3P4W standalone wind/solar hybrid system. The control signals of a standalone hybrid system are designed under phase a of linear/nonlinear load disconnected at t = 2.7 seconds. Figure 11.4(a) Shows the control signals of DROGI-FLL control algorithm $i_{a1}, i_{b1}, \omega_g, i_{a2}, i_{b2}, i_p, V_t$ and u_{pa} where, $i_{a1}, i_{a2}, i_{b1}, i_{b2}, \omega_g, i_p, V_t$ and u_{pa} are the fundamental estimated in-phase current of ROGI-I & ROGI-II, fundamental estimated quadrature current of ROGI-I & ROGI-II, center frequency, peak value of extracted component of load current, amplitude of generator voltage and in-phase unit vector of phase a. Figure 11.4(b) shows the internal control signals of control algorithm are $u_{qa}, i_{sh1}, i_{sh2}, i_q, i_{Lfa}, i_{Lfq}, i_{pv}$ and i_{pa} where the signals are the estimated quadrature unit vector, sample and hold circuit current, AC PI controller current, extracted active/quadrature load current component, solar PV current, and DC PI controller current. Figure 11.4(c) shows the internal control signal of control algorithm are $i_{Ld}, i_{Lq}, i^*_{Ld}, i^*_{Lq}$ and i^*_{gabc} where the control signals represent the extracted active component of load current and quadrature current, extracted reference source current. The control signal gives rapid dynamic response, fast convergence speed to the DROGI-FLL control algorithm, and removes the disturbing harmonic component under worst load condition.

11.4.2 Dynamic Performance of Wind/Solar Hybrid System under Varying Wind Speed/Fixed Solar Intensity Feeding Linear/Nonlinear Loads

The dynamic performance of a 3P4W standalone wind/solar hybrid system under varying wind speed and fixed solar intensity feeding linear/nonlinear load is shown in Figures 11.5. The wind speed varies from 14–18 m/s at t = 2.7 seconds and solar intensity and temperature are fixed at 900 ω / m^2 and 25°C. Due to decline in wind speed, the generated power is reduced then the battery system is to provide real power support to the system and the compensator current is slightly increasing at t = 2.7 sec. As a result, the battery charging current is reduced to meet load requirements while maintaining active power. The excess generated power transfers to the battery for power leveling. As a result, when the generator and solar PV are operating at constant power, the battery power increases. The controller maintains the generated voltage, generator current, load current, and system frequency as constant. Figure 11.5(b) shows the solar intensity level changes; so does the output power of the SPV. When the generator power is kept constant while the solar intensity increases, the extra power is fed to the battery, and the battery enters charging mode. Similarly, as solar intensity decreases, the battery discharges to feed the required load. As a result, the controller maintains the power balance even when the solar intensity changes dynamically.

11.4.3 Dynamic Performance of Wind/Solar Hybrid System under Varying Solar Intensity/Fixed Wind Speed Feeding Linear/Nonlinear Loads

The dynamic performance of a standalone wind/solar hybrid system under varying solar intensity and fixed wind speed feeding linear/nonlinear load is shown in Figure 11.6. The wind speed is fixed at 18 m/s and the solar intensity and temperature are vary from 600–900 ω / m^2 and 20–25°C. Due to reduction in solar intensity, the generator current, battery voltage, and battery current are slightly reduced and compensator current is slightly increased. The DROGI-FLL–based VSC controller maintains the generated voltage, load current and frequency are constant. Figure 11.6(b) shows that due to change in solar intensity at t = 2.7 seconds, the PV current, solar PV power, and duty cycle are reduced. If the solar intensity is reduced, the PV current is also reduced as the solar intensity decreases, resulting in a

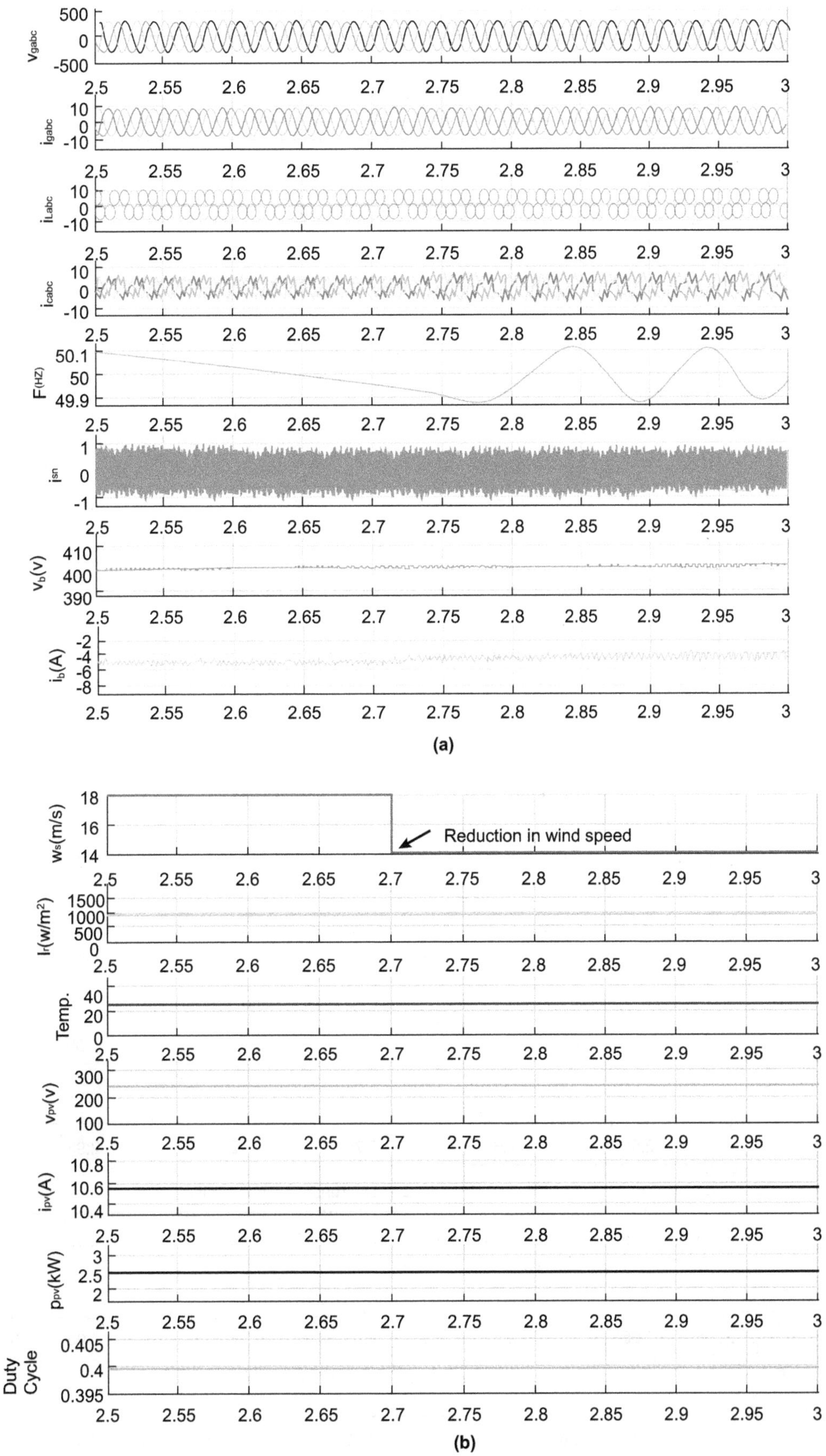

FIGURE 11.5 (a) Dynamic response of wind/solar hybrid system; (b) Variables of SPV system and wind speed under varying wind speed and fixed solar intensity under feeding linear/nonlinear load.

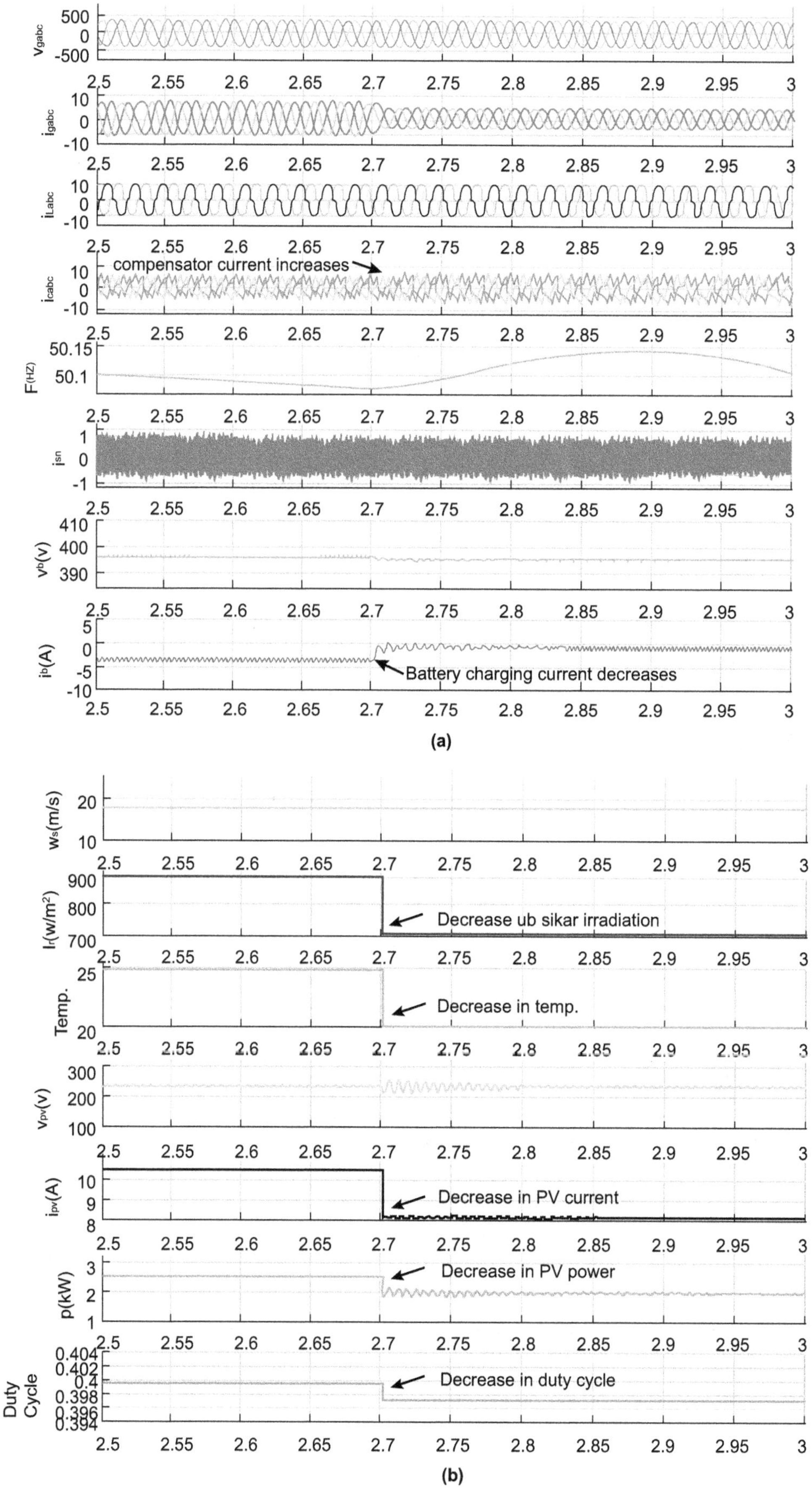

FIGURE 11.6 (a) Dynamic response of wind/solar hybrid system; (b) Variables of SPV system and wind speed under varying solar intensity and fixed wind speed feeding linear/nonlinear load.

reduction in total generated power. The load, on the other hand, is fixed, so the battery is used to provide the necessary power. As a result, the battery current decreases, and the battery switches from charging to discharging mode. When the connected linear/nonlinear load is fixed, the generated solar power drops instantly; as a result, the battery current discharges (changes from charging to discharging mode) to maintain the required power balance between source and load.

11.4.4 Dynamic Performance of Wind/Solar Hybrid System under Fixed Wind Speed/Solar Intensity and Feeding Varying Linear/Nonlinear Loads

Figure 11.7 depicts the dynamic performance of a standalone hybrid system with varying linear/nonlinear load and fixed wind speed/solar intensity. Phase a of linear/nonlinear load is disconnected at t = 2.65–2.85 seconds, then the generator current is remain sinusoidal and balanced. During this duration, the compensator current and battery current is increasing and load current is decreasing. The generated voltage, neutral current, and frequency are all kept constant by the controller. As results of the removals of load one phase a, the load requirement is reduced and the excess power that was previously supplied to the load is now fed to the battery. Hence, the battery current (+ ve direction) increases.

11.4.5 Power Balance Analysis of Standalone Hybrid System

Figure 11.8 shows the simulation result of power balance analysis of standalone hybrid system under different operating conditions. Figure 11.8(a) shows that the controller stabilized the generated power, load power, and solar PV power under fixed wind speed/solar intensity feeding linear/nonlinear load. Figure 11.8(b) shows that during

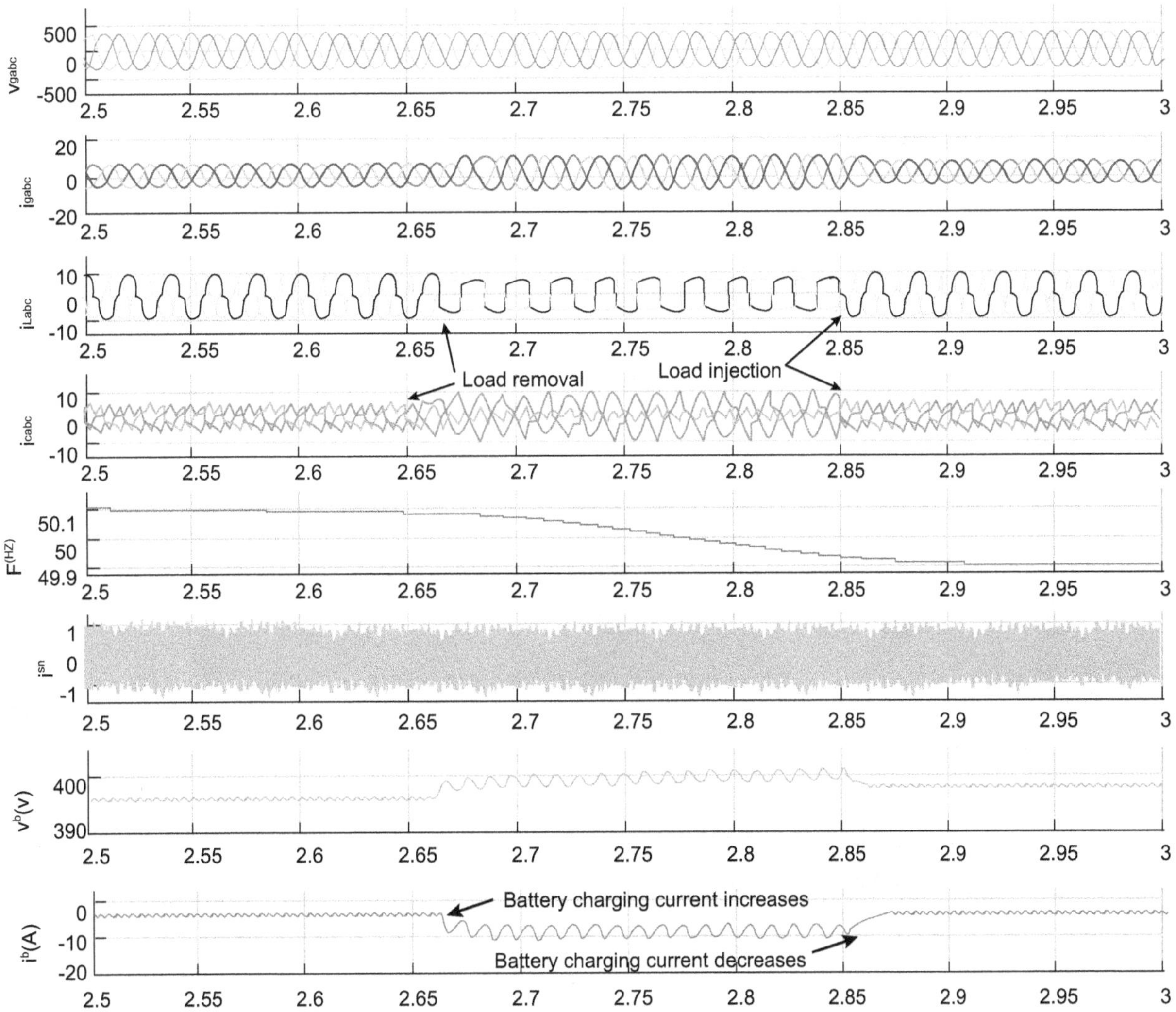

FIGURE 11.7 Dynamic response of wind/solar hybrid system under fixed wind speed/solar intensity feeding varying linear/nonlinear load.

decline in wind speed, the controller maintains the system frequency/voltage and the balance power is provided by the battery system. Figure 11.8(c) shows that the reduction in generated power, solar power, and load power is constant due to decline in solar intensity at t = 2.7 seconds. Figure 11.8(d) shows the rise in generated power and reduction in load power at t = 2.65–2.85 seconds due to disconnection of load of phase a. During removal of load, the extra power

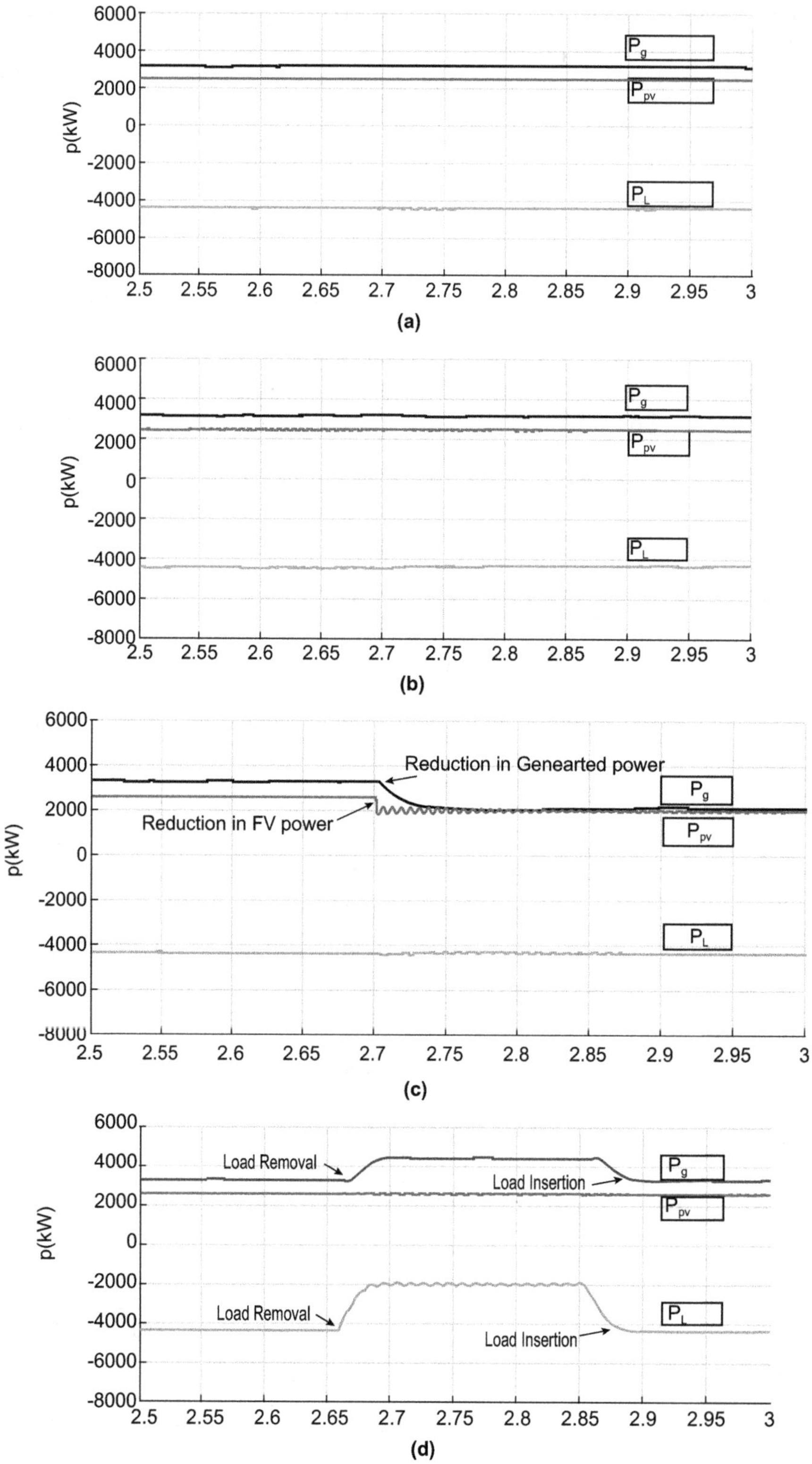

FIGURE 11.8 Power balance analysis of standalone hybrid system: (a) fixed wind speed/solar intensity feeding linear/nonlinear load; (b) varying wind speed and fixed solar intensity feeding linear/nonlinear load; (c) varying solar intensity and fixed wind speed feeding linear/nonlinear load; (d) fixed wind speed/solar intensity feeding linear/nonlinear load.

supply is fed to the battery for charging. The DROGI-FLL–based controller maintains the power under varying load conditions/solar intensity/wind speed. The battery system with VSC is used to provide active/reactive power support and consume the extra power during less load demand.

Table 11.1 represents the power balance of standalone hybrid system under different operating conditions. The DROGI-FLL–based controller maintains the power balance between the source and the load.

11.4.6 Waveform and Harmonic Analysis of Standalone Hybrid Systems

Figure 11.9 shows the waveform and harmonic spectrum of standalone hybrid system feeding linear/nonlinear load.

TABLE 11.1
Generated Power, Load Power, and PV Power under Different Operating Conditions of Standalone Hybrid Systems

S.No.	Parameters	Generated Power (p_g)	Load Power (p_L)	PV Power (p_{pv})
1	Fixed wind speed/solar intensity feeding linear/ nonlinear load	3.2 kW	4.4 kW	2.5 kW
2	Declining wind speed and fixed solar intensity feeding linear/nonlinear load	3.2 kW	4.27 kW	2.5 kW
3	Fixed wind speed and decline solar intensity feeding linear/nonlinear load	2.06 kW	4.4 kW	1.93 kW
4	Fixed wind speed/solar intensity, when load is disconnected of phase a at t = 2.65–2.85 seconds	4.34 kW	1.98 kW	2.5 kW

The proposed 3P4W standalone hybrid system gives good response under nonlinear load even when one- or two-phase load is disconnected. The THD of the generated voltage, generator current, and load current are 0.64%, 4.17% and 26.45%, respectively. The THD of the generator current is becomes below than 5%, which is under theIEEE-519 standard.

Table 11.2 represents the THD analysis of generated voltage, generator current, and load current per phase under nonlinear load of the standalone hybrid system.

11.5 SUMMARY

In this chapter, a proposed standalone hybrid system with a 3P4W load has been connected with a SPV and a wind driven SEIG. DROGI-FLL–based control has been implemented for control of the voltage/frequency and improves the power quality of the system. The DROGI-FLL–based control algorithm successfully extracts the amplitude of fundamental load current component, frequency, and phase angle of load current. To harness maximum power from the SPV, the INC-based MPPT control varies the duty cycle of the boost converter under changing conditions of solar intensity. A battery system, which is interfaced directly to the DC bus, manages the power balance under the changes loading condition and variable power generation. The control algorithm with VSC yields fast convergence, harmonics eradication, reactive power, enhanced current source imbalance rejection ability, and balancing of load, and the fourth leg of VSC is used to provide neutral current compensation. The simulation results shows the dynamic response of 3P4W standalone wind/solar hybrid system and the control algorithm with VSC is able to regulate system voltage/frequency, and improves the system power quality at varying wind speed and changing solar intensity feeding varying linear/nonlinear load. The THDs of generated voltage, generator current, and load current are found well within the limits of the IEEE-519 standard.

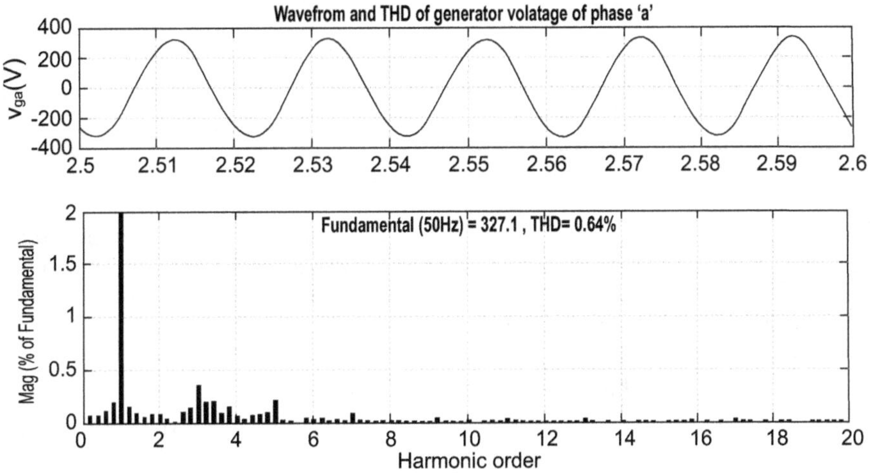

(a)

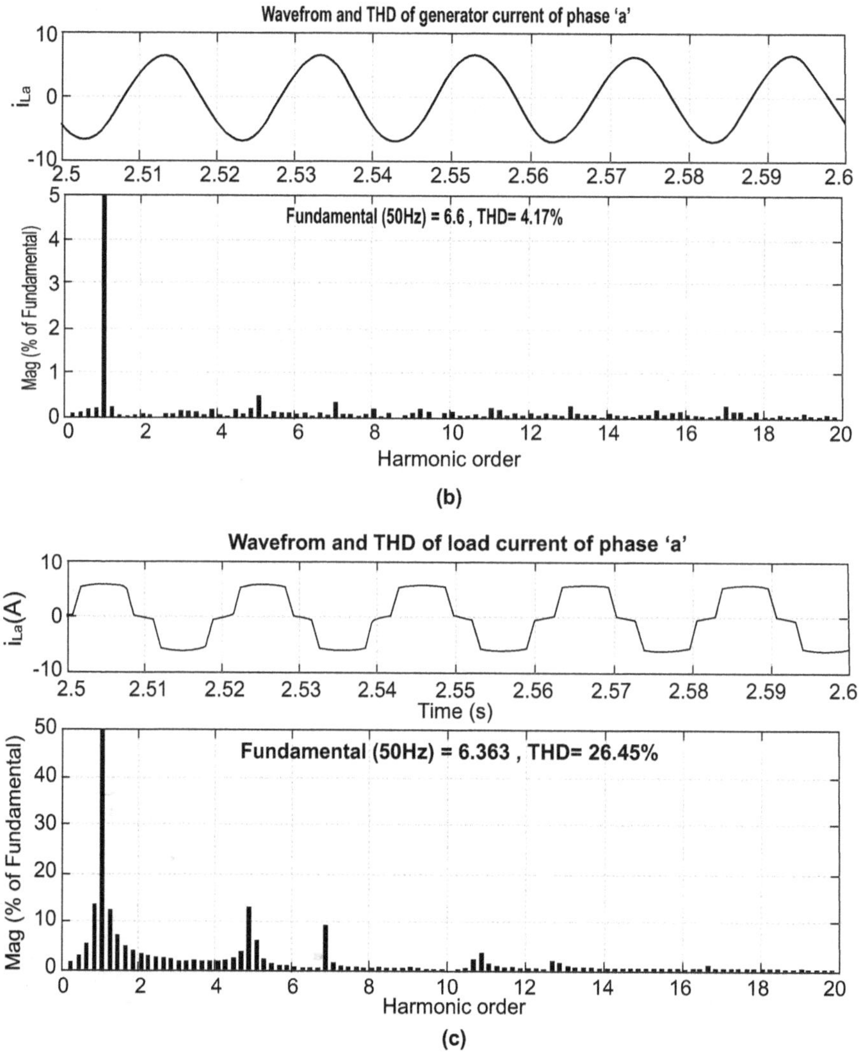

FIGURE 11.9 Waveform and harmonic analysis of: (a) generated voltage; (b) generator current; (c) load current.

TABLE 11.2
Harmonic Analysis of All the Phases from Simulation

S. No.	Phase	Values of THD		
		v_g, %	i_g, %	i_L, %
1	A	0.64	4.17	26.45
2	B	0.67	4.20	26.85
3	C	0.67	4.22	27.01

11.6 EXAMPLES

11.6.1 Solved Problems

1. For a 400 V battery capacity of 7.5 Ah (initially full charge) that is supplying a load of 2,000 Ω, find the state of charge (SoC) of the battery after six hours.

Solution

Given $V_{OC} = 400$ V, $\mu_n = 7.5$ Ah, time (Δt) = 6 hours

Battery is initially full charge i.e. SoC (t − 1) = 100%

Load current (i_L) = Battery current (i_B) = i(t)

$$= \frac{\text{Battery Voltage}(v_B)}{\text{Load Resistance}(R_L)} = \frac{400}{2000} = 0.2\ \text{A}$$

Battery SoC at time instant 't':

$$\text{SoC (t)} = \text{SoC (t−1)} + \frac{i(t)}{\mu_n} \times \Delta_t \times 100 = 100 + \frac{-(0.2)}{7.5} \times 6 \times 100$$

$= 100 - 16 = 84\%$

2. Calculate the power output from a wind turbine with a blade length of 60 m; wind velocity (v_w) is 14 m/s, air density of 1.225 kg/m^3, and power coefficient (c_p) is 0.5.

Solution

Area of wind turbine (A) = $\pi r^2 = \pi \times 50 \times 50 = 7850\ \text{m}^2$

Power output of wind turbine (P) = $\frac{1}{2}\rho A v_w^3 c_p$

$= \frac{1}{2} \times 1.225 \times 7850 \times (14)^3 \times 0.5$

$= 6.596$ MW

3. A three-phase voltage source inverter is supplied from a 600 V DC source. For a star-connected resistive load of 30 Ω per phase, calculate the load power in kW for 120° device conduction.

Solution

Supplied voltage (V_{oc}) = 600 V

Resistive load (R_L) = 20 Ω/phase

RMS value of phase voltage (V_{ph}) = $0.4082 \times V_{dc}$

$= 244.92$ V

$$\text{Load power} = \frac{3v_{ph}^2}{30} = \frac{3 \times (244.92)^2}{30}$$

$= 5.99$ kW

11.6.2 Unsolved Problems

1. The input voltage (v_{in}) is 315 V, duty cycle (d) is 0.25, and the value of inductor ripple current is 10% of the input current. The switch frequency of the boost converter is 20 kHz. Calculate the value of inductor of the boost converter.
2. A wind generator with the following details is installed at a wind site: upward wind velocity is 20 m/s, downward wind velocity is 9 m/s, length of blade is 20 m, and efficiency of generator is 90%. Determine the following:
   a. The input power to wind turbine.
   b. The output power of wind turbine.
   c. The output of generator.
   d. What is the change in the wind turbine, if we increases length of blade to 30 meters?
3. If a PV array has a maximum power output of 10 w under irradiance level of 900 ω/m², determine:
   a. What must the irradiance must be to achieve a power output of 17 w?
   b. Would you expect the open circuit voltage to increase or decrease?

REFERENCES

[1] V. Narayanan, Seema Kewat and Bhim Singh, "Control and implementation of a multifunctional solar PV-BES-DEGS based microgrid" IEEE Transactions on Industrial Electronics, Aug. 2020, doi:10.1109/TIE.2020.3013740.

[2] F. dubuisson, M. Rezkallah, Ambrish Chandra, Maarouf saad, Marco Tremblay and Hussein Ibrahim, "Control of hybrid wind-diesal standalone microgrid for water treatment system application" IEEE Transactions on Industry Applications, vol. 55, no. 6, Nov.–Dec. 2019, doi:10.1109/TIA.2019.2938727.

[3] P. Satish Kumar, R.P.S. Chandrasena, V. Ramu, G.N. sreenivas and K.V.S.M. Babu, "Energy management system for small scale hybrid wind solar battery based microgrid" IEEE access, vol. 8, pp. 2169–3536, doi:10.1109/ACCESS.2020.296405.

[4] V. Narayanan, Seema Kewat and Bhim Singh, "Standalone PV-BES-DG based microgrid with power quality improvement" 2019 IEEE International Conference on Environment and Electrical Engineering and 2019 IEEE Industrial and Commercial Power Systems, Europe, Aug. 2019, Italy, doi:10.1109/EEEIC.2019.8783251.

[5] Priyank Shah, Ikhlaq Hussain and Bhim Singh, "Multi-resonant FLL-based control algorithm for grid interfaced multifunctional solar energy conversion system" IET Science, Measurement & Technology, vol. 12, no., Jan. 2018, pp. 49–62, doi:10.1049/iet-smt.2017.0096.

[6] Seema and Bhim Singh, "Grid synchronization of WEC-PV-BES based distributed generation system using robust control strategy" IEEE Transactions on Industry Applications, vol. 56, no. 6, Nov.–Dec. 2020, doi:10.1109/TIA.2020.3021060.

[7] Menxi Xie, Huiqing wen, Canyan Zhu and Yong Yang, "DC offset rejection improvement in single-phase SOGI-PLL algorithms: methods review and experimental evaluation" IEEE Access, vol. 5, doi:10.1109/ACCESS.2017.2719721.

[8] Priyank Shah, Ikhlaq Hussain and Bhim Singh, "Fuzzy logic based FOGI-FLL algorithm for optimal operation of single-stage three-phase grid interfaced multifunctional SECS" IEEE Trans. on Industrial Informatics, vol. 14, no. 8, Aug. 2018, doi:10.1109/TII.2017.2786159.

[9] Ashutosh K. Giri, Sabha Raj Arya, Rakesh Maurya and B. Chittibabu, "Control of VSC for enhancement of power quality in off-grid distributed power generation" IET Renewable Power Generation, vol. 14, no. 5 pp. 771–778, Apr. 2020, doi:10.1049/iet-rpg.2019.0497.

[10] S. Sombir and Madhusudan Singh, "Voltage and frequency control of self excited induction generator integrated with PV system" IECON, the 46th Annual Conference of the IEEE Industrial Electronics Society, 978-1-7281-5414-5/20, doi:10.1109/IECON43393.2020.9254608.

[11] Ujjwal Kumar Kalla, Bhim Singh, S. S. Murthy, Chinmay Jain and Krishna Kant, "Adaptive sliding mode control of standalone single-phase microgrid using hydro, wind and solar PV array based generation" IEEE Transactions on Smart Grid, vol. 9, no. 6, Nov. 2018, doi:10.1109/TSG.2017.2723845.

[12] Bhasker Rao R, Narsa Reddy Tummuru and B. L. Lingaiah, "Photovoltaic-wind and hybrid energy storage integrated multi-source converter configuration for DC microgrid applications" IEEE Trans. on Sustainable Energy, vol. 12, no. 1, Jan. 2021, doi:10.1109/TSTE.2020.2983985.

[13] Geeta Pathak, Bhim Singh and B. K. Panigarhi, "Wind-hydro microgrid and its control for rural energy system" IEEE Transactions on Industry Applications, vol. 55, no. 3, May–June 2019, doi:10.1109/TIA.2019.2897659.

[14] Ashutosh K. Giri, Sabha Raj Arya and Rakesh Maurya, "Hybrid order generalized integrator based control for VSC to improve the PMSG operation in isolated mode" ICPC2T 2020, pp. 373–378, doi:10.1109/ICPC2T48082.2020.9071498.

[15] Saeed Golestan, Josep M. Guerrero, J. C. Vasquez, A. M. Abusorrah and Yusuf Al-Turki, "A study on three-phase FLLs" IEEE transactions on power electronics, vol. 34, no. 1, Jan. 2019, doi:10.1109/TPEL.2018.2826068.

[16] Ashutosh K. Giri, Sabha Raj Arya, Rakesh Maurya and B. Chitti Babu, "Power quality improvement in standalone SEIG based distributed generation system using lorentzian norm adaptive filter" IEEE Trans. on Industry Applications, vol. 54, no. 5, Oct. 2018, doi:10.1109/TIA.2018.2812867.

[17] Md. Rasheduzzaman and J. W. Kimball, "Modeling and tuning of an improved delayed signal cancellation PLL for microgrid application" IEEE Trans. on Energy Conversion, vol. 34, no. 2, June 2019, doi:10.1109/TEC.2018.2880610.

[18] Abhishek Kumar, Seema Kewat, Bhim Singh and Rashmi Jain, "CC-ROGI-FLL based control for grid tied photovoltaic system at abnormal grid conditions" IET Generation, Transmission & Distribution, vol. 14, no. 17, 04 Sept. 2020, pp. 3400–3411, doi:10.1049/iet-gtd.2019.0765.

[19] Ashutosh K. Giri, Sabha Raj Arya, Rakesh Maurya and Ramakanta Mehar, "Variable learning adaptive gradient based control algorithm for voltage source converter in distributed generation" IET Renewable Power Generation, Oct. 2018, doi:10.1049/iet-rpg.2018.5213.

[20] Ashutosh K. Giri, Sabha Raj Arya, Rakesh Maurya and B. Chittibabu, "Features of power quality in single phase distributed power generation using adaptive nature vectorial filter" IEEE Transactions on Power Electronics, vol. 33, no. 11, pp. 9482–9495, nov. 2018, doi:10.1109/TPEL.2017.2789209.

[21] Duy C. Huynh and Matthew, "Development and comparison of an improved incremental conductance algorith for tracking the MPP of a solar PV panel" IEEE Trans. on Sustainable Energy, vol. 7, no. 4, Oct. 2016, doi:10.1109/TSTE.2016.2556678.

[22] Sombir Kundu, Madhusudan Singh and Ashutosh K. Giri, "Design and control of a standalone wind—solar system with SEIG feeding linear/nonlinear Loads" Proceedings of symposium on power electronic and renewable energy systems control, Lecture notes in electrical engineering, vol. 616, Springer, doi:10.1007/978-981-16-1978-6_28.

[23] Rohini Sharma, Seema Kewat and Bhim Singh, "Robust 3IMPL control algorithm for power management of SyRG/PV/BES-based distributed islanded microgrid" IEEE Trans. on Industrial Electronics, vol. 06, no. 10, Oct. 2019.

[24] Saeed Golestan, J. M. Guerrero, J. C. Vasquez, A. M. Abusorrah nd Yusuf Al-Turki, "Harmonic linearization and investigation of three-phase parallel-structured signal decomposition algorithms in grid-connected applications" IEEE Trans. on Power Electronics, pp. 0885–8993, Apr. 2021, doi:10.1109/TPEL.2020.3021723.

[25] Sabha Raj Arya, Mittel M. Patel, Sayed Javed Alam and Ashutosh K. Giri, "Phase lock loop-based algorithms for DSTATCOM to mitigate load created power quality problems" International transactions on electrical energy systems, Aug. 2019, doi:10.1002/2050-7038.12161.

[26] Abhishek Ranjan, seema Kewat and Bhim Singh, "DSOGI-PLL based solar grid interfaced system for alleviating power quality problems" 2019 National Power Electronics Conference (NPEC), 978-1-7281-4428-3/19, Mar. 2020, doi:10.1109/NPEC47332.2019.9034851.

[27] Ashutosh K. Giri, Sabha Raj Arya, Rakesh Maurya and B. Chiiti Babu, "VCO-less PLL control-based voltage source converter for power quality improvement in distributed generation system" IET Electric Power Applications, vol. 13, issue 8, pp. 1114–1124, Mar. 2019, doi:10.1049/iet-epa.2018.5827.

12 Intelligent Proportional Integral-Sliding Mode Control for Standalone Variable-Speed Wind Energy Conversion Systems

Anjana Jain, Saravanakumar Rajendran, K. Deepa, Ashutosh K. Giri, Sabha Raj Arya, Biswajeet Rout, and Prabhakaran K. K.

CONTENTS

List of Symbols

Kp	Proportional gain
Ki	Integral gain
v	Wind speed
ρ	Density of air
A′	Area swept by blades of turbine
λ′	Wind turbine's tip-speed ratio
$C_p{}'$	Wind turbine's power coefficient
β′	Pitch angle of blades
$P_{turbine}$	Turbine power
ω_r	Rotor rotational speed
r	Wind turbine radius
θ_e	Electrical angle
T_{mech}	Torque of turbine
T_{gen}	Electromagnetic torque of generator
P	Pairs of pairs
ω_m	Rotor mechanical speed
ω_e	Rotor electrical speed
J	Moment of inertia
B	Coefficient of viscous friction
R_s	Resistance of winding of stator
L_d	Direct-axis stator inductance
L_q	Quadrature-axis stator inductance
ϕ_m	Flux-linkage
V_{sd}	Direct-axis stator voltage
V_{sq}	Quadrature-axis stator voltage
I_{sd}	Direct-axis stator current
I_{sq}	Quadrature-axis stator current
V	Terminal voltage
I_{Load}	Load current
I_{gen}	Generator current
$I_{Battery}$	Battery current
I_{ga}	Generator phase-a current
I_{gb}	Generator phase-b current
I_{gc}	Generator phase-c current
P_{Load}	Load power
P_{gen}	Generator power
$P_{Battery}$	Battery power

12.1 INTRODUCTION

Wind energy is viable to replace conventional sources. However, for efficacy of wind energy conversion systems (WECS), use of gear-box in doubly-fed induction generator (DFIG) is one of the limitations that lead to use of direct coupled (to wind turbine) permanent magnet synchronous generators (PMSG) [1–4]. For this scheme to work in an efficient manner, control of frequency and terminal voltage at load end is needed [2], and this is the primary area of study and research in the presented work. Neural-network based frequency and voltage control (VF) for standalone operation of WECS comprises of battery energy storage system (BESS) and PMSG is discussed in [3]. The battery is one necessary element required to absorb the difference between power of generator and load in standalone systems [5, 6]. DSOGI-PLL (phase locked loop with dual second order generalized integrator) is applied during

DOI: 10.1201/9781003229124-12

disturbed grid to achieve orthogonal signals with harmonic and offset rejection [7, 8]. PI and PID (proportional–integral–derivative) are the conventional controllers, simple and with easy implementation. Sensitivity towards non-linearity, parameters variation, weak dynamic response, and poor rejection of disturbances are its major drawbacks [9–11]. In classical PID, ensuring good performances with disparate plants and guaranteeing a suitable adaptation for time-dependent plant is difficult. The urgency of complex mathematical modeling of partially known or unknown system's dynamics is eliminated in model-free-control (MFC), which is modelled in a very simple manner and incurs low computational cost. MFC is proven to be an efficient nonlinear robust technique, and it is updated continuously by using online-estimation-techniques. In i-PID controllers, without any modeling and identification procedure, a strongly nonlinear and time-dependent system's unknown dynamics is considered. Overshoots and undershoots are avoided. Better behavior and efficiency of i-PID are discovered compared to classical PID [12]. MFC provides smooth control variable and is more robust with respect to noises, but complicated derivation of noise is required for SMC [13]. MFC and event triggered control with reduced complexity is proposed for motion control with added robustness of model-free control to the event-based scheme [14]. Analysis of stability of a MF intelligent PI (iPI) controller for flexible-joint and link-manipulator with better trajectory-tracking performance is presented in [15, 16]. Transient-swings in the manipulator during tracking of train of pulses are reduced and external disturbances are handled using i-PI. Also, noise and disturbances from the manipulator's dynamics are removed. The challenges associated with MFC in data-driven control are analyzed in [17] and compared with model-based control. An MFC is presented for optimal control of a hydroelectric power plant's water level control [18]. Two model-free SMC techniques for control of tracking-error dynamics are discussed and compared against model-free iPI controller in [19]. These schemes show robustness against the disturbances and variations in parameters. Design process of integral-SMC with PI-type sliding surface for power control of the WECS based on PMSG is given in [20]. This control includes the reference jump and allows it to gain nominal performance during model-uncertainty. This control provides good track of the smooth reference-command.

The main contributions of this chapter are as follows: (1) design of iPI-SMC for control of frequency and voltage for PMSG-BESS–based WECS under varying wind conditions; (2) detailed analysis for comparison of iPI-SMC with PI control for varying working environment; and (3) validation of the performance of the proposed iPI-SMC through OP4510 HIL real-time simulator. A bi-directional VSC is utilized for regulation of frequency and voltage, along with power balance and stability. The various loads considered to validate the effectiveness and robustness of iPI-SMC controller are: (1) balanced 3-φ linear load; (2) balanced 3-φ non-linear load; and (3) unbalanced 3-φ load.

The chapter organization is as follows: PMSG and WT mathematical modelling is given in Section 12.2. A short review of MFC and the control scheme, along with description of these systems, is given in Section 12.3. Simulation analysis, and results and discussions, are presented in Section 12.4. In Section 12.5, validation of proposed iPI-SMC through test results using op4510 RT HIL simulator is presented. Section 12.6 provides the inference drawn in conclusion of the chapter.

12.2 MATHEMATICAL MODEL OF WIND TURBINE AND PMSG

Mathematical modelling of PMSG and WT are given as follows.

12.2.1 Wind Turbine Modelling [7]

The conversion of the wind's kinetic energy into mechanical power is accomplished by the WT. Eq. (12.1) gives the wind's aerodynamics-power:

$$P_{turbine} = P_w' C_p' = \frac{1}{2}\rho A v^3 C_p'\left(\lambda', \beta'\right) \tag{12.1}$$

$$C_p'\left(\lambda', \beta'\right) = C_1\left(\frac{C_2}{\lambda_i} - C_3\beta' - C_4\right)e^{\frac{-C_5}{\lambda_i}} + C_6\lambda' \tag{12.2}$$

$$\lambda_i = \left(\frac{1}{\lambda' + 0.08} - \frac{0.035}{\beta'^3 + 1}\right) \tag{12.3}$$

$$\lambda' = \frac{\omega_r * r}{v} \tag{12.4}$$

Coefficient: $C_1 = 0.5176, C_2 = 116, C_3 = 0.4, C_4 = 5, C_5 = 21, C_6 = 0.0068$.

12.2.2 Modelling of PMSG

Using WT, electrical power is achieved after converting the mechanical power of wind. Eqns. (12.5–12.8) represent the modelling of PMSG:

$$\frac{d\omega_m}{dt} = \left(\frac{1}{J}\right)\left(T_{mech} - T_{gen}\right) \tag{12.5}$$

Variation in rotor-flux is nil, as permanent magnets are used. Eqns. (12.6) and (12.7) show the stator voltages of *d-q* axes:

$$V_{sd} = R_s I_{sd} + L_d\frac{dI_{sd}}{dt} - \omega_e L_q I_{sq} \tag{12.6}$$

$$V_{sq} = R_s I_{sq} + L_q\frac{dI_{sq}}{dt} + \omega_e L_d I_{sd} + \omega_e\phi_m \tag{12.7}$$

Eq. (12.8) presents the electromagnetic torque.

$$T_e = \left(\frac{3}{2}\right)P\left(\phi_m I_{sq} - \left(L_d - L_q\right)I_{sd}I_{sq}\right) \tag{12.8}$$

$L_d = L_q$ for non-salient rotor.

12.3 DESCRIPTION OF THE SYSTEM AND CONTROL SCHEME

Frequency and voltage are varying during load or wind-velocity variations, and these can be maintained constant by applying a proper controller. Figure 12.1 presents the system configuration and description of control equations. PMSG is connected to a horizontal-axis WT. PMSG is supplying the load. At the terminals of generator, BESS is connected via voltage source converter (VSC). Load receives power from BESS during slow wind velocity, and BESS stores it after receiving from PMSG at high wind velocity. Load-current references are developed with iPI-SMC controller. Use of DSOGI-PLL is employed for frequency-measurement. Control-signals are achieved and given to VSC.

Control objectives: As a result of varying wind speed and/or load in WECS, frequency and voltage are not stable. Hence, the aim of this iPI-SMC–controlled VSC is the regulation of frequency and terminal voltage, and providing adequate response in varying operating environments.

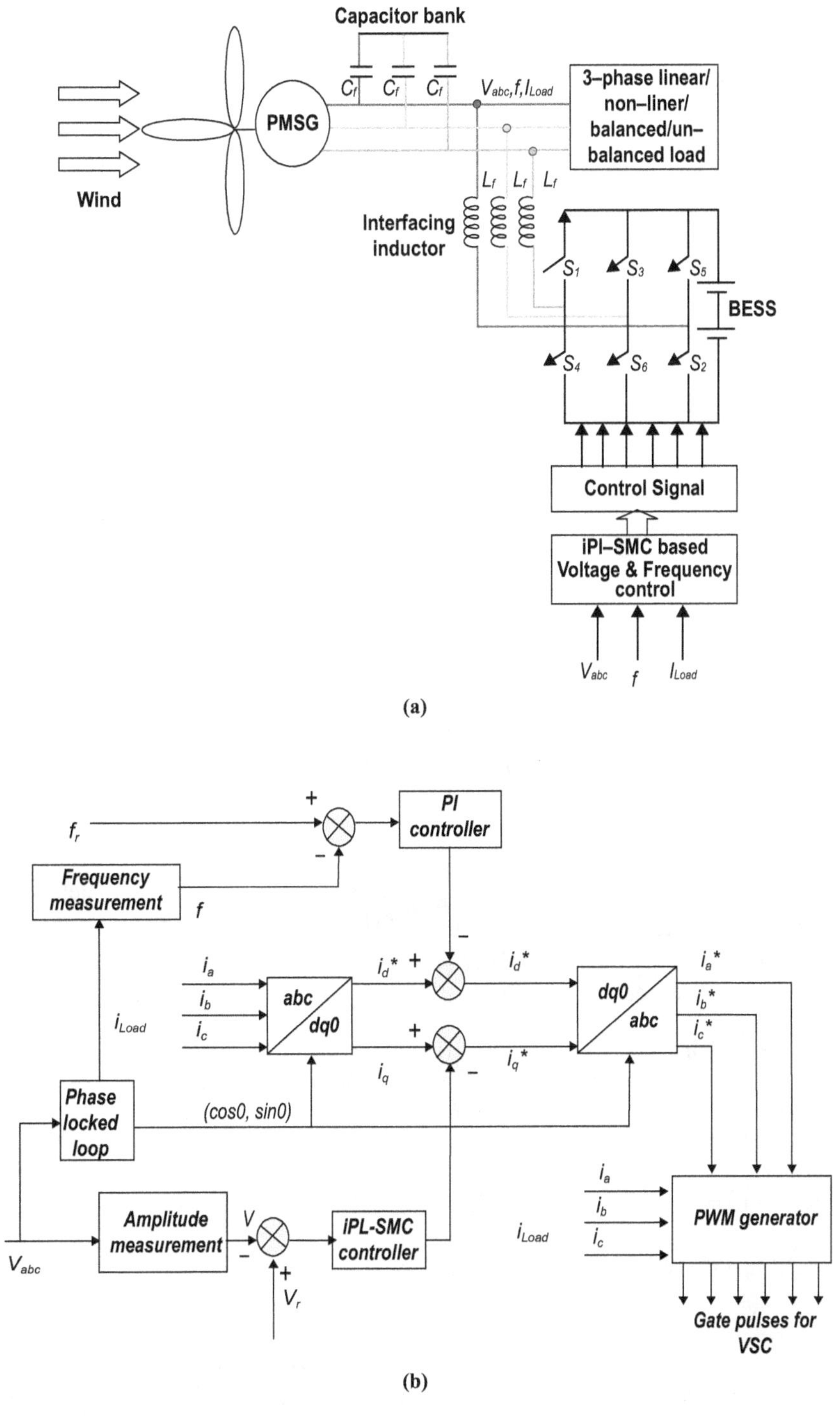

FIGURE 12.1 (a) Configuration of the system; (b) description of the controller.

12.3.1 A Review of MFC

In MFC [12], a finite-dimensional linear or non-linear differential equation (which is either partially known or totally unknown) approximately governs the input-output behavior of the system within its operating range. As an approximation, the input and output are considered as mono-variable. A finite-dimensional SISO (single input single output) system is described by Eq. (12.9):

$$E\left(t, y, \dot{y}, \ldots\ldots\ldots y^{(i)}, u, \dot{u}, \ldots\ldots\ldots u^{(k)}\right) = 0 \qquad (12.9)$$

where E = smooth function of its arguments, y=controlled output-signal, t = continuous variable, and u = control signal

Assuming an integer n, $0 < n \leq i, \dfrac{\partial E}{\partial y^{(n)}} \not\equiv 0$:

$$y^{(n)} = \left(t, y, \dot{y}, \ldots\ldots\ldots y^{(i)}, u, \dot{u}, \ldots\ldots\ldots u^{k}\right) \qquad (12.10)$$

By setting:

$$E = F + \alpha u \qquad (12.11)$$

where $F = y^{(n)} - \alpha u$, $\alpha > 0$ is a constant; it could carry nonlinearities, which impact the un modelled-dynamics of system.

12.3.2 Intelligent PI (iPI)

To get the desired behavior, for $n = 1$, the control law for iPI [19] is by Eq. (12.12):

$$u = -\frac{1}{\alpha}\left(\widehat{F} - \dot{y}_r + K_p e + K_i \int e dt\right) \qquad (12.12)$$

K_p & K_i are the usual tuning constants of PI controller with a transfer function of $C(s) = \left(K_p + \dfrac{K_i}{s}\right)$. Here, $\widehat{F}$ is the estimate of $F(t)$, y_r is the reference output trajectory. The error (e) is defined by Eq. (12.13). Here:

$$e = y - y_r \qquad (12.13)$$

A low-pass filter is utilized to generate the estimate of $\dot{y}$ and is presented by $\hat{\dot{y}}$. Also, a filtered derivative of reference y_r is generated and presented by $\dot{y}_r$. Hence:

$$\widehat{F} = \hat{\dot{y}} - \alpha u \qquad (12.14)$$

Estimation error (e_{est}) of $F(t)$, whose value can be neglected, is given by Eq. (12.15):

$$e_{est} = \left(\dot{y} - \hat{\dot{y}}\right) = \left(F - \widehat{F}\right) \qquad (12.15)$$

From Eqs. (12.12), (12.14), and (12.15):

$$\alpha u = \left(-\hat{\dot{y}} + \alpha u + \dot{y}_r - K_p e - K_i \int e\, dt + \dot{y} - \dot{y}\right)$$

$$\alpha u = \left(\left(\dot{y} - \hat{\dot{y}}\right) - \left(\dot{y} - \dot{y}_r\right) + \alpha u - K_p e - K_i \int e\, dt\right)$$

$$e_{est} = \dot{e} + K_p e + K_i \int e\, dt \qquad (12.16)$$

12.3.3 iPI-SMC and Design for the Proposed System

In variable-speed WECS, continuous change in wind velocity and/or load will cause disturbances, which in turn disturb the voltage and frequency profile. Without modeling, system disturbances and unknown nonlinearities can be housed by model-free iPI control. SMC insure the robustness against the system's disturbances and variations with stable tracking of error dynamics. So, a controller based on iPI-SMC is proposed variable-speed WECS (Figure 12.2). Voltage error is presented by Eq. (12.17):

$$e = \left(V - V_r\right) \qquad (12.17)$$

where V_r= reference voltage and V = terminal voltage

By adding an augmented control signal u_{aug} to Eq. (12.12), Eq. (12.18) gives the control law for iPI-SMC:

$$u = -\frac{1}{\alpha}\left(\widehat{F} - \dot{V}_r + K_p e + K_i \int e\, dt\right) + u_{aug} \qquad (12.18)$$

Eq. (12.19) presents the arrangement of closed-loop-system:

$$e_{est} + \alpha u_{aug} = \dot{e} + K_p e + K_i \int e\, dt \qquad (12.19)$$

x_1 and x_2 are the stated variables:

$$x_1 = \int e\, dt \qquad (12.20)$$

$$x_2 = e \qquad (12.21)$$

Derivative of Eqs. (12.20)–(12.21) are given as follows:

$$\dot{x}_1 = x_2 \qquad (12.22)$$

$$\dot{x}_2 = \dot{e} = e_{est} + \alpha u_{aug} - K_p x_2 - K_i x_1 \qquad (12.23)$$

S is the sliding surface considered for iPI-SMC and given by Eq. (12.24):

$$S = x_1 + T x_2 \qquad (12.24)$$

$T(>0)$ = design parameter to decide the conduct of sliding surface. Lyapunov function presented by Eq. (12.25) specifies the system's stability:

$$v = \frac{1}{2} S^2 \qquad (12.25)$$

For stability of the system:

$$\dot{V} = S\dot{S} \leq 0 \qquad (12.26)$$

Where:

$$\dot{S} = \dot{x}_1 + T\dot{x}_2 = x_2 + T\left(e_{est} + \alpha u_{aug} - K_p x_2 - K_i x_1\right)$$

$$\dot{S} = \left(1 - TK_p\right)x_2 - TK_i x_1 + T\alpha u_{aug} + Te_{est} \qquad (12.27)$$

u_{aug} consists of equivalent-control signal $\left(u_{eq}\right)$ and correction-control signal $\left(u_{correction}\right)$ as follows:

$$u_{aug} = u_{eq} + u_{correction} \qquad (12.28)$$

Ideal condition for sliding-mode stability is $S = 0$. Hence:

$$\dot{S} = 0 \tag{12.29}$$

Then Eqs. (27) and (29) with $u_{aug}=u_{eq}$ gives:

$$u_{eq} = \frac{1}{\alpha T}\left(TK_i x_1 - \left(1 - TK_p\right)x_2 - Te_{est}\right) \tag{12.30}$$

A boundary is adapted to the stability condition in Eq. (12.26). $u_{correction}$ is given by Eq. (12.31):

$$u_{correction} = \left(-\frac{\eta}{\alpha T}\right)sat(S,\varepsilon) = \left(-\frac{\eta}{\alpha T}\right)\begin{cases} -1 \, if \, S < \varepsilon, \\ \frac{S}{\varepsilon} \, if \, |S| \le \varepsilon, \\ 1 \, if \, S > \varepsilon, \end{cases} \tag{12.31}$$

Here, $\eta(>0)$ = convergence-factor and $\varepsilon(>0)$ = thickness of boundary layer. e_{est} is unknown and bounded. It is given by Eq. (12.32):

$$\left|e_{est}\right| \le e_{estmax} \tag{12.32}$$

$$e_{estmax} = \text{upper-boundary of } \left|e_{est}\right|$$

Eqs. (12.28), (12.30), and (12.31) gives:

$$u_{aug} = \frac{1}{\alpha T}\left(TK_i x_1 - \left(1 - TK_p\right)x_2 - Te_{est}\right) + \left(-\frac{\eta}{\alpha T}\right)sat(S,\varepsilon) \tag{12.33}$$

From Eqs. (12.18) and (12.33):

$$u = \frac{1}{\alpha}\left(-\hat{F} + \dot{V}_r - \frac{e}{T} - e_{estmax} - \frac{\eta}{T}sat(S,\varepsilon)\right) \tag{12.34}$$

12.3.4 Stability Proof for iPI-SMC Controller for the Proposed System

Eq. (12.32) presents bounded estimation error. For stability of iPI-SMC based system Eq. (12.27) becomes:

$$\dot{S} = -Te_{estmax} - \eta sat(S,\varepsilon) + Te_{est} \tag{12.35}$$

$$\dot{S} = S\left(e_{est} - e_{estmax}\right)T - S\eta sat(S,\varepsilon) \tag{12.36}$$

Proof: To establish the sliding-mode's stability condition, the following cases are studied for possible values of $S(t)$.

Case 1: $\left|S(t)\right| \le \varepsilon$; State-vector resides in boundary layer.

$$u_{correction} = \left(-\frac{\eta}{\alpha T}\right)\frac{S}{\varepsilon} \tag{12.37}$$

$$S\dot{S} = \left(S\left(e_{est} - e_{estmax}\right)T - S\eta\frac{S}{\varepsilon}\right) = \left(-\frac{S^2\eta}{\varepsilon}\right) + S\left(e_{est} - e_{estmax}\right)T \tag{12.38}$$

$S\dot{S}$ is leads to stability if $S\dot{S}$ is –*ve*. Hence:

$$\left(-\frac{S^2\eta}{\varepsilon}\right) > S\left(e_{est} - e_{estmax}\right)T \tag{12.39}$$

Case 2: $\left|S(t)\right| > \varepsilon$ State-vector is out of boundary layer. Then:

$$u_{correction}\left(-\frac{\eta}{\alpha T}\right)(1) \tag{12.40}$$

$$S\dot{S} = \left(S\left(e_{est} - e_{estmax}\right)T - S\eta\right) \tag{12.41}$$

$S\dot{S}$ is heading towards stability, if $S\dot{S} = -ve$. Then:

$$(S\eta) > S\left(e_{est} - e_{estmax}\right)T \tag{12.42}$$

12.3.4.1 Switching Variable for Steady State

S_∞ is a switching variable under steady state, and satisfies $\dot{S} = 0$, which approaches its boundary layer when:

$$sat(S_\infty, \varepsilon) = \frac{S_\infty}{\varepsilon} \tag{12.43}$$

Then, $\dot{S}_\infty = 0$ leads to:

$$T\left(e_{est\,\infty} - e_{estmax}\right) - \left(\eta\frac{S_\infty}{\varepsilon}\right) = 0 \tag{12.44}$$

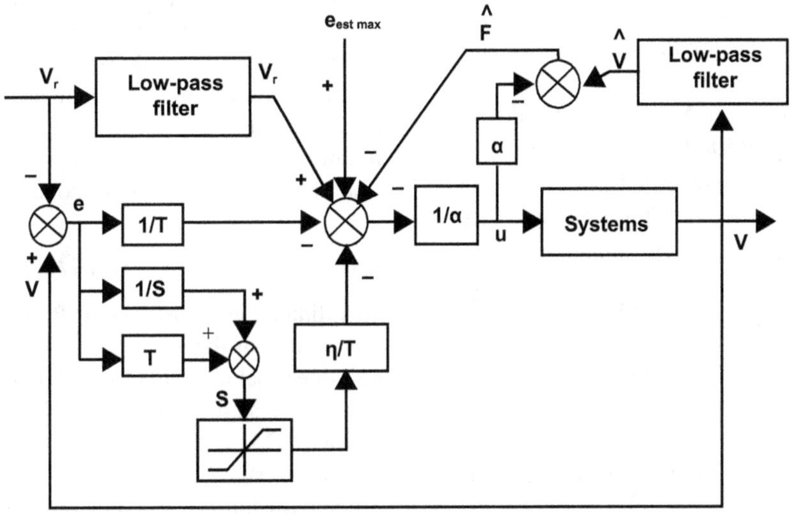

FIGURE 12.2 Control structure of iPI-SMC for the proposed system.

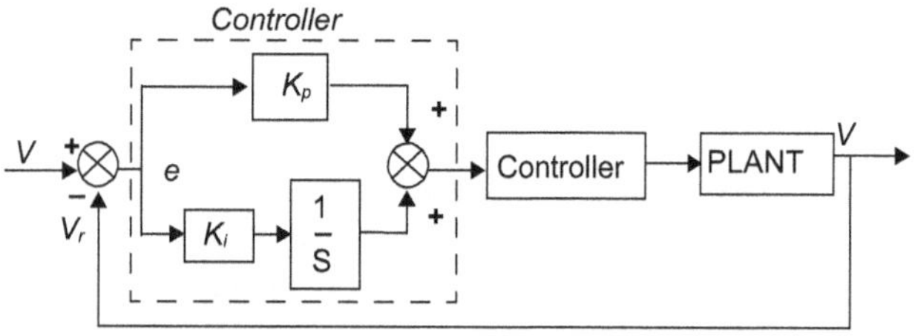

FIGURE 12.3 Structure of PI-based voltage controller.

$$S_\infty = \left(T\frac{\varepsilon}{\eta}\right)\left(e_{est\,\infty} - e_{estmax}\right) \tag{12.45}$$

Here, $e_{est\,\infty}$ is the estimation error under steady state.

12.3.5 PI Controller

The difference among the reference and measured voltage is supplied to the PI controller, and output of PI is utilized to generate signals for the converter. Figure 12.3 is showing the block diagram of the PI controller for voltage control.

12.3.5.1 Terminal Voltage Computation

Phase voltages (V_{sa}, V_{sb}, V_{sc}) are sensed at PCC (point of common coupling), and the amplitude of source voltage is obtained using Eq. (12.46):

$$V_t = \sqrt{\left(\frac{2}{3}\left(V_{sa}^2 + V_{sb}^2 + V_{sc}^2\right)\right)} \tag{12.46}$$

12.3.5.2 Computation of Reference of d-Axis Load Current

Frequency error $f_e(n)$ is achieved by difference among reference and measured frequencies and given to frequency-loop's PI controller. At n^{th} sampling state:

$$f_e(n) = f_r(n) - f(n) \tag{12.47}$$

$$I_d^*(n) - I_d(n-1) + output\ of\ frequency\ loop\ PI\ controller \tag{12.48}$$

where $f_r(n)$ = reference frequency, $f(n)$ = measured frequency

12.3.5.3 Computation of q-Axis Component of Load-Current Reference

Terminal voltage error e(n) is achieved by difference among measured and reference terminal voltages and given to voltage-loop's iPI-SMC controller. At n^{th} sampling state:

$$e(n) = V(n) - V_r(n) \tag{12.49}$$

$$I_q^*(n) = I_q(n-1) + output\ of\ voltage\ loop\ iPI - SMC\ controller \tag{12.50}$$

where $V_r(n)$ = reference terminal voltage, $V(n)$ = measured terminal voltage

12.4 SIMULATION RESULTS AND DISCUSSION

To validate proposed iPI-SMC controller's performance, elaborated analysis based on simulation is performed in MATLAB/Simulink environment under many different work conditions which is then compared with performance of the PI controller. Case I exhibits the dynamic and steady-state response of the controller for varying wind velocity. Two wind profiles are considered: step and trapezoidal. Case II verifies the controller's robustness and capability of power management during various loading. Case III demonstrates controller's response during 1-φ and 3-φ fault at the terminals. In Case IV, performance of controller is discussed under disturbance. The controller's effectiveness evaluation is performed via following parameters: (i) ISE (integral of square of error); (ii) control signal's mean; (iii) ISV (integral of control signal's squared value); and (iv) TV (total-variance) of control signal.

Case I: Steady-State and Dynamic System Performance

The system's performance during steady-state and dynamic conditions with its capability of power management among load, PMSG, and BESS is tested in this case study under varying wind velocity. The (i) simulation starts with 10m/s wind velocity. A 6kW, 100VAR linear load is coupled to the terminals. Wind velocity is changed to 12m/s at t = 0.6 seconds and to 13m/s at t = 1.2 seconds. Figure 12.4(a) shows the terminal voltage for PI and iPI-SMC controllers. From this figure, it can be viewed that iPI-SMC provides faster settling with minimum overshoot during initial transients. Also, transitions between the variable wind velocities for proposed iPI-SMC controller are smother and faster than PI with the least overshoot. Figure 12.4(b) represents the power management among load, generator, and BESS. This figure shows that, if generated power less than the load requirement, during 0 < t < 1.2 seconds, BESS supplies additional power required by load, and for 1.2 < t < 1.8 sec, power generated is larger than load demand; BESS receives surplus power. In (ii), a trapezoidal wind profile is considered which varies from 10–13 m/s. Figure 12.4(c) presents the wind-velocity profile and terminal voltage, and Figure 12.4(d) demonstrates the management of power among load, generator, and BESS for PI and iPI-SMC controllers. It can be observed from these figures that iPI-SMC gives smoother transitions than PI controller.

Case II: Robustness Test During Various Loading

To validate the robustness of iPI-SMC, various load conditions are considered: 1) 3-φ linear balanced load; 2) 3-φ non-linear balanced load; and 3) unbalanced load. Also, the system is tested for 3-φ fault. Various loads are presented in Table 12.1. The controller's capacity for management of power is verified for all the mentioned working conditions. Simulation starts with wind velocity of 13m/s with load-1.

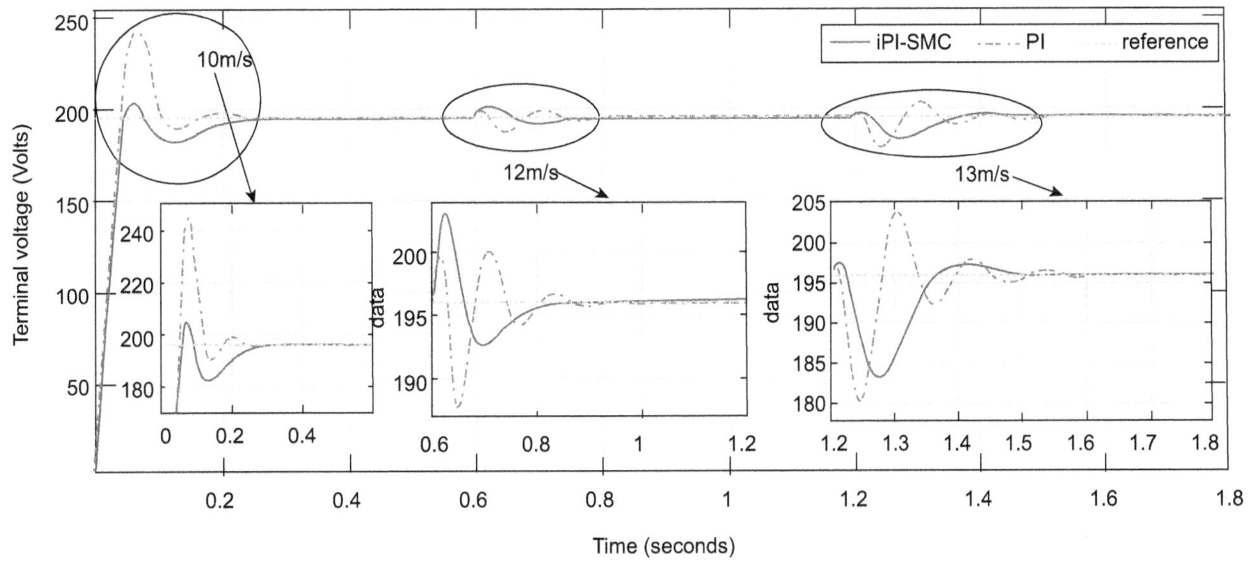

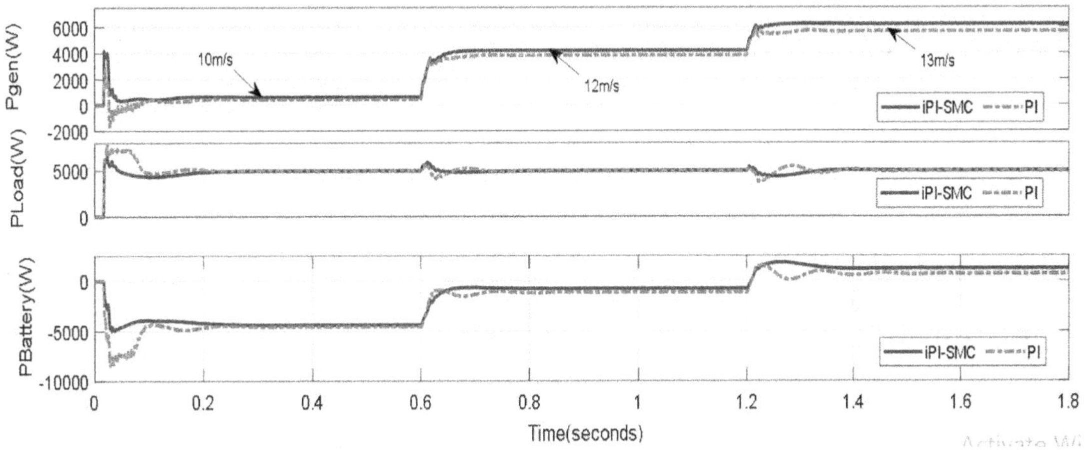

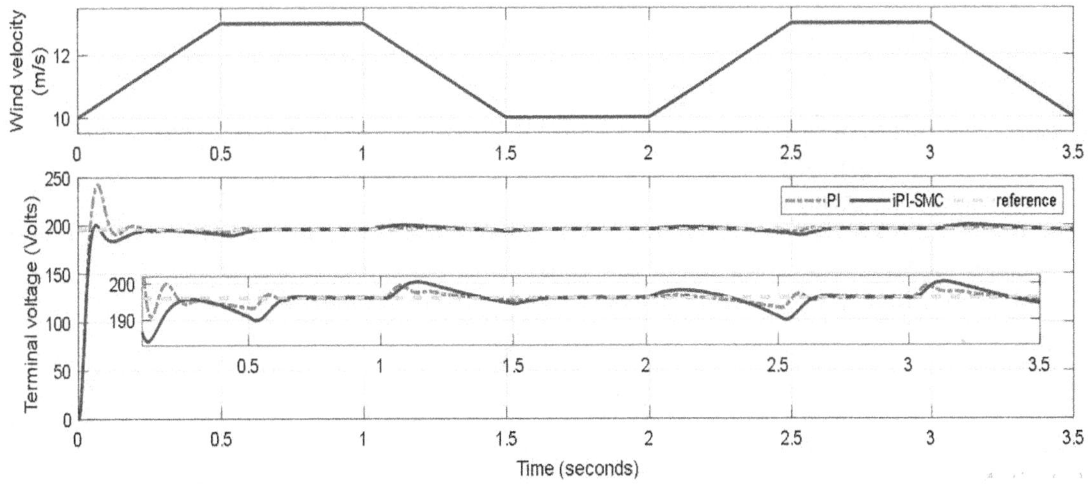

FIGURE 12.4 Terminal voltage and power management (a)–(b) of iPI-SMC and PI among PMSG, load, and BESS for step wind profile; (c) terminal voltage and power management of iPI-SMC and PI among PMSG, load, and BESS for trapezoidal wind profile.

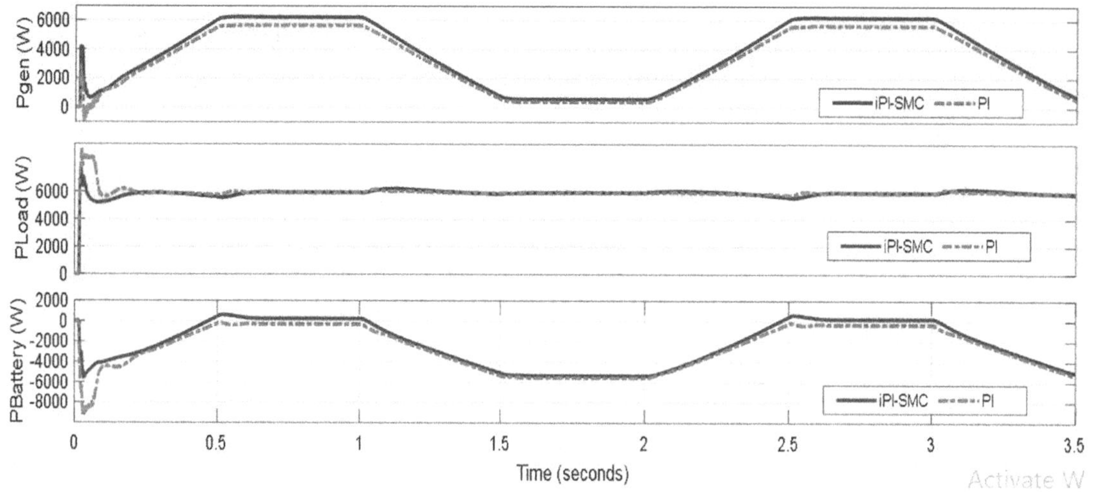

FIGURE 12.4 (Continued)

TABLE 12.1
Different loads

	Load	kW, VAR	Time
Load-1	3-φ balanced linear load	3.5kW, 100VAR	0–0.6 seconds
Load-2	3- φ balanced non-linear load	Load-1 + 1.5kW, 100VAR	0.6–1.2 seconds
Load-3	3-φ balanced non-linear load	Load-2 + 1.5kW, 100VAR	1.2–1.8 seconds
Load-4	3-φ unbalanced non-linear load	Unbalanced load-3	1.8–2.5 seconds

At t = 0.6 seconds and 1.2 seconds, load-2 and load-3 are added at the generator terminals. At t = 1.8 seconds, load-4 is connected to the generator. Figure 12.5(a) presents voltage at terminals for PI and iPI-SMC controllers. The figure exhibits that terminal voltage for iPI-SMC is setting faster than PI. Management of power among load, generator, and BESS is presented in Figure 12.5(b). A smooth and faster transition can be observed for iPI-SMC in comparison to PI during load variations. Figure 12.5(c) displays the voltage and load current during load variation.

Case III: Robustness Test during L-G and LLL-G Fault at the Terminals

Both 3-φ (LLL-G) and 1-φ (L-G) faults are established during 0.8 < t < 0.85 seconds and 1.5 < t < 1.55 seconds, respectively. Figure 12.6 presents the terminal voltage for PI and iPI-SMC during fault condition. It can be viewed from this figure that iPI-SMC performs well during both the faults and settles faster than PI.

Case IV: Effectiveness of iPI-SMC during Controller's Disturbance

Controller's disturbance is introduced to confirm the effectiveness of iPI-SMC. At t = 0.6 seconds, disturbance is introduced to the controllers. Both iPI-SMC and PI controller's performance under the disturbances of 25%, 35%, 50%, and 70% is presented by Figure 12.7(a)–(d), respectively. It is clear from these figures that iPI-SMC is able to accommodate the disturbances with faster settling and minimum oscillation as compared to PI under 25% and 35% disturbances; iPI-SMC performs well and settles faster, but PI is unstable for 70% and 50% disturbances.

Case V: Controllers Performance Assessment

Assessment of controller's performance is executed via following performance metrics:

1. $ISE = \int_0^t e^2 \, dt$ (51)
2. Control signal's mean
3. Control signal's ISV $= \int_0^t u^2 dt$ (52)
4. Control signal's TV $= \sum_{i=1}^{n} \left|(u_{i+1} - u_i)\right|$ (53)

ISE presents error's mean square value, and it is the measure of tracking performance. Quantity of controllers work done is presented by control signal's mean. Consumption of energy by the controller is measured by ISV of control signal. TV is the measure of smoothness of the control signal. The high and low values of TV express the maximum and minimum consumption of control-input. The controller's complexity also is indicated by TV. Assessment of both iPI-SMC and PI step and trapezoidal wind profiles is demonstrated by Table 12.2. It can be viewed from Table 12.2 that iPI-SMC exhibits superior performance to that of PI for all the performance metrices; iPI-SMC gives minimum ISE, which ensures the dynamic tracking. The values of ISV and TV are found lesser for iPI-SMC than PI, which shows the smoothness and lesser energy consumption by iPI-SMC. As compared with PI, the mean of control signal for iPI-SMC is minimum, which conveys that the amount of work done by iPI-SMC is lesser.

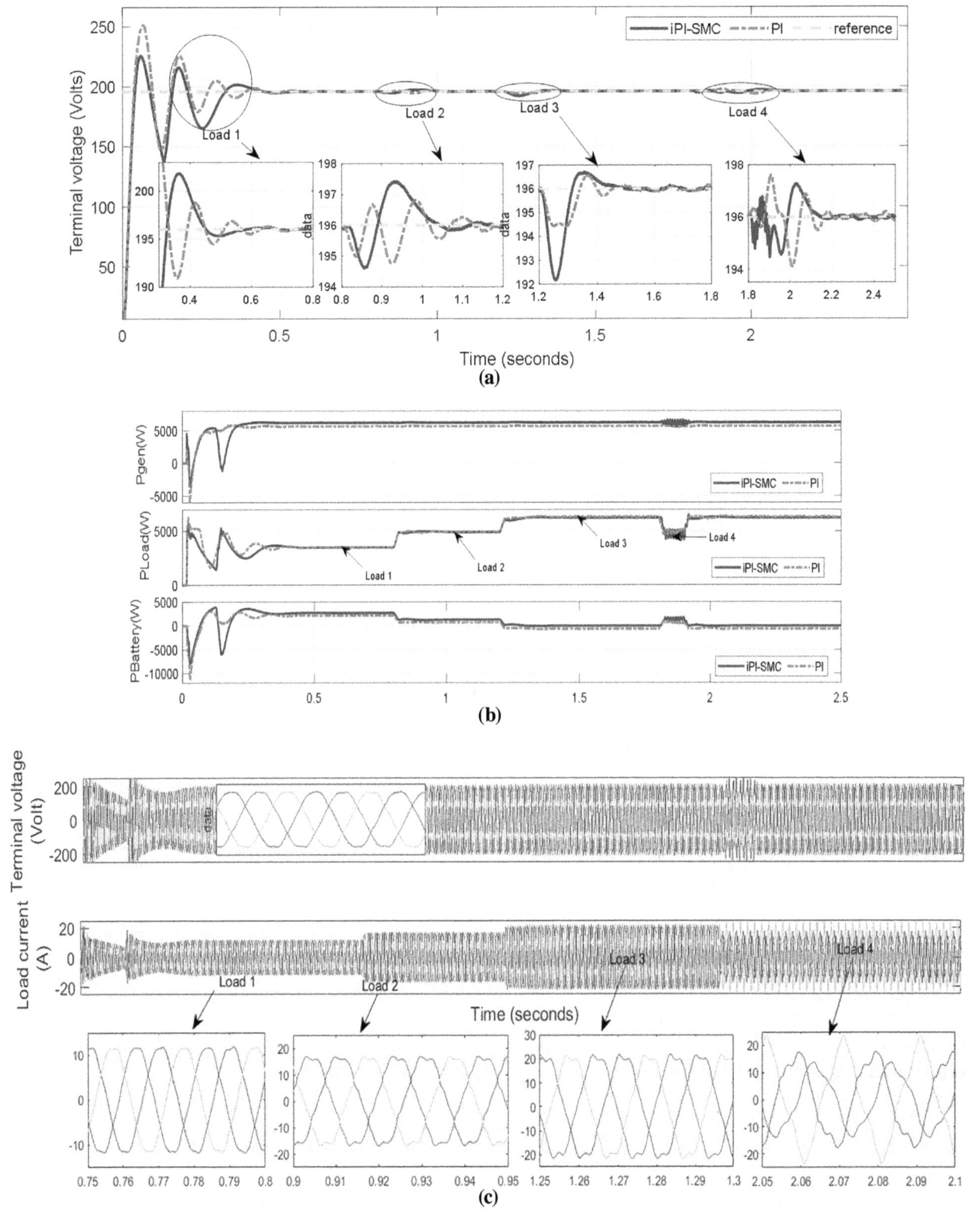

FIGURE 12.5 For different loading (a)–(b) Terminal voltage and power management of iPI-SMC & PI between PMSG, load, and BESS (c) 3-φ terminal voltage and load currents of iPI-SMC.

12.5 TEST RESULTS AND DISCUSSION

Presented iPI-SMC control strategy is verified experimentally in Opal-RT 4510 (Kintex-7 FPGA, 325T, 326,000 logic cells and Intel Xeon E3 v5 CPU 3.5GHz processor), HIL-RTS, in which both plant and controller are fixed in single RTS. A digital real-time oscilloscope (Tektronix DPO2014 C013143) was employed for recording of results. Amplification gain of 1/700, 1/4, and 1/20 is used for power, current, and voltage measurements, respectively. Figure 12.8(a) presents the schematic of HIL setup for the proposed

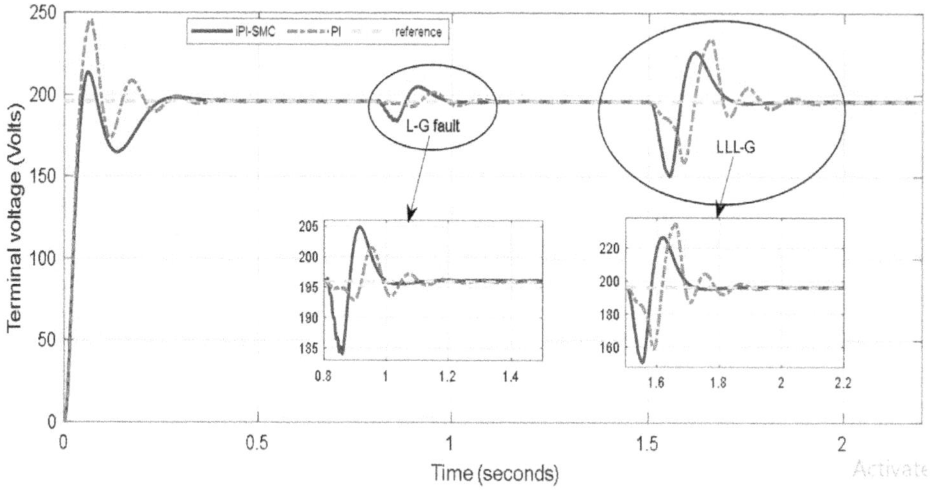

FIGURE 12.6 Terminal voltage for iPI-SMC and PI during L-G and LLL-G fault conditions (time in seconds).

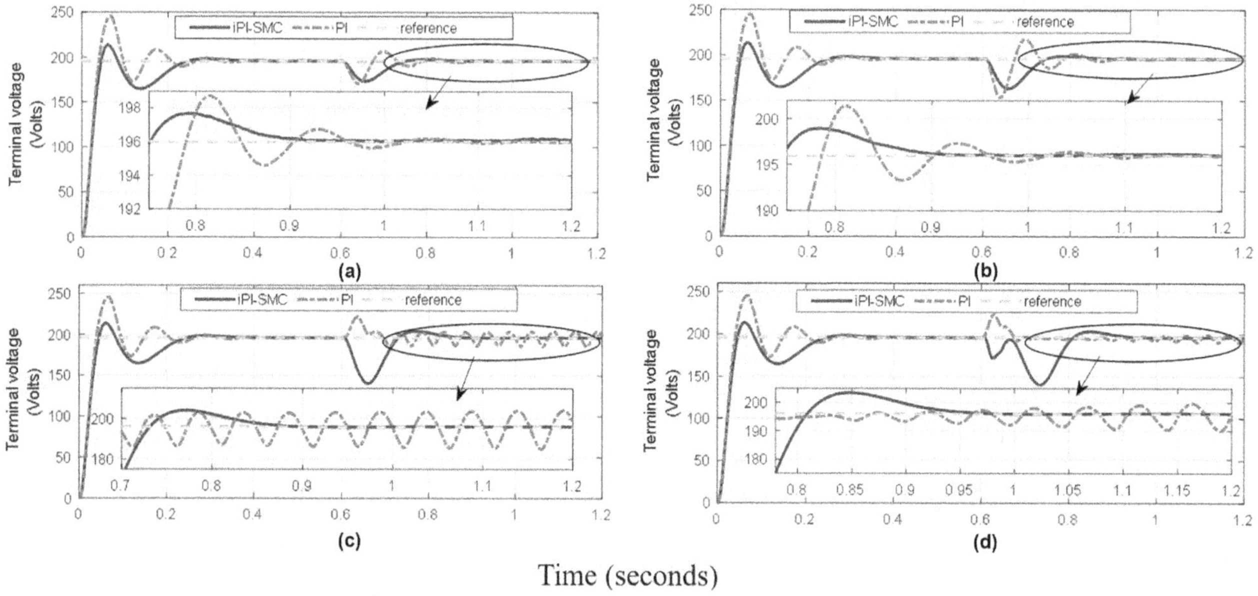

FIGURE 12.7 comparison of iPI-SMC and PI during: (a) 25%; (b) 35%; (c) 50%; (d) 70% disturbance.

TABLE 12.2
Performance Assessment of Controllers

Wind Profile	Controller	ISE	Mean of Control Signal	ISV of Control Signal	Total-Variance of Control Signal
Step	iPI-SMC	678.5	30.34	640.8	5.39×10^{-4}
	PI	699.6	33.56	848.1	8.90×10^{-4}
Trapezoidal	iPI-SMC	690.8	30.99	1.86×10^{3}	1.0×10^{-4}
	PI	734.2	33.08	1.94×10^{3}	7.47×10^{-4}

system, consisting of i) Opal-RT as RTS; ii) PC as programming host to execute the proposed system's MATLAB/Simulink model in Opal-RT; and iii) the router for connecting the setup devices in the same subnetwork. The Opal-RT 4510 is connected to Tektronix digital real-time oscilloscope (DPO2014 C013143). Figure 12.8(b) shows the Opal-RT HIL RTS setup for system execution [21] in the RT-lab using Opal-RT 4510. The results have been recorded during

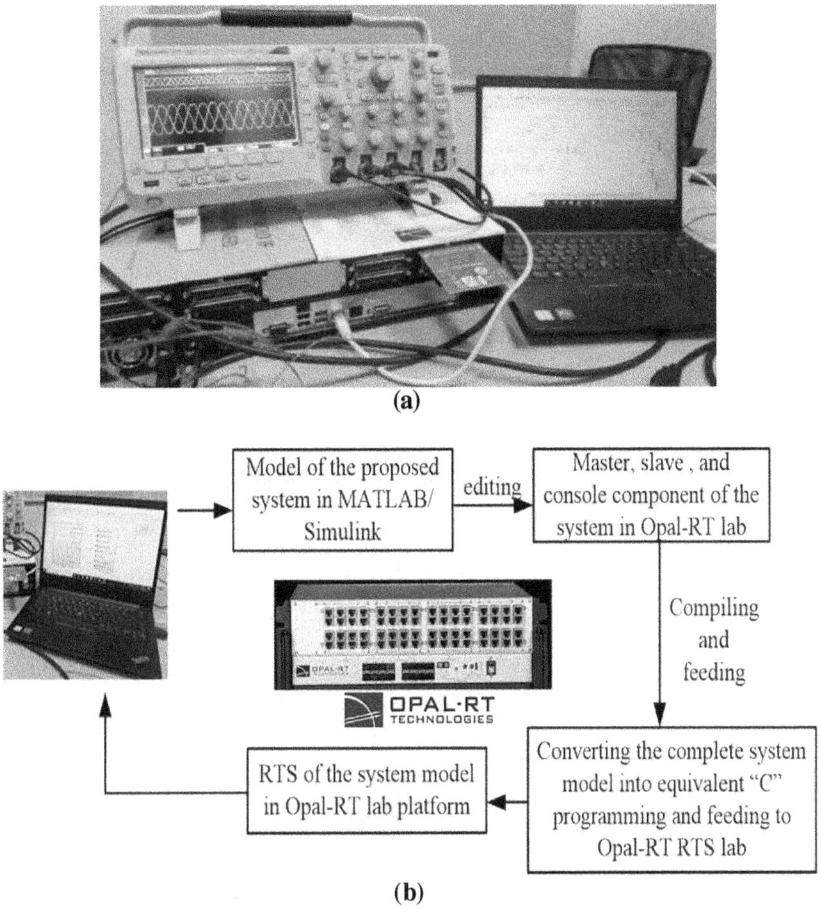

FIGURE 12.8 (a) Schematic of the HIL setup for the proposed system; (b) Opal-RT HIL RTS setup for system execution.

varying wind velocity and load. The following section discusses dynamic behavior of the proposed control method.

12.5.1 Performance of iPI-SMC Controller during Variable Wind Velocity

iPI-SMC controller is executed in real time on FPGA Kintex7A for 10m/s, 12m/s, and 13m/s wind speeds, along with a linear load (100VARs, 5000W). Figure 12.9(a) shows that the controller allows voltage at terminals to settle very fast to its reference during changes in the wind velocity. Figure 12.9(b) presents the 3-φ load voltage. From this figure, it can be observed that constant balanced voltage is obtained in spite of wind velocity variations. Generator current for phases a–c are given in Figure 9(c)–(e). From these figures, it can be realized that PMSG is successfully increasing the power generation with respect to increase in wind velocity. Balance of power among load, generator, and battery is demonstrated by Figure 12.9(f), and it can be viewed from this figure that during under power generation (10m/s and 12m/s), the battery is contributing for extra load demand and during over power generation (13m/s), the surplus power is given to battery after supplying the load demands. From these analyses, it could be surmise that response of the presented iPI-SMC is reasonable to take care of wind velocity variation.

12.5.2 Performance of iPI-SMC Control during Variable Load

iPI-SMC controller is executed in real time on FPGA Kintex7A for varying load at 13m/s of wind velocity. Figure 12.10(a) depicts terminal voltage for load variations. It is observed that voltage at terminals quickly stabilizes to value of reference with minimum oscillation. Balance of power among load, generator, and battery is depicted in Figure 12.10(b). It is observed that, for load 1, the generator can provide power in surplus to battery. While, battery is contributing for increased demand in load for loads 2–3. Figure 12.10(c) presents generator current during fixed wind velocity. From this figure, it is observed that generator current is constant. Variations in load current are seen in Figure 12.10(d) during load variations. Figure 12.10(e) shows battery current variation for change in load. From Figure 12.10(d)–(e), it is observed that current in battery is varying due to load current variation during constant generation. Also, load current imbalance can be seen in Figure 12.10(f). These test results can be used to infer that the

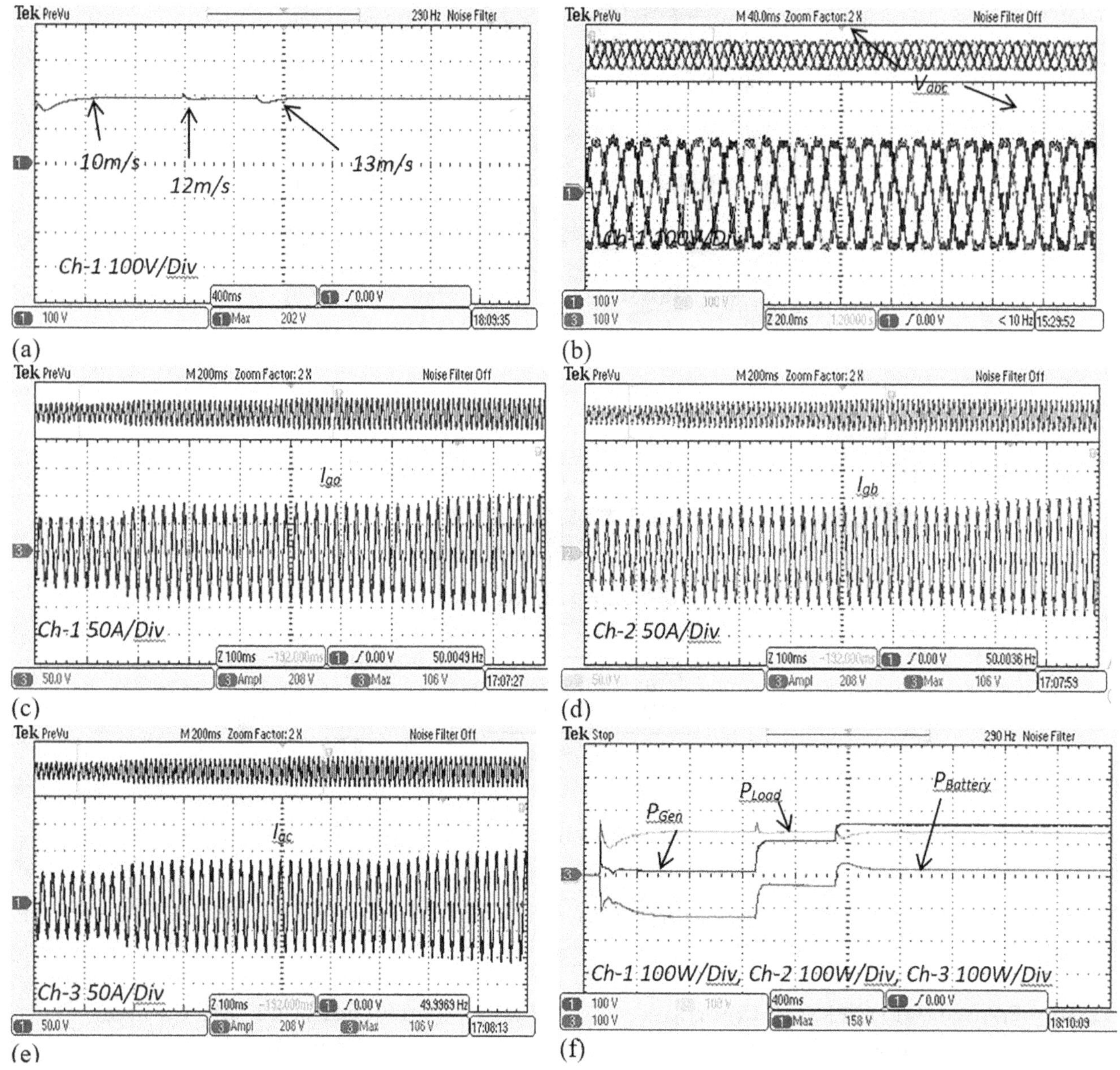

FIGURE 12.9 For varying wind velocity: (a) voltage at terminals; (b) three-phase load voltage; (c) generator current for phase a; (d) generator current for phase b; (e) generator current for phase c; (f) balance of power among load, battery, and generator.

response of the presented iPI-SMC controller is satisfactory under dynamic loads.

12.6 CONCLUSION

The iPI-SMC–based control strategy is applied for WECS connected to PMSG and BESS for voltage and frequency control. The efficacy of the presented control methodology is compared with PI for various work conditions like varying load and wind velocity, load imbalance, L-G and LLL-G fault at generator terminal, and disturbance in the controller. Simulation results demonstrate that the presented iPI-SMC control methodology offers reasonable tracking dynamics and robustness against controller disturbances. Finally, the performance matrices like ISE, control signals ISV, TV, and mean ensure that the performance of iPI-SMC is relatively far more effective than PI. The proposed controller is able to serve a good power management capability among BESS, load, and PMSG under all the varying working states. The viability of presented iPI-SMC controller is validated through OP4510 HIL real-time simulator, and the performances are found satisfactory in variable wind velocity and load conditions.

12.7 SOME WORKED EXAMPLES

Example 1

A 2MW, 690V, 11.25Hz salient-pole PMSG is used in a standalone WECS whereby the number of pole pairs is 30. An RL load is connected to the generator. Following are the parameters for load: $R_L = 0.19125\Omega$, $L_L = 1.3104mH$.

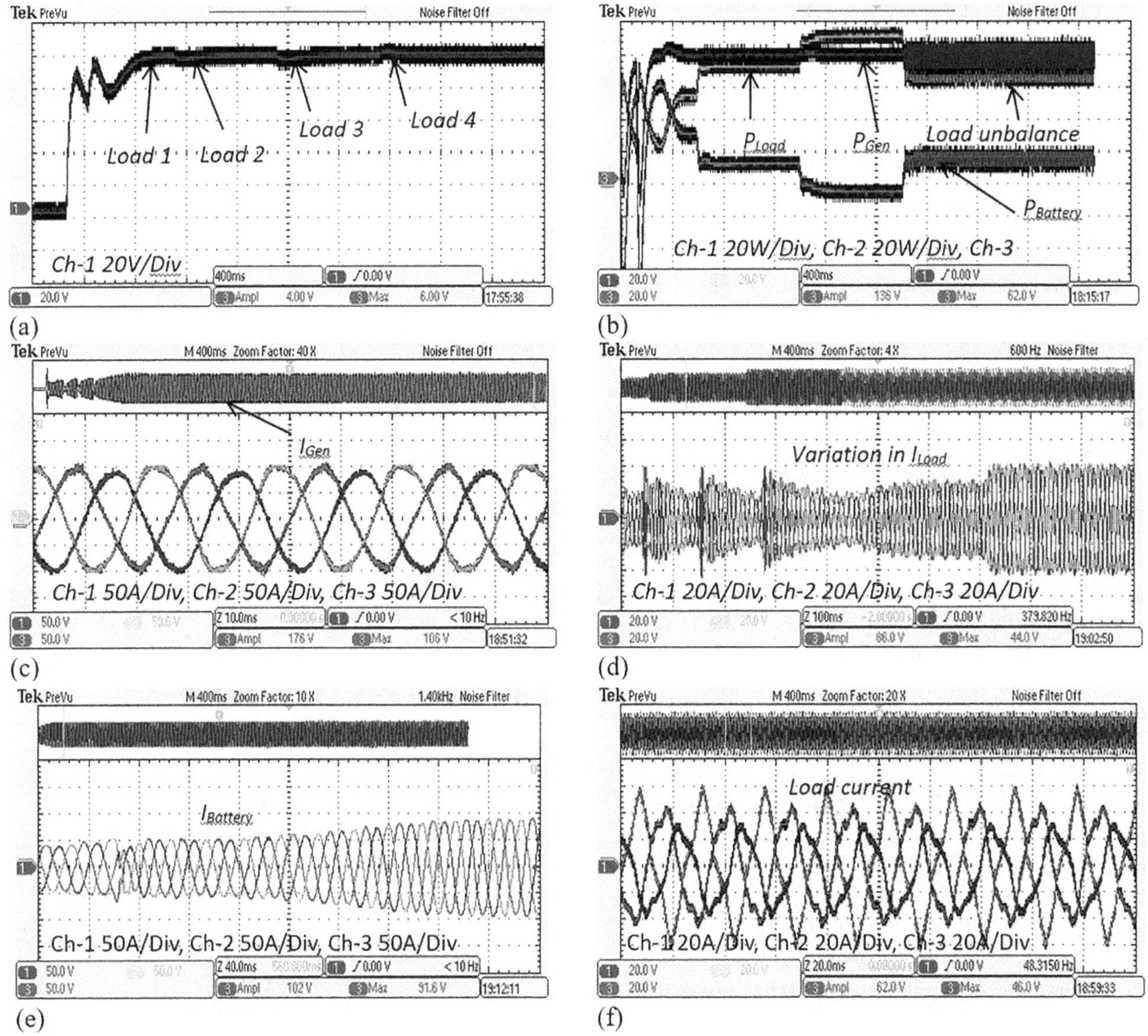

FIGURE 12.10 For varying load: (a) voltage at terminals; (b) balance of power among load, battery, and generator; (c) generator current; (d) load current; (e) battery current; (f) imbalance in load current.

With 12m/s wind speed, the generator works at rotor speed of 1.0pu. Find the rotor's electrical and mechanical speeds, and value of load impedance. (Rated rotor speed = 22.5rpm. Neglect stator core loss and rotational losses.)

Solution

Rated rotor mechanical speed:

$$w_{m,R} = 22.5\times\left(2\pi/60\right) = 2.356\text{rad / sec}$$

Rotor mechanical speed:

$$w_m = w_{m,R}\times w_{m,pu} = 2.356\times 1.0 = 2.356\ rad\ /\ sec$$

Rotor electrical speed:

$$w_r = w_m \times p = 2.356\times 30 = 70.69\ rad\ /\ sec$$

Load impedance:

$$\overline{Z}_L = R_L + jX_L = 0.19125 + j92.626\times 10^{-3}$$
$$= 0.2125 < 25.84°\Omega$$

Example 2

A single channel boost converter is employed for a 600kW, 690V, 50Hz PMSG WECS. The generator's power is transferred via boost converter to a grid of 690V/50Hz using a two-level voltage source inverter and diode rectifier. Capacitance C and Inductance L_x of the boost converter are 2300μF and 270mH, respectively. The boost converter's switching frequency = 2000Hz. The inverter sets the output voltage v_0 of boost converter to 1220V. Operation of the generator is based on MPPT method. The active power of stator is proportional to the cube of rotor speed. PMSG is working at 0.9pu rotor speed at the given wind velocity, and stator's line-to-line voltage is 640V. Determine the input voltage and power to boost converter.

Solution

The input voltage to boost converter:

$$V_i = \frac{3\sqrt{2}}{\pi} \times V_{LL} = 1.35 \times 640 = 864\text{V}$$

At 0.9 pu rotor speed, input power to boost converter:

$$P_i = P_{m,R} \times \left(\omega_{m,pu}\right)^3 = 600 \times 10^3 \times (0.9)^3 = 437.4 \times 10^3 W$$

Example 3

Consider a 2.0MW, 690V, 9.75Hz nonsalient-pole PMSG WECS, whereby the number of pole pairs is 26. The rated mechanical torque is 848.826kNm, and rated rotor speed is 22.5rpm. The generator is control is based on ZDC and MPPT strategies. The generator works at 1.0pu rotor speed with 12m/s wind speed. Determine the torque and mechanical power of generator.

Solution

The rotor mechanical speed:

$$w_m = w_{m,R} \times w_{m,pu} = 22.5 \times (2\pi / 60) \times 1.0 = 2.356 \; rad / sec$$

The rotor electrical speed:

$$w_r = w_m \times P = 2.356 \times 26 = 61.26 \; rad / sec$$

The generator mechanical torque at 1.0 pu rotor speed:

$$\begin{aligned} T_m = T_e = T_{m,R} \times \left(\omega_{m,pu}\right)^2 &= 848.83 \times 10^3 \times (1.0)^2 \\ &= 848.83 \times 10^3 \; N.m \end{aligned}$$

The rated mechanical power:

$$\begin{aligned} P_{m.R} = w_{m,R} \times T_{m,R} &= 22.5 \times (2\pi / 60) \times 848.83 \times 10^3 \\ &= 2000 \times 10^3 W \end{aligned}$$

At 1.0 pu rotor speed, generator's mechanical power:

$$P_m = P_{m.R}\left(w_{m,pu}\right)^3 = 2000 \times 10^3 \times (1)^3 = 2000 \times 10^3 W$$

Example 4

A 2.45 MW, 4000V, 53.3Hz non-salient pole PMSG is employed in a CSC-interfaced WECS whereby rated rotor speed and mechanical torque are 400rpm and 58.4585kNm, respectively. The generator works with an MPPT strategy at 0.9pu rotor speed at 10.8m/s wind speed. The inverter and rectifier filter capacitor values are 213μF and 149μF, respectively. Determine the power and mechanical torque of the generator if controlled by ZDC strategy.

Solution

The rotor mechanical speed:

$$w_m = w_{m,R} \times w_{m,pu} = 400 \times (2\pi / 60) \times 0.9 = 37.7 \; rad / sec$$

The rotor electrical speed:

$$w_r = w_m \times P = 37.7 \times 8 = 301.59 \; rad / sec$$

The generator mechanical torque at 1.0 pu rotor speed:

$$T_m = T_e = T_{m,R} \times \left(\omega_{m,pu}\right)^2 = 58458.5 \times (0.9)^2 = 47351.4 N.m$$

The rated mechanical power:

$$\begin{aligned} P_{m.R} = w_{m,R} \times T_{m,R} &= 400 \times (2\pi / 60) \times 58458.5 \\ &= 2448.7 \times 10^3 W \end{aligned}$$

At 1.0pu rotor speed, mechanical power of generator:

$$P_m = P_{m.R}\left(w_{m,pu}\right)^3 = 2448.7 \times 10^3 \times (0.9)^3 = 1785.1 \times 10^3 W$$

12.8 DISCUSSION QUESTIONS

1. What is the necessity of the model-free controller in a standalone wind energy system?
2. Why are error and integration of the error considered for forming the siding surface?
3. What is the significance of T in sliding surface?
4. What kind of disturbance is applied to validate the control performance?
5. Why not model-free i-PID structures for control of a standalone wind energy system?
6. What is the significance of the control signal's mean over the control signal's TV?
7. How are filters designed in the proposed control algorithms?

12.9 ACKNOWLEDGMENT

This work was supported by Opal-RT Technology, Bengaluru, India by extending help for taking test results for proposed iPI-SMC controller for PMSG-based WECS.

REFERENCES

[1] SN Bhadra, D Kastha, S Banerjee, *Wind Electrical Systems*. Oxford University Press, 1st edition, 2013.
[2] CN Bhende, S Mishra, Siva Ganesh Malla, "Permanent magnet synchronous generator-based standalone wind energy supply system", *IEEE Trans. on Sustainable Energy*, Vol. 2(4), pp. 361–373, 2011.
[3] M Cheng, Y Zhu, "The state of the art of wind energy conversion systems and technologies: A review", *Energy Convers. Manag.*, Vol. 88, pp. 332–347, 2014.
[4] V Yaramasu, B Wu, PC Sen, S Kouro, Narimani M, "High-power wind energy conversion systems: State-of-the-art

and emerging technologies", *Proceedings of the IEEE*, Vol. 103(5), pp. 740–788, 2015.

[5] L Zhang, Y Li, "Optimal energy management of wind-battery hybrid power system with two-scale dynamic programming", *IEEE Trans. on Sustainable Energy*, Vol. 4(3), pp. 765–773, 2013.

[6] MA Tankari, MB Camara, B Dakyo, Lefebvre, "Use of ultra-capacitors and batteries for efficient energy management in wind—diesel hybrid system", *IEEE Trans.on Sustainable Energy*, Vol. 4(2), pp. 414–424, 2013.

[7] Anjana Jain, R Saravanakumar, "Comparative analysis of DSOGI-PLL & adaptive frequency loop-PLL for voltage and frequency control of PMSG-BESS based Hybrid Standalone WECS", *8th International Conference on Power and Energy Systems*, Sri Lanka, 2019.

[8] Pedro Rodriguez, Alvaro Luna, Raul Santiago, Munoz-Aguilar, "A stationary reference frame grid synchronization-system for three-phase grid-connected power converters under adverse grid conditions", *IEEE Trans. on Power Electronics*, Vol. 27(1), pp. 99–112, 2012.

[9] P Cominos, N Munro, "PID controllers: Recent tuning-methods and design to specification", *IEEE Proceeding-Control Theory Application*, Vol. 149, pp. 46–56, 2002.

[10] T Chaiyatham, I Ngamroo, "Optimal fuzzy gain scheduling of PID controller of superconducting magnetic energy storage for power system stabilization", *Int. J. Innov. Comput. Inf. Control.*, Vol. 9(2), pp. 651–666, 2013.

[11] R Teodorescu, F Blaabjerg, "Flexible control of small wind turbines with grid failure detection operating in stand-alone and grid-connected mode", *IEEE Trans. on Power Electronics*, Vol. 19(5), pp. 1323–1332, 2004.

[12] Michel Fliss, Cedric Join, "Model free control and intelligent PID controllers: Towards a possible trivialization of nonlinear control", *15th IFAC Symposium on System Identification*, Saint-Malo, France, 2009.

[13] Samer Riachy, Michel Fliess, Cedric Join, "High-order sliding modes and intelligent PID controllers: First steps toward a practical comparison", *18th World Congress, The International Federation of Automatic Control (IFAC2011)*, Italy, 2011.

[14] Jing Wang, Hugues Mounier, Arben Cela, Silviu-Iulian Niculescu, "Event driven intelligent PID controllers with applications to motion control", *18th IFAC World Congress International Federation of Automatic Control (IFAC2011)*, Italy, 2011.

[15] John T Agee, Selcuk Kizir, Zafer Bingul, "Intelligent proportional-integral (iPI) control of a single link flexible joint manipulator", *J. Vib. Control.*, Vol. 21(11), pp. 2273–228, 2015.

[16] John T Agee, Zafer Bingul, Selcuk Kizir, "Tip trajectory control of a flexible-link manipulator using an intelligent proportional integral (iPI) controller", *Trans. Inst. Meas. Control.*, Vol. 36(5), pp. 673–682, 2014.

[17] Zhong-Sheng Hou, Zhuo Wang, "From model-based control to data-driven control: Survey, classification and perspective", *Inf. Sci.*, Vol. 235, pp. 3–35, 2013.

[18] MA Jama, H Noura, A Wahyudie, A Assi, "Enhancing the performance of heaving wave energy converters using model-free control approach", *Renew. Energy*, Vol. 83, pp. 931–941, 2014.

[19] Radu-Emil Precup, Mircea-Bogdan Radac, Raul-Cristian Roman, Emil M. Petriu, "Model-free sliding mode control of nonlinear systems: Algorithms and experiments", *Inf. Sci.*, Vol. 381, pp. 176–192, 2017.

[20] Rachid Errouissi, Ahmed Al-Durra, "A Novel PI type Sliding Surface for PMSG-based wind turbine withimprovedtransient performance", *IEEE Trans. on Energy Conversion*, Vol. 33(2), pp. 834–844, 2018.

[21] Mahendra Bhadu, Nilanjan Senroy, Indra Narayan Kar, Gayathri Nair Sudha, "Robust linear quadratic gaussian-based discrete mode wide area power system damping controller", *IET Gener. Transm. Distrib.*, Vol. 10(6), pp. 1470–1478, 2016.

13 Lithium-Ion Batteries
Characteristics, Modelling and Application

Ujjval B. Vyas, Varsha A. Shah and Athul Vijay P. K.

CONTENTS

13.1 INTRODUCTION

The escalation in greenhouse gas emissions due to intense energy demands from all sectors has prompted researchers to study and implement zero-emission technologies. The concept of zero-emission technologies mainly concentrates on utilising electricity as a source of energy for the end-user and generating electricity using renewable energy systems. This adoption of zero-emission technologies can be easily observed in the transportation sector as the conventional internal combustion engine vehicles are being replaced by electric vehicles (EVs), and with the electrification of locomotives across the world [1]. Apart from these, research is being carried to utilise energy storage devices and provide auxiliary power in aircraft and ships to reduce emissions [2]. This process of electrification causes a considerable increase in demand on the grid. In the present scenario, as of February 2021, around 61.5% of the energy generated is through thermal sources and 24.5% is through renewable sources, the majority is through solar and wind [3]. Electrification in the transportation sector alone is not sufficient, as the tailpipe emission of a vehicle is just transferred from a distributed location to a concentrated area at thermal plants. Thus, for a substantial reduction in greenhouse gas emissions, renewable energy systems have to be adopted [4].

Adoption of renewable energy systems has various challenges due to the uncertain nature of both the sources and loads. These disadvantages can be solved by adopting power electronic intelligence at edge network (PINE) as proposed in [5]. In PINE, each power electronic device at the users' end acts as an intelligent system communicating with the other converters. The power electronic converters, when connected with an energy storage device, can provide many more services. Advances in the area of batteries have allowed their utilisation in the automobile industry, and as they are controlled by power electronic converters, they act as a PINE without endurance of additional installation cost. The development of smart grids has paved the way for vehicle-to-grid (V2G) technology, which is a prime example of a PINE system. This V2G technology allows bidirectional flow of power offering numerous services to grid such as peak load shaving, power grid regulation [6], reactive power compensation [7] and spinning reserve [8].

V2G technology is classified as unidirectional V2G (UV2G—utilises the communication network between the grid and vehicle to control the rate of charging of battery) and bidirectional V2G (BV2G—the power flow is vehicle to grid and from grid to vehicle) [9]. UV2G services prevent system instability, voltage drop issues, spinning reserves and grid overload, since the battery is considered as dynamic load. Power utilities can also avoid overloading [10].

BV2G services improve the reliability and sustainability of the grid, since the battery is considered as both load and an energy storage; this enables the energy exchange between the grid and the battery. There are three emerging concepts for V2G technologies: V2H (vehicle to home), V2V (vehicle to vehicle) and V2G. Functions such as reactive power support, peak load shaving, frequency regulation and voltage regulation are also provided. The BV2G also provides renewable energy support for the grid. The

DOI: 10.1201/9781003229124-13

large energy storage devices present in the distributed grid networks can also provide such ancillary services [11, 12].

EVs and energy storage devices in distributed generations and micro grid systems connected to grid can be utilised for ancillary services, but as per the literature, Li-ion batteries degrade and degradation factors include capacity throughput, SOC, temperature and variation in magnitude of both charging and discharging [13]. The physical degradation can be quantified as capacity fade (effects the range of EV) and power fade (effects the internal resistance of the battery, thereby decreasing the power capability and efficiency of vehicle). A proper assessment of the battery degradation is of utmost importance, as a trivial capacity and power damage at initial cycles becomes more prominent at later stage of the operation [14]. These battery degradations can be quantified via voltage response of cell and techniques such as incremental capacity and differential voltage. Storing cell at higher SOC and temperature causes capacity and power fade.

The usage of batteries only for ancillary purposes at a constant power is proven to be detrimental, as it limits the life of the battery packs to five years. The impact of delayed charging is trivial at room temperature [15]. However, it could be significant in tropical countries that have relatively warmer climates such as India, where the effect of high temperature on battery degradation is further pronounced.

The initial premise of ancillary services from the energy storage devices was to maximise the profits for the prosumers, but with utilising an intelligent technique, such cycling of power leads to increased degradation. The more profitable scenario is to account for battery degradation, and this could be achieved by intelligent control techniques [16]. To implement this, first a battery management system (BMS) with higher computing power and memory has to be developed such that machine learning algorithms can be implemented to accurately predict the state of the energy storage device. The accuracy of the BMS algorithms depends on the battery models. Therefore, an understanding of battery modelling and the methodology to obtain these models is discussed in this chapter.

13.1.1 Terminology of Batteries

In literature, the battery performance has been explained by certain factors that are described in what follows [17].

Cell, Module and Pack

A cell is a singular entity encasing the various battery components such as anode, cathode, electrolyte, pressure value and many others. A combination of the cells in either series or parallel is known as a module. Generally, the most common module available, especially in our laptops, is 4S3P: four cells in series and three series legs in parallel. The combinations may vary depending on the chemistry employed by the manufacturer and his requirement. A combination of modules forms a battery pack, commonly observed in EVs and large power applications such as electric grids.

Cut-Off Voltage (V_{off})

Cut-off voltage is defined as the minimum terminal voltage of the battery cell terminal allowed while discharging.

Nominal Voltage (V_n)

The cell's nominal voltage is measured after it discharging 50% of its total energy.

Maximum Voltage (V_{max})

The maximum voltage is defined as the maximum terminal voltage until which the cell can be charged.

Ampere Hour (Ah)

Ampere Hour (Ah) represents the amount of charge present in the battery. It is defined as the current that a battery can deliver for one hour before the battery voltage reaches the cut-off voltage.

True Capacity (CAh)

A battery reaches V_{off} with a reduced charge expenditure, especially for higher currents. It is due to the higher voltage drop across the internal impedance for higher discharge currents. Thus, true capacity is defined as the battery capacity in Ah extracted with the discharging current is set to a low value such that the voltage drop across the internal impedance is essentially zero.

Capacity Loss

A battery experiences degradation in both rest and operational states. This degradation is represented by a capacity loss of the battery that will not be available in the future.

State of Health (SoH)

SoH is defined as the ratio of total actual capacity after capacity loss to the full capacity at the initial stage.

State of Charge (SoC)

SoC is defined as the ratio of capacity remaining in the battery to the battery's rated true capacity. It represents present energy in the cell that is altered by both charging and discharging.

Depth of Discharge (DoD)

DoD is defined as the ratio of total Ah extracted to the true capacity (C_{Ah}) of the battery.

C-Rate

C-rate is a dimensionless figure representing the ratio of the discharge current to the battery's actual capacity. For example, a 2Ah battery discharged with 3A current is discharged at 1.5C-rate.

Power Density

Power density is defined as the maximum amount of power that is stored per kilogram. It facilitates the comparison of batteries. The higher power density represents the battery provides high power at a lower weight.

Gravimetric Energy Density

The gravimetric energy density of a battery represents energy that a battery contains compared to its weight.

Volumetric Energy Density

The volumetric energy density of a battery is a measure of how much energy a battery contains compared to its volume.

State of Power (SoP)

SoP is defined as the maximum power that can be extracted to the rated full power.

Cycle Life

Cycle life is defined as the total number of cycles, or the number of times a battery can be charged and discharged until its capacity fades to a point beyond which it cannot be used.

Columbic Efficiency

Columbic efficiency is defined as a battery's ratio of charge it can deliver completely to the total charge it holds initially.

13.1.2 Chemical Properties of Batteries

Electrical energy is stored in batteries as chemical energy and is utilised for supplying current at the time of requirement. The rechargeable cells, also known as secondary cells, are used for storage purposes. A battery cell is available in different sizes and shapes, from compact cylindrical cells to pouch-shaped cells. The output voltage and energy depend on the numbers and composition of the cell. Secondary batteries utilise different materials like lead, nickel, acid, lithium and alkaline. Most of the materials utilised in battery manufacturing are highly active and poisonous materials. Even after removing these materials from batteries, many of these substances inside a battery are toxic to some extent [18]. Thus, care must be taken to dispose of the battery after utilisation. The following section explains the various types of batteries available in the market.

Lead-Acid Batteries

The lead-acid battery is among the utmost established battery technologies, as it originated around the mid-1800s. It consists of lead-based positive and metallic oxide negative electrodes, immersed in an electrolyte solutions such as diluted sulphuric acid [19]. The redox equation governing it is shown in Eq. (13.1):

$$\text{Redox equation:}\quad Pb + PbO_2 + 2H_2SO_4 \leftrightarrow 2PbSO_4 + 2H_2O \tag{13.1}$$

There are two types of lead-acid batteries: flooded lead-acid and sealed valve-regulated lead-acid (VRLA) solutions. A flooded lead-acid battery is rarely utilised due to monthly maintenance and replacement of distilled water, despite its low installation cost. Furthermore, its operation needs to be performed at adequately ventilated locations due to the emission of flammable gases. The VRLA battery was introduced to reduce the maintenance cost as it uses a valve that reduces water loss. Despite these improvements, the low discharge limit (<20%), low energy density, low cycle life and slow charging rate prevent lead-acid batteries in high energy requirements. They are widely utilised in applications when cost-effectiveness, reliability and abuse tolerance are critical but lifetime and energy density are irrelevant [20–22].

Ni-Cd Batteries

The nickel-cadmium battery is a secondary battery based on nickel oxide hydroxide as a positive electrode and metallic cadmium as a negative electrode. Ni-Cd batteries are utilised in various applications, such as low-power portable to large ventilated applications used for reserve power [23]. The redox equation defining the Ni-Cd battery is shown in Eq. (13.2):

$$\text{Redox Equation:}\quad Cd + 2NiO\,(OH) + 2H_2O \leftrightarrow Cd(OH)_2 + 2Ni(OH)_2 \tag{13.2}$$

Compared with other rechargeable cells, Ni-Cd cells have better cycle life and capacity, and operate at low temperatures and high discharge rates. Higher materials costs and a high self-discharge rate hinder their application with high energy and power applications. Ni-Cd cells are widely used in portable devices with low energy requirements [24].

Ni-MH Batteries

Ni-MH battery consists of a metallic cadmium-based negative electrode, a nickel-oxyhydroxide based positive electrode immersed in electrolyte with alkaline nature, usually potassium hydroxide. The Ni-MH battery solves the issues of toxic nature and the high cost of nickel-cadmium (Ni-Cd) battery. It is an environmentally friendly battery with higher power and energy density that is less prone to memory effects. The Ni-MH battery notably become famous for EVs and Hybrid EVs (HEVs) during the early 2000s. However, it has several disadvantages: high self-discharge rate, low columbic efficiency and less cycle life. Apart from these, it has a low ability to tolerate overcharge and fast charging. During rapid charging, the excessive amount of heat is generated builds hydrogen gas and ruptures the cells, leading to significant capacity decay. Thereby, the charging algorithms have to be accurately designed [25–27]. The redox reaction is shown in Eq. (13.3):

$$\text{Redox Equation:}\quad MH + NiO(OH) \leftrightarrow M + Ni(OH)_2 \tag{13.3}$$

Na Batteries

The aforementioned batteries operate at room temperature; therefore, those batteries are not suited for applications in high-temperature regions. For these applications, certain batteries known as molten salt batteries have been introduced. Among these, batteries based on sodium known as

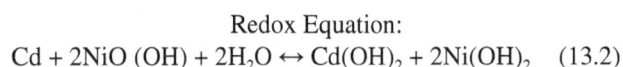

Na batteries are predominately studied. A battery based on liquid sodium (Na) and sulphur (S) has a high efficiency of charge/discharge, high energy density and long cycle life, and is manufactured from economical raw materials. High operating temperatures around 300–350°C and the corrosive nature of the sodium polysulphides makes them appropriate for stationary energy storage applications. As the cost of materials is low, the cell becomes more economical with increasing size. To properly maintain the battery temperature between 270°C and 350°C so that the electrodes remain in a molten state, independent heaters are required.

Li-Ion Batteries

Lithium-ion batteries are among the recent battery technologies commercially available, but their ubiquitous nature in our technology is due to their various advantages. The comparative data of lithium-ion batteries with other batteries in Table 4.1 gives evidence of its superiority. The lithium-ion chemistry-based batteries involve a lithium-based metal oxide as cathode and a graphite anode. The Li-ion cells have an remarkable 200-plus Wh/kg energy density, a decent power density with more than 95% of columbic efficiency. The traditional Li-ion battery has utilised cobalt-based lithium oxide as a cathode. It has several shortfalls that include short cycle life due to significant degradation and the toxic nature of cobalt. Also, cobalt-based lithium-ion batteries pose a fire safety risk if endured through mechanical or electrical stress [28]. Thus, various chemical properties are studied as a cathode material for Li-ion–based batteries, along with a generalised working principle for Li-ion batteries as discussed in Section 13.2.

13.2 LI-ION BATTERIES—WORKING PRINCIPLES AND VARIOUS CHEMICAL PROPERTIES

This section provides a basic understanding of the structure of the Li-ion battery, along with a generalised working principle. Based on the type of metallic oxide, the lithium-ion batteries are classified and a comparison is laid out.

13.2.1 Principles of Li-Ion Batteries

A battery is a collection of electrochemical cells that convert chemical energy into electrical power. An electrochemical cell consists of anode, cathode, electrolyte and separator arranged, as shown in Figure 13.1.

Li-ion batteries, like many other batteries, also work on the principle of redox reaction. A redox reaction consists of

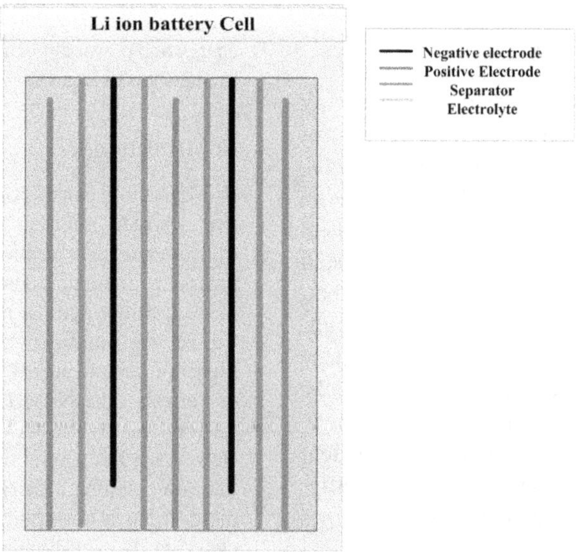

FIGURE 13.1 Lithium-ion battery cell with layers of anode, cathode and separator [30].

TABLE 13.1
Comparison of Parameters of Different Types of Secondary Batteries [29]

Type	Nominal Voltage (V)	Specific Energy (Wh/kg)	Specific Power (W/kg)	Columbic Efficiency	Operating Temperature (°C)	Cycle Life
Nickel metal hydride	1.2	216–432	250–1000	66	30	500–1000
Lead-acid	2	35–50	180	50–92	3–20	500–1500
Nickel-Cadmium	1.2	108–180	100	65–80	20–30	>5000
Li-Ion	3.7	360–950	~250–~340	90–99	-10–65	400–1200
Sodium (Na) Batteries	3.7	120–260	90–230	80–90	270–350	4500

two stages: oxidation and reduction. In the oxidation stage, an electrode releases an electron to the external circuit. The electrode at which oxidation occurs is referred to as an anode. In the reduction stage, an electrode known as a cathode collects the electrons through an external circuit from the anode. During the charging, the positive electrode becomes a cathode during discharge and behaves as an anode.

In contrast, during discharge, the negative electrode becomes an anode and behaves as a cathode during charge [30]. As per convention, the terminal designations are based on discharge operation. The electrolyte conducts the ions between the cathode and anode of a cell, whereas a separator is an insulated layer penetrable by ions yet capable of avoiding a potential short circuit between the electrodes. The electrodes are immersed in an electrolyte as shown in Figure 13.2. When a the cell is connected to a load, the oxidation reaction occurs at the negative electrode, while reduction occurs at a positive electrode. This process is shown below in Eq. (13.4)–(13.6), where M, Y represents the elements and z,k,n represents natural numbers and e represents electrons.

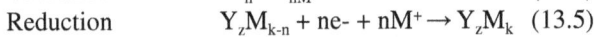

Oxidation	$M_n \rightarrow {}_{nM}{}^{+} + ne^{-}$	(13.4)
Reduction	$Y_zM_{k-n} + ne- + nM^{+} \rightarrow Y_zM_k$	(13.5)
Redox Reaction	$M_n + Y_zM_{k-n} \rightarrow Y_zM_k$	(13.6)

Similarly, the redox reactions while charging are shown in Eq (13.7) to Eq (13.9)

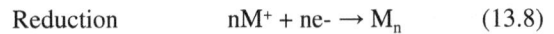

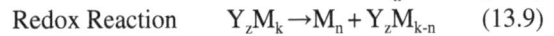

Oxidation	$Y_zM_k \rightarrow Y_zM_{k-n} + ne- + nM^{+}$	(13.7)
Reduction	$nM^{+} + ne- \rightarrow M_n$	(13.8)
Redox Reaction	$Y_zM_k \rightarrow M_n + Y_zM_{k-n}$	(13.9)

13.2.2 Comparison of Various Li-ion Batteries

Many kinds of lithium-ion batteries are employed in high-power applications. The most widely adopted Li-ion battery contains a carbon anode (negative electrodes). The positive electrode material is a lithium-based metal such as LFP, LMO, NCA, NCM, etc. Advantages and disadvantages of the most common Li-ion batteries are presented in the following.

Lithium Cobalt Oxide Batteries

Lithium cobalt oxide ($LiCoO_2$)-based batteries are the most mature Li-ion battery chemistry; their anode is based on graphite, while the cathode is based on lithium cobalt oxide. Lithium cobalt oxide is easy to prepare and batteries based on this cathode material are used mainly for consumer electronics because of their good electrical performance and high energy density [31]. The drawbacks of this chemistry are related to safety issues, since the cathode material is highly reactive and has poorer thermal stability than other cathode materials.

Lithium Manganese Spinel Batteries

Lithium manganese spinel ($LiMn_2O_4$)-based batteries have a lower cost and higher performance in comparison to the lithium cobalt oxide-based batteries. On the other hand, they have a lower specific capacity than the lithium cobalt

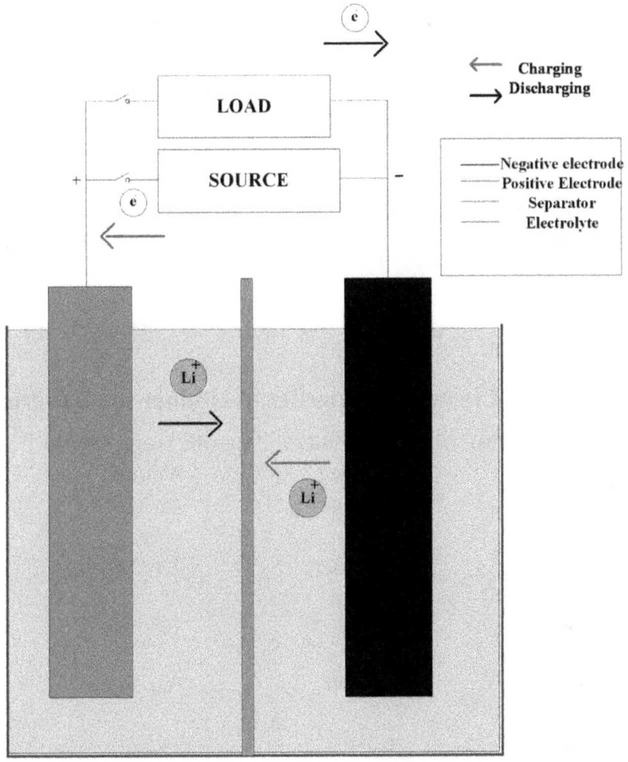

FIGURE 13.2 Internal working of cell [30].

oxide batteries (20—30% lower) [32]. Furthermore, they have an increased safety than previous developed Li-ion batteries (based on cobalt and nickel) and a cycle lifetime of 1,000 cycles. Even though they have increased safety, lithium manganese spinel batteries are susceptible to thermal runaway if overcharged or discharged with current rates which are too high [33, 34].

Lithium Nickel Cobalt Aluminium Oxide Batteries

Lithium nickel cobalt aluminium oxide ($LiNiCoAlO_2$) -based batteries, also referred to as NCA batteries, are characterised by relative high specific power and energy densities and reasonable long cycle lifetime (e.g. 2,000–3,000 cycles) [26, 27]. The main disadvantages of this chemistry are related with their high cost and reduced safety.

Lithium Nickel Manganese Cobalt Oxide Batteries

Lithium nickel manganese cobalt oxide ($LiNi_xMn_xCo_xO_2$)-based batteries, also referred to as NMC batteries, provide a compromise between high capacity and high current [28]. Furthermore, Li-ion batteries based on NMC cathode have similar lifetime characteristics as NCA-based batteries; however, they show better safety than all the aforementioned battery chemical properties [26].

Lithium Iron Phosphate Batteries

Lithium iron phosphate ($LiFePO_4$)-based batteries (with anode based on graphite carbon and cathode based on lithium iron phosphate) have become extremely attractive for various applications because of their inherent safety and relatively low cost. Moreover, lithium iron phosphate–based batteries can have long cycling lifetime (e.g. more than 7,000 full cycles) and high charge and discharge current rates (e.g. up to 5C) [28, 35]. The main disadvantage of lithium iron phosphate batteries is their reduced energy density in comparison to other Li-ion chemical properties.

Lithium Titanate Oxide Batteries

The lithium titanate oxide ($Li_4Ti_5O_{12}$) material represents an appealing and competitive candidate for new types of anodes in Li-ion batteries [28]. Thus, Li-ion batteries based on this anode material have emerged in the market. Batteries based on lithium titanate oxide anodes and lithium manganese spinel cathodes are characterised by very long calendar and cycle lifetimes (e.g. more than 15 years and 10,000 cycles, respectively) and are considered to be the most safe Li-ion batteries that are currently on the market [36]. The main disadvantages of lithium titanate oxide–based batteries are their high cost and reduced energy density [37]. Comparison of lithium-ion battery chemical properties is shown in Table 13.2.

13.3 DESIGN OF BATTERY PACKS FOR POWER SYSTEM APPLICATIONS

The battery packs are extensively utilised in high-power applications such as a storage units for renewable energy storage to supply active-power EV batteries that provide various V2G services such as voltage regulation, up/down frequency regulation, reactive power support, spinning reserves, valley filling, peak shaving, load following, and energy balance in the grid. For all these applications, based on the requirement, the design of the battery pack varies. The design methodology of battery packs can be divided into two steps. Initially, the designer has to determine the total number of cells required and their connections in series and parallel. The latter step involves designing a BMS for the proper and safe functioning of cells. This section provides a basic understanding of the initial steps through numerical examples.

Numerical Question 1

A 100kAh battery pack has to be designed that has to be able to provide a 100A discharge current with a terminal voltage of 440V. A battery cell has the following data:

Maximum terminal voltage: 4.2V
Minimum terminal voltage: 2.75V
Nominal Voltage: 3.7V
Capacity of cell: 2.6Ah

TABLE 13.2
Comparison of Parameters of Different Chemical Properties of Lithium-Ion Batteries [29]

Battery Chemistry	Nominal Voltage (Volts)	Power Density (W/kg)	Specific Energy Density (Wh/Kg)	Cycle Life	Max Temperature (OC)
Lithium, iron, phosphate (LFP)	3.6	200–240	90–160	2,000–5,000	40
Lithium, iron, nickel, manganese, cobalt	3.5	240–260	150–220	1,000–2,000	40
Lithium, manganese spline oxide	3.8	200–230	100–150	300–400	40
Lithium, nickel, cobalt, aluminium oxide	3.4	250–350	200–260	1,000–1,500	40
Lithium titanate oxide	2.4	320–400	50–80	3,500–7,000	60
Lithium cobalt oxide	3.6	220–260	50–200	500–1,000	40

Determine the number of cells to be connected in series, parallel and total number of cells for:

Case 1: Discharge rate 2C
Case 2: Discharge rate 1C
Case 3: Discharge rate 0.5C

Solution

Based on the provided data, the total number of cells required to be connected in series is obtained by:

$$ceil\left(\frac{Terminal\ Voltage\ of\ the\ Pack}{Minimum\ voltgae\ of\ the\ cell}\right) = \frac{440}{2.75} = 160$$

Number of cells required to be connected in parallel is obtained by:

$$ceil\left(\frac{Ah\ Capacity\ of\ the\ Pack}{Ah\ capcity\ of\ the\ cell}\right) = \frac{100}{2.6} = 39$$

Total number of cells is obtained by:

$$(Number\ of\ cells\ in\ series)\left(Number\ of\ cells\ in\ parallel\right) = 160 \times 39 = 6290$$

The number cells apart from the Ah and voltage required also depends on the maximum discharge current, based on the number of parallel cells and discharge rate of each cell. This calculated discharge current has to be greater than or equal to the required discharge current of 100A. If the computed discharge current is not greater than the necessary discharge current, recalculate the number of cells in parallel.

Case 1: Discharge Current Rate of 2C

$$Calculated\ Discharge\ current\ kA = \left(Number\ of\ parallel\ cells\right)\left(maximum\ discharge\ rate\right) = 39 \times 2 \times 2.6 = 202.86A$$

In this case, the calculated discharge is greater than the required maximum discharge; therefore, the number of cells required for Case 1 is 6,290.

Case 2: Discharge Current Rate of 1C

$$Calculated\ Discharge\ current\ kA = \left(Number\ of\ parallel\ cells\right)\left(Maximum\ Discharge\ rate\right) = 39 \times 1 \times 2.6 = 101.43A$$

In this case, the calculated discharge is greater than the required maximum discharge; therefore, the number of cells required for Case 2 is 6,290.

Case 3: Discharge Current Rate of 0.5C

$$Calculated\ Discharge\ current\ kA = \left(Number\ of\ parallel\ cells\right)\left(Maximum\ Discharge\ rate\right) = 39 \times 0.5 \times 2.6 = 50.71A$$

In this case, the calculated discharge is less than the required maximum discharge; therefore, the cells required for Case 3 has to recalculated.

Number of parallel cells

$$ceil\left(\frac{Maximum\ discharge\ Current\ of\ the\ Pack}{Maximum\ discharge\ current\ of\ the\ cell}\right) = \frac{100}{(0.5)(2.6)} = 77$$

As the obtained 77 cells in parallel are greater than previous 39, both Ah and discharge current conditions have been fulfilled. Therefore, for Case 3, the total of cells required is 77 × 160 = 12,320.

Based on this example, the number of cells can be reduced by the following:

- Higher terminal voltage levels of a single cell that reduces the number of series cells.
- Higher discharge/charge current rate.
- Higher capacity of a single cell.

The total number of cells in large battery packs may vary from hundreds to thousands, due to certain manufacturing defects and the impurities in the raw materials of cells. These differences do not affect the battery pack design but create issues in the battery pack's safe operation. These issues are tackled by employing a BMS. The following section describes the BMS and lays out its functionalities and various algorithms utilised to perform these functionalities.

13.4 BATTERY MANAGEMENT SYSTEMS

The BMS is an essential component in the design of battery pack. It has two objectives: the main objective is to protect and maintain healthy operation of cells, and the other objective is to provide basic data such as the SoC, temperature, SoH, state of power and other relevant data to the appropriate user for proper decision making [38]. The main functions regarding the protection of cells are:

- To protect the battery pack cells from overcharging, overheating and other faults such as polarity reversal and others.
- To identify and isolate the cells under fault conditions.

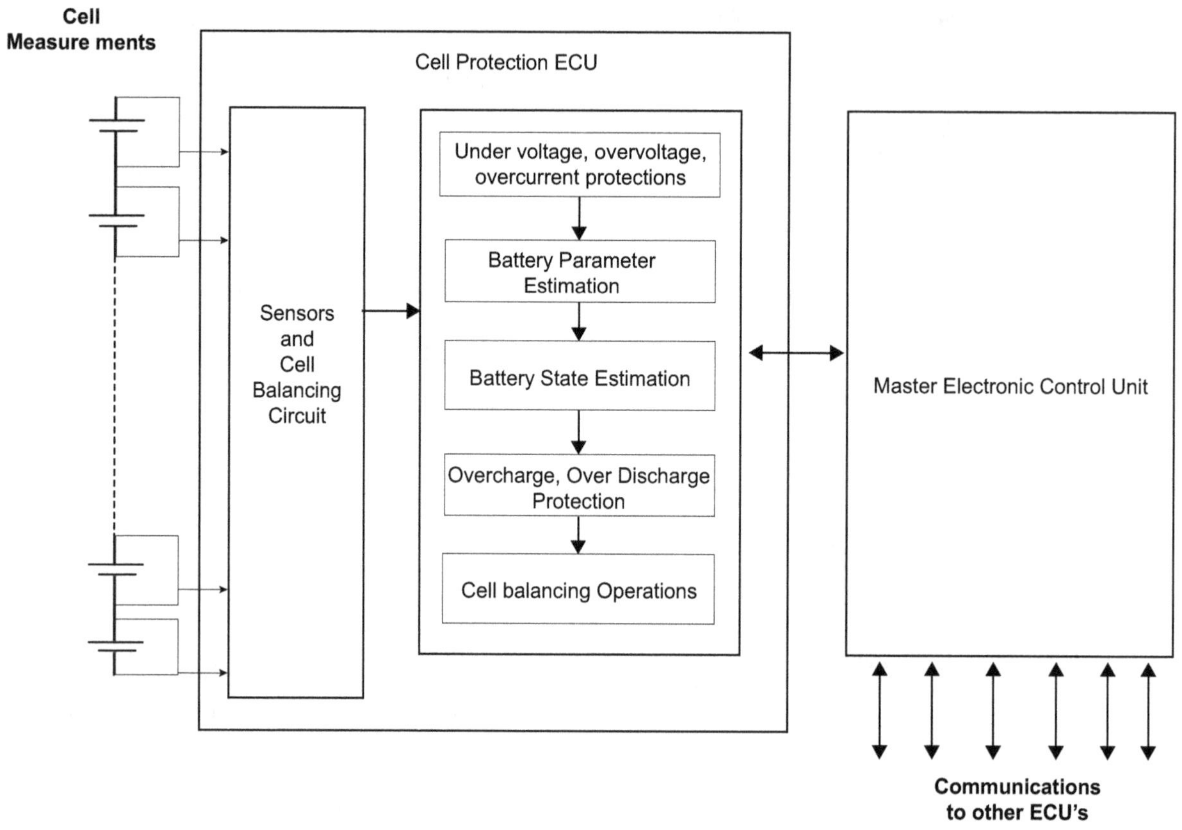

FIGURE 13.3 Schematic layout of the battery management system [39, 40].

- To properly operate the cells such that their life is elongated without compromising the system's performance.

The basic structure of a BMS is shown in Figure 13.3. A BMS is a consortium of several electronic control units (ECU). These ECU are specially designed for BMS applications and generally have a sophisticated microprocessor that processes the data of voltage, current and temperature of each cell through analog-to-digital converters (ADCs). Depending upon the number of ADC and the total number of cells, along with the capability of the designed ECU, the total number of ECU may vary. For a large battery pack, as the total number of cells is much higher, many ECU are required [39]. These ECU operates in tandem with each other through a master ECU as shown in Figure 13.3. This master ECU, based on the data from other ECU, provides information to the decision-making unit generally present outside the battery unit and varies based on the application. For example, in an EV application, this information is shared to the dashboard for the information of drive, to other control units of EV, with manufacturers to assess the performance of batteries, and many more. In a power system–based application, this critical information is shared to the either a localised control unit (such as substations) or a centralised control unit (such as power regulators at the generation end).

The cell voltage, current and temperature are measured through sensors and are passed on to the ECU allotted for those particular cells. Based on the terminal voltage, a decision is made to continue the functioning of cell or to disconnect the cell if it is overcharged or over discharged. The data from the sensors are further passed through parameter identification of the cell. The parameters identified are utilised to estimate SoC and SoH. The difference in SoC provides a base for cell-balancing algorithms, based on the kinds of cell-balancing circuit employed for an appropriate action. The following are among the techniques that are employed at each stage.

13.4.1 Parameter Identification Algorithms

Parameter identification consists of two distinguished stages. In initial stage, a relationship between open circuit voltage and SoC is obtained experimentally for various temperatures. These data are saved either in the form of a look-up table or in the form of an empirical equation obtained by curve fitting technique [41]. The parameters are determined by two methods, either as an online or offline method.

13.4.1.1 Offline-Based Method for Parameter Identification

The offline-based methods of parameter identifications depends on the type of model. The details of the models are discussed extensively in the following sections. The battery models depends on the required accuracy, simplicity of solving the equations, memory required to store the data and cost of ECU [42]. In the offline-based method, the parameters are fitted to the experimentally obtained data. Offline-based methods store data in two forms: one in the form of table, and the other in form an equation that represents the various parameters relationship that is obtained by curve fitting. The parameters are obtained by one of two methods.

- A basic calculation based method based on the voltage and current data obtained from experimentation. This method is simplest of all and can only be employed to develop an electrical circuit–based model up to first order [17].
- The second method is based on the optimisation technique whereby the unknown parameters are considered as the variables and based on the boundary conditions that are determined by the user. These variables are subjected to the objective function that is created, based on the model. The general principle is to calculate the terminal voltage for an input current through the model and obtain error between calculated value and experimental value. The objective function is designed to minimise the error. In literature many optimisation algorithms—such as genetic algorithm [43], differential evolution [44], particle swarm optimisation [45], teaching learning-based optimisation [45] technique and many more—have been employed.

The advantage of these techniques is that the computational burden on the ECU is reduced as the complicated computations are carried in a laboratory setup and just the parameter values are loaded in ECU either in form of a look-up table or in the form of an equation that describes its behaviour based on a particular property. The main disadvantage of this method is that it requires enormous experimental data to properly predict the parameter values with consideration of degradation effect. As the number of experiments increase, this method becomes more complicated if the usage of memory has be kept in check. Apart from the huge dataset required to accurately estimate the parameters, it is difficult to create the exact conditions that a battery might be subjected in to its environmental conditions.

13.4.1.2 Online-Based Method for Parameter Identification

Online-based parameter identification is based on estimation algorithms [46]. The batter models have to be represented in a linearised form as shown by Eq. (13.10). The battery model represented by this form is referred as the auto-regressive exogenous model [47].

$$y(t) = \emptyset(t)^T \theta \tag{13.10}$$

where, $y(t)$ is the output vector, in this case terminal voltage, and $\emptyset(t)^T$ is a transposed vector known as linear regression vector and the elements in the $\emptyset(t)^T$ are known as repressors; the variable θ is a parameter vector formed from the parameters that are required to be determined

This equation is then subjected to estimation algorithms such as recursive least square [48], recursive least square with forgetting factor [49], weighted recursive least square with forgetting factor [50], and many other parameter estimation algorithms. In these algorithms, the general principle is to provide an initial value of the parameters. Based on these initial values, the output vector is calculated from the Eq. (13.10). This output vector is subtracted from the experimentally obtained data. Based on the error, the assumed initial value of the parameters are changed such that error is minimised. After few iterations, the error is minimised and the algorithm is continued until the parameters are required.

13.4.2 State of Charge Estimation Algorithms

The parameters determined through parameter identification are used to determine the various state parameters that provide working knowledge of the battery. One such parameter is SoC, which is determined by many methods depending on the accuracy required and the computational capability of ECU. They are mainly defined as coulomb-counting method, look-up table–based method, model-based method, fuzzy-based method and neural-based methods that are explained briefly as follows.

- *Coulomb-counting method:* The coulomb-counting method is based on the basic definition of SoC, which in terms of ratio is the total number of coulombs utilised to the total charge in terms of coulombs available from the battery [51]. The mathematical representation for the coulomb-counting method is shown in Eq. (13.11). This is either subtracted from or added to the initial SoC value. The simple calculations have prompted its extensive use for small-size batteries. The main issue with this technique is error that arises due to uncertainty in the initial value. This uncertainty arises due to rest condition and self-discharge of the cell. Thus many other methods are adopted for SoC estimation.

$$Z_n = Z_{n-1} + \int_{t_0}^{t_1} \frac{i(t)*t\,dt}{CAh} \tag{13.11}$$

where, Z_n is the SoC at instant t_1, Z_{n-1} is the SoC at instant t_0, i is the current at the terminal voltage and CAh is true capacity of the cell in ampere-seconds

- *Look-up table–based method:* The battery parameters such as open-circuit voltage, resistance and capacitance are functions of SoC. These relationships are obtained from extensive experimentation and are represented in the form of look-up table. If the impedance parameters are considered to represent the relationship, an online parameter estimation method is required to accurately predict the parameters and then a look-up table is used to estimate the SoC [52]. If the open-circuit voltage is considered as the base parameter for the relationship, a combination of look-up table and coulomb-counting methods is employed [53]. The initial SoC in coulomb-counting method at the rest time is obtained from the look-up table, thereby solving the issue. The main disadvantage of this method is the enormous amount of memory required to store the look-up table as the battery degrades and all the parameters are changed.
- *Model-based estimation method:* This method is more reliable than other methods when used in tandem with online parameter determination. In this method, a discrete state space model is developed form the battery model. These models are subjected to the state estimation algorithms such as Kalman filter, extended Kalman filter [54], unscented Kalman filter [53], particle observer and many more [55]. The accuracy of the estimation depends on two factors: accuracy of the battery model, and estimator algorithm employed. During the formation of the discrete state space model, it has to be noted that SoC should be one of the states. Depending on the model employed and its complexity, an appropriate algorithm is employed. The consideration of modelling and measurement error is one of the main advantages of this method, as it provides accurate estimation of SoC.
- *Soft computing–based method:* All the previously described methods require either a basic or precise knowledge of the battery and its characteristics to develop an accurate model, but soft computing techniques such as neural networks and fuzzy systems do not require knowledge of the battery characteristics. In this method, a neural network mode, a fuzzy-based model or a neuro-fuzzy–based model is developed, based on extensive experimentation [56]. These models are developed in such a manner that the inputs are obtained from sensor and fed to the models to provide an estimated SoC. For more accurate SoC estimation, a large dataset is required that consumes a lot of time for experimentation. Apart from this, these methods require a highly sophisticated ECU.

13.4.3 State of Health Estimation Algorithm

The SoH estimation can also be regarded as a lifecycle predictor. The SoH estimation methods are similar to the estimation of the SoC. For this, a life model is developed. In literature, many lifecycle models are presented. The following Eq. (13.12)–(13.15) are the types of models developed [57].

$$\text{Toshiba ageing model } Q_{loss} = 1.544 \times 10^7 e^{\left(\frac{-40498}{8.3143T}\right)t} \quad (13.12)$$

$$\text{Bloom model } Q_{loss} = Ae^{\left(\frac{-E_a}{RT}\right)t^z} \quad (13.13)$$

$$\text{Matsushima model } Q_{loss} = K_f\sqrt{t} \quad (13.14)$$

$$\text{Arrhenius model } Q_{loss} = \left(ae^{\frac{\alpha}{T}} + bI^{\beta} + c\right)n_c^{\left(le^{\frac{\alpha}{T}} + mI^{\delta} + f\right)} \quad (13.15)$$

where, Q_{loss} is the capacity loss, t is time in months, T is temperature in kelvin, I is current in amperes and A, a, b, c, l,α, β, f, R, E_a and K_f, are constants obtained by fitting from experimentation data

The capacity of the cell at Nth cycle is obtained by subtracting the capacity loss Q_{loss} from the previous capacity. C_{N-1} can be rewritten as

$$C_N = C_{N-1} - Q_{loss} \quad (13.16)$$

The Eq. (13.16) is incorporated in the battery model and the estimation algorithms utilised for SoC estimation [57].

13.4.4 Cell-Balancing Techniques

The cell-balancing technique is required especially for the large battery packs, as each cell does not have the same internal resistance and other parameters. Charging and discharging currents creates a change in SoC in the cells connected in series. This difference, if not addressed properly through cell balancing, can create the overcharging and discharging of a single cell when others are still in operation. This cell-balancing technique is mainly classified as the following.

- *Passive cell balancing:* In the passive cell-balancing method, resistors are used to dissipate the energy of a cell with higher SoC level. This is carried by one of two modes. Connecting resistors in parallel with the cells is known as fixed resistor method, as the value of resistance in parallel with the cells is same. The employment of this method continuously discharges the cells and the battery is completely drained out after some time, even if it is kept in ideal mode. This is countered by adding electronic switches that are operated by the controller, and this methodology is known as variable resistor cell-balancing method [58, 59].
- *Active cell balancing:* The main drawback of the passive cell balancing method is the loss of energy due to resistors. This issue is solved by employing active elements that store energy such as capacitors and inductors to store the excess energy from one cell and to transfer the same to other cell without losing much energy. This method is known as

active cell-balancing technique. One of the main issues with this technique is the large size of capacitors and inductors, along with the requirement of a sophisticated controller to properly assess and make a decision [60].

13.5 BATTERY MODELLING

Li-ion batteries are increasingly considered in different stationary and dynamic applications such as power system–based applications and EV-based applications. In order to operate BMS and assess various aspects of technical feasibility for such applications, precise knowledge about their performance behaviour is mandatory. This chapter focuses mainly on the modelling of the selected LFP/C-based battery cells, as they are very much required for the algorithms in BMS.

Li-ion battery performance models should be able to predict with high accuracy the electrical performance (mainly the voltage) of the battery cell at different operating conditions; by operating conditions, it is meant the sum of the conditions given by the load current, temperature and SoC. A precise Li-ion battery cell performance model allows to run simulations in order to improve the battery cell design, verify different operational strategies for the Li-ion ESS and optimise the size of the Li-ion ESS. Thus, by relying on models which are able to estimate accurately the behaviour of the Li-ion battery cells, time- and cost-demanding laboratory experimental tests are minimised or even avoided. Performance modelling of Li-ion batteries could by realised by following different approaches, depending on the final objective of the model. These approaches, categorised by different degrees of complexity, are divided in three main types, as discussed in [61]. Each of this modelling approaches is briefly discussed in the following sections.

13.6 TYPES OF BATTERY CELL MODELLING

The Li-ion battery modelling is categorised into three types: electrochemical models, equivalent circuit models and soft computing–based models.

Electrochemical models are based on the concepts of electrochemistry that aim to incorporate key behaviours of batteries, therefore often achieving higher accuracy. These models are suitable for simulating the distributed chemical reactions in the electrodes and electrolytes [62, 63]. These models are generally represented by partial differential equations that consist of a large number of unknown variables. The complexity of these partial differential equations often requires solving platforms with large memory and high computation capability. During the solving of these models, they face the issue of over-fitting due to their significant number of parameters, therefore rendering its practical application in BMS questionable.

Soft computing–based models developed on the basis of neural networks and fuzzy systems have also been widely studied, but the amount of data required to train the neural network is enormous and the experimentations required for creating such data sets are time-consuming [21, 48].

Equivalent electric circuit models (EECM) are battery models based on lumped components. These models are not highly accurate, but the reduced computational capability required makes them a preferred choice for BMS systems. They have been considered mainly for the purpose of BMS development. They are lumped models with relatively few parameters. A detailed explanation of these models with their governing equations are provided in [64] and the following section. Therefore, EECM are the main focus of this chapter.

13.7 REVIEW OF EQUIVALENT CIRCUIT BATTERY MODELS

In literature, various models based on the lumped parameters such as combined model and simple R model are presented. The resistance-capacitance (RC) network–based equivalent circuit models, such as the first-order RC [65], second-order RC [66] and third-order RC models [67], are widely studied. Dedicated elements depicting the battery hysteresis behaviour were added to the RC models, such as the first-order RC model with hysteresis and second-order RC model with hysteresis [18].

13.7.1 Combined Model

The combined model is based on the relationship between V_{oc} (open-circuit terminal voltage of the battery) and SoC. In this model, the relationship is obtained by combination of Unnewehr universal model, Shepherd model and Nernst model [68]. The electrical representation is shown in Figure 13.4. The equations that represent the relationship between SoC (Z) and V_{oc} is given by Eq. (13.17)–(13.20).

$$Z_n = Z_{n-1} + \int_{t_1}^{t_0} \frac{i(t) * t\,dt}{C_N} \tag{13.11}$$

Shepherd model: $V_{oc} = K_1/Z$ (13.17)
Unnewehr universal model: $V_{oc} = K_2 Z$ (13.18)
Nernst model: $V_{oc} = K_3 \ln(Z) + K_4 \ln(1-Z)$ (13.19)
Combined model: $V_{oc} = K_1/Z + K_2 Z + K_3 \ln(Z) + K_4 \ln(1\text{-}Z)$ (13.20)

where, V_{oc} represents open-circuit terminal voltage, Z represents SoC and K_1, K_2, K_3 and K_4 are constants

This method is rarely used due to high error quantity and in practice the relationship between V_{oc} and Z is obtained by curve-fitting techniques.

13.7.2 Simple Model

The simple battery cell model shown in Figure 13.5 is an extension of the combined model, but the relationship between the V_{oc} and SoC is obtained by curve fitting techniques and a resistance that represents the ionic diffusion through the separator and is a function of temperature and degradation of the separator [18]. This resistance is a combination of resistance (R_0) due to charge transfer (R_{ct}), bulk resistance (R_b) and solid stat interface (R_{sei}). It is

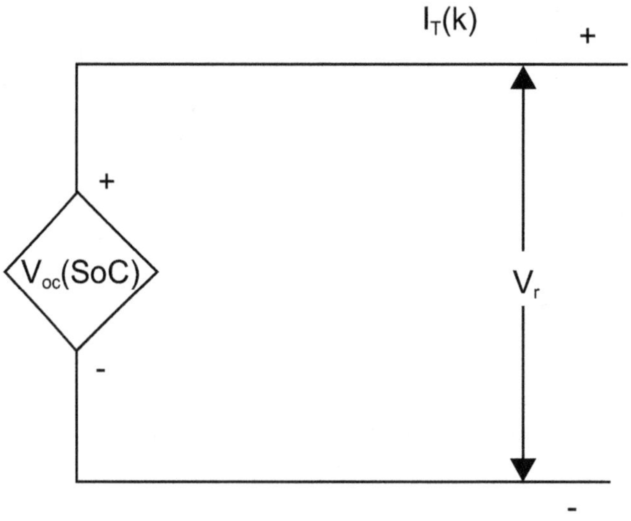

FIGURE 13.4 Combined model equivalent representation [68].

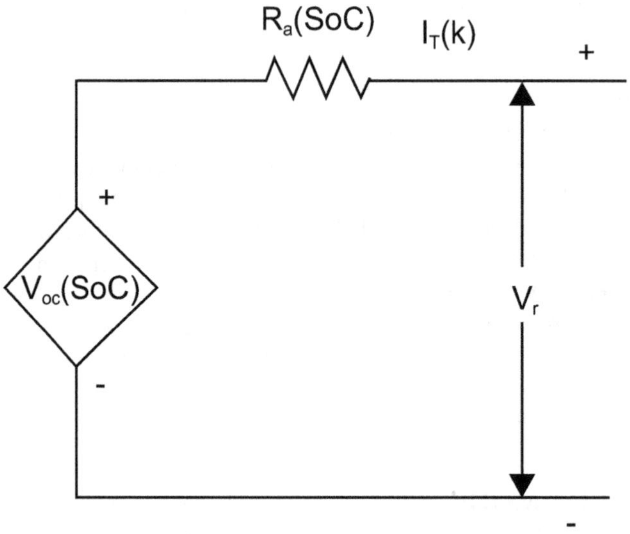

FIGURE 13.5 Simple R model [69].

observed that the resistance R_{ct} is temperature dependent; the equation representing the relationship is shown in Eq. (13.21)–(13.22).

$$R_0 = R_{ct} + R_b + R_{sei} \tag{13.21}$$

$$\log\left(\frac{1}{R_{ct}}\right) = \frac{E_a}{2.303RT} + A \tag{13.22}$$

where E_a is activation Energy, T is absolute temperature, R is gas constant, A is constant obtained through experimentation

13.7.3 Randle's Model

The preceding model is appropriate for static systems whereby a continuous current is either applied or drawn. For a vehicle load, the current is never continuous and the effect of capacitance between anode and cathode comes into picture. The separator acts as dielectric material. In the physical system, if a current is applied, the electron and Li-ion move from anode to cathode and the sudden removal of current drops the terminal voltage, and at a slower pace, charge redistribution is observed. This charge redistribution occurs until the potential difference across the anode and cathode is at such a value that the ions and electrons are at standstill [70]. As shown in the Figure 13.1, each cell has a lot of layers of cathode, anode and separator. This is modelled by Randle's circuit and has a Warburg impedance in parallel with C_{dl} (represents double layer capacitance), as shown in Figure 13.6.

In this Randle's model, the Warburg impedance is represented by infinite RC circuit as shown in Figure 13.3(b).

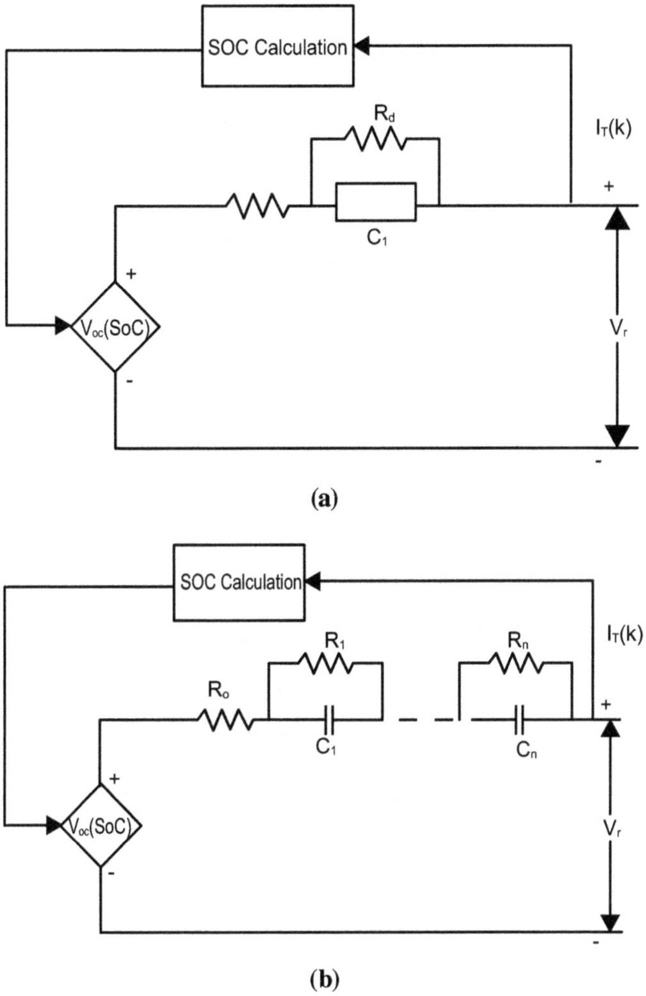

FIGURE 13.6 Electrical equivalent circuit of: (a) Randle's model; (b) Infinite RC model [70].

Since the infinite RC model is not practically implementable the circuit has been simplified as first-order RC, second-order RC and third-order RC. It has been observed in literature that as the order of the system goes on increasing the error decreases but at orders more than four the change in error minimal. So a well approximate model can be obtained by first-order RC and second-order RC networks. The equivalent circuit model with first-order RC and second-order RC has been shown in Figure 13.7(a) and Figure 13.7(b) respectively.

Mathematical Modelling of First-Order RC [17]

For KVL equation, we have

$$V_t(t) = V_{OC}(Z(t)) - V_{C_1}(t) - V_{R_o}(t) \tag{13.23}$$

$$V_{C_1}(t) = R_1 i_{R_1}(t) \tag{13.24}$$

$$V_{R_0}(t) = R_0 i(t) \tag{13.25}$$

By applying KCL, $i_{R_1}(t) + i_{C_1}(t) = i(t)$ (13.26)

$$i_{C_1}(t) = C_1 \frac{dV_{C_1}}{dt} \tag{13.27}$$

Substitute Eq. (13.27) in Eq. (13.26):

$$i_{R_1}(t) + C_1 \frac{dV_{C_1}}{dt} = i(t) \tag{13.28}$$

$$i_{R_1}(t) + R_1 C_1 \frac{di_{R_1}(t)}{dt} = i(t) \tag{13.29}$$

$$\frac{di_{R_1}(t)}{dt} = \frac{1}{R_1 C_1}\left[i(t) - i_{R_1}(t)\right] \tag{13.30}$$

By solving the preceding equation and discretising by considering kΔt = t, the following equation is obtained:

$$i_{R_1}[k+1] = i_{R_1}[k] e^{\left(\frac{-\Delta t}{R_1 C_1}\right)} + i[k](1 - e^{\left(\frac{-\Delta t}{R_1 C_1}\right)} R_1 C_1) \tag{13.31}$$

Discretising the Eq. (13.23) and Eq. (13.30):

$$V_t[k] = V_{OC}(Z[k]) - V_{C_1}[k] - V_{R_o}[k] \tag{13.32}$$

$$V_{C_1}[k] = R_1 i_{R_1}[k] \tag{13.33}$$

$$V_{R_0}[k] = R_0 i[k] \tag{13.34}$$

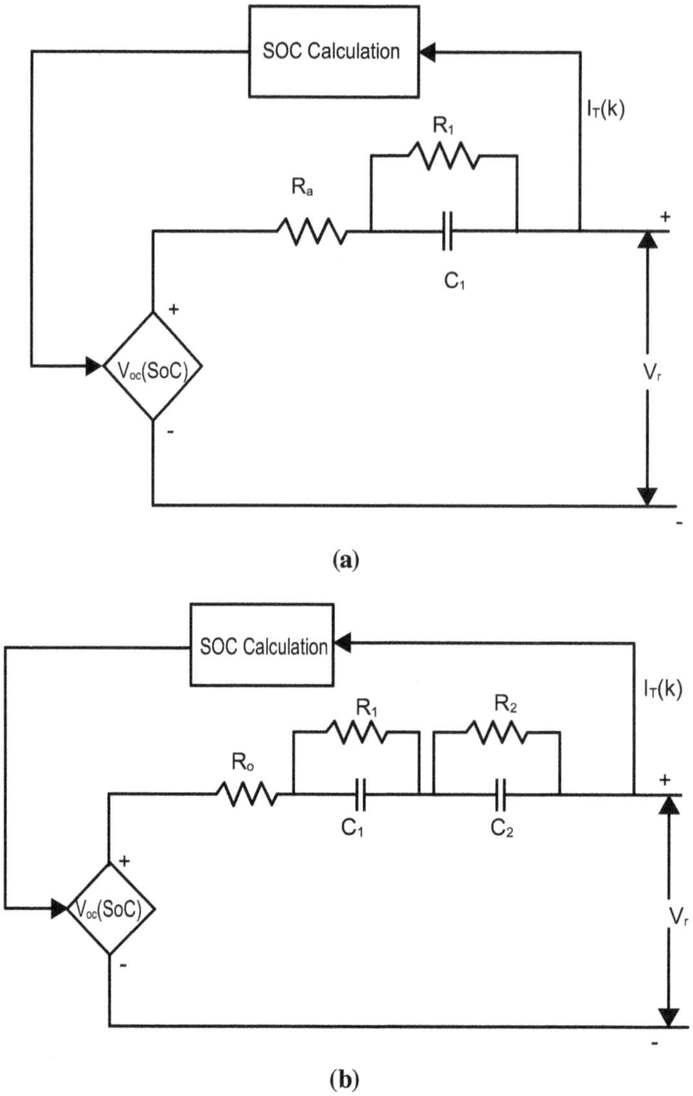

FIGURE13.7 Electrical equivalent circuit of: (a) first-order RC model; (b) second-order RC model [17].

Combing the equation for SoC estimation, Eq. (13.11), Eq. (13.31) and Eq. (13.32):

$$z[k+1] = z[k] - \frac{\Delta t}{C}\eta i[k] \quad (13.35)$$

$$i_{R_1}[k+1] = i_{R_1}[k]e^{\left(\frac{-\Delta t}{R_1 C_1}\right)} + i[k](1 - e^{\left(\frac{-\Delta t}{R_1 C_1}\right)}R_1 C_1) \quad (13.36)$$

$$V_t[k] = V_{oc}(Z[k]) - R_1 i_{R_1}[k] - R_o i_{R_o}[k] \quad (13.37)$$

Mathematical Model of Second-Order RC Model [17]

The equations for second-order RC model are derived in similar manner as in the governing equations are:

$$z[k+1] = z[k] - \frac{\Delta t}{C}\eta i[k] \quad (13.38)$$

$$i_{R_1}[k+1] = i_{R_1}[k]e^{\left(\frac{-\Delta t}{R_1 C_1}\right)} + i[k](1 - e^{\left(\frac{-\Delta t}{R_1 C_1}\right)}R_1 C_1) \quad (13.39)$$

$$i_{R_2}[k+1] = i_{R_2}[k]e^{\left(\frac{-\Delta t}{R_2 C_2}\right)} + i[k](1 - e^{\left(\frac{-\Delta t}{R_2 C_2}\right)}R_2 C_2) \quad (13.40)$$

$$V_t[k] = V_{OC}(Z[k]) - R_1 i_{R_1}[k] - R_2 i_{R_2}[k] - R_o i_{R_o}[k] \quad (13.41)$$

13.7.4 Hysteresis Model

The path followed by the charging curve is different than the discharge graph. This hysteresis voltage is dependent on the SoC and the equation obtained by curve-fitting technique. The first-order RC and second-order RC with hysteresis is shown in Figure 13.8(a) and Figure 13.8(b), respectively.

The equations representing hysteresis model for first-order RC model [17] are:

$$z[k+1] = z[k] - \frac{\Delta t}{C}\eta i[k] \quad (13.42)$$

$$i_{R_1}[k+1] = i_{R_1}[k]e^{\left(\frac{-\Delta t}{R_1 C_1}\right)} + i[k](1 - e^{\left(\frac{-\Delta t}{R_1 C_1}\right)}R_1 C_1) \quad (13.43)$$

$$V_t[k] = V_{oc}(Z[k]) - R_1 i_{R_1}[k] - R_o i_{R_o}[k] + MH(Z) \quad (13.44)$$

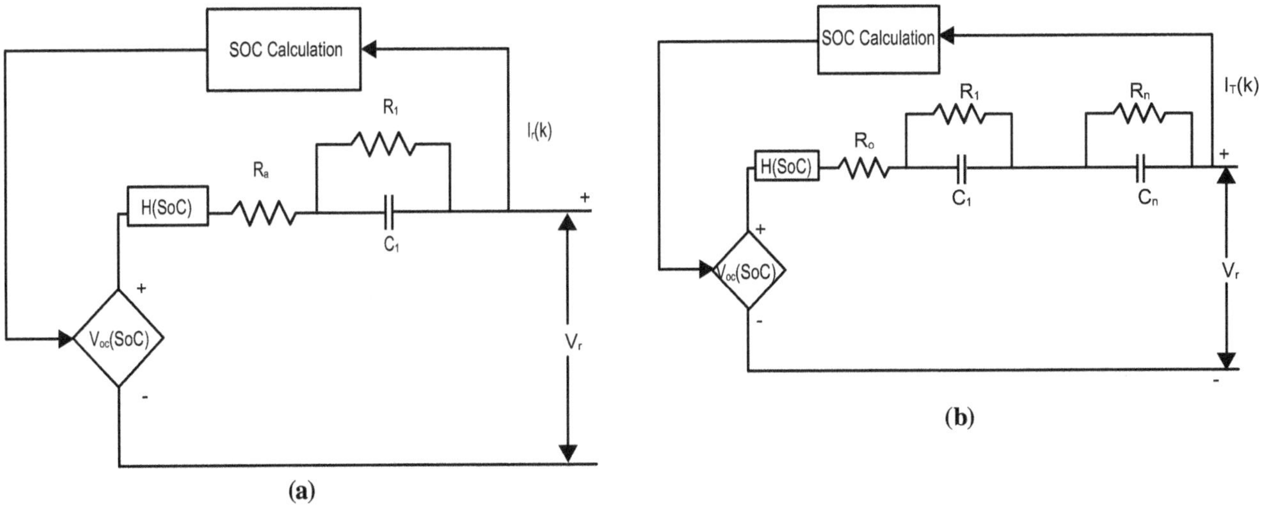

FIGURE13.8 Electrical equivalent circuit of: (a) first-order RC model with hysteresis; (b) second-order RC model with hysteresis [17].

For second-order RC model with Hysteresis [17], the equations are:

$$z[k+1]=z[k]-\frac{\Delta t}{C}\eta i[k] \tag{13.45}$$

$$i_{R_1}[k+1]=i_{R_1}[k]e^{\left(\frac{-\Delta t}{R_1C_1}\right)}+i[k](1-e^{\left(\frac{-\Delta t}{R_1C_1}\right)}R_1C_1) \tag{13.46}$$

$$i_{R_2}[k+1]=i_{R_2}[k]e^{\left(\frac{-\Delta t}{R_2C_2}\right)}+i[k](1-e^{\left(\frac{-\Delta t}{R_2C_2}\right)}R_2C_2) \tag{13.47}$$

$$V_t[k]=V_{oc}(Z[k])-R_1i_{R_1}[k]-R_2i_{R_2}[k] -R_oi_{R_o}[k]+MH(Z) \tag{13.48}$$

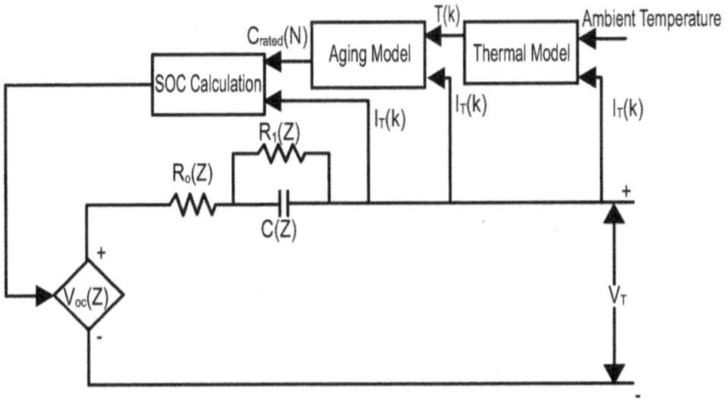

FIGURE 13.9 Battery model adopted for the simulation study.

13.7.5 SoC-Dependent Electro-Thermal-Degradation Battery Model

The battery model employed in this chapter is can be categorised as electrical model, thermal model and ageing model. The Figure 13.9 describes a complete battery model that includes analogous electrical representation of chemical reactions, ageing model and thermal model. The electrical model adopted in this work is a first-order RC model.

Electrical Model

The electrical model had been developed by pulse testing methodology. The parameter identification has been carried by the data obtained from test. The Eq. (13.49) represents coulomb-counting methodology adopted for SoC Calculation. In Eq. (13.49), the terms z, C[N],ηi[k] denotes SoC, capacity at Nth cycle, Columbic efficiency and terminal current at kth sample, respectively.

$$z[k+1]=z[k]-\frac{\Delta t}{C[N]}\eta i[k] \tag{13.49}$$

$$i_{R_1}[k+1]=i_{R_1}[k]e^{\left(\frac{-\Delta t}{R_1([k])C_1([k])}\right)}+i[k]\left(1-e^{\left(\frac{-\Delta t}{R_1([k])C_1([k])}\right)}\right) \tag{13.50}$$

$$V_t[k]=V_{oc}([k])-R_1([k])i_{R_1}[k]-R_o([k])i_{R_o}[k] +MH([k]) \tag{13.51}$$

Thermal Model

This work adopts a temperature model for the battery from [71]. Refs [72] and [73] have considered a linear system that comprises two states, T_{core} and T_{air}, and reference [74] uses the nonlinear heat transfer equation with a single state. A simulation study shows that the dynamics of T_{air} have negligible effect on the temperature rise. Therefore, the a simple temperature model, provided in [71] represented by Eq. (13.52)–(13.55) has been adopted.

$$T[k]=T[k-1]-\left(f\left(T[k-1]-T_a\right)\right)+\left(b[k]i[k]^2\right) \tag{13.52}$$

$$f = \frac{1}{mC_{cell} R_{eff}} = 3.8069 \times 10^{-4} \tag{13.53}$$

$$R_{eff} = \frac{1}{hA} \tag{13.54}$$

$$b[k] = \frac{R_o[k] + R_1[k]}{mC_{cell}} = \frac{R_o[k] + R_1[k]}{0.045 * 995.2} \tag{13.55}$$

Ageing Model

The battery ageing model utilised for the ageing analysis is based on a general Arrhenius equation that calculates the capacity fade associated with the both charging and discharging cycle. In literature, Meintz et al. [75] through experimental data defined the battery capacity degradation as a function of temperature and time. Keil and Jossen [11] further developed an ageing model by incorporating the C-rate, the total Ah and temperature. Suri and Onori [76] included the effect of SoC in the model, and the combined model is shown by Eq. (13.56)–(13.64).

$$DoD[k] = 1 - z[k] \tag{13.56}$$

$$Ah[k] = N.DoD[k].Ah_{rated} \tag{13.57}$$

$$C_{rate}[k] = \frac{i[k]}{Ah_{rated}} \tag{13.58}$$

$$E_a[k] = 31500 - 152.5C_{rate}[k] \tag{13.59}$$

$$B[k] = \begin{cases} 2896.6z[k] + 7411.2 \text{ if } z[k] \le 0.45 \\ 2694.5.z[k] + 6022.2 \text{ if } z[k] > 0.45 \end{cases} \tag{13.60}$$

$$Q_c[k] = B[k] e^{\frac{-E_a[k]}{(8.314T[k])}} Ah^{0.55} \tag{13.61}$$

$$Q_{cyc}[N] = \sum Q_c[k] \tag{13.62}$$

$$C_{\%}[N+1] - C_{\%}[N] - Q_{cyc}[N] \tag{13.63}$$

$$C[N+1] = 2.5 * \frac{C_{\%}[N]}{100} \tag{13.64}$$

13.8 METHODOLOGY FOR BATTERY MODELLING

The methodology followed for battery modelling in literature is shown in Figure 13.10. This battery modelling can be divided into static and dynamic testing. The static test provides the relationship between the V_{oc} and SoC, while the dynamic testing provides the values for parameters. This obtained data is utilised to construct a model as shown in Figure 13.10.

13.8.1 Static Testing

A cell's open-circuit voltage is a static function of its SoC and temperature. All other aspects of a cell's performance are dynamic. Thus, separate experiments have to be performed to collect experimental data for the SoC-versus-OCV relationship and for the dynamic relationship. In this section, experiments to determine the OCV relationship have been discussed. Before the cell test begins, the cell must be fully charged. Then, the cell is very slowly discharged to a minimum operating voltage while continuously measuring cell voltage and accumulated ampere-hours discharged. The cell is rested for at least a day so that enough relaxation time is provided for charge redistribution, then the cell is very slowly charged and constant current–constant voltage (CC-CV) charging scheme has been adopted. In literature, the pulse-charging scheme has been adopted and it is observed that the SoC-vs.-V_{oc} graph has a missing data from 95–100 since the test is stopped as the voltage reaches the V_{max}, but due to charge redistribution, a false voltage is sensed. As the supply is removed, it is observed that the voltage across terminal reduces. It is hypothesised that the capacitance layer at the terminal end has been fully charged. Once the supply is removed, the charge in terminal capacitors is distributed among the other capacitors. Thus, to fully charge a battery, a constant current is applied until voltage is V_{max} then a constant voltage V_{max} is applied and the current is brought to zero [17].

The purpose of the slow rate is to minimise the excitation of the dynamic parts of the cell model. We desire to keep the cell in a quasi-equilibrium state at all times. So, we generally use a C/20 rate, which is a compromise between the desire to

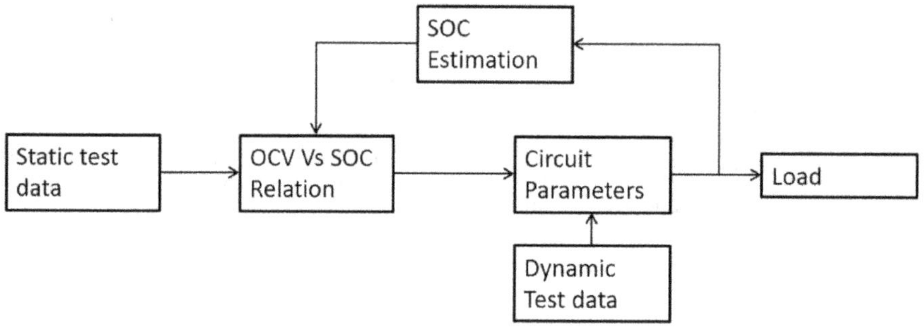

FIGURE 13.10 Flow diagram of battery model [17].

have zero current to have a true equilibrium and the practical realisation that the test will already require on the order of 40 hours to complete with a C/30 discharge followed by a C/30 charge. The collected data allow us to create a curve of "discharge OCV" versus discharge ampere-hours and a curve of "charge OCV" versus charge ampere-hours. With appropriate conversion between ampere-hours and SoC, we then have discharge OCV and charge OCV versus SoC. These curves are different because:

- The test is not conducted at zero current, so there will still be some ohmic and diffusion polarisation.
- Hysteresis is present in the cell voltage response.

It is assumed that the deviation from true OCV to the discharge OCV and charge OCV curves is equal in magnitude, so an approximate the cell's true OCV as the average of these two values.

13.8.2 Dynamic Testing

Dynamic testing is done to obtain the values of lumped parameters of each module. This dynamic testing can be categorised as frequency-based testing and time-based testing.

Frequency-Domain-Based Testing

In frequency-based testing, an instrument known as impedance spectroscope is utilised to obtain the impedance spectroscopy of the cell. Electrochemical impedance spectroscopy has two modes of experimentation, namely galvanostatic and potentiostatic modes. In galvanostatic mode, the energy storage device under investigation is subjected to a small AC current and its voltage response is recorded both in terms of magnitude and phase change. Based on the both input and output data, the impedance of energy storage device is determined with discrete Fourier transforms (DFTs). The impedance of the energy storage device is a function of SoC and temperature. As the average value of an AC signal is zero, the battery is neither charged nor discharged. Thus, the cell is subjected to potentiostatic mode of operation that utilises an AC signal superimposed with a small DC signal; thereby, it either charges or discharges the cell. Thus, a huge set of data is required to analyse the impedance at various SoC and temperature [41–43]. This method is generally an offline testing method and is used for benchmarking purposes. Dynamic testing is done to obtain the values of lumped parameters of each module. This dynamic testing can be categorised as frequency-based testing and time-based testing. After the spectroscopy has been obtained for all required SoC values, the constructed model is subjected to same test and in a simulation environment with initial parameter values. The impedance response is then compared with actual response, and the parameter values are adjusted such that the error is minimised.

Time-Domain-Based Testing

In time-based testing, a DC pulse is applied either for charging and discharging the cell for a certain duration, and then the pulse is removed. An example of single-pulse discharge testing is shown in Figure 4.3. At the instant when the discharge current pulse is removed, at time 20 minutes, and considering Eq. (13.65), the instantaneous change in voltage must be equal to the instantaneous change in current multiplied by the series resistance R_0 because the capacitor voltage cannot change instantly, and SoC is not changing when current is zero [17]. Thus,

$$\Delta V_0 = R_0 \Delta i \tag{13.65}$$

Δi is change in current during the test, and ΔV_0 change in voltage; therefore, the value:

$$R_0 = |\Delta V_0/\Delta i| \tag{13.66}$$

Then, the steady-state change in voltage can approximate by the value around time 60 minutes. The overall steady-state voltage change can be found from Eq. (13.67) to be:

$$\Delta V_\infty = (R_0 + R_1)\Delta i \tag{13.67}$$

again with signs computed so that R_0 and R_1 are both positive, knowing that the capacitor voltage will converge to zero in steady-state. Thus the value of R_1 is obtained by:

$$R_1 = |\Delta V_\infty/\Delta i| - R_0 (13.68)$$

Finally, the pulse response converges to a value close to steady-state in about four time constants of the R-C circuit, whereby the time constant of the exponential decay is:

$$4\tau = R_1 C_1 \tag{13.69}$$

In literature, there are many other dynamic testing procedures involving continuous charging and discharging pulses. Based on the voltage pattern, the parameters values are estimated either by usage of recursive least square algorithm, recursive least square algorithm with forgetting factor, Kalman filter or any multi-objective optimisation technique.

13.9 DEVELOPMENT OF TEST BENCH

Based on the literature on the testing methodology of the cell, modelling this work is based on time-based modelling. A test bench setup has been developed to conduct the static and dynamic test. The schematic of the developed test bench setup is shown in Figure 13.11, and the test bench setup has been shown in Figure 13.12.

The specifications of the test bench setup, along with Li-ion battery used, is shown in Table 13.3.

13.9.1 Experimentation of Static Test

The procedure for static testing is outlined in what follows.

- Measure the voltage of the cell, and if it is below V_{max}, charge the cell until the terminal voltage under rest condition reaches V_{max}.

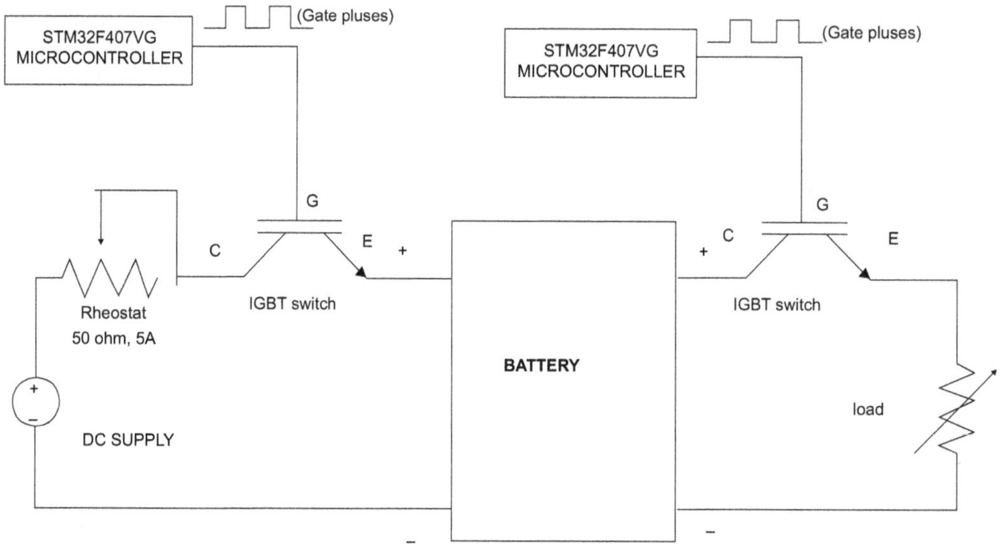

FIGURE 13.11 Test bench setup of the testing system.

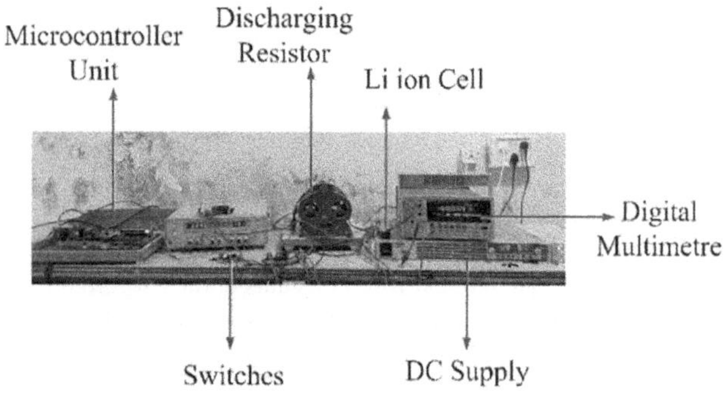

FIGURE 13.12 Developed test bench setup of the system.

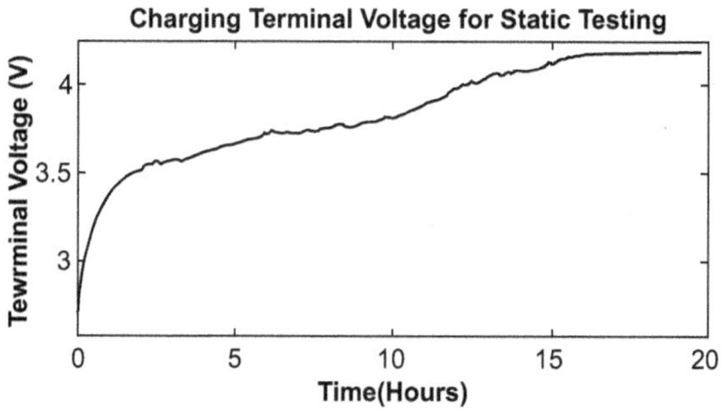

FIGURE 13.13 Battery terminal voltage V time in hours for discharge test.

TABLE 13.3
Specifications of Test Setup [77]

Equipment	Parameter	Specifications
Li-ion battery	Nominal voltage (V)	3.6V
	Capacity (Ah)	2.5Ah
	Minimum cut-off voltage (V)	2.75V
	Maximum cut-off voltage (V)	4.2V
Programmable DC power supply	Rated maximum DC voltage	60V
	Rated maximum DC current	45A

- Discharge the cell by connecting a resistance of 33Ω such that the current is 0.13A; i.e., at the rate of C/20 until V_{min} is obtained.
- Continuously measure the terminal voltage and current using a DMM under a three-wire two-measurement scheme.
- Plot the graphs of voltage and current against time obtained through discharging of battery. The obtained graphs are shown in Figure 13.13 and Figure 13.14.
- At the value of V_{min}, let the battery rest for at least for two hours so that the effect of charge redistribution is negated.

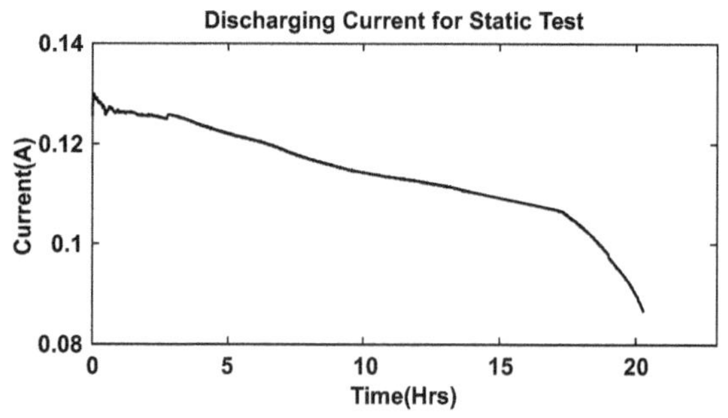

FIGURE 13.14 Battery current V time in hours for discharge test.

- Charge the cell by connecting a constant current source such that the current is 0.13A; i.e., at the rate of C/20 until V_{max} is obtained, and then a constant voltage of V_{max} is applied until current reaches zero.
- Continuously measure the terminal voltage and current using a DMM under three-wire two-measurement scheme.
- At the value of V_{min}, let the battery rest for at least for two hours so that the effect of charge redistribution is negated.
- Plot the graphs of voltage and current against time obtained through charging of battery. The obtained graphs are shown in Figure 13.15 and Figure 13.16.

13.9.2 Experimentation of Single-Pulse Testing

The procedure for single-pulse testing is outlined in what follows.

- Measure the voltage of the cell, and if it is below V_{max}, charge the cell until the terminal voltage under rest condition reaches V_{max}.
- Discharge the cell by connecting a resistance of 33Ω such that the current is 0.13A; i.e., at the rate of C/20 for ten minutes.
- Continuously measure the terminal voltage and current using a DMM under a three-wire two-measurement scheme.
- At the time t = 20 minutes, remove the pulse and let the battery rest for at least for 45 minutes so that the effect of charge redistribution is negated.

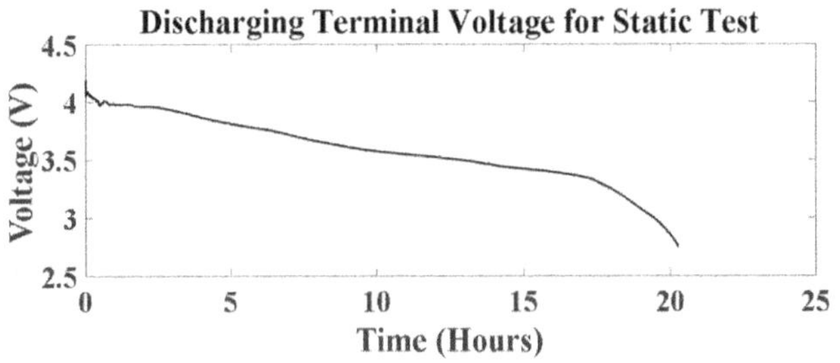

FIGURE 13.15 Battery terminal voltage V time in hours for discharge test.

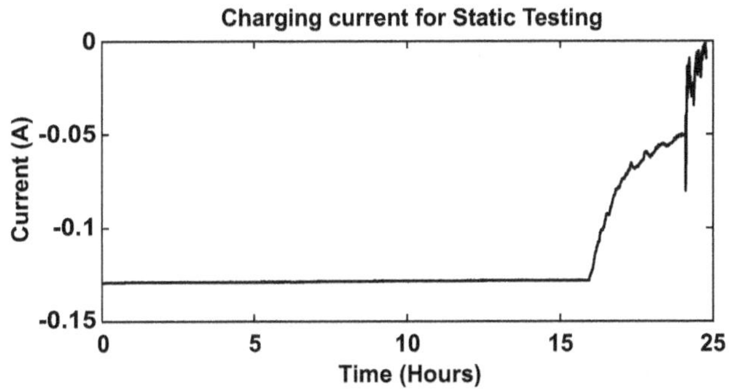

FIGURE 13.16 Battery current V time in hours for charge test.

- Charge the cell by connecting a constant current source such that the current is 0.13A; i.e., at the rate of C/20 for ten minutes, and let the battery rest for at least for 45 minutes so that the effect of charge redistribution is negated.
- Continuously measure the terminal voltage and current using a DMM under a three-wire two-measurement scheme.
- Plot the graphs of pulse charging and discharging of the battery. The graphs are shown in the Figure 13.17 and Figure 13.18.

Numerical Question 2

In a single-pulse testing, the battery initially is at a voltage of 4.09V. A pulse of discharging current of 0.126A has been applied. When the terminal voltage reaches 3.949V, the pulse is removed. The terminal voltage suddenly rises to 4.003V and after a rest of 40 minutes, the terminal voltage reaches 4.007V. Find the internal parameters R_o, R_1 and C_1. The resistance of wire connecting the battery is 0.4Ω.

Solution

The sudden change in voltage when pulse is removed is obtained by $\Delta V_0 = 0.054V$

The sudden change in current is $\Delta i = 0.126A$

The total resistance combining wire and the cell is $R_0 + R_w = |\Delta V_0/\Delta i| = 0.4\ \Omega$

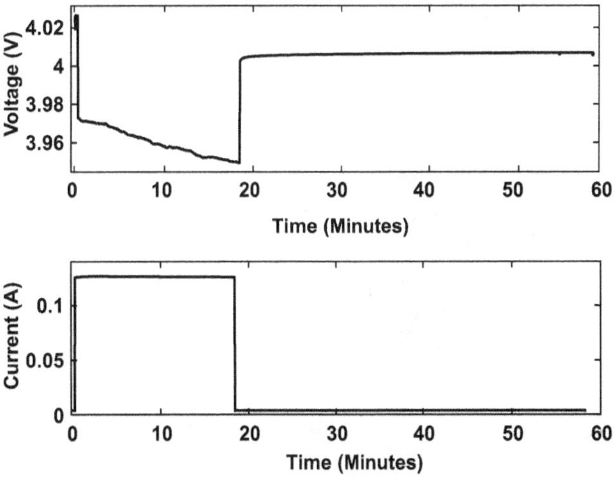

FIGURE 13.17 Single-pulse test data for discharging.

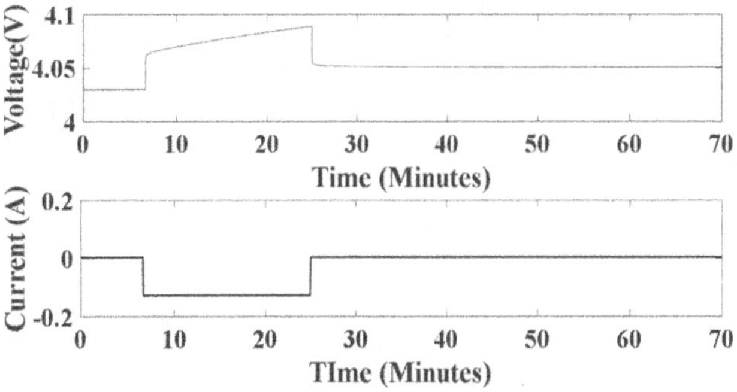

FIGURE 13.18 Single-pulse test data for charging.

The wire resistance provided is $R_w = 0.4\Omega$

Thus internal resistance R_1 is 0.04 Ω

The voltage difference for the steady-state at the end of 45 minutes and the moment pulse is removed is $\Delta V_\infty = 0.058V$

The total resistance $R_1 + R_0 + R_w = |\Delta V_\infty/\Delta i| = 0.46\Omega$

The internal resistance $R_1 = 0.02\Omega$

The time constant is $4\tau = 2{,}400$ seconds

Therefore, the capacitance C_1 is 30kF

Thus, the required internal parameters are $R_0 = 0.04\Omega$, $R_1 = 0.02$ Ohms and capacitance C_1 is 30 kF.

Numerical Question 3

In a single-pulse testing, the battery initially is at a voltage of 4.03V. A pulse of charging current of 0.126A had been applied. When the terminal voltage reaches 4.089V, the pulse is removed. The terminal voltage suddenly drop to 4.055V, and after a rest of 40 minutes, the terminal voltage reaches 4.051V. Find the internal parameters R_o, R_1 and C_1. The resistance of wire connecting the battery is 0.2Ω.

Solution

The sudden change in voltage when pulse is removed is obtained by $\Delta V_0 = 0.034V$

The sudden change in current is $\Delta i = 0.126A$

The total resistance combining wire and the cell is $R_0 + R_w = |\Delta V_0/\Delta i| = 0.26\Omega$

The wire resistance provided is $R_w = 0.4\Omega$

Thus internal resistance R_1 is 0.06Ω

The voltage difference for the steady-state at the end of 45 minutes and the moment pulse is removed is $\Delta V_\infty = 0.038V$

The total resistance $R_1 + R_0 + R_w = |\Delta V_\infty/\Delta i| = 0.30\Omega$

The internal resistance $R_1 = 0.04$ ohms

The time constant is $4\tau = 2{,}400$ seconds

Therefore the capacitance C_1 is 30kF

Thus the required internal parameters are $R_0 =$ 0.04ohms, $R_1 = 0.02$ Ohms and capacitance C_1 is 60 kF.

The two numerical questions were calculated for the same cell, but the experiments were conducted at different initial terminal voltages. There is a change in value of calculated parameters with change in SoC. Therefore, multi-step testing is designed to obtain the parameters at various intervals of SoC.

13.9.3 Experimentation of Multi-Step Testing

- The terminal voltage of the battery initially was brought to 4.2V at rest state.
- Then, the battery was discharged with a C/7 rate for 15 minutes from 4.2V (SoC at 100%) and rested for 45 minutes.
- This step is repeated until the rest voltage reaches at 2.75V (SoC at 0%). The battery was rested for two hours and then was charged with a CC-CV strategy.
- Initially in CC mode, the battery was charged with C/7 current for 15 minutes and a rest time of 45 minutes was provided.
- The same procedure was carried out in the CV stage until the current reached 0.1C and the terminal voltage during the rest period reaches 4.2V.
- The charging and discharging data form the experimentations are shown in Figure 13.19 and Figure 13.20, respectively. In Figure 13.19(b) and

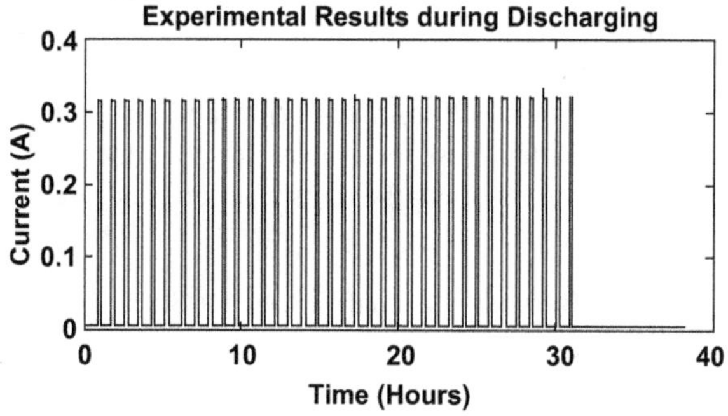

(a)

FIGURE 13.19 Experimental results during discharging: (a) discharging current; (b) terminal voltage.

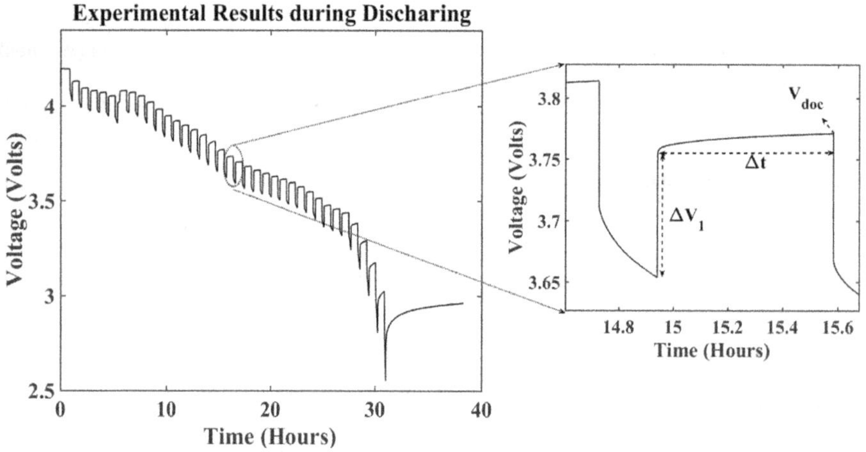

(b)

FIGURE 13.19 (Continued)

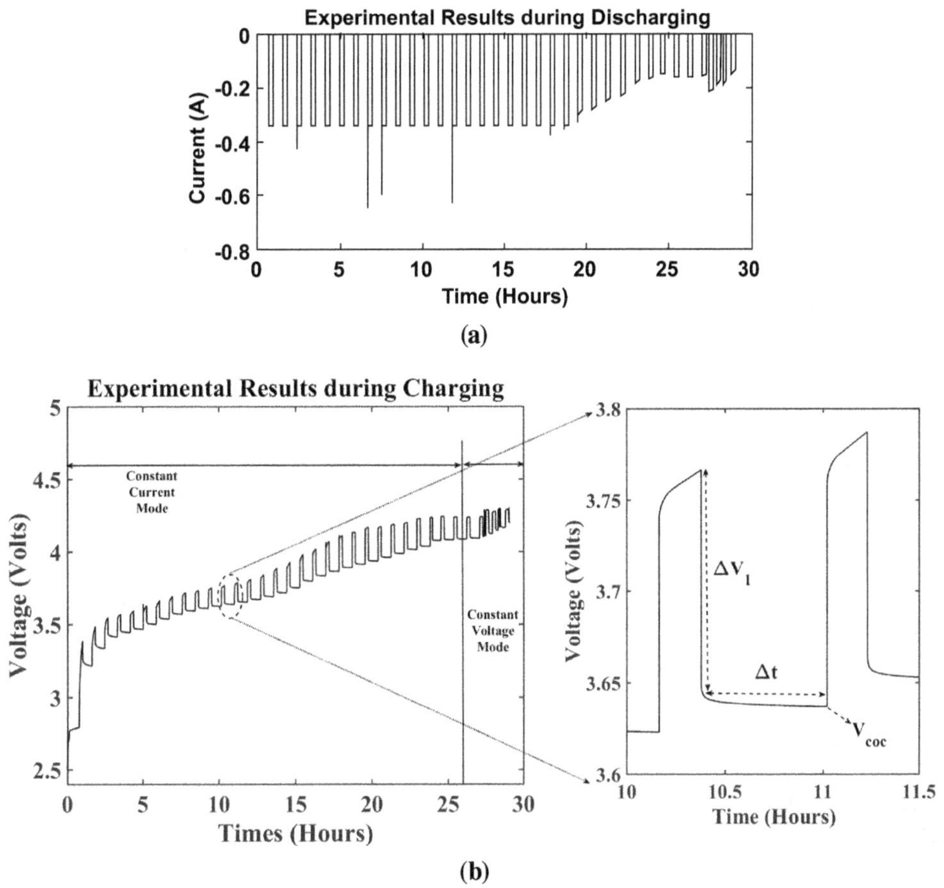

FIGURE 13.20 Experimental results during charging: (a) charging current; (b) terminal voltage.

Figure 13.20(b), an enlarged version of a one-hour cycle of the test is shown, and the open-circuit voltage and voltage drops (ΔV_1 and ΔV_2) are described.

Based on the obtained experimental results, the open-circuit voltage during charging and discharging open-circuit voltages are identified. The methodology to determine the parameters in single-pulse testing is utilised to obtain the parameters such as resistances R_O, R_1 and capacitance C_1. Figure 13.21(a) shows the open-circuit voltage obtained from the charging part of experiment, while Figure 13.21(b) shows open-circuit voltage for discharging. Based on the equations provided, calculating the internal resistances and capacitance for all the pulses has been obtained. Figure 13.22. shows the variation of the internal parameters with SoC.

The average of open-circuit voltage for charging and discharging is calculated and is represented in Figure 13.23. Based on the data average open-circuit voltage, the hysteresis is then obtained and is shown in Figure 13.24. The hysteresis phenomenon is considered with a common Eq. (13.70) for both charging and discharging hysteresis data, where I is terminal current.

$$H = sign(I)\big(f(Z)\big) \tag{13.70}$$

These obtained parameters are functions of SoC and the relationship of these parameters as shown in Eq.

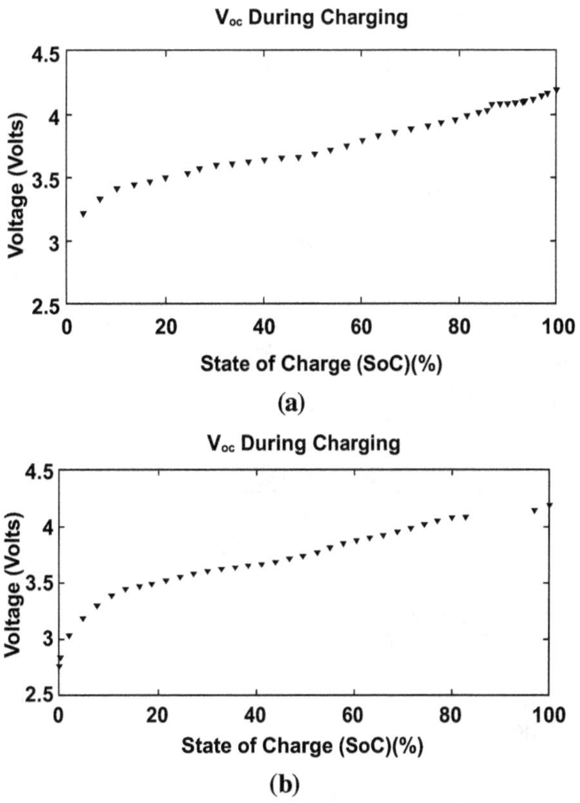

FIGURE 13.21 Open-circuit voltage obtained from experimentation: (a) V_{oc} during charging; (b) V_{oc} during discharging.

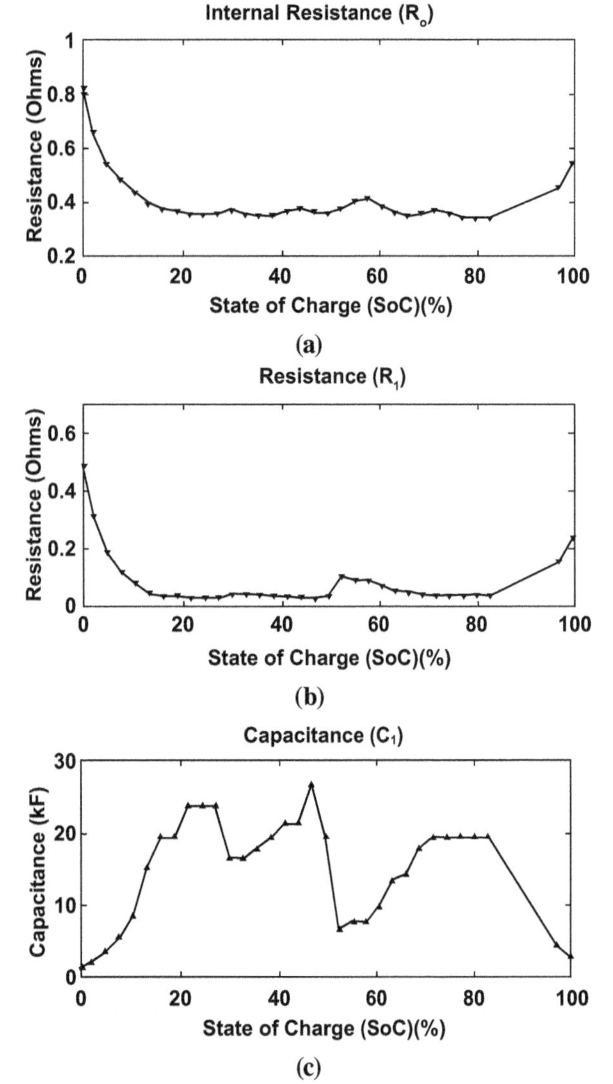

FIGURE 13.22 Parameters of the lithium-ion cell: (a) internal resistance (R_o); (b) Resistance (R_1); (c) capacitance (C).

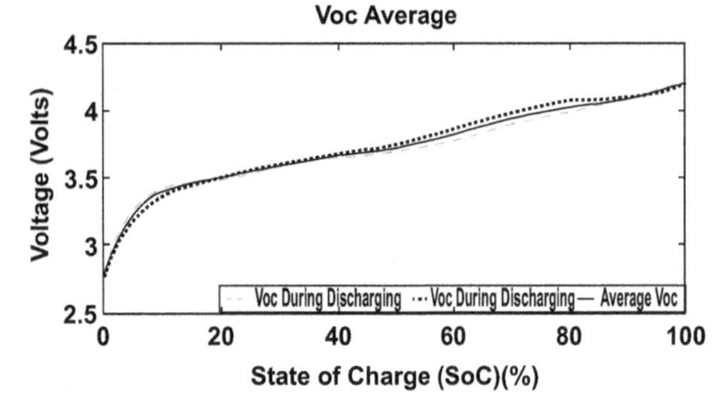

FIGURE 13.23 Plot of V_{oc} during charging, V_{oc} during discharging and Average V_{oc} with experimental data.

(13.71)–(13.78) is determined by applying curve fitting technique.

$$z[k+1]=z[k]-\frac{\Delta t}{C[N]}\eta i[k] \tag{13.71}$$

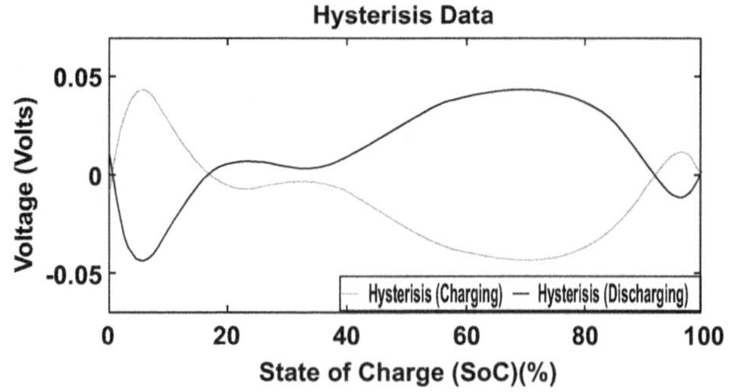

FIGURE 13.24 Hysteresis data.

$$\begin{aligned} V_{oc}([k]) = & -454.2425(z[k])^{8}+2029.4(z[k])^{7} \\ & -3741.1(z[k])^{6}+3684.1(z[k])^{5} \\ & -2101(z[k])^{4}+705.6841(z[k])^{3} \\ & -136.0894(z[k])^{2}+14.6430(z[k]) \\ & +2.7645 \end{aligned} \tag{13.72}$$

$$\begin{aligned} H([k]) = & -24.0442(z[k])^{8}+1011.2(z[k])^{7} \\ & -1757.3(z[k])^{6}+1630.3(z[k])^{5} \\ & -864.752(z[k])^{4}+259.82(z[k])^{3} \\ & -40.7872(z[k])^{2}+2.6082(z[k]) \\ & -0.0123 \end{aligned} \tag{13.73}$$

$$\begin{aligned} R_{o}([k]) = & \,283.032(z[k])^{8}-1212.908(z[k])^{7} \\ & +2169.959(z[k])^{6}-2093.401(z[k])^{5} \\ & +1179.54(z[k])^{4}-396.7483(z[k])^{3} \\ & +79.2972(z[k])^{2}-9.053(z[k]) \\ & +0.8376 \end{aligned} \tag{13.74}$$

$$\begin{aligned} R_{1}([k]) = & \,292.6524(z[k])^{8}-1305.12(z[k])^{7} \\ & +2424.113(z[k])^{6}-2419.836(z[k])^{5} \\ & +1403.596(z[k])^{4}-480.881(z[k])^{3} \\ & +95.4744(z[k])^{2}-10.2678(z[k]) \\ & +0.5084 \end{aligned} \tag{13.75}$$

$$\begin{aligned} C_{1}(z[k]) = & \,219831.221(z[k])^{8}-838371.69(z[k])^{7} \\ & +1306552.05(z[k])^{6}-10794590.51(z[k])^{5} \\ & +512207.27(z[k])^{4}-139086.95(z[k])^{3} \\ & +18938.93(z[k])^{2}-604.1638(z[k]) \\ & +9.5813 \end{aligned} \tag{13.76}$$

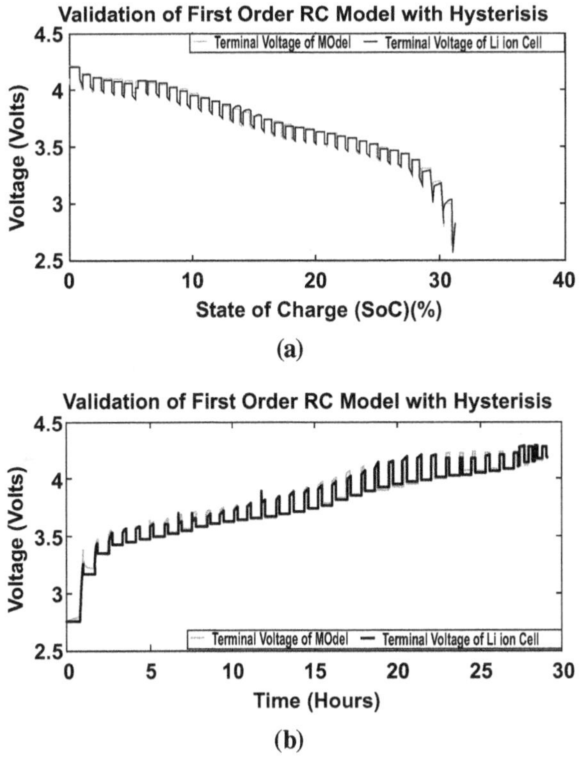

FIGURE13.25 Validation of first-order RC with hysteresis model: (a) discharging; (b) charging.

$$i_{R_1}[k+1] = i_{R_1}[k]e^{\left(\frac{-\Delta t}{R_1([k])C_1([k])}\right)} + i[k](1 - e^{\left(\frac{-\Delta t}{R_1([k])C_1([k])}\right)}) \quad (13.77)$$

$$V_t[k] = V_{oc}([k]) - R_1([k])i_{R_1}[k] - R_o([k])i_{R_o}[k] + MH([k]) \quad (13.78)$$

Based on this equation, the terminal voltage is calculated for both charging and discharging, and the error is calculated by root mean square error. The validation for SoC-based first-order RC model with charging and discharging currents is shown in Figure 13.25.

13.10 SIMULATION FOR ASSESSING BATTERY AGEING

The simulation for the battery ageing is carried by providing a charge/discharge cycle to the cell. The basic structure of the battery system with DC/DC converter is shown in Figure 13.26. A battery is connected to a bi-directional DC/DC converter controlled by a control algorithm. In this algorithm, the current and voltage reference is provided wither by BMS of the battery for charging and from the grid utility for discharging. The methodology adopted to analyse the performance of charging strategy in terms of charging time, temperature rise and capacity fade is shown

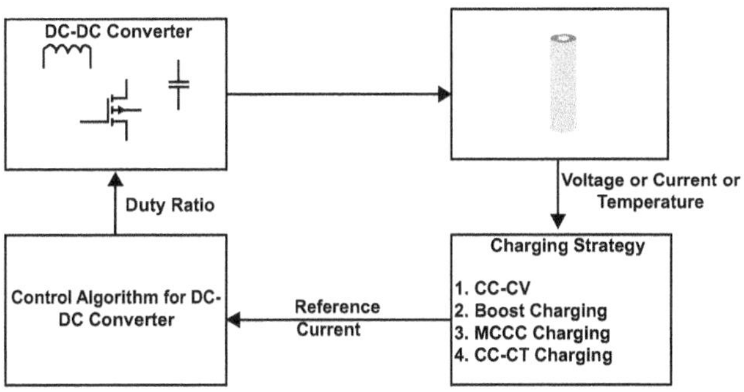

FIGURE 13.26 Basic structure of a battery system with a DC/DC converter.

in Figure 13.27. Initially, upper and lower SoC limits for charging and discharging of the battery is determined. The analysis of the charging strategy is carried over a charge/discharge cycle up to 350 cycles as the manufacturer states that the battery starts rapid degradation after 300 cycles. In this methodology, the battery is either charged by a fast-charging strategy or a standard-charging strategy, as stated by the manufacturer. The decision on the charging with fast-charging strategy is selected based on the number of cycles. In this study, the developed methodology is simulated for three times, and each time, fast charging is selected by differences between fast charging cycles.

In the initial round, every cycle adopts fast charging, and in the second round, every fifth cycle adopts fast-charging strategy. Finally, in the third round, every tenth cycle adopts fast charging. The battery when subjected to standard charging is charged by 0.2C current in constant-current mode and once the terminal voltage reaches maximum allowable voltage, the battery is charged by constant voltage until the battery current reaches 0.01C.

In the adopted methodology, the battery is discharged by a UDDS discharging current that has a random current profile [77]. This study can be replicated with different discharging patterns. This UDDS discharging current is obtained from UDDS drive cycle and vehicle dynamics. The calculations of UDDS drive cycle and vehicle dynamics provide the power required by the motor through an efficiency model of the vehicle power of the battery. An assumption of constant-terminal voltage provides UDDS discharging current. In this method of discharge, care is taken to not violate the upper limit of 4.2V and lower limit of 2.75V of the cell during discharging.

The CC-CV strategy is employed for fast charging of the battery. The battery is charged by a 5A current until the terminal voltage reaches 4.2V. Later, the battery is charged by constant voltage mode. In this report, the battery is charged by having three different SoC ranges: 0–00%, 10–90% and 20–80% SoC. The temperature rise is noted with consideration of having or not having a cooling system.

Figure 13.28(a) provides terminal voltage, open-circuit voltage and SoC for charging with 0–100% SoC. Figure 13.28(b) provides temperature and current data for charging. Table 13.4 provides the result of both CC-CV fast

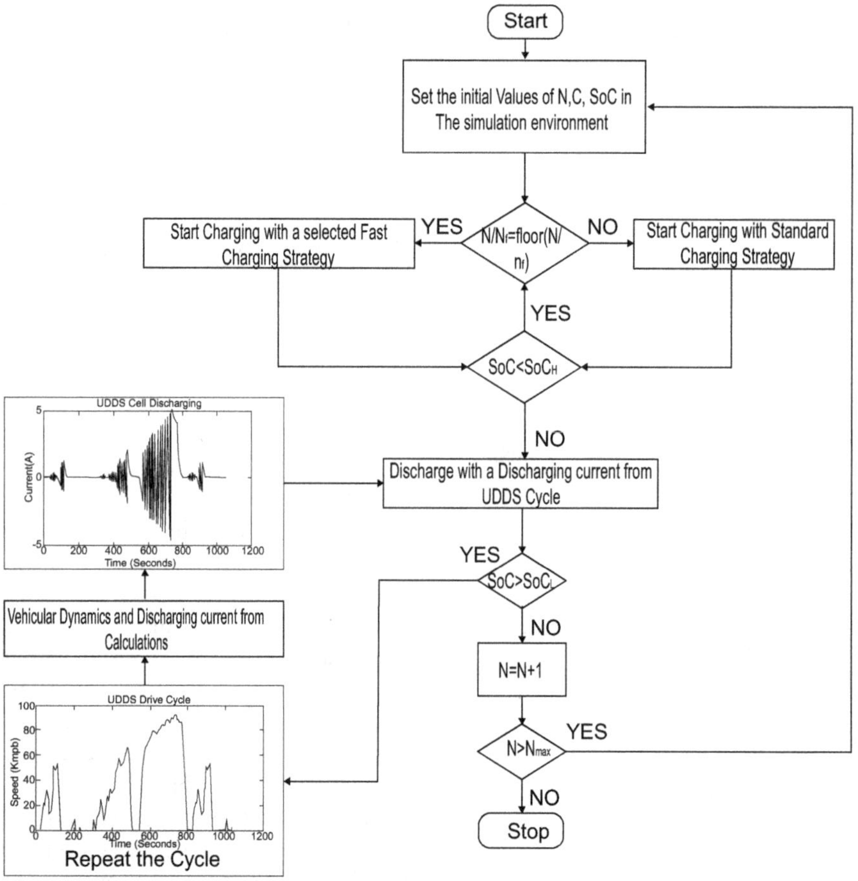

FIGURE 13.27 Flowchart depicting methodology adopted for the study.

TABLE 13.4
Performance of CC-CV Charging Strategies

	Fast Charging (2C)			Standard Charging (0.2C)		
SoC (Charging)	**Charging Time (Minutes)**	**Temperature Rise (oC) (without Cooling)**	**Temperature Rise (oC) (with Cooling)**	**Charging Time (Minutes)**	**Temperature Rise (oC) (without Cooling)**	**Temperature Rise (oC) (with Cooling)**
0–100	105.21	23.18	17.68	333	2	1
10–90	31.55	21.69	16.98	243	1.74	0.92
20–80	19.5	19.8	15.81	171	1.64	0.71

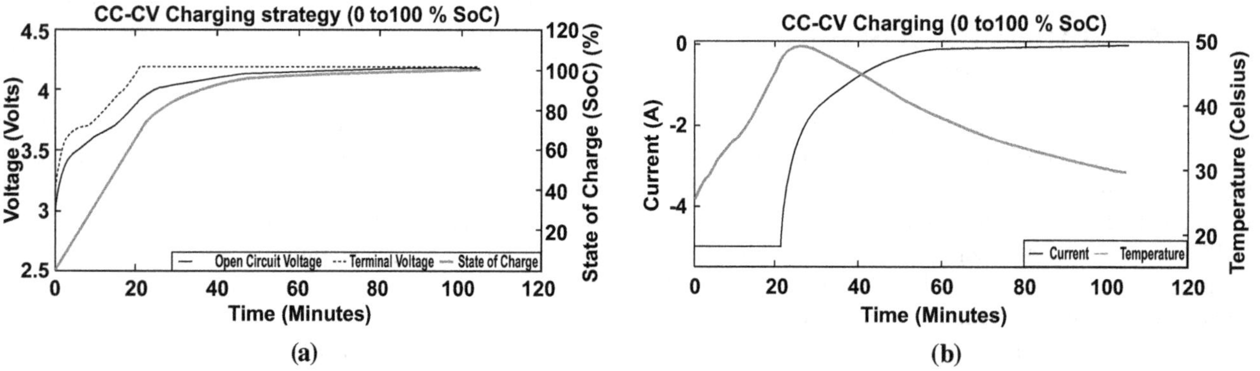

FIGURE 13.28 Waveform for CC-CV charging strategy: (a) terminal voltage, OCV, SoC; (b) temperature, current.

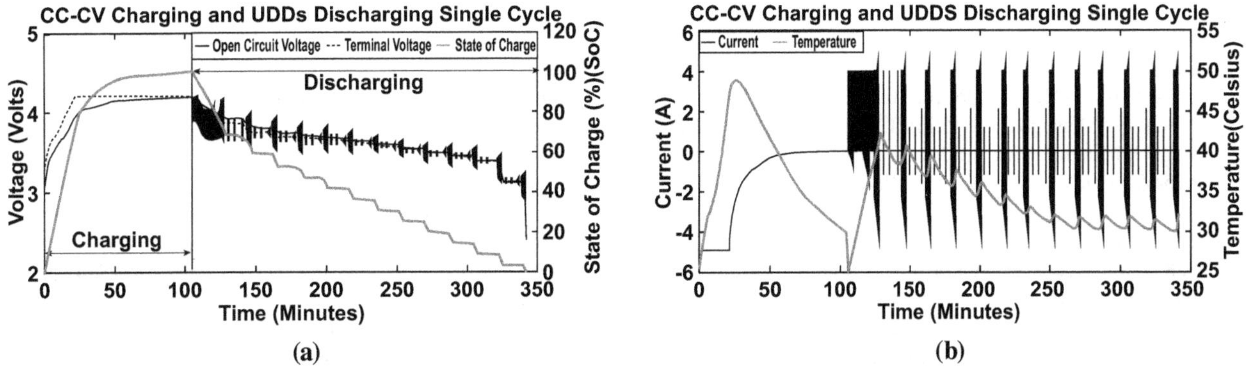

FIGURE 13.29 Charge/discharge cycle for CC-CV charging strategy: (a) terminal voltage, OCV, SoC; (b) temperature, current.

TABLE 13.5
Capacity Analysis of CC-CV Charging

Cycle Number for Fast Charging	Capacity Remaining (350 Cycles)
All	99.941
5	99.9521
10	99.9538

charging and standard charging. The charging time for 20–80% SoC interval is 19.5 minutes with a temperature rise of 19.8°C without cooling and 15.81°C with cooling. The temperature rise is higher for 10–90% SoC and 0–100% SoC due to higher charging current. The rise in temperature should be near to standard charging to reduce the degradation effect. The charge/discharge cycle of CC-CV charging is shown in Figure 13.29(a) and Figure 13.29(b). Table 13.5 provides the details for capacity remaining after 350 cycles. Based on the data, battery degradation is highest when battery is fast charged every cycle.

13.11 CONCLUSION

The battery packs required in power system applications are enormous in size, with high cell numbers. These cells are operated in a safe operating region due to various algorithms adopted in BMS. Apart from safe operation, BMS also communicates with the smart grid. The decision-making component of the smart grid should be aware about the various battery conditions and states for effective utilisation of the battery. It can only be achieved by understanding various algorithms and the battery models utilised for the estimations, as the error in modelling is high. The same is reflected in the estimation algorithms; thus, a proper study of types of batteries their characteristics and modelling is required. Therefore, this chapter provides a brief discussion regarding the various types of batteries. Later, a brief discussion regarding the BMS algorithms shows the importance of battery modelling, as most of the algorithms in BMS depend on them. A brief discussion on the methodology of modelling through experimentation. A SoC-based first-order RC model is developed as an example. The RMSE error for the developed model when validated with the multi-step experiment is 2.4×10^{-4}. Based on the developed model, a simulation on the effect of charge/discharge cycle to understand the battery degradation, the battery when subjected to higher currents degrades the battery. Smart grid operations have to take this into account for the benefit of both the utility and the prosumer.

REFERENCES

[1] IEA, *Global EV Outlook 2019,* 2019. IEA, Paris, https://www.iea.org/reports/global-ev-outlook-2019.

[2] B. Sarlioglu and C. T. Morris, "More electric aircraft: review, challenges, and opportunities for commercial transport aircraft," *IEEE Trans. Transp. Electrif.*, vol. 1, no. 1, pp. 54–64, 2015, doi:10.1109/TTE.2015.2426499.

[3] "Power Sector at a Glance ALL INDIA | Government of India | Ministry of Power." https://powermin.gov.in/en/content/power-sector-glance-all-india (accessed May 26, 2021).

[4] F. Ahmad, M. S. Alam, and M. Asaad, "Developments in xEVs charging infrastructure and energy management system for smart microgrids including xEVs," *Sustainable Cities and Society*. vol. 35, Nov. 2017, pp. 552–564, doi:10.1016/j.scs.2017.09.008.

[5] H. M. Chou, L. Xie, P. Enjeti, and P. R. Kumar, "Power electronics intelligence at the network edge (PINE)," *2017 IEEE Energy Convers. Congr. Expo. ECCE 2017*, vol. 2017-Janua, pp. 5214–5221, 2017, doi:10.1109/ECCE.2017.8096876.

[6] Y. Li *et al.*, "Optimal scheduling of isolated microgrid with an electric vehicle battery swapping station in multi-stakeholder scenarios: A bi-level programming approach via real-time pricing," *Appl. Energy*, vol. 232, pp. 54–68, Dec. 2018, doi:10.1016/j.apenergy.2018.09.211.

[7] D. Li, A. Zouma, J. T. Liao, and H. T. Yang, "An energy management strategy with renewable energy and energy storage system for a large electric vehicle charging station," *eTransportation*, vol. 6, p. 100076, Nov. 2020, doi:10.1016/j.etran.2020.100076.

[8] X. Yan, B. Zhang, X. Xiao, H. Zhao, and L. Yang, "A bidirectional power converter for electric vehicles in V2G systems," in *Proceedings of the 2013 IEEE International Electric Machines and Drives Conference, IEMDC 2013*, 2013, pp. 254–259, doi:10.1109/IEMDC.2013.6556261.

[9] Y. Gao, X. Zhang, Q. Cheng, B. Guo, and J. Yang, "Classification and review of the charging strategies for commercial lithium-ion batteries," *IEEE Access*, vol. 7, pp. 43511–43524, 2019, doi:10.1109/ACCESS.2019.2906117.

[10] Mcdavis A. Fasugba, and Philip T. Krein, "Gaining vehicle-to-grid benefits with unidirectional electric and plug-in hybrid vehicle chargers," in *2011 IEEE Vehicle Power and Propulsion Conference*, 2011, pp. 1–6. IEEE Xplore, https://doi.org/10.1109/VPPC.2011.6043207

[11] P. Keil and A. Jossen, "Charging protocols for lithium-ion batteries and their impact on cycle life-An experimental study with different 18650 high-power cells," *J. Energy Storage*, vol. 6, pp. 125–141, 2016, doi:10.1016/j.est.2016.02.005.

[12] K. Clement-Nyns, E. Haesen, and J. Driesen, "The impact of charging plug-in hybrid electric vehicles on a residential distribution grid," *IEEE Trans. Power Syst.*, vol. 25, no. 1, pp. 371–380, Feb. 2010, doi:10.1109/TPWRS.2009.2036481.

[13] E. Redondo-Iglesias, P. Venet, and S. Pelissier, "Modelling lithium-ion battery ageing in electric vehicle applications—calendar and cycling ageing combination effects," *Batteries*, vol. 6, no. 1, p. 14, Feb. 2020, doi:10.3390/batteries6010014.

[14] M. Jafari, A. Gauchia, S. Zhao, K. Zhang, and L. Gauchia, "Electric vehicle battery cycle aging evaluation in real-world daily driving and vehicle-to-grid services," *IEEE Trans. Transp. Electrif.*, vol. 4, no. 1, pp. 122–134, 2017, doi:10.1109/TTE.2017.2764320.

[15] P. Keil and A. Jossen, "Aging of lithium-ion batteries in electric vehicles," *Dissertation*, vol. 7, no. 1, pp. 41–51, 2017, doi:10.3390/wevj7010041.

[16] K. Uddin, M. Dubarry, and M. B. Glick, "The viability of vehicle-to-grid operations from a battery technology and policy perspective," *Energy Policy*, vol. 113, no. October 2017, pp. 342–347, 2018, doi:10.1016/j.enpol.2017.11.015.

[17] G. L. Plett, *Battery Management Systems, Volume 1: Battery Modeling Battery Modeling,* 2015. Artech House, London.

[18] A. Fotouhi, D. J. Auger, K. Propp, S. Longo, and M. Wild, "A review on electric vehicle battery modelling: From Lithium-ion toward Lithium-Sulphur," *Renew. Sustain. Energy Rev.*, vol. 56, pp. 1008–1021, Apr. 2016, doi:10.1016/j.rser.2015.12.009.

[19] S. V. Rajani, V. J. Pandya, and V. A. Shah, "Experimental validation of the ultracapacitor parameters using the method of averaging for photovoltaic applications," *J. Energy Storage*, vol. 5, pp. 120–126, Feb. 2016, doi:10.1016/j.est.2015.12.002.

[20] H. Budde-Meiwes *et al.*, "A review of current automotive battery technology and future prospects," *Proc. Inst. Mech. Eng. Pt. D J. Automobile Eng.*, vol. 227, no. 5. pp. 761–776, May 2013, doi:10.1177/0954407013485567.

[21] C. C. Chan, E. W. C. Lo, and S. Weixiang, "Available capacity computation model based on artificial neural network for lead-acid batteries in electric vehicles," *J. Power Sources*, vol. 87, no. 1, pp. 201–204, 2000, doi:10.1016/S0378–7753(99)00502–9.

[22] Y. Cao, R. C. Kroeze, and P. T. Krein, "Multi-timescale parametric electrical battery model for use in dynamic electric vehicle simulations," *IEEE Trans. Transp. Electrif.*, vol. 2, no. 4, pp. 432–442, Dec. 2016, doi:10.1109/TTE.2016.2569069.

[23] S. Y. Park *et al.*, "A universal battery charging algorithm for Ni-Cd, Ni-MH, SLA, and Li-Ion for wide range voltage in portable applications," in *PESC Record—IEEE Annual Power Electronics Specialists Conference*, 2008, pp. 4689–4694, doi:10.1109/PESC.2008.4592708.

[24] K. Pourabdollah, "Development of electrolyte inhibitors in nickel cadmium batteries," *Chem. Eng. Sci.*, vol. 160, pp. 304–312, Mar. 2017, doi:10.1016/j.ces.2016.11.038.

[25] M. Anderman, "The challenge to fulfill electrical power requirements of advanced vehicles," *J. Power Sources*, vol. 127, no. 1–2, pp. 2–7, 2004, doi:10.1016/j.jpowsour.2003.09.002.

[26] M. Chen and G. A. Rincón-Mora, "Accurate electrical battery model capable of predicting runtime and I-V performance," *IEEE Trans. Energy Convers.*, vol. 21, no. 2, pp. 504–511, Jun. 2006, doi:10.1109/TEC.2006.874229.

[27] A. Fotouhi, D. J. Auger, K. Propp, and S. Longo, "Electric vehicle battery parameter identification and SOC observability analysis: NiMH and Li-S case studies," *IET Power Electron.*, vol. 10, no. 11, pp. 1289–1297, Sep. 2017, doi:10.1049/iet-pel.2016.0777.

[28] N. Nitta, F. Wu, J. T. Lee, and G. Yushin, "Li-ion battery materials: Present and future," *Mater. Today*, vol. 18, no. 5, pp. 252–264, 2015, doi:10.1016/j.mattod.2014.10.040.

[29] H. Ibrahim, A. Ilinca, and J. Perron, "Energy storage systems-characteristics and comparisons," *Renew. Sustain. Energy Rev.*, vol. 12, no. 5. pp. 1221–1250, Jun. 2008, doi:10.1016/j.rser.2007.01.023.

[30] White Paper, *Introduction to Lithium Polymer Battery Technology*. Jauch Quartz GmbH & Jauch Battery Solutions, 2018, https://www.jauch.com/downloadfile/5c5050fa5b6510e9a8ad76299baae4e53/white_paper_introduction_to_lipo_battery_technology_11-2018_en.pdf

[31] G. Sikha, P. Ramadass, B. S. Haran, R. E. White, and B. N. Popov, "Comparison of the capacity fade of Sony US 18650 cells charged with different protocols," *J. Power Sources*, vol. 122, no. 1, pp. 67–76, Jul. 2003, doi:10.1016/S0378-7753(03)00027-2.

[32] Y. Huang *et al.*, "Lithium manganese spinel cathodes for lithium-ion batteries," *Adv. Energy Mater.*, vol. 11, no. 2., p. 2000997, Jan. 01, 2021, doi:10.1002/aenm.202000997.

[33] J. S. Kim, K. Kim, W. Cho, W. H. Shin, R. Kanno, and J. W. Choi, "A truncated manganese spinel cathode for excellent power and lifetime in lithium-ion batteries," *Nano Lett.*, vol. 12, no. 12, pp. 6358–6365, Dec. 2012, doi:10.1021/nl303619s.

[34] C. M. Julien and M. Massot, "Lattice vibrations of materials for lithium rechargeable batteries I. Lithium manganese oxide spinel," *Mater. Sci. Eng. B Solid-State Mater. Adv. Technol.*, vol. 97, no. 3, pp. 217–230, Feb. 2003, doi:10.1016/S0921–5107(02)00582–2.

[35] L. Liao *et al.*, "Effects of temperature on charge/discharge behaviors of LiFePO 4 cathode for Li-ion batteries," *Electrochim. Acta*, vol. 60, pp. 269–273, 2012, doi:10.1016/j.electacta.2011.11.041.

[36] B. Nykvist and M. Nilsson, "Rapidly falling costs of battery packs for electric vehicles," *Nat. Clim. Chang.*, vol. 5, no. 4, pp. 329–332, 2015, doi:10.1038/nclimate2564.

[37] K. Zaghib *et al.*, "Safe and fast-charging Li-ion battery with long shelf life for power applications," *J. Power Sources*, vol. 196, no. 8, pp. 3949–3954, 2011, doi:10.1016/j.jpowsour.2010.11.093.

[38] Q. Lin, J. Wang, R. Xiong, W. Shen, and H. He, "Towards a smarter battery management system: A critical review on optimal charging methods of lithium ion batteries," *Energy*, vol. 183, pp. 220–234, 2019, doi:10.1016/j.energy.2019.06.128.

[39] "BQ2954 data sheet, product information and support | TI.com." www.ti.com/product/BQ2954 (accessed May 26, 2021).

[40] M. U. Ali, A. Zafar, S. H. Nengroo, S. Hussain, M. J. Alvi, and H. J. Kim, "Towards a smarter battery management system for electric vehicle applications: a critical review of lithium-ion battery state of charge estimation," *Energies*, vol. 12, no. 3. MDPI AG, p. 446, Jan. 30, 2019, doi:10.3390/en12030446.

[41] W. D. Widanage, et al. "Design and use of multisine signals for Li-ion battery equivalent circuit modelling. Part 2: model estimation," *Journal of Power Sources*, vol. 324, Aug. 2016, pp. 61–69. DOI.org (Crossref), https://doi.org/10.1016/j.jpowsour.2016.05.014

[42] Lokesh Gurjer, et al. "Detailed modelling procedure for lithium-ion battery using thevenin equivalent," in *2019 IEEE International Conference on Electrical, Computer and Communication Technologies (ICECCT)*, 2019, pp. 1–6. IEEE, DOI.org (Crossref), https://doi.org/10.1109/ICECCT.2019.8869224.

[43] M. J. Rothenberger, D. J. Docimo, M. Ghanaatpishe, and H. K. Fathy, "Genetic optimization and experimental validation of a test cycle that maximizes parameter identifiability for a Li-ion equivalent-circuit battery model," *J. Energy Storage*, vol. 4, pp. 156–166, Dec. 2015, doi:10.1016/j.est.2015.10.004.

[44] G. Yang, "Battery parameterisation based on differential evolution via a boundary evolution strategy," *J. Power Sources*, vol. 245, pp. 583–593, Jan. 2014, doi:10.1016/j.jpowsour.2013.06.139.

[45] X. Hu, S. Li, and H. Peng, "A comparative study of equivalent circuit models for Li-ion batteries," *J. Power Sources*, vol. 198, pp. 359–367, Jan. 2012, doi:10.1016/j.jpowsour.2011.10.013.

[46] Y. Xu *et al.*, "State of charge estimation for lithium-ion batteries based on adaptive dual Kalman filter," *Appl. Math. Model.*, 2020, doi:10.1016/j.apm.2019.09.011.

[47] H. Ben Sassi, F. ERRAHIMI, and N. ES-Sbai, "State of charge estimation by multi-innovation unscented Kalman filter for vehicular applications," *J. Energy Storage*, vol. 32, p. 101978, Dec. 2020, doi:10.1016/j.est.2020.101978.

[48] X. Hu, S. E. Li, and Y. Yang, "Advanced machine learning approach for lithium-ion battery state estimation in electric vehicles," *IEEE Trans. Transp. Electrif.*, vol. 2, no. 2, pp. 140–149, 2016, doi:10.1109/TTE.2015.2512237.

[49] S. Zhang, X. Guo, and X. Zhang, "An improved adaptive unscented kalman filtering for state of charge online estimation of lithium-ion battery," *J. Energy Storage*, vol. 32, p. 101980, Dec. 2020, doi:10.1016/j.est.2020.101980.

[50] M. Verbrugge, "Adaptive, multi-parameter battery state estimator with optimized time-weighting factors," *J. Appl. Electrochem.*, vol. 37, no. 5, pp. 605–616, May 2007, doi:10.1007/s10800-007-9291-7.

[51] I. Baccouche, S. Jemmali, A. Mlayah, B. Manai, and N. E. Ben Amara, "Implementation of an improved Coulomb-counting algorithm based on a piecewise SOC-OCV relationship for SOC estimation of Li-ion battery," *Int. J. Renew. Energy Res.*, vol. 8, no. 1, pp. 178–187, 2018.

[52] R. Xiong, J. Cao, Q. Yu, H. He, and F. Sun, "Critical review on the battery state of charge estimation methods for electric vehicles," *IEEE Access*, vol. 6, pp. 1832–1843, 2017, doi:10.1109/ACCESS.2017.2780258.

[53] Y. Tian, B. Xia, W. Sun, Z. Xu, and W. Zheng, "A modified model based state of charge estimation of power lithium-ion batteries using unscented Kalman filter," *J. Power Sources*, vol. 270, pp. 619–626, 2014, doi:10.1016/j.jpowsour.2014.07.143.

[54] G. L. Plett, "Extended Kalman filtering for battery management systems of LiPB-based HEV battery packs—Part 3. State and parameter estimation," *J. Power Sources*, vol. 134, no. 2, pp. 277–292, Aug. 2004, doi:10.1016/j.jpowsour.2004.02.033.

[55] M. A. Hannan, M. S. H. Lipu, A. Hussain, and A. Mohamed, "A review of lithium-ion battery state of charge estimation and management system in electric vehicle applications: challenges and recommendations," *Renew. Sustain. Energy Rev.*, vol. 78, pp. 834–854, Oct. 2017, doi:10.1016/j.rser.2017.05.001.

[56] M. Charkhgard and M. Farrokhi, "State-of-charge estimation for lithium-ion batteries using neural networks and EKF," *IEEE Trans. Ind. Electron.*, vol. 57, no. 12, pp. 4178–4187, Dec. 2010, doi:10.1109/TIE.2010.2043035.

[57] Y. Wang *et al.*, "A comprehensive review of battery modeling and state estimation approaches for advanced battery management systems," *Renewable and Sustainable Energy Reviews*, vol. 131. Elsevier Ltd, p. 110015, Oct. 01, 2020, doi:10.1016/j.rser.2020.110015.

[58] Jian Cao, et al. "Battery balancing methods: a comprehensive review," in *2008 IEEE Vehicle Power and Propulsion Conference*, 2008, pp. 1–6. IEEE Xplore, https://doi.org/10.1109/VPPC.2008.4677669.

[59] A. K. M. A. Habib, M. K. Hasan, M. Mahmud, S. M. A. Motakabber, M. I. Ibrahimya, and S. Islam, "A review: energy storage system and balancing circuits for electric vehicle application," *IET Power Electron.*, vol. 14, no. 1, pp. 1–13, Jan. 2021, doi:10.1049/pel2.12013.

[60] K. M. Lee, Y. C. Chung, C. H. Sung, and B. Kang, "Active cell balancing of Li-Ion batteries using LC series resonant circuit," *IEEE Trans. Ind. Electron.*, vol. 62, no. 9, pp. 5491–5501, Sep. 2015, doi:10.1109/TIE.2015.2408573.

[61] X. Wang *et al.*, "A review of modeling, acquisition, and application of lithium-ion battery impedance for onboard battery management," *eTransportation*, vol. 2, no. 7, pp. 1–21, Feb. 2021, doi:10.1016/j.etran.2020.100093.

[62] E. Samadani *et al.*, "Empirical modeling of lithium-ion batteries based on electrochemical impedance spectroscopy tests," *Electrochim. Acta*, vol. 160, pp. 169–177, Apr. 2015, doi:10.1016/j.electacta.2015.02.021.

[63] S. K. Chung, A. A. Andriiko, A. P. Mon'Ko, and S. H. Lee, "On charge conditions for Li-ion and other secondary lithium batteries with solid intercalation electrodes," *J. Power Sources*, vol. 79, no. 2, pp. 205–211, Jun. 1999, doi:10.1016/S0378-7753(99)00058-0.

[64] Jorge V. Barreras, et al. "An improved parametrization method for Li-ion linear static equivalent circuit battery models based on direct current resistance measurement," in *2015 International Conference on Sustainable Mobility Applications, Renewables and Technology (SMART)*, 2015, pp. 1–9. IEEE, DOI.org (Crossref), https://doi.org/10.1109/SMART.2015.7399223

[65] X. Lai *et al.*, "A comparative study of global optimization methods for parameter identification of different equivalent circuit models for Li-ion batteries," *Electrochim. Acta*, vol. 295, pp. 1057–1066, 2019, doi:10.1016/j.electacta.2018.11.134.

[66] Y. Huangfu, J. Xu, D. Zhao, Y. Liu, and F. Gao, "A novel battery state of charge estimation method based on a super-twisting sliding mode observer," *Energies*, vol. 11, no. 5, 2018, doi:10.3390/en11051211.

[67] C. Zhang, K. Li, S. McLoone, and Z. Yang, "Battery modelling methods for electric vehicles—a review," *2014 Eur. Control Conf. ECC 2014*, pp. 2673–2678, 2014, doi:10.1109/ECC.2014.6862541.

[68] S. L. Wang *et al.*, "Open circuit voltage and state of charge relationship functional optimization for the working state monitoring of the aerial lithium-ion battery pack," *J. Clean. Prod.*, vol. 198, pp. 1090–1104, Oct. 2018, doi:10.1016/j.jclepro.2018.07.030.
[69] T. Feng, L. Yang, X. Zhao, H. Zhang, and J. Qiang, "Online identification of lithium-ion battery parameters based on an improved equivalent-circuit model and its implementation on battery state-of-power prediction," *J. Power Sources*, vol. 281, pp. 192–203, May 2015, doi:10.1016/j.jpowsour.2015.01.154.
[70] P. L. Moss, G. Au, E. J. Plichta, and J. P. Zheng, "An electrical circuit for modeling the dynamic response of li-ion polymer batteries," *J. Electrochem. Soc.*, 2008, doi:10.1149/1.2999375.
[71] A. Abdollahi *et al.*, "Optimal battery charging, Part I: minimizing time-to-charge, energy loss, and temperature rise for OCV-resistance battery model," *J. Power Sources*, vol. 303, pp. 388–398, 2016, doi:10.1016/j.jpowsour.2015.02.075.
[72] A. A. Pesaran, "Battery thermal models for hybrid vehicle simulations," *J Power Sources*, vol. 110, no. 2, pp. 377–382, Aug. 2002, doi:10.1016/S0378-7753(02)00200-8.
[73] X. Lin *et al.*, "A lumped-parameter electro-thermal model for cylindrical batteries," *J. Power Sources*, vol. 257, pp. 12–20, Jul. 2014, doi:10.1016/j.jpowsour.2014.01.097.
[74] A. A. Pesaran and S. D. Burch, "Thermal performance of EV and HEV battery modules and packs prepared under FWP HV71," 1997. https://www.nrel.gov/docs/legosti/old/23527.pdf
[75] A. Meintz *et al.*, "Enabling fast charging—vehicle considerations," *J. Power Sources*, 2017, doi:10.1016/j.jpowsour.2017.07.093.
[76] G. Suri and S. Onori, "A control-oriented cycle-life model for hybrid electric vehicle lithium-ion batteries," *Energy*, vol. 96, pp. 644–653, Feb. 2016, doi:10.1016/j.energy.2015.11.075.
[77] Samsung SDI Co., "Lithium-ion rechargeable cell for power tools—datasheet," *INR18650-25R datasheet*, no. 1, pp. 0–16, 2014.

Unsolved Questions

1. A 4kAh battery pack has to be designed that has to be able to provide a 100A discharge current with a terminal voltage of 400V. A battery cell has the following data

   Maximum terminal voltage: 4.2V
   Minimum terminal voltage: 2.75V
   Nominal Voltage: 3.7V
   Capacity of cell: 2.6Ah

   Determine the number of cells to be connected in series, parallel and total number of cells for

   - Case 1: Discharge rate 4C
   - Case 2: Discharge rate 1.5C
   - Case 3: Discharge rate 1C

   Solutions

   Case-1: 14 parallel, 146 series and total 2,044
   Case-1: 34 parallel, 146 series and total 4,964
   Case-1: 50 parallel, 146 series and total 7,300

2. A Li ion battery of 2,500mAh has an internal resistance of 90 milliohms, resistance of RC network is 20 milliohms with capacitance of 17,000F. Consider a first-order RC model if the terminal voltage at kth instant is 3.5V and 1.2A. Calculate the open circuit voltage at kth instant.

   Solution

   3.36 V

3. The open circuit voltage in a battery follows the pattern of $S^2 + 0.5S + 2.75$ where S is SoC in ratio. Consider the parametric data for the battery as in Question 2. A constant current of 5 A is applied to the battery for five minutes at initial SoC of 10%. Calculate the initial terminal voltage and terminal voltage at the end of five minutes. Assume that columbic efficiency is 1.

   Solution

   3.36V at start, and 3.4976 V at end

4. The initial temperature of a battery is at ambient temperature of 25 °C at kth instant. Based on the parameters provided in Question 2 and Eq. (13.52)–(13.55), calculate the temperature at (k + 1)th and (k + 2)th instant.

   Solution

   25.0042 °C and 25.0084 °C

5. In a single-pulse testing, the battery initially is at a voltage of 3.59 V. A pulse of discharging current of 1.026 A had been applied. When the terminal voltage reaches 3.549V, the pulse is removed. The terminal voltage suddenly rises to 3.613 V, and after a rest of 40 minutes, the terminal voltage reaches 3.628 V. Find the internal parameters R_o, R_1 and C_1. The resistance of wire connecting the battery is 0.4Ω.

   Solution

   Internal resistance = 62.37 mΩ, R_1 = 14.62mΩ, C1 = 41.036 kF

6. Derive an auto regressive exogenous model for first-order RC Model of Li-ion battery (Refer. [49]).

   Answer:

$$E_t[k] = P_1 E_t[k-1] + P_2 i[k] + P_3 i[k-1]$$

Where

$$P_1 = a = e^{\left(\frac{-\Delta t}{R_1 C_1}\right)}$$
$$P_2 = -R_o$$
$$P_3 = aR_o - [1-a]R_1$$

14 A Complete Battery Management System of Li-Ion Batteries in Electric Vehicle Applications

J. Saikrishna Goud and Kalpana R.

CONTENTS

14.1 INTRODUCTION

Renewable energy sources have become extremely popular in recent years due to increasing environmental concerns and the shortage of fossil fuels. Every field and application utilizing energy is switching to an alternate source of energy instead of fossil fuel. In the current world of the automotive sector, the electric vehicle (EV) has found immense importance due to the same. Electric cars are making big waves in the automobile world. These noise-free, pollution-free and high-performance vehicles are expected to make their internal combustion engine counterparts obsolete by 2025. Moreover, the Indian market for EVs is growing exponentially and it is expected to be around \$50 billion (Rs. 3.7 lakh crore) by the year 2030 [1].

The powerhouse of EVs is the battery pack which consists of battery cells connected in a combination of series and parallel to produce the power required to run EVs. The commonly used battery for EV application is the lithium-ion (Li-ion) battery, owing to its better performance as compared to the other available battery compositions. The term Li-ion battery refers to a family of different chemical properties such as lithium cobalt oxide ($LiCoO_2$), lithium iron phosphate ($LiFePO_4$ or LFP), lithium nickel cobalt aluminum oxide ($LiNiCoAlO_2$ or NCA), lithium manganese oxide ($LiMn_2O_4$ or LMO) and lithium nickel manganese cobalt oxide ($LiNiMnCoO_2$ or NMC). A comparison chart of these Li-ion chemical properties in terms of cycle life, cost per cycle, safety, thermal performance, specific energy and specific power is presented in Figure 14.1.

The major advantages of Li-ion batteries as compared to their counterparts are high specific energy and specific power, higher voltage per cell, low self-discharge rate and long life. Despite having many advantages, Li-ion batteries are intolerant to overcharge and deep discharge. The charging of battery cells at elevated temperatures with high charging rates affects it adversely. A similar effect can be observed for the discharging process at elevated temperatures with high discharging rates. Eventually, all these factors contribute toward premature failure, accelerated aging of the battery and reduced life as compared to any other component of an EV system. Therefore, this results in frequent replacement of the battery; hence, the recurring cost of the EV increases.

To overcome the premature failure and accelerated aging of the battery, a battery management system (BMS) is required. BMS is essentially an embedded system for the safe operation of the battery stack. The primary function of BMS is to monitor the parameters such as voltage, current and temperature to protect the battery from damage and to protect the safety of the EV operator. Second, BMS should be effective to estimate the state of charge (SoC) and state of health (SoH). Moreover, BMS should indicate the safe operating points and battery states (i.e., SoC and SoH) on the EV dashboard. Furthermore, the important functionality of BMS is to equalize the voltage/SoC of serially connected cells in a stack to enhance the battery life. To design a BMS (or) to develop SoC and SoH estimation techniques, it is important to know the battery run time (Ah vs. time) characteristics and battery aging mechanism. Therefore, modeling of a Li-ion cell will be discussed in the next section.

14.2 MODELING OF LI-ION CELL

In literature, two types of modeling techniques are available: electrochemical and electrical. In this study, electrical circuit modeling is discussed. Figure 14.2 shows the

DOI: 10.1201/9781003229124-14

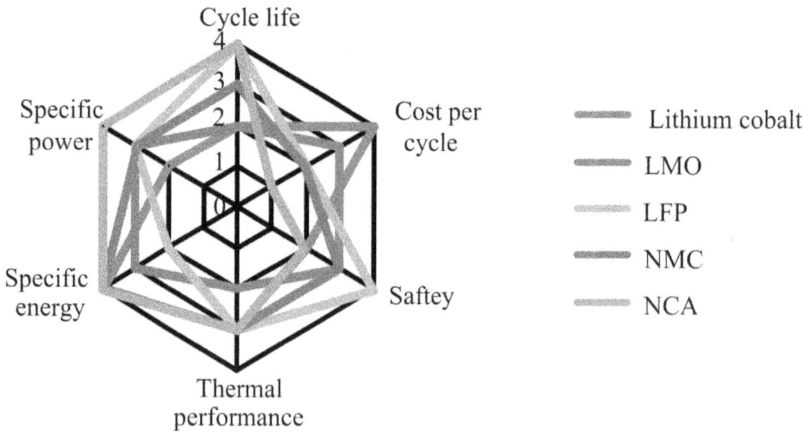

FIGURE 14.1 Comparison of different lithium-ion chemical properties.

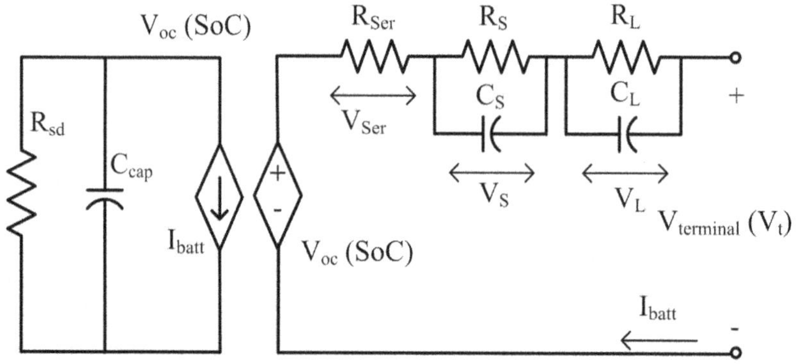

FIGURE 14.2 Equivalent circuit of a lithium-ion cell.

equivalent circuit of a Li-ion cell [2–3] where V_{oc} is the open circuit voltage and it depends on battery SoC.

SoC is the percentage of charge available in the battery with respect to rated capacity. The two parallel branches of R and C represent the transient behavior; R_{Ser} is the DC resistance and is responsible for the voltage drop. Calendar life losses of the battery are modeled using a resistor R_{sd}. Calendar loss is the degradation of battery life, when it is in idle condition. The capacitor C_{cap} determines the capacity of the cell.

In this study, modeling of LFP battery is discussed, the parameters V_{oc}, R_{Ser}, R_S, R_L, C_S and C_L of the LFP battery are expressed as:

$$V_{OC} = -0.5863 \times e^{-21.9 \times SoC} + 3.414 + 0.1102 \times SoC - \left(0.1718 \times e^{\frac{-0.008}{1-SoC}}\right) \quad (14.1)$$

$$R_{Ser} = 0.1298 \times SoC^4 - 0.2892 \times SoC^3 + 0.2273 \times SoC^2 - 0.07216 \times SoC + 0.0898 \quad (14.2)$$

$$R_S = -0.0108 \times e^{-11.03 \times SoC} + 0.01827 - 6.462 \times 10^{-3} \times SoC \quad (14.3)$$

$$C_S = 0.01697 \times SoC^3 - 1.007 \times 10^{-3} \times SoC^2 + 1.408 \times 10^{-3} \times SoC - 3.897 \times 10^{-2} \quad (14.4)$$

$$R_L = -0.295 \times e^{-20 \times SoC} + 0.04722 - 0.0242 \times SoC \quad (14.5)$$

$$C_L = 2.13 \times 10^6 \times SoC^6 - 6.007 \times 10^6 \times SoC^5 + 6.271 \times 10^6 \times SoC^4 - 2.958 \times 10^6 \times SoC^3 + 5.998 \times 10^5 \times SoC^2 - 3.102 \times 10^4 \times SoC + 2.232 \times 10^3 \quad (14.6)$$

By using the aforementioned equations, a 4Ah single cell LFP battery is modeled as an example. The V_{oc} and V_t characteristics during discharge and run time characteristics during charging are shown in Figure 14.3. Figure 14.3(a) describes the V_{oc} and V_t characteristics at different discharge C-rates. In battery terminology, C-rate describes the rate at which battery is charging or discharging. For example, 1C rate of a 4Ah battery means that a fully charged 4Ah battery can be completely discharged with 4Amps in one hour.

Figure 14.3(b) shows the terminal voltage profile of a LFP battery at 1C charge rate. During the battery course of usage, there exists two type of capacity losses (capacity fade)

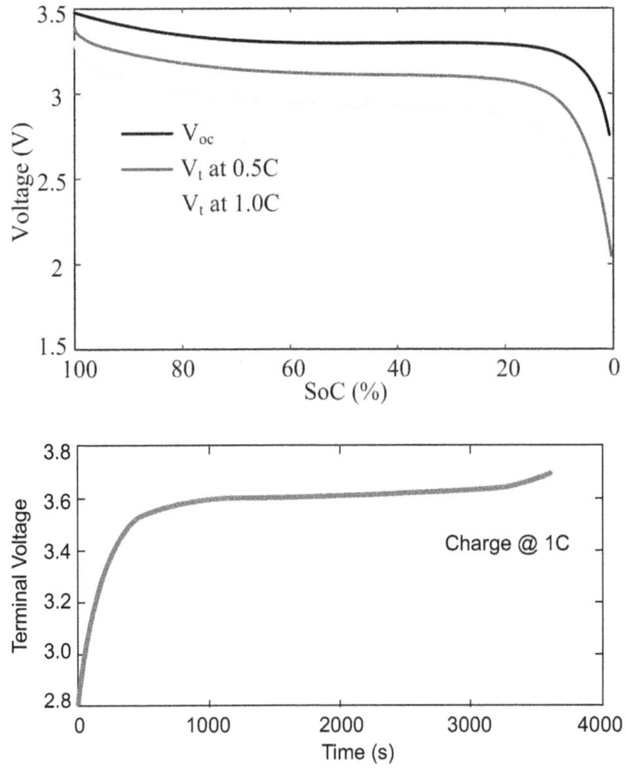

FIGURE 14.3 (a) V_{oc} and V_t characteristics discharge rates of 0.5C and 1C; (b) runtime characteristics during charging mode.

such as calendar loss and cycle loss. Typically, the calendar losses are minimal and can be ignored if the battery is operated regularly. From the aging characteristics of LFP cell reported in [4–6], it is evident that large depth of discharge (DoD), charging and discharging the battery with high currents and elevated ambient temperatures of the battery accelerates the aging (i.e., capacity degradation). Hence, the capacity degradation due to operating the battery at extreme temperatures and large DoDs are mathematically modeled and incorporated to the LFP cell equivalent circuit.

The capacity degradation (i.e., change in C_{cap} in Figure 14.2) due to the effect of DoD is modeled as follows [5, 6]:

$$\Delta C_{cap} = 30330 \times exp\left(\frac{-31500}{8.314 \times T}\right) \times \left(N \times DoD \times 2\right)^{0.552} \quad (14.7)$$

where, ΔC_{cap} is the degraded capacity, T is temperature in Kelvin and N represents the number of cycles that has been already

The effect of temperature on battery aging is modeled using following equations (14.5–14.6):

$$\Delta C_{cap} = 0.1825 \times exp\left(\frac{-1324.65}{T}\right) \times N^{0.5878} \quad (14.8)$$

The equations (14.7)–(14.8) determine the capacity loss due to the effect of aging parameters DoD and temperature. Therefore, the effective usable capacity is expressed as:

$$C_{cap} = C_{cap-rated} - \left(\Delta C_{cap\_DoD} + \Delta C_{cap\_Temp}\right) \quad (14.9)$$

where, $C_{cap-rated}$ is the rated capacity of the battery specified by the manufacturer

The equations (14.8)–(14.9) are developed using curve-fitting technique with the battery accelerated aging test data. The effect of DoD and operating temperature on battery run time over a period of time is shown in Figure 14.4 and Figure 14.5. Figure 14.4(a) shows the battery run time characteristics at different cycle numbers and the battery temperature is maintained at 35°C. From Figure 14.4(a), it is evident that as the cycle number increases, battery run time decreases. Figure 14.4(a) shows the capacity degradation due to the effect of temperature.

From Figure 14.4(b), it is evident that battery ages at a faster rate when it is operating at elevated temperatures. Similarly, Figure 14.5(a) shows the battery run time characteristics at different cycle numbers and the DoD is maintained at 100%. Figure 14.5(b) shows the effect of DoD on battery capacity. From Figure 14.5(b), it is observed that higher DoD degrades the battery faster. Moreover, it can also be observed that the effect of temperature on battery

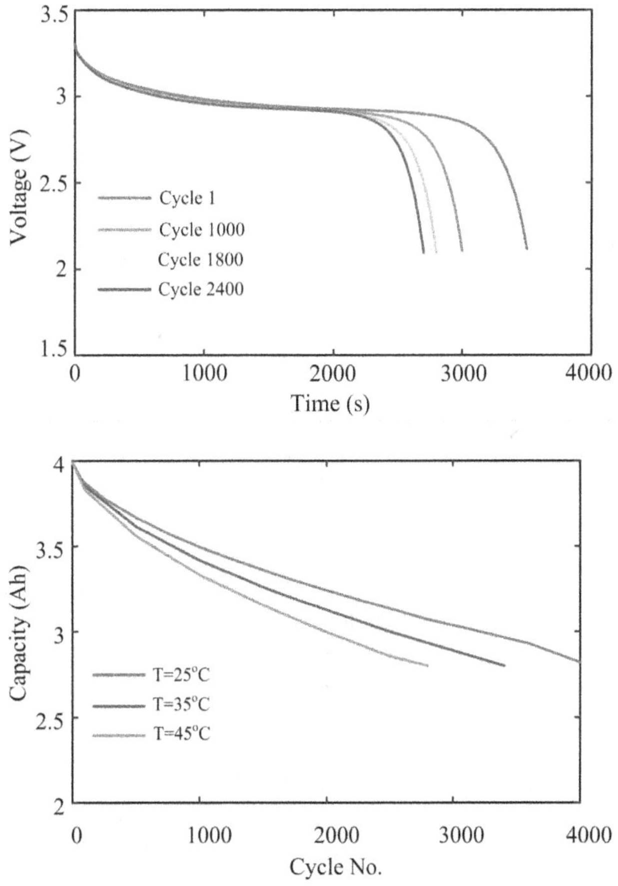

FIGURE 14.4 (a) Battery runtime characteristics at different cycle numbers and at operating at 35°C; (b) capacity degradation of LFP battery at different cycle numbers and operating at different temperature conditions.

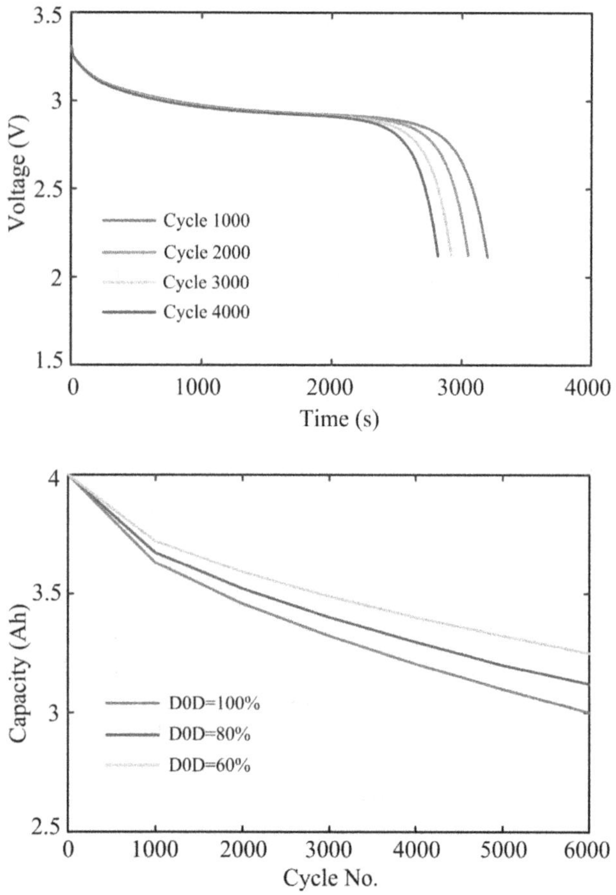

FIGURE14.5 (a) Battery runtime characteristics at different cycle numbers and operating at 100% DoD; (b) Capacity degradation of LFP battery at different cycle numbers and operating at different DoD conditions.

lifecycle is higher than the effect of DoD. Effect of other battery parameters viz. charging voltage, charging and discharging currents, can also be modeled from the accelerated aging test data and can be incorporated to the equivalent circuit.

14.3 STATE OF CHARGE ESTIMATION

State of charge (SoC) is an index which describes the amount of charge available with respect to the rated capacity of the battery. Accurate SoC estimation protects the battery from overcharge and deep discharge, and also acts as a fuel indicator similar to conventional vehicles. Different kinds of SoC estimation techniques are presented in literature; these techniques can be classified into four categories: conventional methods [7–9], adaptive model–based techniques [10–13], learning algorithm–based techniques [14–16] and non-linear observer–based techniques [17–18].

Open circuit voltage technique [7], coulomb-counting technique [8] and impedance spectroscopy technique [9] fall under the category of conventional techniques; these techniques measure the battery's physical parameters such as voltage, current and impedance/resistance to estimate the SoC. Kalman filter [10], extended Kalman filter [11], unscented Kalman filter [12] and sigma-point Kalman filter [13] fall under the category of adaptive model based techniques. These techniques require the equivalent circuit model of the cell. Therefore, the modeling accuracy of equivalent circuit determines the accuracy of SoC estimation.

Neural-network [14], fuzzy-logic [15] and support-vector machine [16] techniques fall under the category of learning algorithm–based techniques. These techniques require large sets of test data to estimate the SoC. Sliding-mode observer [17] and proportional integral observer [18] techniques fall under non-linear observer category. These techniques are complex to implement among all. Therefore, conventional techniques will be discussed in this chapter because they are simple to implement in real-time applications.

14.3.1 OPEN-CIRCUIT VOLTAGE TECHNIQUE

Open circuit voltage (OCV) of the battery is measured to estimate the SoC, since there exists a relationship between OCV and SoC of all rechargeable batteries. For example, lead acid batteries have an approximately linear relation between OCV of the battery and SoC. Therefore, by measuring OCV of the battery, SoC can be estimated. However, in case of Li-ion batteries, the SoC vs. OCV relation is nonlinear (Figure 14.6), hence making it difficult to model the SoC vs. OCV relation accurately. For measuring the open circuit voltage, battery requires adequate relaxing time to reach the equilibrium state from operating state. Moreover, the duration of relaxation time varies depending on environmental conditions. Therefore, this technique is not possible to implement when a vehicle is under running conditions, and this method is mostly used in laboratories.

14.3.2 COULOMB-COUNTING TECHNIQUE

Coulomb-counting technique is the most popular technique. It uses battery current integration to estimate the

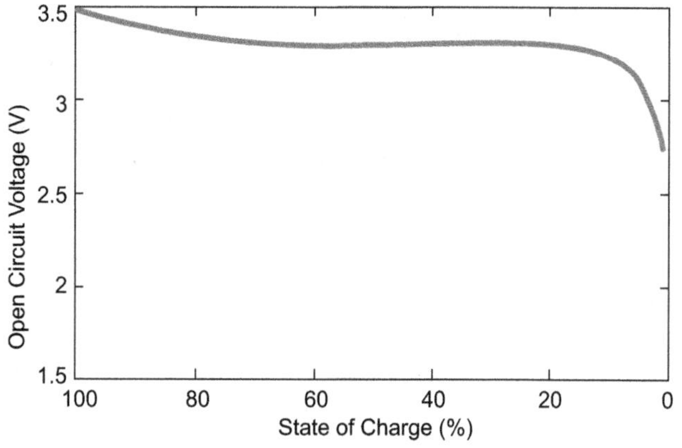

FIGURE 14.6 SoC vs. OCV characteristics of LFP battery.

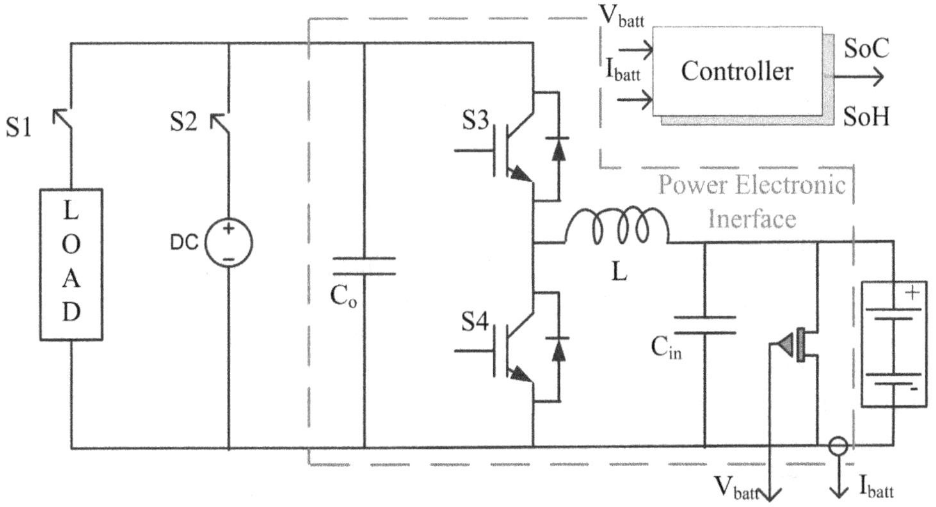

FIGURE 14.7 Schematic of the electronic circuit to implement SoC and SoH techniques [19].

SoC. The following equation describes the SoC estimation using coulomb counting:

$$SoC = SoC_o \pm \left(\int \frac{I_{batt}}{Rated\,Capacity} dt \right) \times 100 \quad (14.10)$$

where, SoC_0 is the SoC measured when battery is connected for the first time, and it is determined using V_{oc} vs. Ah characteristics of the battery; Rated Capacity will be specified in the manufacturer data sheet; I_{batt} is the battery current

Figure 14.7 depicts the typical schematic of the electronic circuitry required to implement the coulomb-counting technique, wherein the power electronic interface is used to charge and discharge the battery, and S1 and S2 are the switches changes the battery operating mode, i.e., charging or discharging mode. V_{batt} and I_{batt} are the sensors which give the readings of battery terminal voltage and charge/discharge currents. A flowchart of the coulomb-counting technique to estimate the SoC is described in Figure 14.8. The controller (Figure 14.7) measures the V_t and I_{batt} to estimate the SoC. SoC_0 in Eq. (14.10) is an important parameter which determines the estimation accuracy. Controller measures the SoC_0 using the Ah-V_{oc} characteristics (Figure 14.6) of a fresh battery as given in a manufacturer datasheet. In EV applications, measuring V_{oc} is impractical since the battery cannot be isolated during its operation. Therefore, V_t is used instead of V_{oc} to estimate SoC_0. Therefore, there will be an estimation error since V_t is used in place of V_{oc}. Figure 14.9 shows the MATLAB/Simulink results of SoC of a 4Ah LFP battery SoC characteristics for 4A discharge current.

14.4 STATE OF HEALTH ESTIMATION

The capacity fading of the battery depends on various stress factors namely overcharge, DoD, temperature, charge and discharge rates. However, understanding the aging mechanism of a battery is complex, as battery aging does not depend on a single stress factor but on a number of factors whose interactions causes the battery aging. Effect of these factors on battery age is modeled and discussed in Section 14.2.

As the battery ages, a solid electrolyte interface (SEI) layer forms between the electrodes of the battery, increasing the battery's impedance while reducing its usable capacity. Moreover, with the battery aging, loss of lithium takes place in the electrolyte [20]. Knowing the effect of these stress factors and knowing the health of the battery in advance helps to enhance the cycle life of the battery, and thus it prevents the premature failure of the battery. The term SoH expresses the health status of the battery by comparing it with the characteristics of a new battery of the same rating. However, it is not simple to measure the SoH as it is not a direct measurement.

With battery aging, impedance and resistance increase; also, the capacity of the battery is reduced. Therefore, by monitoring aforementioned parameters, SoH can be estimated. Electrolytic impedance spectroscopy (EIS) is the benchmark SoH estimation technique which gauges change in impedance to estimate SoH. The following equation is used to estimate SoH through impedance measurement [19]:

$$SoH = \frac{Z_{EoL} - Z_i}{Z_{EoL} - Z_{New}} \times 100[\%] \quad (14.11)$$

SoH can be estimated by measuring the DC resistance of the battery [21]. Variation in DC resistance of the battery can be determined by observing the terminal voltage by discharging or charging with a pulsed current as described in Eq. (14.12). Further, with the measured DC resistance, SoH can be estimated similar to Eq. (14.11).

$$R_i = \frac{\Delta V}{\Delta I} \quad (14.12)$$

where in Eq. (14.11)–(14.12), R_i and Z_i are the resistance and impedance measured at ith cycle; Z_{EoL} and Z_{new} are

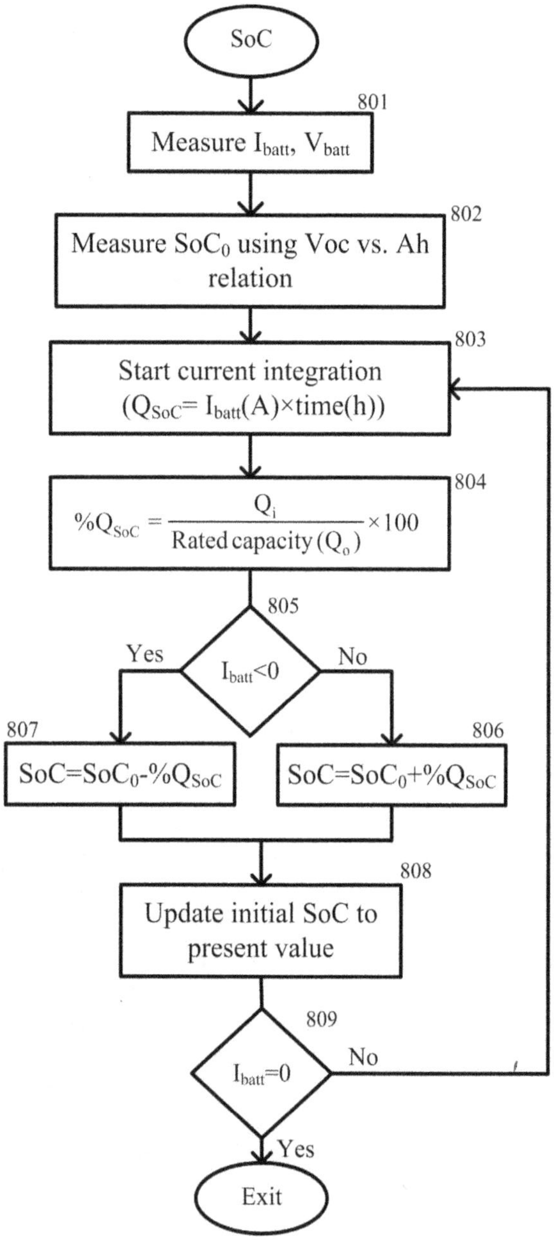

FIGURE 14.8 Flowchart of the coulomb-counting technique to estimate SoC [19].

impedances at end of life and when battery was new, respectively; ΔI is the magnitude of the current pulse, and ΔV is the variation in voltage

Figure 14.10 shows the terminal voltage variation at different cycle numbers (i.e., at different ages) for the applied current pulse. Therefore, by observing these variations in terminal voltages, SoH can be estimated. However, estimation accuracy of this technique is less compared to an EIS test. Data fitting is another type of SoH estimation technique, which uses the large set of resistance/impedance growth or capacity degradation data and a characteristics map is formed from the data. Further, by comparing the real-time battery parameters in every charge/discharge cycle with the characteristic map, SoH can be obtained [22, 23].

Similar to the SoC estimation, Kalman filter (KF) is also used to estimate SoH. KF technique is very popular and an old approach for estimating the state of a dynamic system. Typically, KF technique is a set of mathematical equations that consist of predictor and corrector type of estimators. KF technique works in two stages: in stage one, the predictor estimates the current output variable; in stage two, the estimated variable will get updated to achieve high estimation accuracy. The major advantage of KF technique is that it provides dynamic error boundary for estimated values.

KF technique works well for the linear systems; however, battery characteristics are non-linear, and hence to improve the estimation accuracy, extended Kalman filter technique [24] and unscented Kalman filter [25] are introduced. All the Kalman filtering techniques involve huge matrix operations, and require accurate models of battery aging mechanisms. Therefore, the computational and implementation complexity of these techniques are high. In [26], a unique SoH estimation of LFP cells is proposed. This method uses differential voltage (DV) curves for SoH estimation because DV curves have the ability to represent the degradation mechanism.

As the battery ages, two kinds of capacity losses take place. Calendar loss is one of the battery capacity losses which arise due to storing the battery for a long time. Cycle loss is the another capacity loss, which is caused by numerous factors, viz. elevated temperatures, charging more than the maximum cut-off voltage, discharging below the lower cut-off voltage, or charging and discharging at high currents. SoH of a battery can be estimated by comparing the change in capacity with the capacity of a new battery. By measuring the capacity change, SoH can be estimated as follows:

$$SoH = \frac{Q_i}{Q_o} \times 100[\%] \tag{14.13}$$

where Q_i is the capacity measured in every cycle, and Q_o is the capacity stated on the battery name plate

To measure the change in capacity (Q_i) of the battery over time to estimate SoH, coulomb-counting technique is presented in [27]. Coulomb counting requires discharge of the battery with a constant current until the battery reaches its cut-off voltage (i.e., 100% DoD). However, in real time applications, discharge current varies—and also, the DoD cannot not be 100% always. In [19], a modified coulomb-counting technique is presented to estimate SoH for variable DoD and variable discharge rate conditions. Therefore, in the following section, modified coulomb-counting technique will be discussed, since this method is simple to implement and works effectively for EV applications.

14.4.1 Modified Coulomb-Counting Technique to Estimate SoH

Figure 14.11 depicts the flowchart of the SoH estimation using modified coulomb counting technique. In this

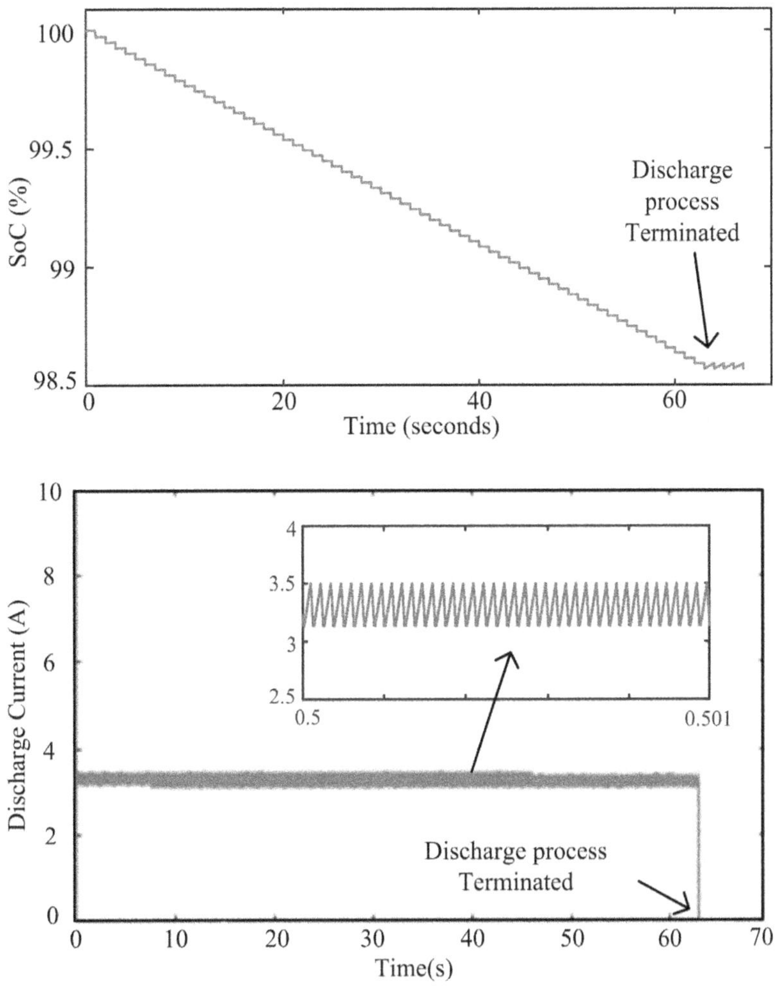

FIGURE 14.9 SoC characteristics for 63 seconds duration of time.

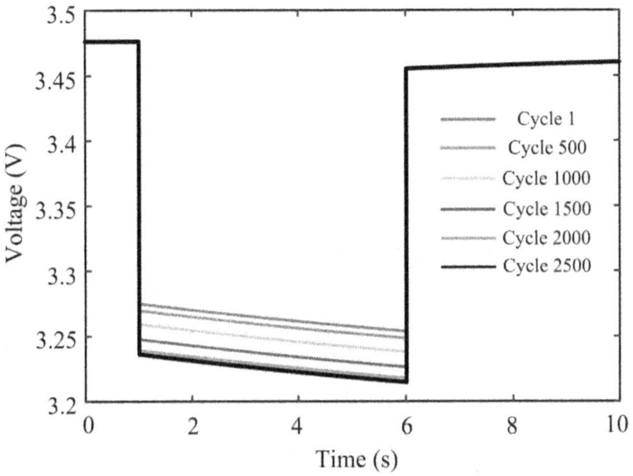

FIGURE 14.10 Terminal voltage variation of LFP battery at different cycle numbers for five seconds duration current pulse.

technique, capacity of the battery is measured in every discharge cycle and compared with capacity of the new battery as described in Eq. (14.13). Capacity of the battery is measured by integrating the discharge current with the time as described in Eq. (14.14):

$$Q_i = \int_{t=0}^{t=end} I_{discharge}\, dt \tag{14.14}$$

In conventional coulomb-counting technique, Q_0 is set to a constant value equal to a rated capacity mentioned in the manufacturer's specifications and that is stored in the memory. Due to use of the constant rated capacity (Q_0), the SoH determined is inaccurate on a number of occasions. For example, the SoH measured when the battery is fully discharged and when the battery is partly discharged shall be different, thereby providing a different SoH when the battery has undergone a same number of charge/discharge cycles. To illustrate the disadvantage further, consider a 2Ah-rated battery operated for 200 cycles which is fully charged then discharged with 1C rate to its cut-off voltage (i.e., 0% SoC) as shown in Figure 14.12(a). From the characteristics shown in Figure 14.12(a), it is evident that the battery is taking 3,380 seconds to discharge completely. Therefore, by integrating the battery discharge current (i.e., 2A) with the discharge time, capacity of the battery is

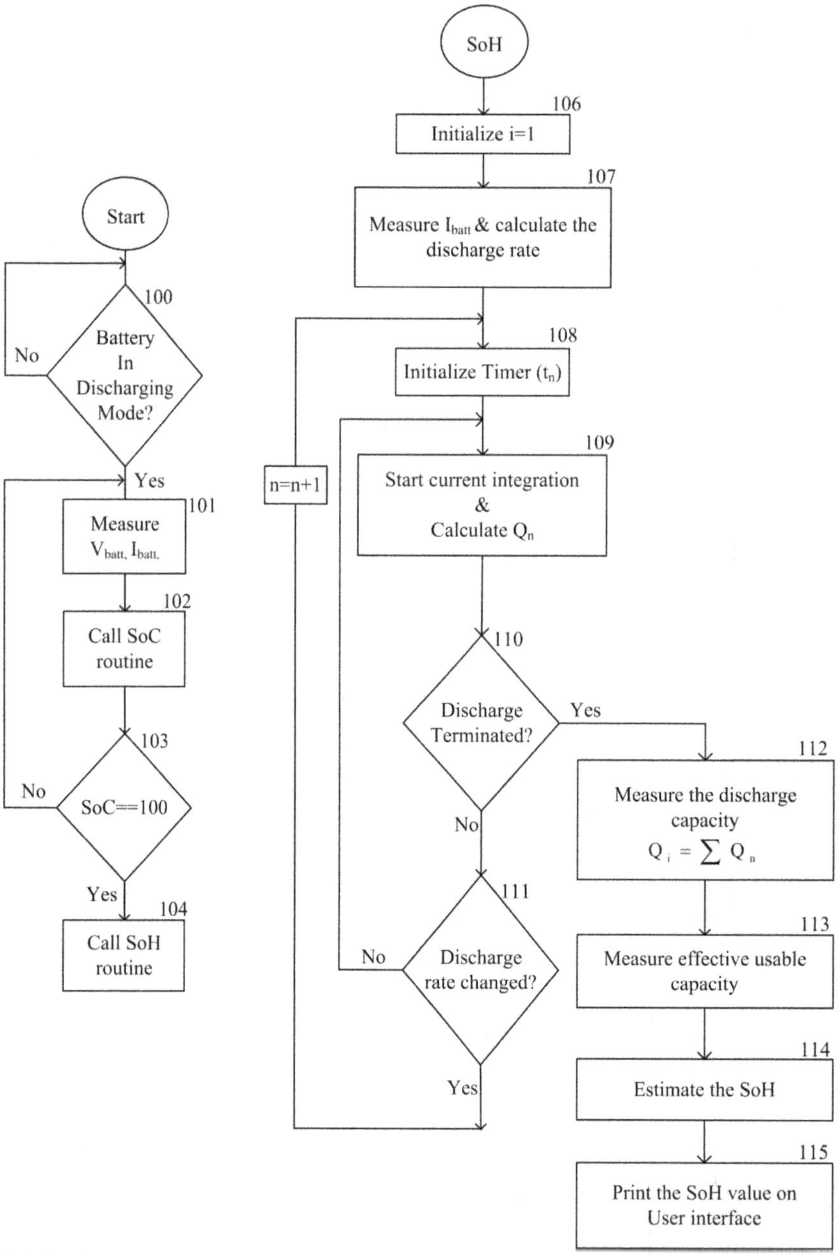

FIGURE 14.11 Flowchart of the modified coulomb-counting technique.

determined as 1.87Ah. Furthermore, using the Eq. (14.13), SoH is measured as 93.88%.

Consider another test case whereby the battery is of the same age as the previous case but the discharge is stopped after 2,180 seconds as shown in Figure 14.12(b), then the capacity—through coulomb counting (i.e., current integration)—is found to be 1.21Ah. In this case, battery capacity is reduced compared to the previous case. The reduced capacity in this case is not because the battery is aged, but it is because of battery is not discharged completely. Therefore, the estimated SoH through coulomb-counting technique (i.e., Eq. [14.13]) gives an incorrect result. Moreover, this technique requires 100% DoD, which is not desirable in many applications.

From the discharge Ah-V characteristics of a 2Ah Li-ion battery shown in Figure 14.13 and from the MARQUE ACL9014 LFP battery datasheet is observed that rated capacity (Q_o) varies with the battery discharge current.

In EV applications, load current varies continuously based on many factors such as road profile, vehicle speed, etc. As a result, discharge current varies with the load. Therefore, using the constant rated capacity (Q_0) in Eq. (14.13) results in incorrect SoH estimation.

In Figure 14.12(b), discharge of the battery is stopped after 2,180 seconds (i.e., when the battery SoC is at 37%) and using coulomb counting, Q_i is calculated, whereas Q_0 is can be calculated from Figure 14.13 (i.e., capacity at 3V). The limitation of change in discharge current during

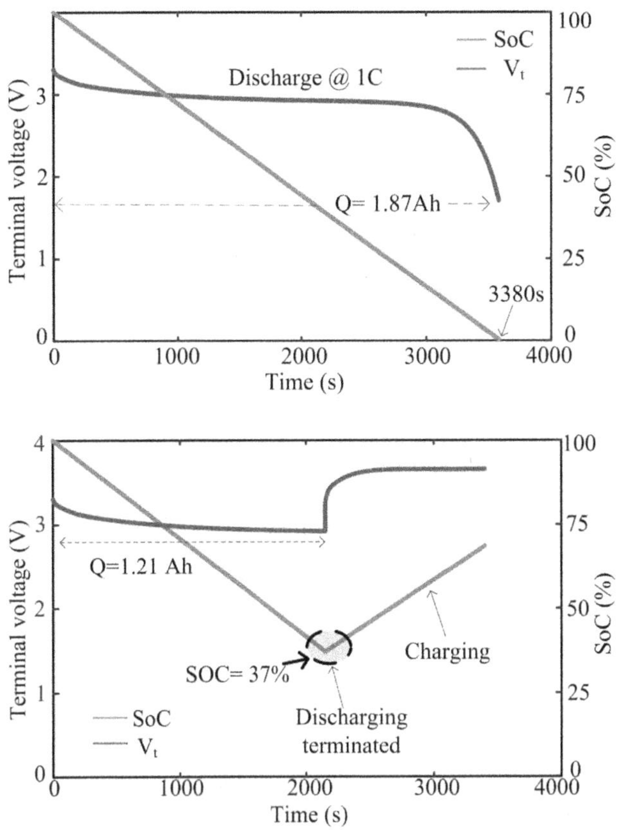

FIGURE 14.12 (a) Complete discharge characteristics of a battery aged 200 cycles; (b) discharge characteristics up to 63% DoD.

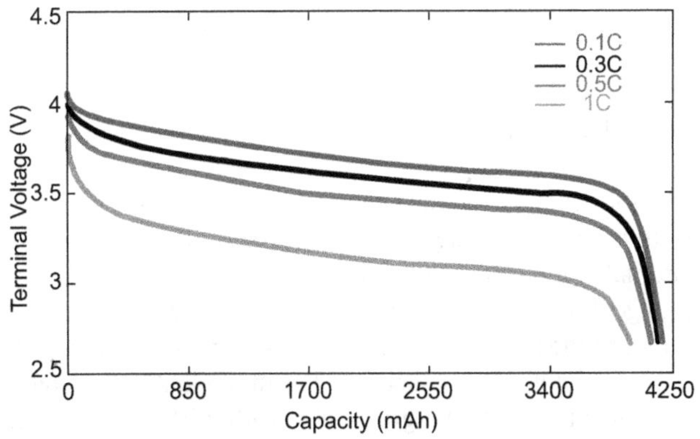

FIGURE 14.13 Ah-V characteristics of an LFP battery at different discharge rates.

the discharge of the battery is addressed in the modified coulomb-counting technique. In modified technique, rated capacity (Q_0) is calculated using a time weightage method as described in Eq. (14.15).

$$Q_o = \frac{(Q_{o1} \times t_1) + (Q_{o2} \times t_2) + (Q_{o3} \times t_3) + \ldots + (Q_{on} \times t_n)}{t_1 + t_2 + t_3 + \ldots + t_n} \times (1 - SoC_{termination}) \tag{14.15}$$

where Q_{o1}, Q_{o2}, . . . Q_{on} are capacities for different charge/discharge rates reported in the data sheet of the battery; t_1, t_2, . . . t_n are the time spent by the battery in each discharge rate; and $SoC_{termination}$ is the SoC of the battery at which discharge process stopped

In the Eq. (14.15), the term $(1 - SoC_{termination})$ overcomes the variable DoD limitation and the other term $\frac{(Q_{o1} \times t_1) + (Q_{o2} \times t_2) + (Q_{o3} \times t_3) + \ldots + (Q_{on} \times t_n)}{t_1 + t_2 + t_3 + \ldots + t_n}$ overcomes the limitation of discharging the battery with different discharge currents (i.e., discharge rates). In the flowchart of the modified coulomb-counting method (Figure 14.11), block 100 verifies whether the battery is in charging or discharging mode. The modified coulomb-counting technique works only when the battery is fully charged, and it should start discharging the load. For estimating the SoC, controller measures the battery current and voltage and calls the SoC routine (Figure 14.8). Once the controller make sure that the battery is fully charged (i.e., 100% SoC), then it calls the modified coulomb-counting SoH estimation routine. Further, the controller (blocks 107–108) determines the rate at which battery is discharging and starts the timer to count the time spent by the battery in the same discharge rate. Furthermore, controller activates the current integration to calculate the Q_i of the battery being discharged.

While the battery is being discharged, if the battery current (i.e., discharge rate) varies, the controller activates the timer to count the time spent by the battery in a new discharge rate; also, it measures the charge delivered to the load in new discharge rate. This process remains the same until the battery discharge stops in the middle or the battery discharges completely. The controller (block 112) measures the capacity (Q_i) of the particular discharge cycle and the controller determines the normalized capacity (i.e. Q_0) of the battery using time weightage method (i.e., Eq. [14.15]) immediately after the discharge process stops. Furthermore, using the Eq. (14.13), the controller estimates the SoH of the battery. Therefore, the accuracy of this technique is very high since it considers the variable DoD and variable discharge currents to estimate SoH. The estimated SoH, along with the SoC details, can be shown on the vehicle's dashboard.

14.4.2 Implementation and Validation of Modified Coulomb-Counting Technique

Figure 14.14 shows the experimental setup to validate the modified coulomb-counting technique. The experimental prototype is the implementation of schematic shown in Figure 14.7. The experimental setup contains of a DC/DC power electronic interface to connect battery and load, digital signal processor (i.e. dSPACE 1104) and a Li-ion battery of 6Ah, 48V rating. The experiments are carried on two sets of batteries of the same rating and different ages. The experimental results of the SoH estimation are shown in Figure 14.15. Figure 14.15(a)–(b) show the charge

FIGURE 14.14 Experimental setup used to estimate SoH.

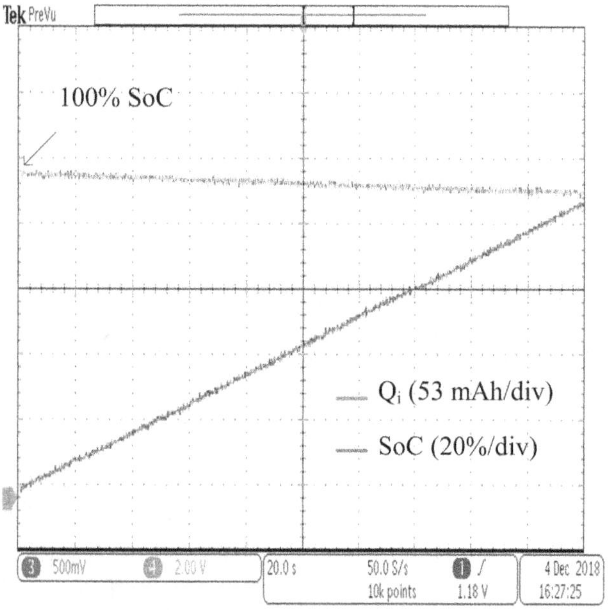

(a)

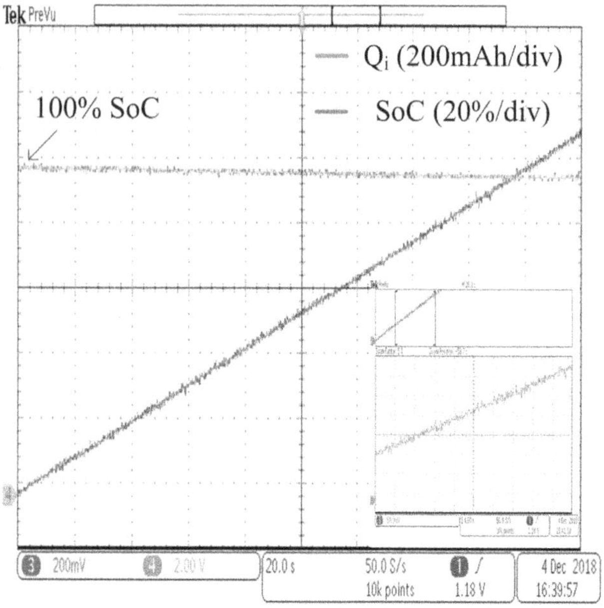

(b)

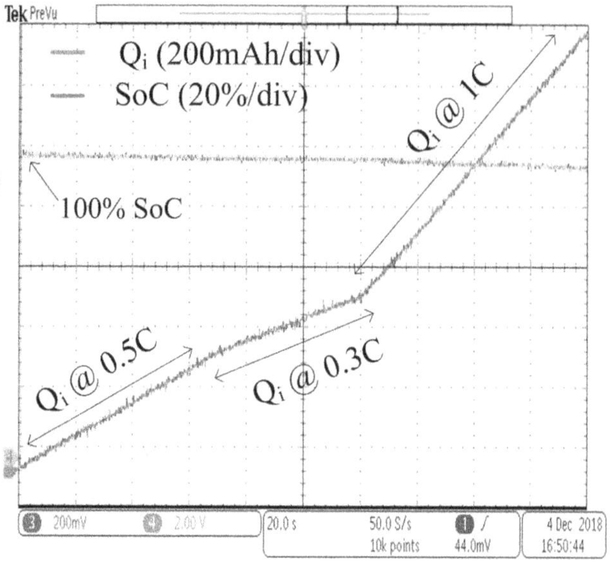

(c)

FIGURE 14.15 Experimental results of SoC and charge delivered to the load: (a) with a 1C discharge rate; (b) with 0.5C discharge rate; (c) with multiple discharge rates.

delivered to the load SoC characteristics when the battery is discharged with 1C and 0.5C. In Figure 14.15, charge delivered to the load and SoC characteristics are shown only for 200s due to oscilloscope constraints. However, the controller has estimated SoH at the end of discharge and displayed on dSPACE control desk.

14.5 CELL BALANCING IN LI-ION BATTERY PACKS

Cell imbalance refers to the situation when series connected individual cells in a battery pack are not at the same voltage/SoC level during the charging and discharging process. With regular charging and discharging of the batteries, the internal cell chemical properties of the individual cells can change. For example, if the internal resistances of the Li-ion cells are measured after a certain time period, the differences in internal resistance of the individual cells would be evident. Upon charging cells with different internal resistances, there would be differences in the time taken for each cell to reach full voltage. Further, charging and discharging these battery packs intensifies the cell imbalances and causes degradation of battery capacity, and it sometimes leads to permanent failure of the battery. Therefore, maintaining cell voltages/SoC levels in a battery pack at an equalized level is significant. To overcome this problem, cell-balancing techniques can be employed to the cells to balance out the cell voltages. The cell-balancing techniques are classified into two categories: passive cell-balancing techniques and active cell-balancing techniques.

Passive cell-balancing techniques convert the excess energy in the cells of a battery pack to heat energy by shunting a resistor. This type of balancing is also called dissipative balancing technique [28]. Fixed-shunt resistor and switched-shunt resistor are the types of passive cell-balancing techniques. Figure 14.16 shows schematics of the passive cell-balancing techniques. The passive cell-balancing circuits are cheap and easy to implement; however, the efficiency of this type of balancing is very poor.

Active cell-balancing techniques transfer the excess energy from a single cell to the entire battery pack or from an excess-energy cell to a lower-energy cell. This type of cell balancing is called non-dissipative cell balancing.

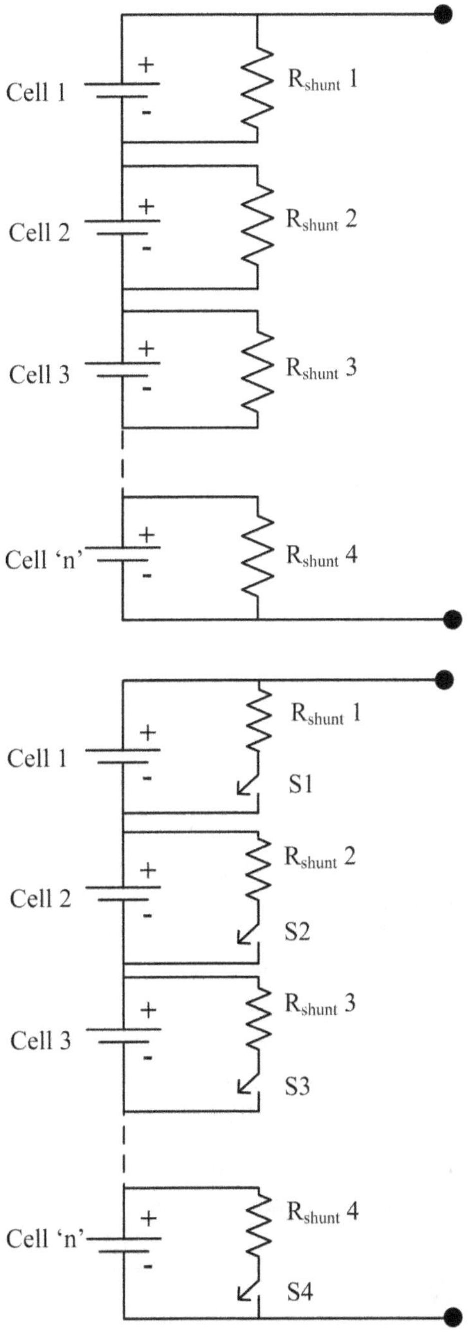

FIGURE 14.16 Passive cell-balancing techniques: (a) fixed shunt resistor; (b) switched shunt resistor cell-balancing circuits [28].

Various active cell-balancing techniques are available in the literature based on the energy transferring component, and they are classified into three categories such as:

1. Switched capacitor based cell balancing
2. Inductor/transformer based cell balancing
3. Converter-based cell balancing

Since the active cell-balancing technique has better efficiency and balancing speed than passive techniques, in this chapter, three techniques are explained from each category.

14.5.1 Switched Capacitor–Based Cell Balancing

A schematic of the switched capacitor–based cell balancing is shown in Figure 14.17. This method requires single capacitor and (n – 1) bi-directional switches wherein, 'n' is the number of series connected cells in a battery pack. In switched capacitor technique, the controller continuously monitors the voltages of series connected cells in a battery pack. Further, the controller selects the higher-energy cell and lower-energy cell for balancing; once the balancing completes, the controller again checks for the cell voltage imbalances. This procedure continues until all the cells in the battery pack attain the same voltage level. The major drawback of this technique is that it takes a long time to equalize the cells compared to other active cell-balancing techniques [29].

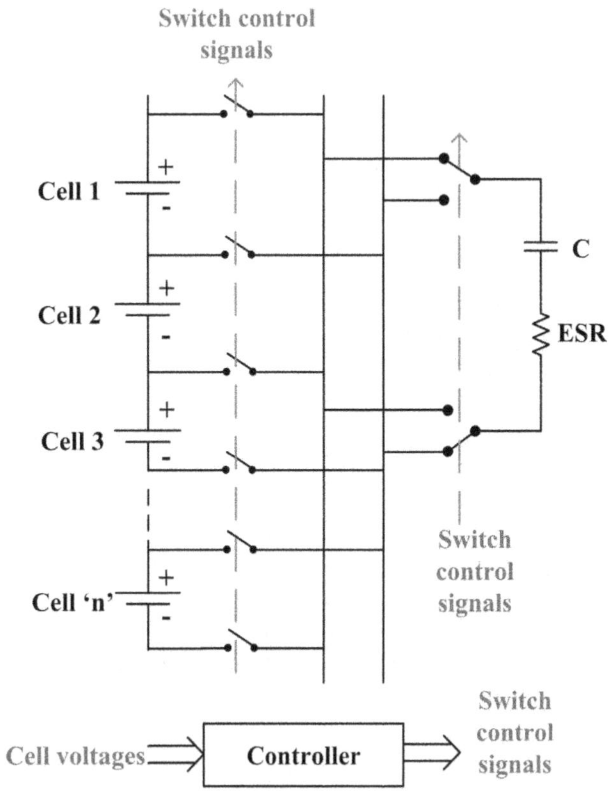

FIGURE 14.17 Switched capacitor–based cell-balancing circuit [29].

14.5.2 Transformer-Based Cell Balancing

The schematic of the transformer-based cell-balancing circuit is shown in Figure 14.18. This technique uses multi-winding flyback transformer; the primary side of the flyback transformer is connected to the entire battery pack, and each cell has a secondary side connected to it. The controller identifies the cell with the lowest voltage, then the corresponding cell secondary winding switch is closed and the primary switch (SM) is closed. Now, the circuit is completed and the power can be sent from the battery pack to the individual cell until that particular cell is not the lowest possible voltage. This type of cell balancing is costly, since it requires a multi-winding transformer. Moreover, the cost increases as the number of series cells in battery pack increases [30].

14.5.3 Converter-Based Cell Balancing

Figure 14.19 shows the circuit diagram of the Ćuk converter–based cell-balancing technique. In this technique, a Ćuk converter is connected between all two adjacent cells of battery pack. The controller continuously monitors the cell voltages and identifies the cell with high voltage. Further, the controller operates the switches (S1–S4) to transfer the energy from high-voltage cells to adjacent cells, and this procedure continues until the battery pack is equalized. This technique takes a long time to balance the cells if the battery pack has more series cells. This type of equalization is highly efficient only for modularized battery packs [31].

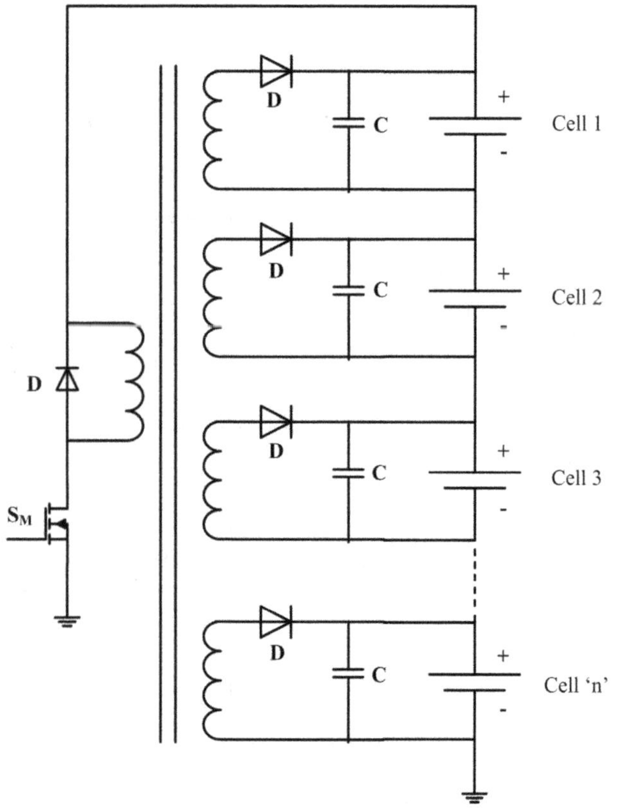

FIGURE 14.18 Flyback transformer–based cell-balancing circuit [30].

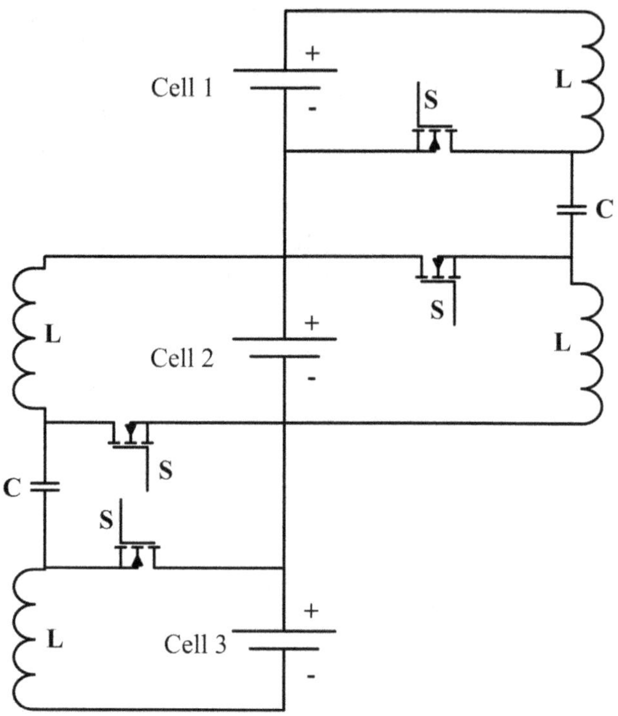

FIGURE 14.19 Ćuk converter–based cell-balancing circuit [31].

14.6 SUMMARY

Use of a battery management system (BMS) in electric vehicle (EV) applications has three main functionalities: state of charge (SoC) estimation, state of health (SoH) estimation and battery cell equalization. Therefore, in this chapter, a conventional coulomb-counting technique to estimate SoC and a modified coulomb-counting technique to estimate SoH are discussed in detail. In addition, hardware implementation of these SoC and SoH techniques, along with the results analysis, are presented which validate the proposed methodology. Moreover, different methods to balance the cells in battery packs are discussed in this chapter.

14.7 NUMERICAL PROBLEMS BASED ON BATTERY MANAGEMENT SYSTEMS

Problem 1

A battery labeled as 25Ah at a C-rate of 0.25C. Calculate the equivalent current (A) and the time required to charge the battery completely.

Solution

Battery capacity = 25Ah
Equivalent current (A) = 25 × 0.25 = 6.25A
Time required to charge the battery = 1/0.25 = four hours

Problem 2

Determine the rating of the lithium iron phosphate (LFP) battery required to power a 3kW load for four hours duration. Also, find the number of series and parallel cells to be connected in the battery pack by considering the specifications of the battery round trip efficiency and charge controller efficiency as 90%, 48V rating of the charge controller, single LFP cell of 3.3V as nominal voltage and 2000mAh capacity.

Solution

Capacity of the battery required for the aforementioned specifications can be calculated as follows [32]:

$$\text{Capacity}(\text{Ah}) = \frac{\text{P} \times \text{N}}{\text{n}_{\text{batt}} \times \text{n}_{\text{charge controller}} \times \text{DoD} \times \text{V}_{\text{batt}}}$$

where, 'P' is the total power of the load, 'N' is the number of hours to power up the battery, nbatt and ncharge controller are efficiency of battery and charge controller, respectively, and they are assumed to be 90%; the maximum DoD is considered as 80% to avoid the deep discharge of the battery, and V_{batt} is the voltage rating of the battery

$$\text{Capacity}(\text{Ah}) = \frac{3000 \times 4}{0.9 \times 0.9 \times 0.8 \times 48} = 385\text{Ah}$$

Number of series and parallel cells required to get a 48V, 385Ah battery pack:

$$\text{Number of series cells required}(\text{N}_\text{s}) = \frac{\text{Battery pack voltage}}{\text{nominal voltage of single cell}}$$

$$\text{Number of series cells required}(\text{N}_\text{s}) = \frac{48}{3.3} = 14.54 \approx 15$$

$$\text{Number of parallel strings required}(\text{N}_\text{P}) = \frac{\text{Battery pack capacity}(\text{Ah})}{\text{N}_\text{s} \times \text{Single cell capacity}(\text{Ah})}$$

$$\text{Number of parallel strings required}(\text{N}_\text{P}) = \frac{385}{15 \times 2} = 12.83 \approx 13$$

$$\text{Number cells in the battery pack} = \text{N}_s \times N_p = 195$$

Problem 3

Figure 14.20 shows the energy characteristics of new European drive cycle (NEDC) with maximum load on vehicle for urban conditions. Using the characteristics, determine the rating of a battery required to drive an EV for a range of 100km. Consider the EV power train of the battery as 48V and the energy required for supplying vehicle auxiliary loads as 8kWh/100km.

Solution

From the NEDC drive cycle energy characteristics, the energy required to drive 100km for maximum load condition is 16kWh.

Auxiliary load energy = 8kWh/100km

Total energy required (Etotal) = 16kWh + 8kWh

Capacity of the battery required for the aforementioned specifications can be calculated as follows [32]:

$$Capacity(Ah) = \frac{E_{total}}{n_{batt} \times n_{power\,train} \times DoD \times V_{batt}}$$

where n_{batt} and $\text{n}_{power\,train}$ are the efficiencies of the battery and power train, respectively, and they are assumed to be 90%; the maximum DoD is considered as 80% to avoid the deep discharge of the battery

$$Capacity(Ah) = \frac{24 \times 1000}{0.9 \times 0.9 \times 0.8 \times 48} = 772Ah$$

Therefore, the rating of a battery required to drive an EV for a range of 100km is found to be 772Ah.

Problem 4

A 37kWh EV has a battery SoC of 40%. Obtain the total expected drive range of the vehicle for NEDC drive cycle shown in Figure 14.20 for maximum load conditions.

Solution

According to NEDC dive cycle energy characteristics, the vehicle requires an energy of 16kWh/100km, and 40% of battery energy corresponds to 14.8kWh of battery energy. Therefore, with the 40% SoC, the vehicle can cover 92.5km.

Problem 5

An EV has a battery rating of 300Ah, 48V, and the maximum allowable DoD of the battery is 80% (i.e., 10% at top and 10% at bottom). The EV consumes 10kWh per 100km. Determine the range of the vehicle.

Solution

Energy rating of the battery (E_{batt}) = 300Ah × 48V = 14.4kWh

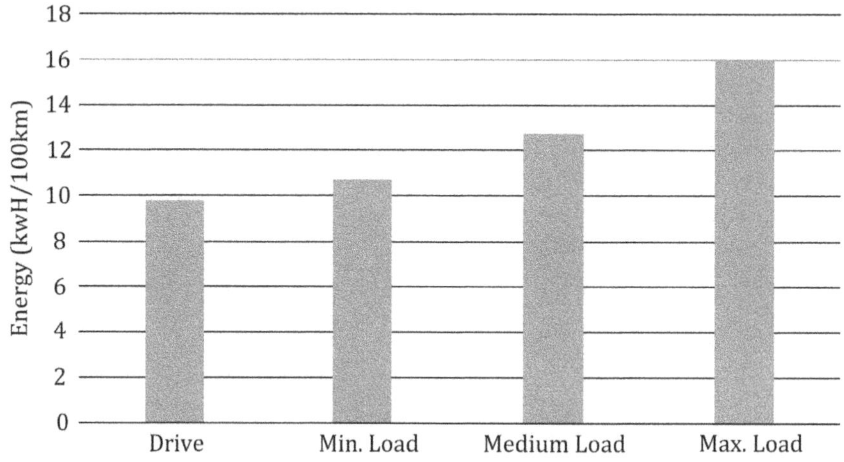

FIGURE 14.20 Energy characteristics of an NEDC drive cycle.

Usable capacity of the battery = DoD × E_{batt} = 11.52kWh

To cover a distance of 100km, the vehicle requires 10kWh of battery energy; therefore, with the available energy, the vehicle can cover a distance of 115.2km.

14.8 PRACTICE QUESTIONS

1. A 30kWh, 300V rated Li-ion battery can be charged from 0–50% SoC in 45 minutes. Determine the charging current, C-rate and charging power required to achieve this.
2. A battery pack consists of 24 cells in series per string and three parallel strings. The internal resistance of each cell is 2mΩ. The battery is fully discharged to the voltage level of 2.7V and subjected to charging process using 50kW charger. Determine the battery charging current, charging voltage and power efficiency of the battery.
3. A Li ion cell has a rated capacity of 3000mAh, a nominal voltage rating of 3.7V and internal resistance of 2mΩ. Determine the maximum deliverable energy of the battery for the discharge rate of 0.5C.
4. An EV has a battery rating of 100kWh and operating with a DoD of 90%. Determine the range of vehicle (in kilometers) if the battery consumes 100Wh/km.
5. Determine the rating of the battery required to cover the range of 150km by considering the WLTC drive cycle. Also, consider 10kWh auxiliary power required throughout the range.

REFERENCES

1. NITI Aayog and Rocky Mountain Institute, Mobilizing Finance for EVs in India: A Toolkit of Solutions to Mitigate Risks and Address Market Barriers, January 2021.
2. L. Lam, P. Bauer and E. Kelder, "A practical circuit-based model for Li-ion battery cells in electric vehicle applications," in Proceedings IEEE 33rd International Telecommunications Energy Conference, 2011, pp. 1–9
3. B. Xu, A. Oudalov, A. Ulbig, G. Andersson and D. S. Kirschen, "Modeling of lithium-ion battery degradation for cell life assessment," IEEE Trans. Smart Grid, vol. 9, no. 2, pp. 1131–1140, Mar. 2018.
4. B. Xu, A. Oudalov, A. Ulbig, G. Andersson and D. S. Kirschen, "Modeling of lithium-ion battery degradation for cell life assessment," IEEE Trans. Smart Grid, vol. 9, no. 2, pp. 1131–1140, Mar. 2018.
5. J. Wang, et al., "Cycle-life model for graphite-LiFePO4 cells," J. Power Sources, vol. 196, no. 8, pp. 3942–3948, 2011.
6. X. Han, M. Ouyang, L. Lu and J. Li, "A comparative study of commercial lithium ion battery cycle life in electric vehicle: Capacity loss estimation," J. Power Sources, vol. 268, pp. 658–669, 2014.
7. K. S. Ng, C. S. Moo, Y. P. Chen and Y. C. Hsieh, "State-of-charge estimation for lead-acid batteries based on dynamic opencircuit voltage," in Proceedings of the 2nd IEEE International Power and Energy Conference (PECon '08), pp. 972–976, Johor Bahru, Malaysia, December 2008.
8. K. Movassagh, S. A. Raihan and B. Balasingam, "Performance analysis of coulomb counting approach for state of charge estimation," 2019 IEEE Electrical Power and Energy Conference (EPEC), 2019, pp. 1–6. doi:10.1109/EPEC47565.2019.9074781
9. M. Coleman, C. K. Lee, C. Zhu and W. G. Hurley, "Stateof-charge determination from EMF voltage estimation: using impedance, terminal voltage, and current for lead-acid and lithium-ion batteries," IEEE Trans. Ind. Electron., vol. 54, no. 5, pp. 2550–2557, 2007.
10. Mori W. Yatsui and Hua Bai. "Kalman filter based state-of-charge estimation for lithium-ion batteries in hybrid electric vehicles using pulse charging." 2011 IEEE Vehicle Power and Propulsion Conference. IEEE, 2011.
11. Rui Xiong, et al. "Evaluation on state of charge estimation of batteries with adaptive extended Kalman filter by experiment approach." IEEE Trans. Veh. Technol. Vol. 62, no. 1, pp. 108–117, 2012.

12. Fengchun Sun, et al. "Adaptive unscented Kalman filtering for state of charge estimation of a lithium-ion battery for electric vehicles." Energy, vol. 36, no. 5, pp. 3531–3540, 2011.
13. Fan Zhang, et al. "State-of-charge estimation based on microcontroller-implemented sigma-point kalman filter in a modular cell balancing system for lithium-ion battery packs." 2015 IEEE 16th Workshop on Control and Modeling for Power Electronics (COMPEL). IEEE, 2015.
14. LiuWang Kang, Xuan Zhao and Jian Ma. "A new neural network model for the state-of-charge estimation in the battery degradation process." Appl. Energy, vol. 121, pp. 20–27, 2014.
15. Li, I-Hsum, et al. "A merged fuzzy neural network and its applications in battery state-of-charge estimation." IEEE Trans. Energy Convers., vol. 22, no. 3, pp. 697–708, 2007.
16. J. N. Hu, et al. "State-of-charge estimation for battery management system using optimized support vector machine for regression." J. Power Sources., vol. 269, pp. 682–693, 2014.
17. Xiaopeng Chen, et al. "Sliding mode observer for state of charge estimation based on battery equivalent circuit in electric vehicles." Aust. J. Electr. Electron. Eng., vol. 9, no. 3, pp. 225–234, 2012.
18. Jun Xu, et al. "The state of charge estimation of lithium-ion batteries based on a proportional-integral observer." IEEE Trans. Veh. Technol., vol. 63, no. 4, pp. 1614–1621, 2013.
19. J. S. Goud, K. R and B. Singh, "An online method of estimating state of health of a li-ion battery," IEEE Trans. Energy Conversion, vol. 36, no. 1, pp. 111–119, March 2021. doi:10.1109/TEC.2020.3008937
20. J. Vetter, P. Nov´ak, M. R. Wagner, C. Veit, K. C. M¨oller, J. Besenhard, M. Winter, M. Wohlfahrt-Mehrens, C. Vogler and A. Hammouche, "Ageing mechanisms in lithium-ion batteries." J Power Sources, vol. 147, no. 1–2, pp. 269–281, 2005.
21. J. S. Goud, R. Kalpana and B. Singh, "Modeling and estimation of remaining useful life of single cell li-ion battery," 2018 IEEE International Conference on Power Electronics, Drives and Energy Systems (PEDES), 2018, pp. 1–5. doi:10.1109/PEDES.2018.8707554
22. T. Sarikurt, M. Ceylan and A. Balikci, "An analytical battery state of health estimation method." In Industrial Electronics (ISIE), 2014 IEEE 23rd International Symposium on, pp. 1605–1609. IEEE, 2014.
23. K. Goebel, B. Saha, A. Saxena, J. R. Celaya and J. P. Christophersen, "Prognostics in battery health management." IEEE Instrumentation & Measurement Magazine, vol. 11, no. 4, 2008.
24. G. L. Plett, "Extended kalman filtering for battery management systems of lipb-based hev battery packs: Part 3. state and parameter estimation." J. Power Sources, vol. 134, no. 2, pp. 277–292, 2004.
25. F. Zhang, G. Liu and L. Fang, "Battery state estimation using unscented kalman filter. In Robotics and Automation, 2009. ICRA'09. IEEE International Conference on, pp. 1863–1868. IEEE, 2009.
26. M. Berecibar, M. Garmendia, I. Gandiaga, J. Crego and I. Villarreal, "State of health estimation algorithm of lifepo4 battery packs based on differential voltage curves for battery management system application." Energy, vol. 103, pp. 784–796, 2016b.
27. K. S. Ng, C. S. Moo, Y. P. Chen and Y. C. Hsieh, "Enhanced Coulomb counting method for estimating state-of-charge and state-of-health of lithium-ion batteries," Appl. Energy, vol. 86, no. 9, pp. 1506–1511, 2009.
28. M. M. Hoque, et al., "Battery charge equalization controller in electric vehicle applications: a review." Renew. Sustain. Energy Rev., vol. 75, pp. 1363–1385, 2017.
29. M. Daowd, M. Antoine, N. Omar, P. Van Den Bossche and J. Van Mierlo, "Single switched capacitor battery balancing system enhancements." Energies, vol. 6, pp. 2149–74, 2013.
30. M. Einhorn, W. Roessler and J. Fleig, "Improved performance of serially connected li-ion batteries with active cell balancing in electric vehicles," IEEE Transactions on Vehicular Technology, vol. 60, no. 6, pp. 2448–2457, July 2011. doi:10.1109/TVT.2011.2153886
31. Lu Xi, Qian Wei and Peng Fang-Zheng, "Modularized buck-boost + Cuk converter for high voltage series connected battery cells." Applied Power Electronics Conference and Exposition (APEC), 2012 Twenty-Seventh Annual IEEE, p. 2272,2278; 2/ 2012.
32. J. S. Goud, S. Tunga, Q. Sultana and S. Bindu, "Li-ion based optimally sized grid interfaced roof top PV system," 2016 3rd International Conference on Electrical Energy Systems (ICEES), 2016, pp. 267–271. doi: 10.1109/ICEES.2016.7510651

15 Comprehensive Analysis of Ultracapacitors

Athul Vijay P. K., Varsha A. Shah, Ujjval B. Vyas and Nikunj Patel

CONTENTS

FIGURE 15.1 Ultracapacitor.

15.1 INTRODUCTION

In electric storage devices, the battery is the most familiar energy storage system. The main motto behind the battery is its huge energy density, lesser weight, and volume. However, the battery has suffered from many limitations such as less power density, less lifecycle, etc. [1]. However, the requirement of power has recently been increased drastically. Hence, an alternative source to support the battery is a must. The ultracapacitor (Figure 15.1) is a prominent energy storage device to support the battery because of its predominant characteristics like high power density, large lifecycle, ability to withstand high temperatures, and high charge-discharge efficiency [2]. Also, the ultracapacitor can be reused if it remains unused for many years, which is not possible in the cases of the battery [3].

The construction of the ultracapacitor is that two electrodes are immersed in an electrolyte with a separator in between. The electrode is of high surface area porous material with a porous size in the range of nanometres. The surface area of the ultracapacitor electrode is much higher than that of the battery [4]. In the 1800s, Helmholtz discovered first-time charge storage at the boundary between a conductor and an ionic solution; later this phenomenon was named a double layer [5]. The double layer has less thickness in comparison with the normal capacitor; also, by using an electrode with a large surface area, the capacitance reached in the range of farad. The power density of supercapacitor is higher than that of batteries, and the energy density is 10–20 times higher than that of electrolytic capacitors for power applications [1]. These capacitors are low-voltage devices. Higher voltages can be achieved by connecting many cells in series like in batteries. Differently than conventional capacitors, supercapacitors make use of electrolytes instead of a dielectric medium, and therefore can store electrochemical energy at the interface between electrodes and the electrolyte. Both liquid and solid electrolytes can be used in supercapacitors. In the electrolyte, mobile charge carriers with a positive charge can accumulate near a negatively charged electrode, while the concentration of negative charge carriers becomes depleted [5]. In this way, a space charge region is developed in the electrolyte near the interface with the electrode and a large electrochemical capacitance is obtained, as shown in Figure 15.2.

In these electrochemical capacitors, the double-layer capacitance can reach high values, depending on the nature of the electrolyte and electrode. In general, the thickness of the space charge layer is negligible compared with the distance between the electrodes. Moreover, the entire interface between the electrodes and electrolytes contributes to the formation of this double layer [5]. Therefore, supercapacitors normally consist of electrodes possessing a very large electrode surface area. A combination of both the double-layer capacitance and the large interfacial area between electrode

DOI: 10.1201/9781003229124-15

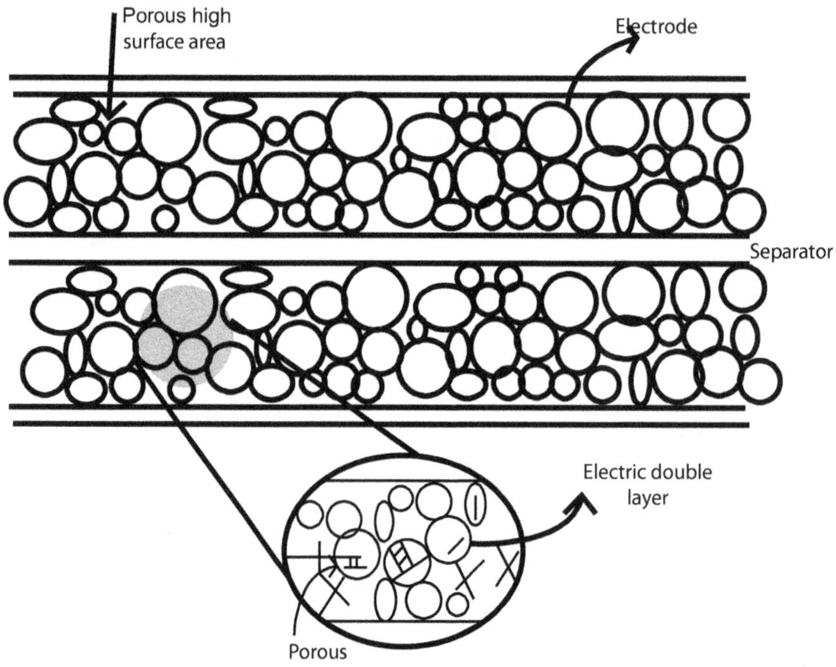

FIGURE 15.2 Schematic of the ultracapacitor [1].

and electrolyte results in supercapacitors within the range of farad compared to a few millifarad for conventional capacitors [5]. The capacitive nature exhibit in the interface between the solid electronic conductor and a liquid ionic conductor is the principle of the double layer. Appying voltage to the ultracapacitor creates a space charge at electrode and electrolyte interface, which is solid/liquid interface and is named as (space charge) electric double layer [5].

The dielectric function of supercapacitors is performed by the electric double layer, which is mainly constituted of solvent molecules. This results in two consequences. First, huge capacitance can be obtained by using electrode materials with high specific surface areas [2]. Second, rated voltage is limited to some volts, depending on the solvent. The polarization due to the build-up of a space charge layer consumes more time than electronic and ionic polarization. Therefore, in contrast with conventional capacitors, which are used in AC applications, supercapacitors are more suitable for DC applications [6].

The ultracapacitor has huge applications in the field of power quality, renewable integration, electrical vehicles, and industrial applications. Hence, a detailed understanding of the ultracapacitor and its working mechanism is mandatory to use the ultracapacitor for the required applications effectively. The main motto of this work is to provide a detailed study on the ultracapacitor concerning its terminal characteristics, material characteristics, modelling, and testing the ultracapacitor in real time as well as with the software. The hardware studies have been carried out with the aid of a BCAP3000F Maxwell ultracapacitor and an STM32M4 microcontroller. The simulation studies have been carried out with the aid of MATLAB/Simulink.

15.1.1 Organization of the Chapter

Apart from the introductory Section 1, this chapter comprises ten sections. Section 15.2 discusses different variants of the ultracapacitor, which is followed by characteristics of the ultracapacitor in Section 15.3. Section 15.4 discusses detailed specifications of the ultracapacitors available in the market. Section 15.5 discussed different materials of ultracapacitors. Section 15.6 discusses modelling of the ultracapacitor, which is followed by the sizing of the ultracapacitor in Section 15.7. Section 15.8 discusses the hardware and simulation studies, which is followed by the application of the ultracapacitor in Section 15.9, conclusion in Section 15.10, and numerical problems in Section 15.11.

15.2 CHARGE STORAGE OF THE ULTRACAPACITOR

Ultracapacitors store energy similar to that of the conventional capacitor as charge separation. In an ultracapacitor, two electrodes are immersed in an electrolyte with a separator in between. The electrodes are made of porous materials, which results in a high surface area. The charge is stored in those porous or the interface between solid interface and electrolyte. Storage of the charge in the ultracapacitor can be possible with the two mechanisms, either with double-layer capacitance or by utilizing pseudocapacitance. An ultracapacitor utilizing a pseudocapacitance mechanism is termed as a hybrid capacitor.

15.2.1 Double-Layer Capacitor

In the double-layer capacitor, the energy is stored as a charge separation occurs at the interface between the solid electrode material and the liquid electrolyte. The ions in the pores are transferred between two electrode terminals via electrolyte in a form of the double layer. Capacitance primarily depends on the surface area of the electrode. Depending on the pore size of the surface electrode, the capacitance, and the discharge, current density will determine the cell voltage depending on the type of electrolyte.

15.2.2 Pseudocapacitance

In an ultracapacitor, no faradic reaction occurs between the electrode and electrolyte. Hence, the capacitance of the ultracapacitor is independent of the voltage. In devices with pseudocapacitance, most of the charges will be transferred at the surface of the electrode. Hence, the interaction between the electrolyte and the electrode results in the faradic reaction and the capacitance will remain voltage-dependent. For the development of an ultracapacitor with pseudocapacitance, three variants of the electrochemical process are there. First is the surface adsorption of ions from the electrolyte. The second is the redox reaction of ions from the electrolyte. The third is doping and undoping of the materials in the electrode [1].

15.2.3 Hybrid Ultracapacitors

Ultracapacitors with one electrode of a double-layer material and the next electrode of material with pseudocapacitance are termed as hybrid capacitors [1]. This is shown in Figure 15.3.

Figure 15.4 and Figure 15.5 show the terminal voltage characteristics and charging current of the ultracapacitor in the double-layer capacitor, as well as a hybrid ultracapacitor. The terminal voltage characteristics of the

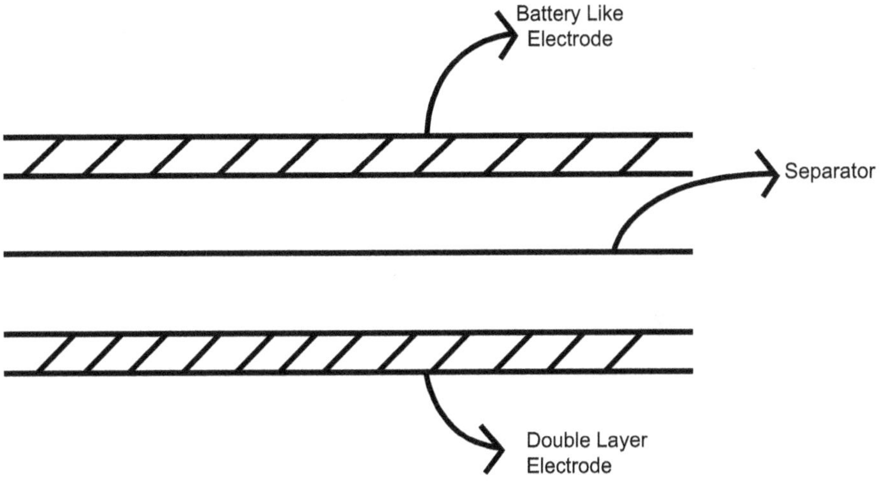

FIGURE 15.3 Hybrid ultracapacitor [1].

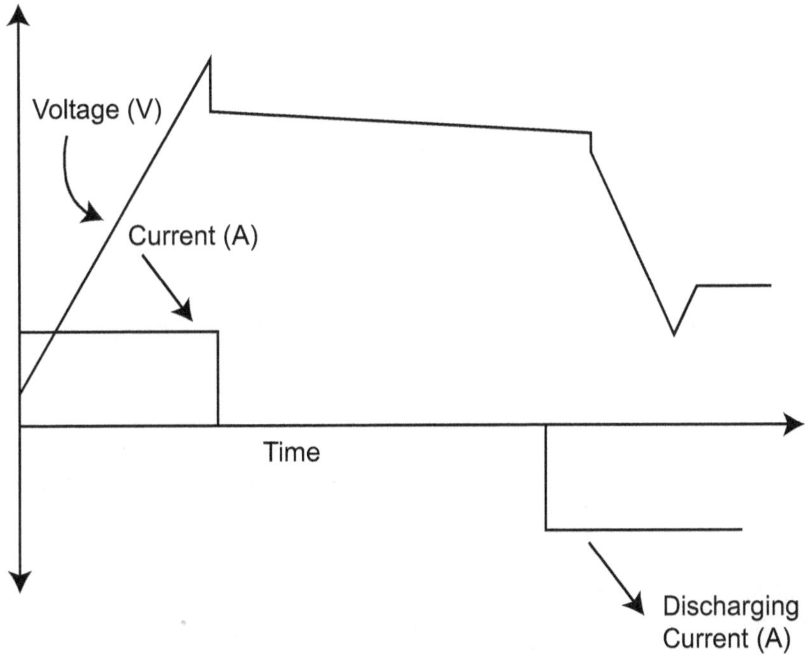

FIGURE 15.4 Characteristics of the double.

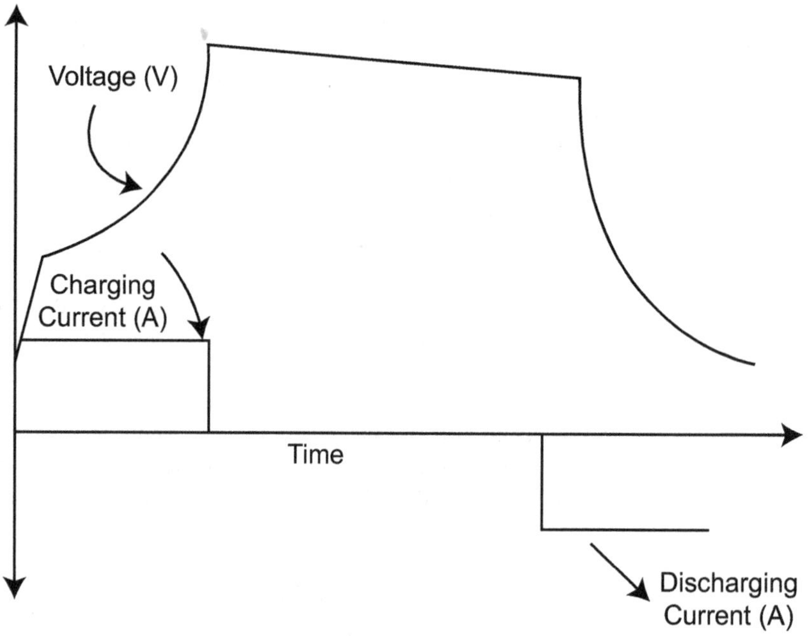

FIGURE 15.5 Characteristics of the hybrid-layer capacitor [1] ultracapacitor [1].

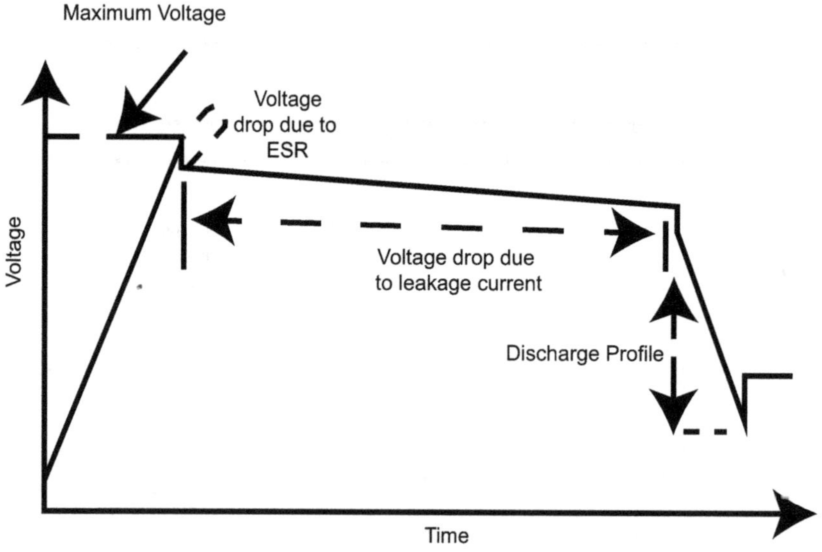

FIGURE 15.6 Terminal voltage of the ultracapacitor [7].

double-layer capacitor will be linear in the charging and discharging period. The terminal voltage characteristics of the hybrid capacitor will be nonlinear in the charging and discharging period.

15.3 CHARACTERISTICS OF THE ULTRACAPACITOR

For a better understanding of the ultracapacitor initially, the reader should have detailed knowledge about the characteristics of the ultracapacitor, which will be discussed in this section.

15.3.1 Terminal Voltage Characteristics

Terminal voltage of the ultracapacitor will increase as the constant current is applied to the ultracapacitor, and has a linear relation with the state of charge, as shown in Figure 15.6. A sudden dip in the voltage can be observed whenever the charging current has been disconnected; this phenomenon is due to the equivalent series resistance of the ultracapacitor. The ultracapacitor can be utilized a maximum of up to 50% of the state of charge, while 75% of the energy will be utilized from the ultracapacitor by that time [8].

15.3.2 Cell, Module, and Pack

A single cell is a complete ultracapacitor with two current leads and a separate compartment holding electrodes, separator, and electrolyte (Figure 15.7). A module is composed of a few cells either by physical attachment or by welding in between cells. A pack of the ultracapacitor is composed of many modules placed in a single envelope [8].

15.3.3 Specific Energy

Specific energy is the amount of energy that can be stored per unit mass. It is expressed in watt-hours per kilogram (Wh/kg) [8].

$$\text{Specific energy} = \frac{0.5 * C * V^2}{3600 * m} \quad (15.1)$$

where c, v and m, are the capacitance, rated voltage, and mass of the single ultracapacitor cell

15.3.4 Specific Power

Specific power is the amount of peak power per unit mass. It is expressed in W/kg [8].

$$\text{Specific Power} = \frac{v^2}{4 * ESR * m} \quad (15.2)$$

where v, ESR, and m are the rated voltage, equivalent series resistance, and mass of the single ultracapacitor cell, respectively [8, 9].

15.3.5 Self-Discharge

Self-discharge is a process of discharging the voltage when the ultracapacitor remains an open circuit if the ultracapacitor opens the circuit at a voltage V and remains open for 24 hours, and if the open-circuit voltage is Vth, the self-discharge voltage is V-Vth. The leakage current is the culprit of the self-discharge. In comparison with the batteries, the self-discharge of the ultracapacitor is huge. However, the merit is that the ultracapacitor can be a chargeback to its original state even though the ultracapacitor remains unused for many years, which is not possible in the batteries [8, 9].

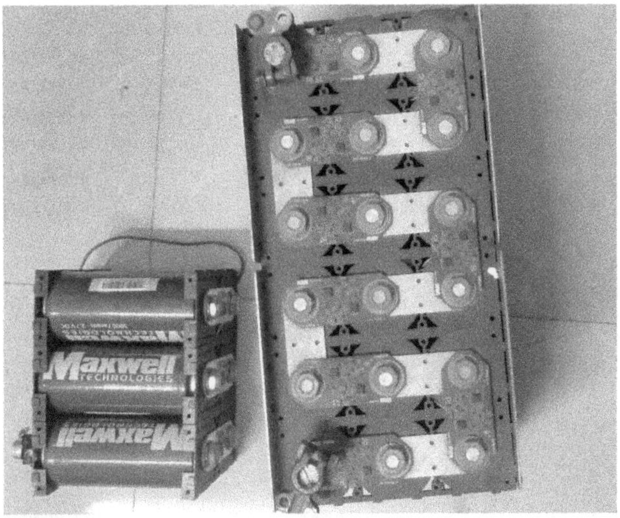

FIGURE 15.7 Ultracapacitor module.

15.3.6 Thermal Resistance

Thermal resistance (R_{th}) is another parameter, which can be calculated as shown in Eq. (15.3):

$$R_{th} = \frac{\Delta T}{ESR * I^2} \quad (15.3)$$

where ΔT is the variation in the temperature from the ambient temperature, and I is the maximum current for a rise in temperature ΔT[8, 9]

15.3.7 State of Charge (SoC)

The remaining capacity of the ultracapacitor is named state of charge (SoC). The SoC of the ultracapacitor has a linear relation with the terminal voltage of the ultracapacitor [8, 9].

$$\text{SoC of the Ultracapacitor} = \frac{\textit{Remaining Capacity of the Ultracapacitor}}{\textit{Rated Capacity of the Ultracapacitor}} \quad (15.4)$$

15.3.8 Depth of Discharge (DoD)

The total amount of capacity that has been discharged from the ultracapacitor indicated by depth of discharge (DoD) [8, 9].

$$\text{DoD} = 1 - \text{SoC} \ (15.5)$$

15.3.9 Lifecycle

The lifecycle is the number of charge-discharge cycles that the ultracapacitor can handle. Unlike the battery, the charge-discharge cycle of the ultracapacitor is very huge [8, 9].

15.4 ULTRACAPACITOR MANUFACTURERS AND SPECIFICATIONS

Major manufactures of the ultracapacitors are Skeleton technologies, AVX Corporation, CDE, Exaton, Eina America, EPCOS, Illinois, KEMET, LICAP, KEMET, Maxwell Technologies, Murata electronics, Nichicon, Ohmite, Panasonic electronics, Rubycon, Seiko Instruments, Skeleton Technologies, Taiyo yuden, TDK Corporation, Tecate Group, United Chemi-Con, Vishay Beyschlag, Wurth Elecktronik, etc. The range of the ultracapacitor is available from 0.082mF to 6,000F in the market. Specifications of the most relevant ultracapacitors available in the market are mentioned in Table 15.1. It has been concluded that Skelcap and Ioxus are much advanced in comparison with other ultracapacitor manufacturers, having very high specific power, high specific energy, high peak current low ESR, and high terminal voltage. The specific power is almost 10–20 times more than the Maxwell ultracapacitor. The CDCL is providing an ultracapacitor of terminal voltage 5.2V, which is almost two times in comparison with the other manufacturers; however, the energy density and lifecycle are less [9–11].

To know about the dependency of various parameters of the ultracapacitor, they have been analyzed by comparing various parameters of the ultracapacitor of various manufacturers concerning the variation of their capacitance. Work has considered the various ultracapacitor from 50–3400F capacitance range.

Figure 15.8 and Figure 15.9 show the variation of the ESR concerning the ultracapacitor capacitance for Maxwell and Skelcap. It has been observed that as the capacitance increases, the ESR will decrease.

Figure 15.10 and Figure 15.11 show the variation in the mass concerning the ultracapacitor capacitance for Maxwell and Skelcap. It has been observed that that ultracapacitors with higher capacitance will have higher mass.

Figure 15.12 and Figure 15.13 show the variation in the specific energy concerning the ultracapacitor capacitance for Maxwell and Skelcap. It has been observed that that ultracapacitors with higher capacitance will have higher specific energy.

Figure 15.14 and Figure 15.15 show the variation in the specific power concerning the ultracapacitor capacitance for Maxwell and Skelcap. It has been observed that that

TABLE 15.1
Ultracapacitor Specifications

Manufacturer	Product	Voltage (V)	Capacitance (F)	Weight (g)	Specific Energy (Wh/kg)	Specific Power (KW/kg)	Imax (A)	Ishort Circuit (A)	ESR (mΩ)	Lifecycle
Maxwell	BCAPO50	2.70	50	12.2	4.1	4.4	37	--------	16	500,000
Maxwell	BCAP0310	2.70	310	60	5.2	6.6	250	1200	2.2	500,000
Maxwell	BCAP0350	2.70	350	60	5.9	4.6	170	840	3.2	1,000,000
Maxwell	BCAP0650	2.70	650	160	4.1	6.8	680	3400	0.8	1,000,000
Maxwell	BCAP1200	2.70	1200	260	4.7	5.8	930	4700	0.58	1,000,000
Maxwell	BCAP1500	2.70	1500	280	5.4	6.6	1150	5700	0.47	1,000,000
Maxwell	BCAP2000	2.70	2000	360	5.6	6.9	1500	7700	0.35	1,000,000
Maxwell	BCAP3000	2.70	3000	510	6	7.9	1900	9300	0.29	1,000,000
Maxwell	BCAP3400	2.85	3400	520	7.4	6.7	2000	10,000	0.28	1,000,000
IOXUS	1RD 1200 K 270 CT	3	1200	290	4.6	12	1300	10,500	0.32	1,000,000
IOXUS	IRD 2000K 270 CT	3	2000	390	5.8	11.4	2000	12,900	0.3	1,000,000
IOXUS	IRD 3000K 270 CT	3	2700	510	6.6	9.6	2700	14,200	0.28	1,000,000
Skelcap	SCA0500	2.85	500	110	5.1	26	600	7500	0.38	1,000,000
Skelcap	SCA750	2.85	750	150	5.8	27	900	8900	0.32	1,000,000
Skelcap	SCA1200	2.85	1200	250	5.4	27	1400	15800	0.18	1,000,000
Skelcap	SCA1800	2.85	1800	340	6	22	2000	17800	0.16	1,000,000
Skelcap	SCA3200	2.85	3200	530	6.8	21	3100	20400	0.14	1,000,000
CDCL	CDLC 122P2R7 K04	2.7	1200	260	4.69	12	1000	4700	0.58	50,0000
CDCL	CDLC 152P2 R7K	3	1500	280	5.43	14	1200	5700	0.47	50,0000
CDCL	CDLC 20212R7 K04	4.2	2000	360	5.64	14	1600	7700	0.35	50,0000
CDCL	CDLC 302P2R7 K04	5.2	3000	510	5.96	12	2200	9300	0.29	50,0000

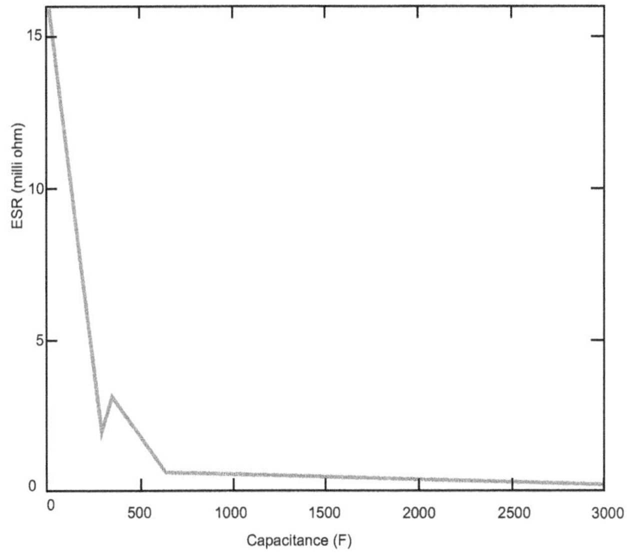

FIGURE 15.8 Capacitance versus ESR plot for Maxwell ultracapacitor.

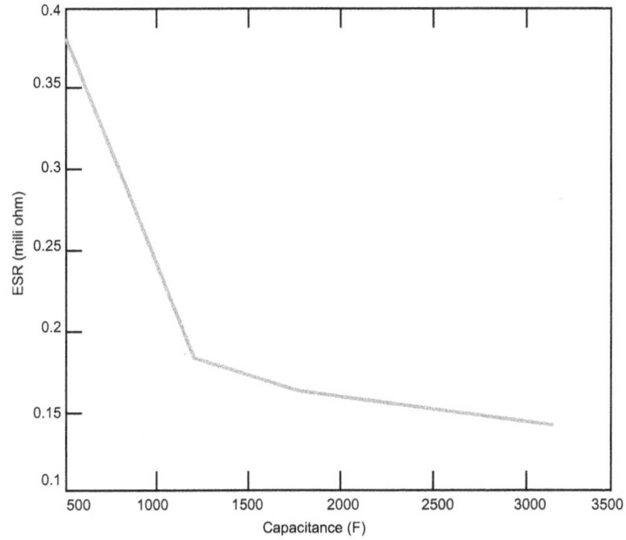

FIGURE 15.9 Capacitance versus ESR plot for Skelcap ultracapacitor.

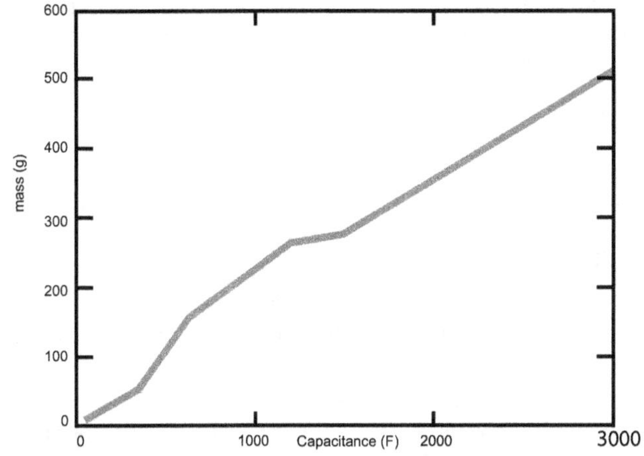

FIGURE 15.10 Capacitance versus mass plot for Maxwell.

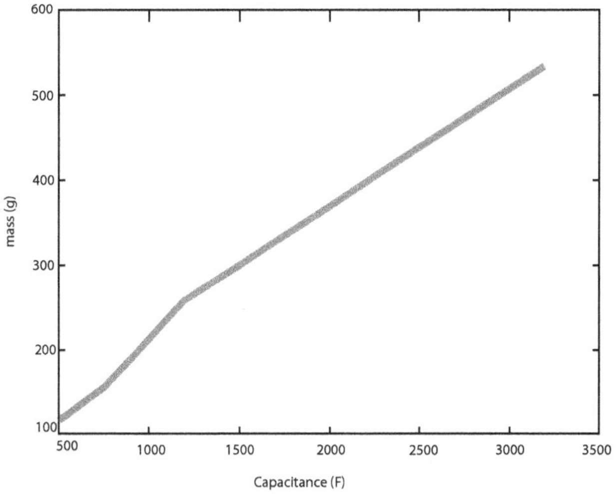

FIGURE 15.11 Capacitance versus mass plot for Skelcap ultracapacitor.

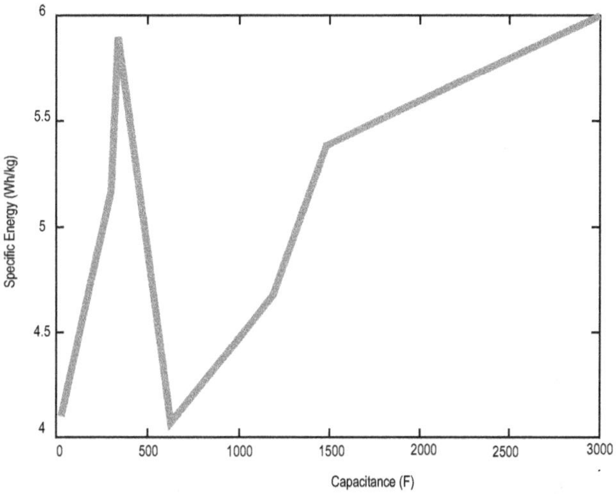

FIGURE 15.12 Capacitance versus specific energy plot for Maxwell ultracapacitor.

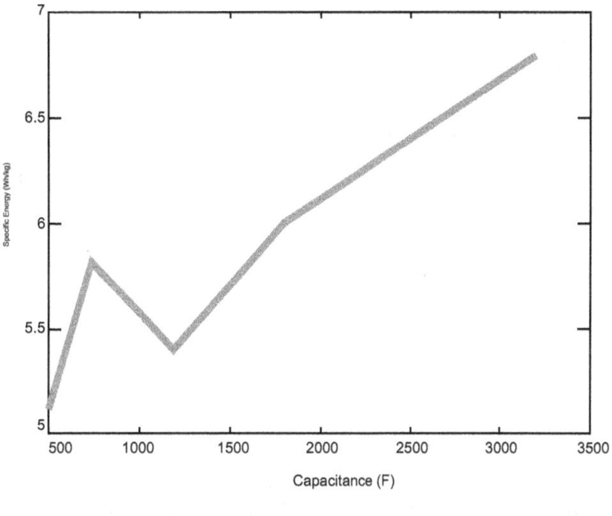

FIGURE 15.13 Capacitance versus specific energy plot for Skelcap ultracapacitor.

ultracapacitors with higher capacitance will have a higher specific power.

Figure 15.16 and Figure 15.17 show the variation in the maximum concerning the ultracapacitor capacitance for Maxwell and Skelcap. It has been observed that that ultracapacitors with higher capacitance will have higher peak current withdrawing capability.

15.5 ULTRACAPACITOR MATERIALS

The key bid for an ultracapacitor is the small energy density. As the energy density of the ultracapacitor is directly proportional to capacitance and the square of its terminal voltage, the energy density can be increased by boosting the terminal voltage or by boosting the capacitance. This is possible by utilizing the ultracapacitor with the electrode material having high capacitance, or with the electrolyte composed of the wide potential pane and with the optimal sizing of the electrode and electrolyte within the ultracapacitor. The wide voltage range, high ion concentration, less viscosity, smaller ion radius, non-toxicity, and lower cost are the vital requirements of the supercapacitor materials. Organic, aqueous, redox character, solid and semi-solid, and ionic liquids are the major electrolytes available for ultracapacitors. Research has shown that aqueous electrolytes are more advanced than organic electrolytes because of lesser resistance, the lesser radius of ion, higher voltage range, and higher capacitance. Also, the manufacturing processes of the organic electrolyte are very complicated [2].

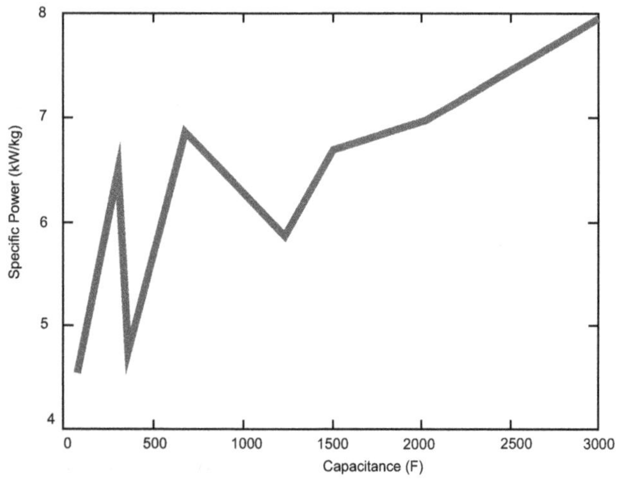

FIGURE 15.14 Capacitance versus specific power plot for Maxwell ultracapacitor.

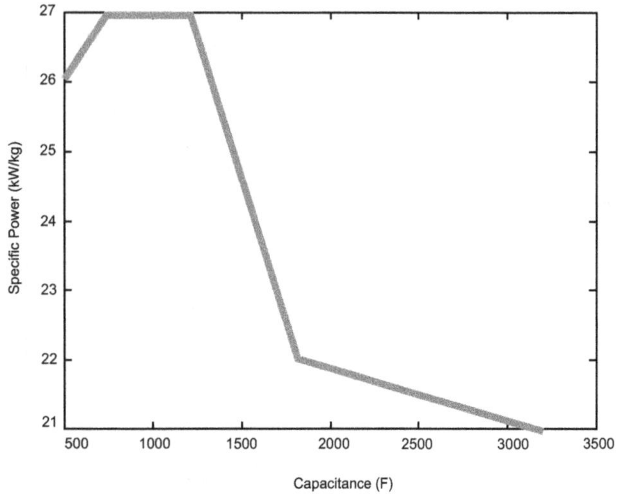

FIGURE 15.15 Capacitance versus specific energy plot for Skelcap ultracapacitor.

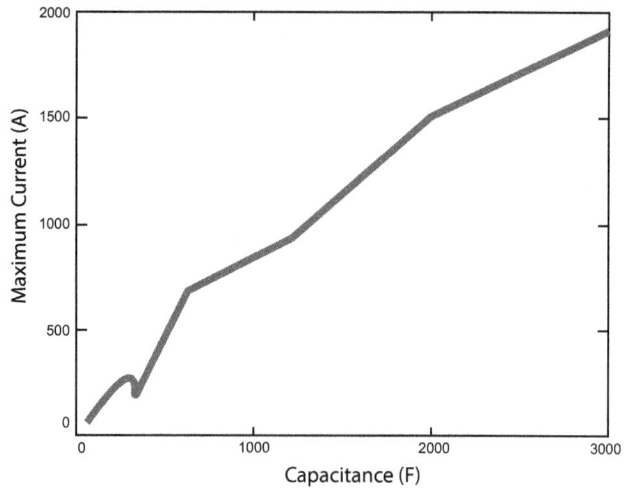

FIGURE 15.16 Capacitance versus maximum current plot for Maxwell ultracapacitor.

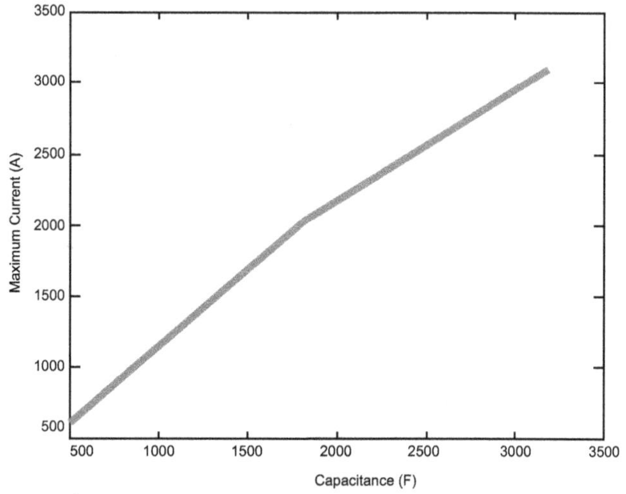

FIGURE 15.17 Capacitance versus maximum current plot for Skelcap ultracapacitor.

The major limitation of the aqueous electrolyte is lower terminal voltage, which is approximately 1.2V [2], [12]. On another side with the organic electrolyte, the voltage range is possible between 2.5V and 2.85V [2], [12]. The organic electrolyte has smaller electrical conductivity. Hence, the power density is less. On another side, the ionic liquid–type electrolytes are well advanced because of higher chemical and thermal stabilities, and low flammability. Also, liquid ions are solvent-free; hence, ion size can be well distinguished. The voltage range of the ion liquid electrolyte ultracapacitor can be obtained up to 4V [2]. The semi-solid electrolyte has recently obtained huge attention because of no potential leakage constraints. The redox character ultracapacitor utilizes pseudocapacitance. Electrodes of ultracapacitors should be of higher thermal stability, higher conductivity, acceptable chemical stability, higher surface area, resistant of corrosion, and lower cost. Also, the material should be capable of pseudocapacitance. With the smaller pore size on the electrode surface, the ultracapacitor can be achieved larger capacitance, as well as smaller ESR. But the terminal voltage will be less. However, for the higher power density, the pore size of the electrode surface should be more [12], which results in a lower energy density. The electrode materials can be classified as follows:

1. Nanostructured carbon-based materials.
2. Conducting polymer-based materials.
3. Metal oxide–based materials.
4. Nanocomposed materials.
5. Carbon material with conducting polymer.
6. Carbon-based nanomaterial with metal oxide-based materials.
8. Metal oxide and conducting polymer-based composites.
9. Metal organic frame work based.
10. Covalent organic material based.
11. Metal nitride based.

With different combinations of electrodes and electrolytes, the specific capacitance, energy density, and power density of the ultracapacitor will vary [13–68]. In [15], the authors show an electrolyte made of copper oxide at AUPD at manganese dioxide core shell whiskers and the electrolyte of 1MKoH, which gives a specific capacitance of 1,376F/g and specific energy of 0.55m Wh/cm^3. In [19], the authors show an electrode of VA-CNT-Graphene with $Ni(OH)_2$/GO/CNTs and with an electrolyte of 2MKo, which gives a specific capacitance of 1,065F/g. In [25], the authors show an electrode of N-CNF/N-CNF and $Ni(OH)_2$ and with an electrolyte of 6MKoH, which gives a specific capacitance of 1,045F/g and specific energy of 51Wh/kg. In [28], the authors show an electrode of GF/Ni foam/$Co(OH)_2$ and electrolyte of 0.08MK3 $Fe(CN)_6$, which provides a specific capacitance of 7,514F/g at 16A/g. In [37], the authors show an electrode of V205Nano sheets and electrolyte of 1MKCL, which provides a specific energy of 635 at 1A/g. In [40], the authors show an electrode of CO_3O_4/polyindole and electrolyte of 1M KoH, which provides a specific energy of 1,805 at 2A/g. In [42], the authors show an electrode of $Ni_x Z_{n1-x}S$ and electrolyte of 3MKoH, which provides specific capacitance of 1,867F/gat 1A/g. In [44], the authors show an electrode of CNT-graphene films and electrolyte of $1MH_2So_4$, which provides specific capacitance of 140F/g at 0.1A/g.

In [45], the authors show an electrode of rGo/carbon-black and electrolyte of PVA/H_2So_4, which provides a specific capacitance of 79 at 1A/g. In [46], the authors show an electrode carbon-black pillared graphene and electrolyte of Nafion, which provides a specific capacitance of 118.5 at 0.1A/g. In [47], the authors show an electrode of functionalized rg0 film and electrolyte of Naflon, which provides a specific capacitance of 118.5F/g at 0.1A/g. In [49], the authors show an electrode of graphene/polyaniline (Pani) composite paper and electrolyte of $1MH_2SO_4$, which provides a specific capacitance of 233 at 2Mv/S. In [50], the authors show an electrode of graphene/Pani hybrid paper and electrolyte of $1MH_2SO_4$, which provides a specific capacitance of 489F/g at 0.4A/g. In [51], the authors show an electrode of NiO-graphene three-dimensional network and electrolyte of 3MkOH, which provides a specific capacitance of 816F/g at 5Mv/s. In [52], the authors show an electrode of Embossed rgo-MnO_2 hybrid films and electrolyte of $1MNo_2So_4$, which provide a specific capacitance of 389F/g at 1A/g, specific energy of 64Wh/kg, and power density of 25W/kg. In [53], the authors show an electrode of PPy/graphene and electrolyte of Kcl, which provides 237F/g at 0.01V/s. In [54], the authors show an electrode of functionalized graphene hydrogel and electrolyte of 1 MH_2So_4, which provides specific capacitance of 441F/g at 1A/g. In [55], the authors show an electrode of three-dimension N and B co-doped graphene hydrogel and electrolyte of $1MH_2So_4$, which provides a specific capacitance of 239 at 1mV/s. In [57], the authors show an electrode of graphene-Pani composite and electrolyte of $1MH_2S0_4$, which provides specific capacitance of 763F/g at 1A/g.

In [58], the authors show an electrode of three-dimensional MnO_2 composite network and an electrolyte of 0.5M Na_2So_4, which provides a specific capacitance of 465F/g at 2mV/S, and specific energy of 8.34Wh/kg and specific power of 94W/Kg. In [59], the authors show an electrode of C0-ALLDH/r60 films and electrolyte 1M Na_2So_4, which provides specific energy of 450F/g at 2mV/s. In [60], the authors show an electrode of graphene-PANI composite and electrolyte of 1M Na_2So_4, which provides a specific energy of 315F/g at 2mV/s. In [61], the authors show an electrode of graphene MnO_2 nanostructured sponges and electrolyte of 1M Na_2So_4, which provides a specific energy of 450F/g at 2mV/s. In [65, 66], the authors show an electrode of CuO2 and electrolyte of 1MKoH, which provided specific energy of 1,079F/g at 1.73A/g, which provide specific energy of 210.9F/g at 0.5A/g. In [67, 68], the authors show an electrode of Co_3O_4 and electrolyte of MKoH which provide specific energy of 410F/g at 0.7A/g. Certain other research is also available for different electrodes or with different electrolytes [16–18, 20–24, 26, 27, 29–36, 38, 39, 41, 43–45, 48, 56, 62–64].

15.6 MODELLING OF THE ULTRACAPACITOR

For detailed analysis of any system, there is required a common platform for engineers, as well as researchers. This platform is named modelling, the detailed mathematical analysis of a system. For the ultracapacitor system, modelling is mandatory to know about the internal characteristics of the system, which helps researchers to know about the dependency of parameters of an ultracapacitor (UC) towards many physical quantities. Hence, the same helps for accurate design prediction and development of control algorithms for UC-based systems. When the modelling comes to the ultracapacitor, it is broadly classified into two domains: frequency and time domain. The frequency/domain models help to achieve the internal electrochemical characteristics of the ultracapacitor in a large frequency range. However, the measurement apparatus (electrochemical impedance spectroscopy [EIS]) required for the frequency domain analysis is much costlier than the time domain analysis; also, the model is more focused on factors affecting UC such as temperature effect, lifecycle, and charge-discharge efficiency rather than real electrical characteristics of the UC. On another side, the time domain analysis models of UC have intelligibly explained the real electrical characteristics of UC [1–3].

15.6.1 Time Domain Modelling of the Ultracapacitor

The time domain modelling of the ultracapacitor basically classified into the following four categories.

1. Electrochemical models.
2. Equivalent circuit models.
3. Intelligent models.
4. Fractional-order models.

15.6.1.1 Electrochemical Models

Initial research in ultracapacitors has been done on the electrochemical models. The electrochemical model considers the real physical reaction that takes place inside the UC with the assistance of partial differential equations. The first model for the ultracapacitor was developed by Helmholtz (Figure 15.18). The author considers that all the charges of the ultracapacitor will be accumulated at the electrode surface, which is very much similar to the conventional capacitor model. Later, Gouy and Chapman incorporated the mobility of the ion due to the dissemination in the Helmholtz model with the aid of the Boltzmann distribution equation (Figure 15.19). Later, Stern combined the Helmholtz model and the Gouy-Chapman model and represented a model with two layers. One layer is of surface charge absorption, and the second is of the diffusion layer (Figure 15.20). The analytic part of the same has been done by combining the Poisson-Boltzmann equation. However, the model ignored the physical size of the ions and represented all the ions as point charges, which is valid only in low electric potential [69, 70].

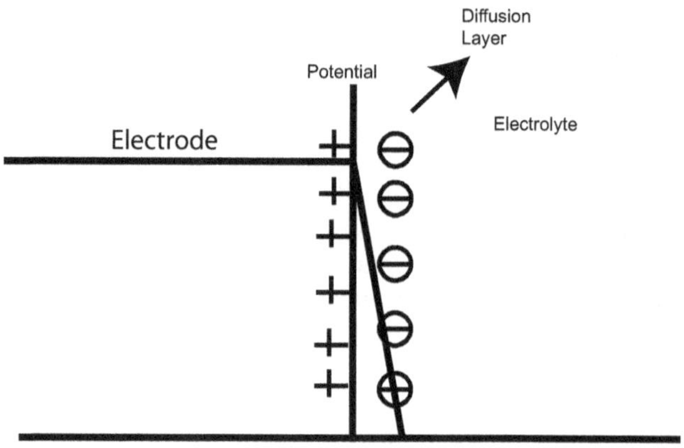

FIGURE 15.18 Helmholtz's ultracapacitor model [69, 70].

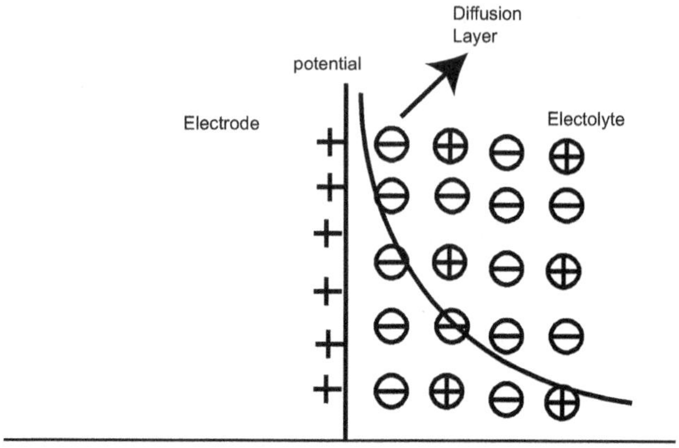

FIGURE 15.19 Gouy and Chapman' model [69, 70].

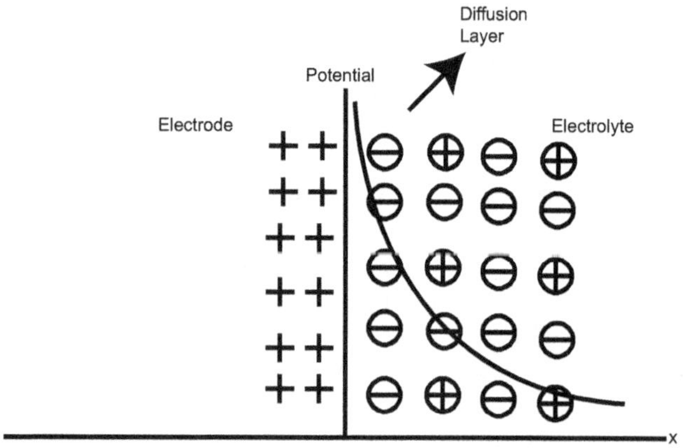

FIGURE 15.20 Stern's EDL model [69, 70].

Later, Bikerman modified the Poisson-Boltzmann equation to incorporate the different sizes of the ions. Still later, Verbrugge and Liu developed a model considering porous electrodes and dilute solutions theory, which is a one-dimensional model. Later, Wang and Pilon developed a three-dimensional model, in which the model included electrolyte dielectric permittivity and its dependency on the applied field, size of the ion, and electrode properties. However, the demerits of the electrochemical models are that most of the parameters are unable to be found in real time and the model is not focused much on the real electrical characteristics of the ultracapacitor [69, 70].

15.6.1.2 Equivalent Circuit Models

The UC models based on equivalent circuits are less complex and have relatively fewer parameters in number compared to all other models, which can be identified accurately for electrical vehicle applications. The classical equivalent circuit model for the UC, which consists of three elements such as capacitor (C), series resistance (Resr), and parallel resistance (Repr) as shown in Figure 15.21. The model parameters of the classical equivalent circuit model potentially can be determined from the manufacturer datasheet. However, for more accurate anticipation of the electrical behaviour of the UC the simple RC branch model is inadequate.

Literature [71, 72] shows UC models that altered the classical equivalent circuit model by excluding the Repr branch and accommodating one more branch of the variable capacitor, for the indication of dependency of the capacitance with cell voltage as shown in Figure 15.22. The variable capacitor value is determined from the look-up table and contains current, temperature, state of charge, and lifecycle as data. However, for the look-up table parameters, there should be peculiar measurement apparatus with high accuracy. Also, the model fails to incorporate the internal porous structure characteristics of the ultracapacitor.

$$C_{tot} = C_0 + C_v v \tag{15.6}$$

where tot is the total capacitance of the cell, C is the variable capacitance in farad/voltage, and V is the voltage

In literature [73], UC is modelled as a ladder circuit, as shown in Figure 15.23. The parameters of ladder circuits are

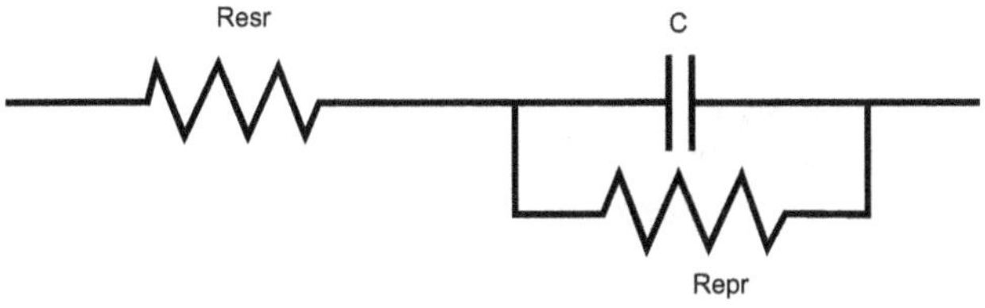

FIGURE 15.21 Simple model of ultracapacitor [72].

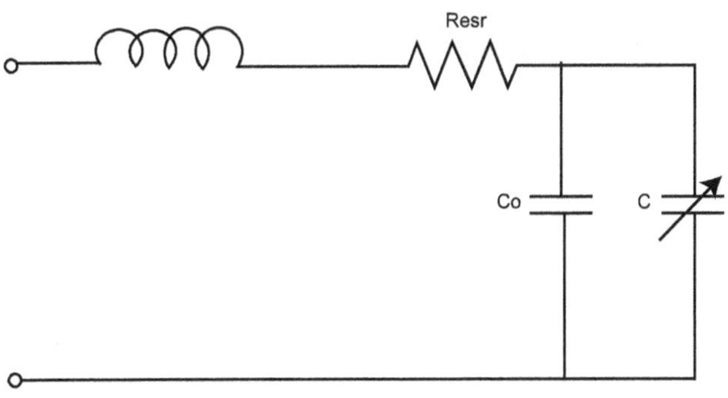

FIGURE 15.22 Equivalent circuit model of UC with additional variable capacitance branch [72].

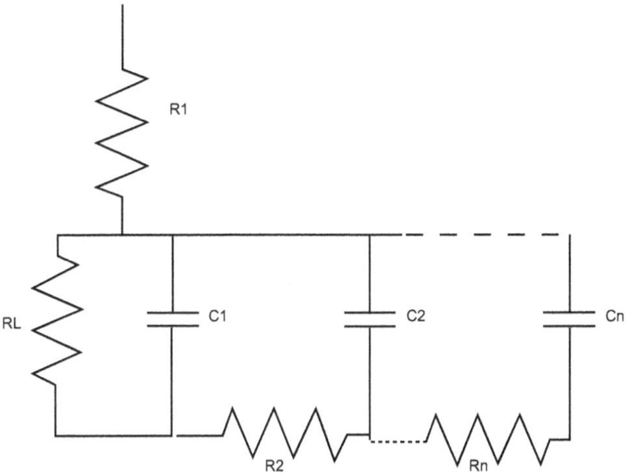

FIGURE 15.23 Ladder circuit for UC [73].

extracted by AC impedance measurements using a computer development programme formed at the University of Twente [74]. The result obtained through the ladder network further compares with the conventional classical model of the UC, and it was found that the classical model of the UC has better anticipation of the UC voltage under slow discharge [74]. Further modification had been done on the ladder circuit by adding a greater number of the RC branch, for the better prediction of the UC characteristics. However, the ladder network is preferred less than the UC model, because of the complex parameter identification methodology and unavailable to identify the number of RC branches required for the accurate prediction of characteristics of the UC.

The transmission line model has further developed for the UC, as shown in Figure 15.24. The main motto to represent the UC model as a transmission line circuit is the porous structure of the electrode material of the UC. The porous structure of the electrode makes a nonlinear relationship in parameters of the UC; hence, resistance and capacitance should be represented as lumped parameters. The parameter identification of the transmission line model is the same as the AC impedance characterization of the ladder network. However, the transmission line model anticipated the behaviour of the UC more efficiently, because of the inclusion of inductance as a parameter in the model and also taking into consideration the current path for the capacitors. However, for the application of UC in EV, the model of UC should be less in branch numbers. The main motto for the reduction of branch parameters of the UC model is because EV UC works as a peak power source; hence, the model should be describing the characteristics of UC for short periods [72].

The Zubieta model developed in early 2000 is considered most desirable for explaining the behaviour of the UC for application in the field of power electronics [7]. The model consists of three branches, as shown in Figure 15.25. The model parameters for each branch have been selected in such a way that each branch has a distinct time constant; hence, charge will accumulate in

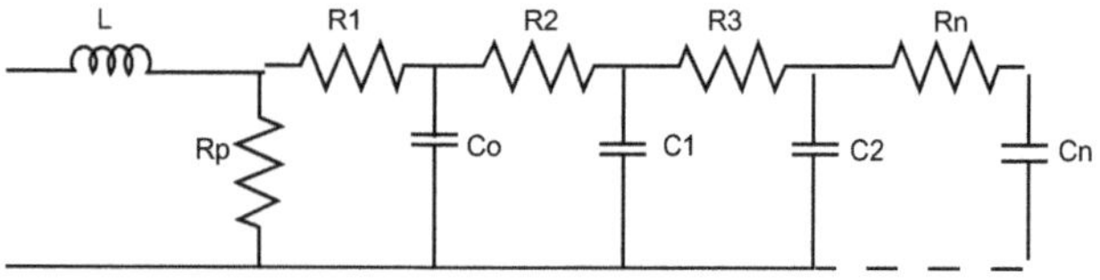

FIGURE 15.24 Transmission circuit for ultracapacitor [72].

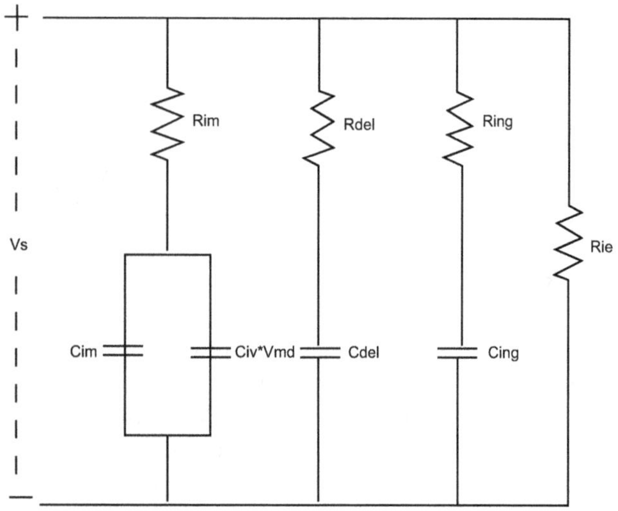

FIGURE 15.25 Three-branch model of the ultracapacitor [7].

the first branch initially during the charging of UC, and later, charge distribution occurs to the second and third branches of the model accordingly.

The parameter identification method Zubieta model is experimental and more complex. The certain vague regarding the chosen time delay of 20ms and voltage of 50mV, which has been used throughout the parameter identification of the model, always leading to certain disarray. Later in [75], the authors modified the methodology more conveniently and in a comparatively easy way for identification of the three-leg model of ultracapacitor by the Zubieta, in which the authors utilized slop changing point of charging time curve as well as ideal time curve for finding the immediate, delayed, and long branch parameters, respectively. Additionally, the author has incorporated the initial state of charge (SoC) of the ultracapacitor as a function of equivalent series resistance in the model parameter methodology; hence, the methodology proposed in [75] demonstrated an excellent internal characteristic of the ultracapacitor for power electronics applications.

15.6.1.3 Intelligent Models

Intelligent models have been utilized for accurate prediction of the demonstration of the ultracapacitor, which mostly consider many parameters for the assessment of accomplishment of the system such as ageing, temperature variations, lifecycle, energy density, power density, ESR, cost, and accuracy in one model, and predict the output, which is not possible in the electrochemical or equivalent circuit models. With the aid of intelligent models such as artificial neural networks, fuzzy logic, the absolute nonlinear porous structure characteristics of the ultracapacitor can be demonstrated.

In [75], the authors developed an artificial neural network–based model which evaluates the effects of various integral characteristics such as specific power, specific energy, SoC, etc., on the ultracapacitor performance, and in which various parameters like surface arrangement, size of the crystal, specific power, and specific energy have been utilized as the input to the system. Each input to the model will have a weight and depend on the propagation of the input from one layer to another layer, and may result in increase or decrease in magnitude. The summation of the input with weighting, along with the history, will propagate to the second layer, and the same procedure will occur in the second layer, as shown in Figure 15.26. In [76], the model in which the rate of current, temperature, the chemistry of the ultracapacitor electrode, and electrolyte and history are considered for the identification of the parameter identification of the model. However, the demerit is the need a for adequate proper training of the elements. Also, the efficiency of the model has a huge dependency on the objective functions, boundary conditions, etc. [77].

15.6.1.4 Fractional-Order Models

A fractional order–based model has been introduced to improve accuracy by introducing calculus of fractional orders for the modelling of the ultracapacitor. The model comprises series, parallel resistance, constant phase element, and Walburg element [78], as shown in Figure 15.27. The parameters of the fractional-order model have been identified with the aid of impedance data. In [79], the authors developed a model based on frequency analysis. In [80], the authors developed the model with the time domain data by a constant current charging. In [81], the authors developed a model by step voltage response. However, all these models have only shown an excellent performance in the laboratory. In real-time conditions with varying load, model accuracy has many variations from the laboratory results.

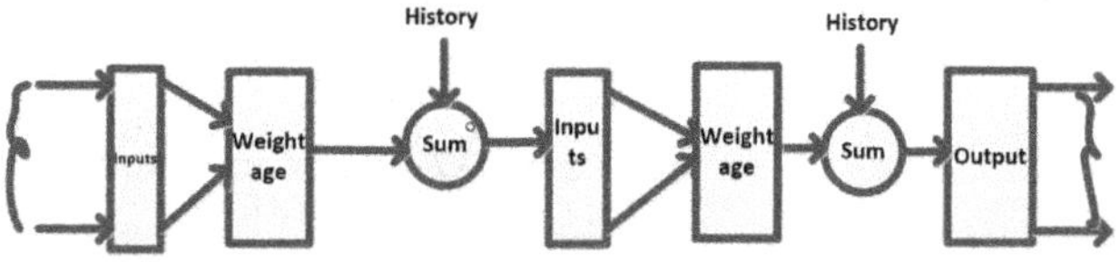

FIGURE 15.26 Intelligent model for ultracapacitor [75].

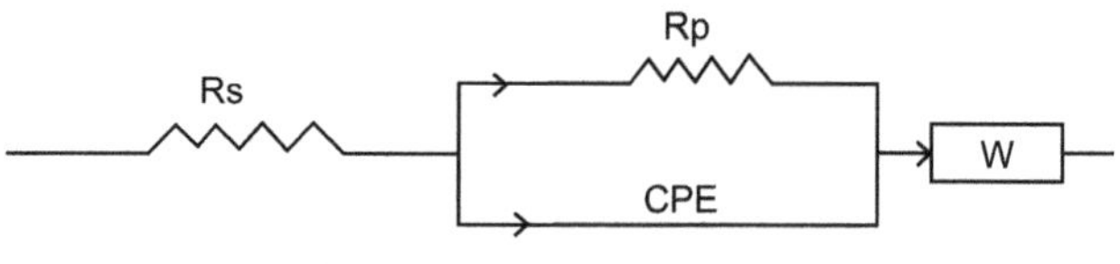

FIGURE 15.27 Fraction order model [78].

15.6.2 Comparison of the Models

TABLE 15.2
Comparison of the Ultracapacitor Models

Model	Merits	Demerits
Electrochemical Michael model	Discusses real physical reaction taking place inside the ultracapacitor	Many parameters are unable to be found in real time
Equivalent circuit model	Describes the electrical behaviour of the ultracapacitor; apt for the power electronics applications.	Absence of physical and ageing considerations in the model.
Intelligent model	Well, and advanced; describes the dependency of many parameters of the ultracapacitor	The efficiency of the model depends on the training of the neurons; adequate training is mandatory for the appropriate functioning of the algorithm
Fractional-order model	Well and accurate determination of parameter using fractional-order equations	Sensitive to the variable load; most of the time, nonlaboratory results show a large variation in comparison with the laboratory results

15.6.3 Thermal Modelling

Thermal analysis on the ultracapacitor is crucial because mostly ultracapacitor operates at high-rate cycling. Hence, adequate heat will be generated in the ultracapacitor cell. Therefore, an appropriate cooling mechanism is mandatory. For the design of the cooling mechanism, there needs to be an accurate prediction of the thermal behaviour of the ultracapacitor at the cell level which helps to improve the life—as well as the efficiency—of the ultracapacitor. Heat generation in the ultracapacitor cell occurs due to the altering of entropy due to ion movement [82]. In the literature, it has been found that the temperature model of the supercapacitor can be broadly classified into two enactments. The first is based on principle models, and the second is on a comprehensive model. Principle models represent the thermodynamic behaviour of the ultracapacitor with the aid of partial differential equations. In [83], the authors developed a heat equation for ultracapacitors by utilizing the finite differential methodology for the understanding of the temperature distribution in concern with the position and time. In [84], the authors determined a methodology for the identification of the central temperature area by utilizing a three-dimensional finite element thermal model.

In the comprehensive model [85, 86], the authors developed an electro-mechanical model in which an electrical model has been coupled with the mechanical model. Those models provide better anticipation of the thermal behaviour of the ultracapacitor in real time. The comprehensive model considers the geometry of the ultracapacitor cell, thermal behaviour of the electrode and electrolyte of the ultracapacitor, etc. [86].

15.7 SIZING OF THE ULTRACAPACITOR MODULE

The sizing of the ultracapacitor is a vital part of an ultracapacitor-based system, and has a huge impact on the cost and efficiency of the system. Sizing of the ultracapacitor implies the determination of the required number of parallel and series capacitors in an ultracapacitor module for the proper functioning of the system. To size, the ultracapacitor there required information about certain parameters as shown in what follows.

1. Maximum operating voltage of the ultracapacitor (V_{max}).
2. Minimum operating voltage of the ultracapacitor (V_{min}). The V_{min} should be the half the value of the V_{max}, because 75% of the energy in the UC will be utilized within the voltage range of $V_{max}/2$.
3. Working voltage (V_W) is the difference between V_{max} and V_{min}.
4. Time duration (T_d) for the discharge of the ultracapacitor from V_{max} to V_{min}.
5. Capacitance of the single capacitor cell (C_{single}).

As shown in Figure 15.28, two factors will influence the performance of the ultracapacitor discharge profile: the first is a resistive component, and the second is the capacitive component, where the resistive component and the capacitive components imply the voltage change in the ultracapacitor due to the ESR and the change in the energy.

$$I = c\,\frac{dV}{dt} \tag{15.7}$$

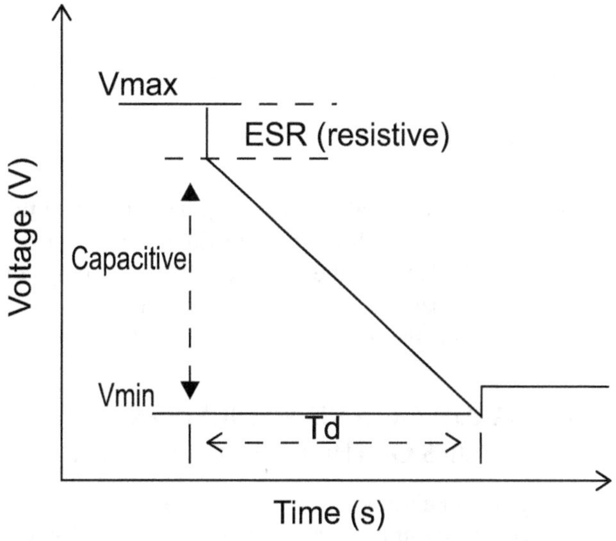

FIGURE 15.28 Discharge profile of the ultracapacitor.

The voltage changes due to the capacitive components can be represented as mention in Eq. (15.8).

$$V = IR \tag{15.8}$$

where I is the average current, c is the capacitance, dv is the change in voltage, and dt is the total duration of time

The voltage changes due to the resistive components can be represented as mentioned in Eq. (15.9).

$$dv = I\frac{dt}{C} + IR \tag{15.9}$$

where V is the voltage and R is the ESR

Total charge-discharge characteristics of the ultracapacitor include both the components as shown in Eq. (15.10).

$$I_{avg} = \frac{I_{max} + I_{min}}{2} \tag{15.10}$$

where I_{avg} is the average current

$$I_{max} = \frac{Power}{V_{min}}, I_{min} = \frac{Power}{V_{max}} \tag{15.11}$$

where $I_{max}, I_{min}, V_{max}, V_{min}$ are maximum and minimum voltage and current, respectively

$$C = \frac{C_{rs} * n_{cpar}}{n_{cser}} \tag{15.12}$$

where C, C_{rs}, n_{cpar}, n_{cser} are capacitance of ultracapacitor module, capacitance of single ultracapacitor single, number of parallel capacitors and series capacitors, respectively

$$n_{cser} = \frac{Rated\ voltage\ of\ the\ UC\ bank\ (V_{max})}{Rated\ voltage\ of\ single\ cell\ (V_{rs})} \tag{15.13}$$

$$R = \frac{R_{rs} * n_{cser}}{n_{cpar}} \tag{15.14}$$

where R, R_{rs}, n_{cpar}, n_{cser} are total resistance, resistance of the single cell, number of capacitances in parallel and series, respectively

Consider a random value of the n_{cpar}, then substitute the equation from Eqs. (15.10)–(15.14) in Eq. (15.9). After the substitution, the value of the dv should be in the range of $V_{max} - V_{min}$. If dv has a large variation, then change the value of n_{cpar} and repeat the step until dv matches [9].

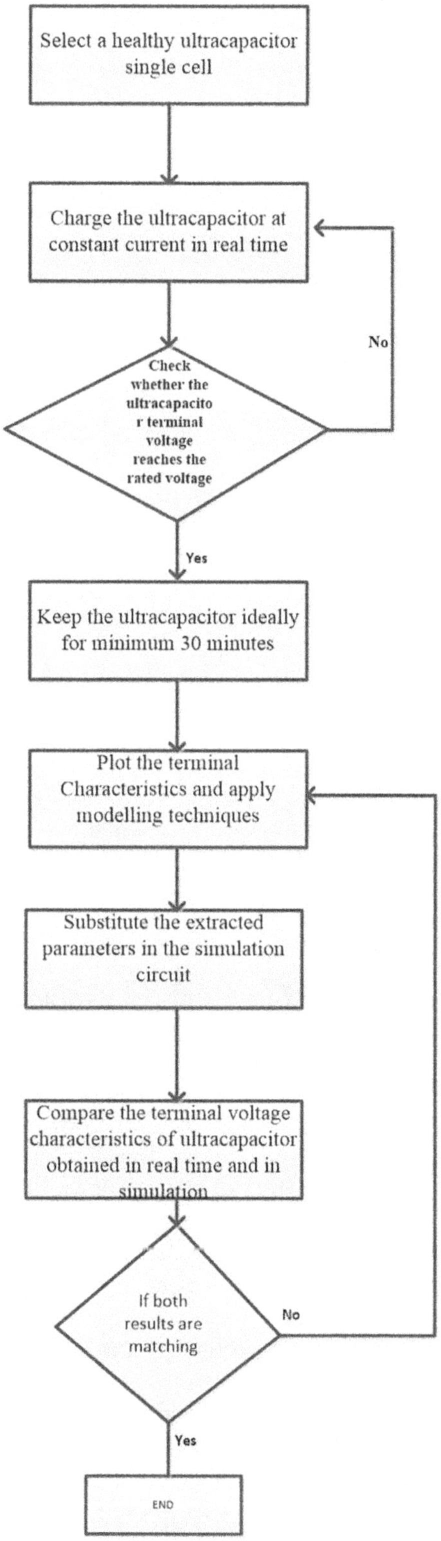

FIGURE 15.29 Flowchart representation of charging procedure and validation of ultracapacitor model.

15.8 HARDWARE AND SIMULATION STUDIES OF THE ULTRACAPACITOR

This section will discuss the detailed studies of ultracapacitor characteristics in hardware and software. Initially, the ultracapacitor should be tested in the hardware with constant current and obtain the terminal voltage characteristics of the ultracapacitor. Then extract the branch parameters with aid of mathematical model techniques. Simulate the model and compare the results obtained in the software as well as hardware. If the results match, the applied modelling technique is appropriate; otherwise, retest the model in

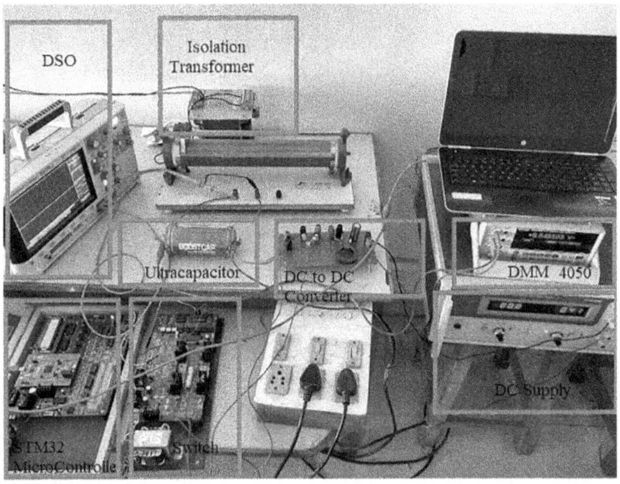

FIGURE 15.30 Real-time test bench.

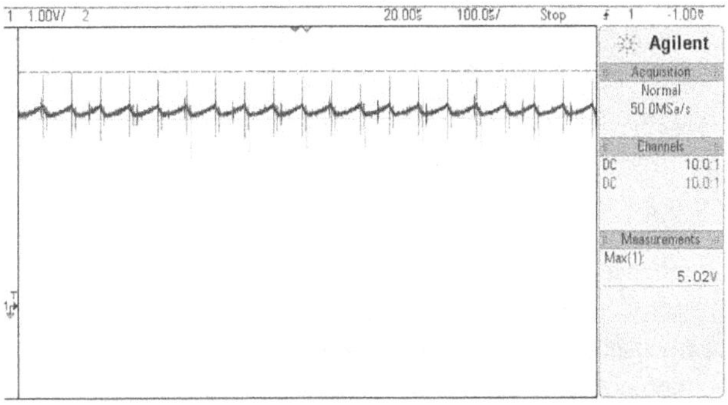

FIGURE 15.31 Charging current for ultracapacitor.

real time or modify the modelling techniques and redo the procedure. Figure 15.29 shows the flowchart representation of this.

15.8.1 Hardware Studies of Ultracapacitors

The UC equivalent circuit parameters are found using real-time charging of the UC at constant current, as shown in Figure 15.30 and Figure 15.31. The test bench consists of Maxwell UCAP 3000 UC, which has capacity of 3000F and rated voltage of 2.7V. A Tektronix DMM 4050 digital multimeter was used to store the measured voltage data from the UC. DC/DC converter provided constant current to charge the UC. A STM32-M4 microcontroller was used to generate required gate pulse to the DC/DC converter.

Once the terminal voltage characteristics of the ultracapacitor were obtained, the energy-based parameter identification methodology proposed in [75] was utilized for the derivation of the ultracapacitor three-leg model parameters (Table 15.3). In [75], the authors divided the obtained charging curve in real time into multiple areas and utilized the slope changing point of the curve to identify the three-leg model parameters of the ultracapacitor.

Figure 15.33 shows the simulation circuit for the three-branch circuit model. Simulation studies have been carried with the aid of MATLAB/Simulink. The parameters of the branch model have been identified with the aid of the energy-based parameter identification methodology introduced in [75]. The charging and discharging have occurred with a 5A current. 5A implies charging and –5A implies discharging of the ultracapacitor.

TABLE 15.3
Derived Parameters of the Three-Leg Ultracapacitor Model Utilizing Energy-Based Parameter Identification Methodology

Parameters	Rim	Cim	Civ	Rmd	Cmd	Rlong	Clong
Value	72mΩ	690F	353F/V	3.41Ω	200F	16Ω	1065F

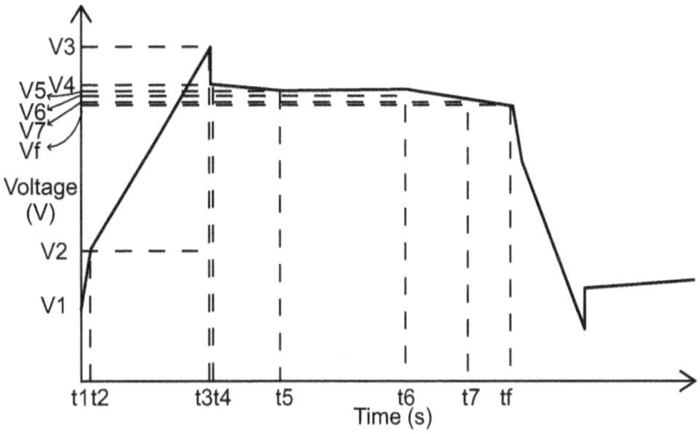

FIGURE 15.32 Simulation studies on ultracapacitor.

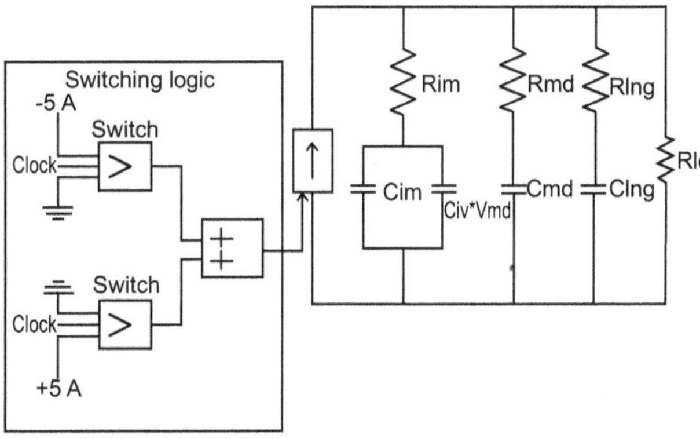

FIGURE 15.33 Simulation circuit for three-branch ultracapacitor model.

Figure 15.34 shows the terminal voltage of the ultracapacitor obtained in the simulation. The ultracapacitor has been kept ideal for 30 minutes for proper transfer of the charges from the immediate branch to the long branch via delayed branch [75].

Figure 15.35 shows the ultracapacitor terminal voltage characteristics in the hardware, as well as software. The figure shows that there has an excellent match between hardware well as software results.

15.9 APPLICATION OF THE ULTRACAPACITOR

The application of the ultracapacitor is immense and broad due to its inherent properties like high power density, large lifecycle, high charge-discharge efficiency, high temperature

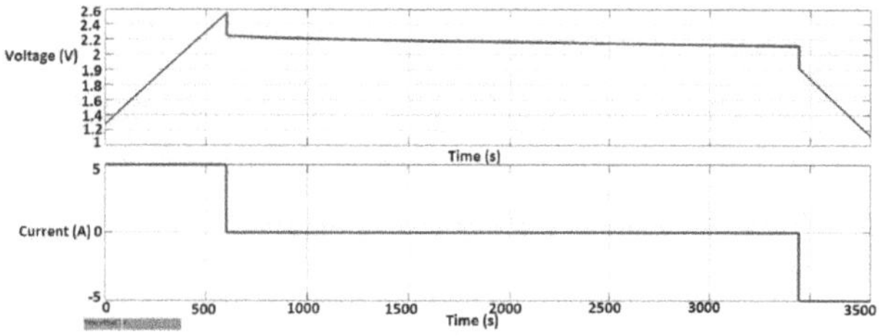

FIGURE 15.34 Terminal voltage and charging current profile of the ultracapacitor in simulation.

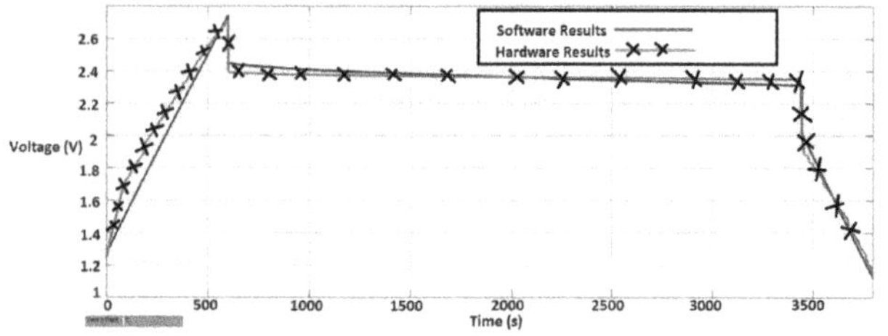

FIGURE 15.35 Terminal voltage characteristics of the ultracapacitor in simulation and hardware.

withstand capabilities, low ESR, etc. The major application field of the ultracapacitor is the uninterrupted power supply (UPS), electrical vehicle, power electronics, renewable integration, power quality applications, etc. A detailed explanation of the same appears in Figure 15.36.

15.9.1 Electrical Vehicles

Ultracapacitors have keen attention in the domain of the electrical vehicle as an auxiliary device for supporting the battery. The vehicle driving power is composed of average power and peak power. The peak power will be of a short duration of time; hence, the design cost of the battery for the same will be huge. Also, it will make the battery life redundant. On the other side, ultracapacitors provide prominent energy storage in the peak power requirement area because of its predominant inherent characteristics of high power density. Hence, in the hybrid source arrangement, the ultracapacitor will supply the required peak power, which will enhance battery life. The Chinese company Sunwin has made an electric bus that has run completely on UC as the primary source. The Sunwin bus with the UC has a mileage of 3–6km and able to charge adequately at each bus stop. Fifty percent of the charge can fill in 30 seconds and 100% can fill within 80 seconds [87].

15.9.2 Power Electronics

In the field of power electronics, in DC/DC converter or DC/AC converter, the DC link capacitance can be replaced by a ultracapacitor [88].

15.9.3 Renewable Energy Integration

Interruption of power will occur during integration of renewable energy sources in the power grid; hence, the ultracapacitor

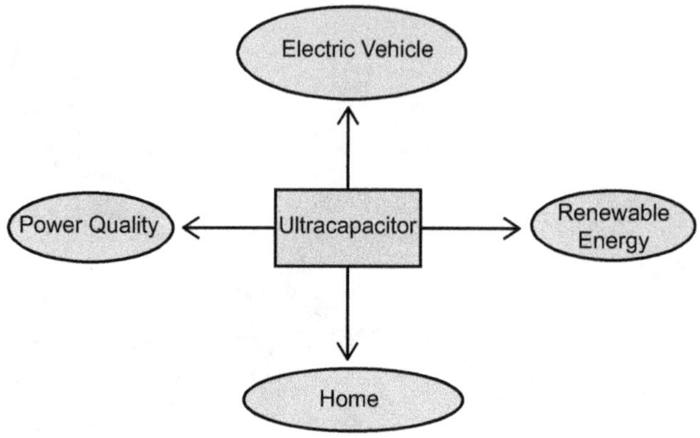

FIGURE 15.36 Application of the ultracapacitor.

can be utilized as a peak power source during the same duration. In [89], the authors have utilized the ultracapacitor as a hybrid energy source—along with the wind turbine—to extenuate the effect of power intermittent. In [90], the authors utilized an ultracapacitor as a clamping capacitance in a three-level inverter in a wind energy changeover arrangement.

15.9.4 Power Quality Applications

Ultracapacitors can be utilized as an energy storage device in the active filters, dynamic voltage restorer, static synchronous compensators, etc., for the mitigation of harmonic current, voltage sag, and voltage swell, since the application is for short periods and the ultracapacitor is apt for the same [91].

15.9.5 Uninterrupted Power Supply (UPS)

Generally, the UPS battery will be utilized as an energy storage system. However, the battery has many disadvantages like lower power density, lower lifecycle, etc.; hence, the ultracapacitor can be utilized as an auxiliary device to support the battery in the UPS system in which ultracapacitor can act as a peak power supply system.

15.10 CONCLUSION

The ultracapacitor has a huge role in the power system and power electronics applications, especially during renewable integration, power quality, electrical vehicles, etc. Since the battery suffers from many limitations such as lower power density, lower lifecycle, higher ESR, etc., studies have proven that ultracapacitors can act either as auxiliary devices to support batteries or as primary sources, depending on the requirements; hence, a detailed study about the ultracapacitor is mandatory. This work has concentrated on doing a comprehensive analysis of the ultracapacitor and its applications. The chapter explained different types of ultracapacitors, different variants of the materials for the ultracapacitor, detailed modelling, and analysis of the ultracapacitor, in-depth sizing of the ultracapacitor module, real-time and software testing of the ultracapacitor, and detailed analysis of the specifications of the ultracapacitors available in the market. Studies have shown that various ultracapacitors of voltage range up to 5.2V are available in the market. However, the ultracapacitor voltage range from 2.5–3V has been widely utilizing for the application. Also, specific energies from 4–7 Wh/kg and specific power from 4–27kW/kg with a lifecycle of 10,00,000 are available. Much research is consistently happening with different electrodes and electrolytes to improve the energy density of the ultracapacitor; hence, in the future, a complete replacement of batteries with ultracapacitors with huge energy density, huge power density, and long lifecycles may be possible.

15.11 NUMERICAL PROBLEMS

15.11.1 Solved Question

Question 1: An ultracapacitor is designed to supply peak power of 10kW for ten seconds into the grid. The selected ultracapacitor bank is of 48V. Calculate the number of parallel and series ultracapacitor cell required. The voltage and capacitance rating of a single ultracapacitor cell is 2.7V and 3,000F. If the cost of the single cell is 500Rs., calculate the approximate cost of the ultracapacitor module.

Solution:

Maximum voltage of the ultracapacitor (Vmax) = 48V

Minimum voltage of the ultracapacitor = Vmax/2 = 24V

Maximum current (Imax) = $\frac{Peak\ Power}{Vmin} = \frac{10000}{24}$ = ~420 A

Minimum current (Imin)= $\frac{Peak\ Power}{Vmax} = \frac{10000}{48}$ = ~240 A

Average current (I) = $\frac{Imax + Imin}{2} = \frac{240 + 420}{2} = 330\,\text{A}$

Time duration (dt) = 10 seconds

Number of ultracapacitor cells in series (nCser) $= \frac{Rated\ Voltage}{Volatge\ of\ a\ single\ cell} = \frac{48}{2.7} = \sim 18$

Assume number of parallel capacitor cell (nCpar) = 1

Rated capacitance of single cell (csing) = 3,000F

Total capacitance (Ctot) = $\frac{csing * nCpar}{nCser} = \frac{3000*1}{18}$ = ~167 farad

Resistance of the single cell (Rsing) = 3 mΩ

Total resistance of the ultracapacitor module (Rmod) $= \frac{Rsing * nCser}{nCpar} = \frac{3\ m\Omega * 18}{1} = 54 m\Omega$

$dv = I\frac{dt}{C} + IR. = 330*\frac{10}{167} + 33 * 54\ m\Omega = \sim 22\ \text{V}$

Hence, the calculated dv is approximately equal to Vmax–Vmin, which is 24 V.

Hence, the ultracapacitor module with 18 cells in series and one cell in parallel will be appropriate for this application.

Total cost of the module = ~ Cost of single cell × number of cells in series × number of cells in parallel.

= 500 × 18 × 1 = ~9000 Rs.

15.11.2 Unsolved Question

An ultracapacitor has been designed to perform as a peak power source during renewable integration. The ultracapacitor has been designed to supply a peak power of a maximum of 10kW for 10 seconds. The number of series cells and parallel cells are 112 and one, respectively, for the ultracapacitor module, and the total capacitance of the ultracapacitor module is 167F. Find out the rated capacitance of the single ultracapacitor cell. Also, calculate the rated voltage value of the ultracapacitor module (rated voltage of the single ultracapacitor cell is 2.7V).

Answer

1. Rated capacitance of single cell = 3,000 F
2. Rated voltage value of the ultracapacitor module = 300V.

REFERENCES

[1] A. Burke, "Ultracapacitors: why, how, and where is the technology", *Journal of Power Sources*, vol. 91, no. 1, pp. 37–50, 2000. Doi:10.1016/s0378-7753(00)00485-7

[2] K. Sharma Poonam, A. Arora and S. Tripathi, "Review of supercapacitors: materials and devices", *Journal of Energy Storage*, vol. 21, pp. 801–825, 2019. Doi:10.1016/j.est.2019.01.010

[3] X. Hu Zhang, Z. Wang, F. Sun and D. Dorrell, "A review of supercapacitor modelling, estimation, and applications: a control/management perspective", *Renewable and Sustainable Energy Reviews*, vol. 81, pp. 1868–1878, 2018. Doi:10.1016/j.rser.2017.05.283

[4] Y. Kumar, S. Rawal, B. Joshi and S. Hashmi, "Background, fundamental understanding and progress in electrochemical capacitors", *Journal of Solid State Electrochemistry*, vol. 23, no. 3, pp. 667–692, 2019. Doi:10.1007/s10008-018-4160-3

[5] W. Schmickler, "Double layer theory", *Journal of Solid State Electrochemistry*, vol. 24, no. 9, pp. 2175–2176, 2020. Doi:10.1007/s10008-020-04597-z

[6] A. Celzard, F. Collas, J. Marêché, G. Furdin and I. Rey, "Porous electrodes-based double-layer supercapacitors: pore structure versus series resistance", *Journal of Power Sources*, vol. 108, no. 1–2, pp. 153–162, 2002. Doi:10.1016/s0378-7753(02)00030-7

[7] L. Zubieta and R. Bonert, "Characterization of double-layer capacitors for power electronics applications", *IEEE Transactions on Industry Applications*, vol. 36, no. 1, pp. 199–205, 2000. Doi:10.1109/28.821816

[8] A. Burke, "Energy Storage energy storage: ultracapacitor ultracapacitor", *Transportation Technologies for Sustainability*, pp. 494–524, 2013. Doi:10.1007/978-1-4614-5844-9_809

[9] Maxwell.com, 2021 [Online]. Available: www.maxwell.com/images/documents/1007239-EN_test_procedures_technote.pdf

[10] "Ultracapacitor—IOXUS", *Ioxus.com*, 2021 [Online]. Available: https://ioxus.com/product-tag/ultracapacitor/?orderby=rating&per_page=30

[11] 2021 [Online]. Available: www.digikey.com/en/product-highlight/s/skeleton-technologies/skelcap-ultracapacitor

[12] M. Abdel Maksoud et al., "Advanced materials and technologies for supercapacitors used in energy conversion and

storage: a review", *Environmental Chemistry Letters*, vol. 19, no. 1, pp. 375–439, 2020. Doi:10.1007/s10311-020-01075-w

[13] X. Lang, A. Hirata, T. Fujita and M. Chen, "Nanoporous metal/oxide hybrid electrodes for electrochemical supercapacitors", *Nature Nanotechnology*, vol. 6, no. 4, pp. 232–236, 2011. Doi:10.1038/nnano.2011.13

[14] G. Zhang, H. Wu, H. Hoster, M. Chan-Park and X. Lou, "Single-crystalline NiCo2O4 nanoneedle arrays grown on conductive substrates as binder-free electrodes for high-performance supercapacitors", *Energy & Environmental Science*, vol. 5, no. 11, p. 9453, 2012. Doi:10.1039/c2ee22572g

[15] Z. Yu and J. Thomas, "Energy storing electrical cables: integrating energy storage and electrical conduction", *Advanced Materials*, vol. 26, no. 25, pp. 4279–4285, 2014. Doi:10.1002/adma.201400440

[16] Y. Wang, X. Zhang, C. Guo, Y. Zhao, C. Xu and H. Li, "Controllable synthesis of 3D NiχCo1−χ oxides with different morphologies for high-capacity supercapacitors", *Journal of Materials Chemistry A*, vol. 1, no. 42, p. 13290, 2013. Doi:10.1039/c3ta12713c

[17] C. Zhou, Y. Zhang, Y. Li and J. Liu, "Construction of high-capacitance 3D CoO@Polypyrrole nanowire array electrode for aqueous asymmetric supercapacitor", *Nano Letters*, vol. 13, no. 5, pp. 2078–2085, 2013. Doi:10.1021/nl400378j

[18] M. Deng et al., "Fabrication of Mn/Mn oxide core—shell electrodes with three-dimensionally ordered macroporous structures for high-capacitance supercapacitors", *Energy & Environmental Science*, vol. 6, no. 7, p. 2178, 2013. Doi:10.1039/c3ee40598b

[19] F. Du, D. Yu, L. Dai, S. Ganguli, V. Varshney and A. Roy, "Preparation of tunable 3D pillared carbon nanotube—graphene networks for high-performance capacitance", *Chemistry of Materials*, vol. 23, no. 21, pp. 4810–4816, 2011. Doi:10.1021/cm2021214

[20] X. Ma, J. Liu, C. Liang, X. Gong and R. Che, "A facile phase transformation method for the preparation of 3D flower-like β-Ni(OH)2/GO/CNTs composite with excellent supercapacitor performance", *Journal of Materials Chemistry A*, vol. 2, no. 32, pp. 12692–12696, 2014. Doi:10.1039/c4ta02221a

[21] S. Wang, S. Shumin, S. Li, F. Gong, Y. Li, Q. Wu, P. Song, S. Fang and P. Wang, "Time and temperature dependent multiple hierarchical $NiCo_2O_4$ for high-performance supercapacitors", *Dalton Transaction of Royal Society of Chemistry,* no. 45, pp. 7469-7475, 2016.

[22] W. Du, Z. Wang, Z. Zhu, S. Hu, X. Zhu, Y. Shi, H. Pang and X. Qian, "Facile synthesis and superior electrochemical performances of CoNi2S4/graphene nanocomposite suitable for supercapacitor electrodes", *Journal of Materials Chemistry A*, vol. 2, no. 25, pp. 9613–9619, 2014. Doi:10.1039/c4ta00414k

[23] G. Xiong, P. He, L. Liu, T. Chen and T. Fisher, "Plasma-grown graphene petals templating Ni—Co—Mn hydroxide nanoneedles for high-rate and long-cycle-life pseudocapacitive electrodes", *Journal of Materials Chemistry A*, vol. 3, no. 45, pp. 22940–22948, 2015. Doi:10.1039/c5ta05441a

[24] H. Chen, S. Zhou and L. Wu, "Porous Nickel Hydroxide—Manganese Dioxide-Reduced Graphene Oxide Ternary Hybrid Spheres as Excellent Supercapacitor Electrode Materials", *ACS Applied Materials & Interfaces*, vol. 6, no. 11, pp. 8621–8630, 2014. Doi:10.1021/am5014375

[25] S. Min, C. Zhao, Z. Zhang, G. Chen, X. Qian and Z. Guo, "Synthesis of Ni(OH)2/RGO pseudocomposite on nickel foam for supercapacitors with superior performance", *Journal of Materials Chemistry A*, vol. 3, no. 7, pp. 3641–3650, 2015. Doi:10.1039/c4ta06233g

[26] S. Min, C. Zhao, Z. Zhang, G. Chen, X. Qian and Z. Guo, "Synthesis of Ni(OH)2/RGO pseudocomposite on nickel foam for supercapacitors with superior performance", *Journal of Materials Chemistry A*, vol. 3, no. 7, pp. 3641–3650, 2015. Doi:10.1039/c4ta06233g

[27] J. Cai, H. Niu, Z. Li, Y. Du, P. Cizek, Z. Xie, H. Xiong and T. Lin, "High-performance supercapacitor electrode materials from cellulose-derived carbon nanofibers", *ACS Applied Materials & Interfaces*, vol. 7, no. 27, pp. 14946–14953, 2015. Doi:10.1021/acsami.5b03757

[28] L. Chen, Y. Hou, J. Kang, A. Hirata and M. Chen, "Asymmetric metal oxide pseudocapacitors advanced by three-dimensional nanoporous metal electrodes", *Journal of Materials Chemistry A*, vol. 2, no. 22, p. 8448, 2014. Doi:10.1039/c4ta00965g

[29] M. Jana et al., "Growth of Ni—Co binary hydroxide on a reduced graphene oxide surface by a successive ionic layer adsorption and reaction (SILAR) method for high performance asymmetric supercapacitor electrodes", *Journal of Materials Chemistry A*, vol. 4, no. 6, pp. 2188–2197, 2016. Doi:10.1039/c5ta10297a

[30] Luo et al., "Three-dimensional graphene foam supported Fe3O4 lithium battery anodes with long cycle life and high rate capability", *Nano Letters*, vol. 13, no. 12, pp. 6136–6143, 2013. Doi:10.1021/nl403461n

[31] X. Sun et al., "Facile synthesis of Co3O4 with different morphologies loaded on amine modified graphene and their application in supercapacitors", *Journal of Alloys and Compounds*, vol. 685, pp. 507–517, 2016. Doi:10.1016/j.jallcom.2016.05.282

[32] M. Li, J. Cheng, J. Wang, F. Liu and X. Zhang, "The growth of nickel-manganese and cobalt-manganese layered double hydroxides on reduced graphene oxide for supercapacitor", *Electrochimica Acta*, vol. 206, pp. 108–115, 2016. Doi:10.1016/j.electacta.2016.04.084

[33] Xu, C. Yang, Y. Xue, C. Wang, J. Cao and Z. Chen, "Facile synthesis of novel metal-organic nickel hydroxide nanorods for high performance supercapacitor", *Electrochimica Acta*, vol. 211, pp. 595–602, 2016. Doi:10.1016/j.electacta.2016.06.090

[34] X. Mu et al., "A high energy density asymmetric supercapacitor from ultrathin manganese molybdate nanosheets", *Electrochimica Acta*, vol. 211, pp. 217–224, 2016. Doi:10.1016/j.electacta.2016.06.072

[35] X. Liu, G. Du, J. Zhu, Z. Zeng and X. Zhu, "NiO/LaNiO3 film electrode with binder-free for high performance supercapacitor", *Applied Surface Science*, vol. 384, pp. 92–98, 2016. Doi:10.1016/j.apsusc.2016.05.005

[36] Y. Liu, N. Wang, C. Yang and W. Hu, "Sol—gel synthesis of nanoporous NiCo2O4 thin films on ITO glass as high-performance supercapacitor electrodes", *Ceramics International*, vol. 42, no. 9, pp. 11411–11416, 2016. Doi:10.1016/j.ceramint.2016.04.071

[37] Y. Chang et al., "Hierarchical Ni 3 S 2 nanosheets coated on mesoporous NiCo 2 O 4 nanoneedle arrays as high-performance electrode for supercapacitor", *Materials Letters*, vol. 176, pp. 274–277, 2016. Doi:10.1016/j.matlet.2016.04.059

[38] M. Pal, R. Rakshit, A. Singh and K. Mandal, "Ultra high supercapacitance of ultra small Co3O4 nanocubes", *Energy*, vol. 103, pp. 481–486, 2016. Doi:10.1016/j.energy.2016.02.139

[39] D. Nagaraju, Q. Wang, P. Beaujuge and H. Alshareef, "Two-dimensional heterostructures of V2O5 and reduced graphene oxide as electrodes for high energy density asymmetric supercapacitors", *Journal of Materials Chemistry A*, vol. 2, no. 40, pp. 17146–17152, 2014. Doi:10.1039/c4ta03731f

[40] W. Yong-gang and Z. Xiao-gang, "Preparation and electrochemical capacitance of RuO2/TiO2 nanotubes composites", *Electrochimica Acta*, vol. 49, no. 12, pp. 1957–1962, 2004. Doi:10.1016/j.electacta.2003.12.023

[41] D. Nagaraju, Q. Wang, P. Beaujuge and H. Alshareef, "Two-dimensional heterostructures of V2O5 and reduced graphene oxide as electrodes for high energy density asymmetric supercapacitors", *Journal of Materials Chemistry A*, vol. 2, no. 40, pp. 17146–17152, 2014. Doi:10.1039/c4ta03731f

[42] W. Yong-gang and Z. Xiao-gang, "Preparation and electrochemical capacitance of RuO2/TiO2 nanotubes composites", *Electrochimica Acta*, vol. 49, no. 12, pp. 1957–1962, 2004. Doi:10.1016/j.electacta.2003.12.023

[43] S. Ratha et al., "Supercapacitors based on patronite—reduced graphene oxide hybrids: experimental and theoretical insights", *Journal of Materials Chemistry A*, vol. 3, no. 37, pp. 18874–18881, 2015. Doi:10.1039/c5ta03221k

[44] R. Raj, P. Ragupathy and S. Mohan, "Remarkable capacitive behavior of a Co3O4—polyindole composite as electrode material for supercapacitor applications", *Journal of Materials Chemistry A*, vol. 3, no. 48, pp. 24338–24348, 2015. Doi:10.1039/c5ta07046e

[45] H. Wang, H. Yi, X. Chen and X. Wang, "Asymmetric supercapacitors based on nano-architectured nickel oxide/graphene foam and hierarchical porous nitrogen-doped carbon nanotubes with ultrahigh-rate performance", *Journal of Materials Chemistry A*, vol. 2, no. 9, pp. 3223–3230, 2014. Doi:10.1039/c3ta15046a

[46] X. Wang, J. Hu, W. Liu, G. Wang, J. An and J. Lian, "Ni—Zn binary system hydroxide, oxide and sulfide materials: synthesis and high supercapacitor performance", *Journal of Materials Chemistry A*, vol. 3, no. 46, pp. 23333–23344, 2015. Doi:10.1039/c5ta07169k

[47] Z. Xing et al., "Ni3S2 coated ZnO array for high-performance supercapacitors", *Journal of Power Sources*, vol. 245, pp. 463–467, 2014. Doi:10.1016/j.jpowsour.2013.07.012

[48] S. Sahoo and J. Shim, "Facile synthesis of three-dimensional ternary ZnCo2O4/reduced graphene Oxide/NiO composite film on nickel foam for next generation supercapacitor electrodes", *ACS Sustainable Chemistry & Engineering*, vol. 5, no. 1, pp. 241–251, 2016. Doi:10.1021/acssuschemeng.6b01367

[49] L. Qiu et al., "Dispersing carbon nanotubes with graphene oxide in water and synergistic effects between graphene derivatives", *Chemistry—A European Journal*, vol. 16, no. 35, pp. 10653–10658, 2010. Doi:10.1002/chem.201001771

[50] Y. Wang et al., "Graphene/carbon black hybrid film for flexible and high rate performance supercapacitor", *Journal of Power Sources*, vol. 271, pp. 269–277, 2014. Doi:10.1016/j.jpowsour.2014.08.007

[51] X. Yang, J. Zhu, L. Qiu and D. Li, "Bioinspired effective prevention of restacking in multilayered graphene films: towards the next generation of high-performance supercapacitors", *Advanced Materials*, vol. 23, no. 25, pp. 2833–2838, 2011. Doi:10.1002/adma.201100261

[52] B. Choi, J. Hong, W. Hong, P. Hammond and H. Park, "Facilitated Ion transport in all-solid-state flexible supercapacitors", *ACS Nano*, vol. 5, no. 9, pp. 7205–7213, 2011. Doi:10.1021/nn202020w

[53] C. Chen et al., "Macroporous 'bubble' graphene film via template-directed ordered-assembly for high rate supercapacitors", *Chemical Communications*, vol. 48, no. 57, p. 7149, 2012. Doi:10.1039/c2cc3

[14] D. Wang et al., "Fabrication of graphene/polyaniline composite paper via in situ anodic electropolymerization for high-performance flexible electrode", *ACS Nano*, vol. 3, no. 7, pp. 1745–1752, 2009. Doi:10.1021/nn900297m.2189k

[54] D. Wang et al., "Fabrication of graphene/polyaniline composite paper via in situ anodic electropolymerization for high-performance flexible electrode", *ACS Nano*, vol. 3, no. 7, pp. 1745–1752, 2009. Doi:10.1021/nn900297m

[55] X. Yan, J. Chen, J. Yang, Q. Xue and P. Miele, "Fabrication of free-standing, electrochemically active, and biocompatible graphene oxide–polyaniline and graphene–polyaniline hybrid papers", *ACS Applied Materials & Interfaces*, vol. 2, no. 9, pp. 2521–2529, 2010. Doi:10.1021/am100293r

[56] X. Cao et al., "Preparation of novel 3D graphene networks for supercapacitor applications", *Small*, vol. 7, no. 22, pp. 3163–3168, 2011. Doi:10.1002/smll.201100990

[57] Y. Cheng, S. Lu, H. Zhang, C. Varanasi and J. Liu, "Synergistic effects from graphene and carbon nanotubes enable flexible and robust electrodes for high-performance supercapacitors", *Nano Letters*, vol. 12, no. 8, pp. 4206–4211, 2012. Doi:10.1021/nl301804c

[58] B. Choi, M. Yang, W. Hong, J. Choi and Y. Huh, "3D macroporous graphene frameworks for supercapacitors with high energy and power densities", *ACS Nano*, vol. 6, no. 5, pp. 4020–4028, 2012. Doi:10.1021/nn3003345

[59] A. Davies et al., "Graphene-based flexible supercapacitors: pulse-electropolymerization of polypyrrole on free-standing graphene films", *The Journal of Physical Chemistry C*, vol. 115, no. 35, pp. 17612–17620, 2011. Doi:10.1021/jp205568v

[60] A. Davies et al., "Graphene-based flexible supercapacitors: pulse-electropolymerization of polypyrrole on free-standing graphene films", *The Journal of Physical Chemistry C*, vol. 115, no. 35, pp. 17612–17620, 2011. Doi:10.1021/jp205568v

[61] Y. Xu, Z. Lin, X. Huang, Y. Wang, Y. Huang and X. Duan, "Functionalized graphene hydrogel-based high-performance supercapacitors", *Advanced Materials*, vol. 25, no. 40, pp. 5779–5784, 2013. Doi:10.1002/adma.201301928

[62] Y. He et al., "Freestanding three-dimensional graphene/MnO2 composite networks as ultralight and flexible supercapacitor electrodes", *ACS Nano*, vol. 7, no. 1, pp. 174–182, 2012. Doi:10.1021/nn304833s

[63] G. Yu et al., "Solution-processed graphene/MnO2 nanostructured textiles for high-performance electrochemical capacitors", *Nano Letters*, vol. 11, no. 7, pp. 2905–2911, 2011. Doi:10.1021/nl2013828

[64] X. Dong, L. Wang, D. Wang, C. Li and J. Jin, "Layer-by-layer engineered co—al hydroxide nanosheets/graphene multilayer films as flexible electrode for supercapacitor", *Langmuir*, vol. 28, no. 1, pp. 293–298, 2011. Doi:10.1021/la2038685

[65] Ge et al., "Facile dip coating processed graphene/MnO2 nanostructured sponges as high performance supercapacitor electrodes", *Nano Energy*, vol. 2, no. 4, pp. 505–513, 2013. Doi:10.1016/j.nanoen.2012.12.002

[66] [14]Y. Horng, Y. Lu, Y. Hsu, C. Chen, L. Chen and K. Chen, "Flexible supercapacitor based on polyaniline nanowires/carbon cloth with both high gravimetric and

area-normalized capacitance", *Journal of Power Sources*, vol. 195, no. 13, pp. 4418–4422, 2010. Doi:10.1016/j.jpowsour.2010.01.046

[67] K. Wang et al., "An all-solid-state flexible micro-supercapacitor on a chip", *Advanced Energy Materials*, vol. 1, no. 6, pp. 1068–1072, 2011. Doi:10.1002/aenm.201100488

[68] H. Lv, Q. Pan, Y. Song, X. Liu and T. Liu, "A review on nano-/microstructured materials constructed by electrochemical technologies for supercapacitors", *Nano-Micro Letters*, vol. 12, no. 1, 2020. Doi:10.1007/s40820-020-00451-z

[69] C. Zhong, Y. Deng, W. Hu, D. Sun, X. Han, J. Qiao and J. Zhang. *Electrolytes for electrochemical supercapacitors*. [s.l.]. CRC Press, 2019.

[70] Brian E. Conway. *Electrochemical supercapacitors: scientific fundamentals and technological applications*. Springer Science & Business Media, 2013.

[71] K. Nakajo et al., "Modeling of a lithium-ion capacitor and its charging and discharging circuit in a model-based design", *Circuits and Systems*, vol. 07, no. 01, pp. 11–22, 2016. Doi:10.4236/cs.2016.71002

[72] G. Feng, R. Qiao and P. Cummings, "Modeling of supercapacitors", *Encyclopedia of Microfluidics and Nanofluidics*, pp. 2282–2289, 2015. Doi:10.1007/978-1-4614-5491-5_1758 [Accessed 1 June 2021].

[73] R. Nelms, D. Cahela and B. Tatarchuk, "Modeling double-layer capacitor behavior using ladder circuits", *IEEE Transactions on Aerospace and Electronic Systems*, vol. 39, no. 2, pp. 430–438, 2003. Doi:10.1109/taes.2003.1207255

[74] R. Spyker and R. Nelms, "Classical equivalent circuit parameters for a double-layer capacitor", *IEEE Transactions on Aerospace and Electronic Systems*, vol. 36, no. 3, pp. 829–836, 2000. Doi:10.1109/7.869502

[75] P. Vijay, V. Shah and V. Shimin, "Energy based equivalent circuit modelling of ultracapacitor considering variation of ESR with OCV", *International Journal of Power and Energy Systems*, vol. 40, no. 2, 2020. Doi:10.2316/j.2020.203-0139

[76] H. Farsi and F. Gobal, "Artificial neural network simulator for supercapacitor performance prediction", *Computational Materials Science*, vol. 39, no. 3, pp. 678–683, 2007. Doi:10.1016/j.commatsci.2006.08.024

[77] A. Eddahech, O. Briat, M. Ayadi and J. Vinassa, "Modeling and adaptive control for supercapacitor in automotive applications based on artificial neural networks", *Electric Power Systems Research*, vol. 106, pp. 134–141, 2014. Doi:10.1016/j.epsr.2013.08.016

[78] P. Saha and M. Khanra, "Equivalent circuit model of supercapacitor for self-discharge analysis—a comparative study", 2016 International Conference on Signal Processing, Communication, Power and Embedded System (SCOPES), 2016, pp. 1381–1386. Doi:10.1109/SCOPES.2016.7955667

[79] N. Bertrand, J. Sabatier, O. Briat and J. Vinassa, "Fractional non-linear modelling of ultracapacitors", *Communications in Nonlinear Science and Numerical Simulation*, vol. 15, no. 5, pp. 1327–1337, 2010. Doi:10.1016/j.cnsns.2009.05.066

[80] A. Dzieliński, D. Sierociuk and G. Sarwas, "Some applications of fractional order calculus", *Bulletin of the Polish Academy of Sciences: Technical Sciences*, vol. 58, no. 4, 2010. Doi:10.2478/v10175-010-0059-6

[81] T. Freeborn, B. Maundy and A. Elwakil, "Measurement of supercapacitor fractional-order model parameters from voltage-excited step response", *IEEE Journal on Emerging and Selected Topics in Circuits and Systems*, vol. 3, no. 3, pp. 367–376, 2013. Doi:10.1109/jetcas.2013.2271433

[82] Y. Dandeville, P. Guillemet, Y. Scudeller, O. Crosnier, L. Athouel and T. Brousse, "Measuring time-dependent heat profiles of aqueous electrochemical capacitors under cycling", *Thermochimica Acta*, vol. 526, no. 1–2, pp. 1–8, 2011. Doi:10.1016/j.tca.2011.07.027

[83] H. Gualous, H. Louahlia and R. Gallay, "Supercapacitor characterization and thermal modelling with reversible and irreversible heat effect", *IEEE Transactions on Power Electronics*, vol. 26, no. 11, pp. 3402–3409, 2011. Doi:10.1109/tpel.2011.2145422

[84] K. Wang, L. Zhang, B. Ji and J. Yuan, "The thermal analysis on the stackable supercapacitor", *Energy*, vol. 59, pp. 440–444, 2013. Doi:10.1016/j.energy.2013.07.064

[85] A. Berrueta, I. San Martín, A. Hernández, A. Ursúa and P. Sanchis, "Electro-thermal modelling of a supercapacitor and experimental validation", *Journal of Power Sources*, vol. 259, pp. 154–165, 2014. Doi:10.1016/j.jpowsour.2014.02.089

[86] W. Sarwar, M. Marinescu, N. Green, N. Taylor and G. Offer, "Electrochemical double layer capacitor electro-thermal modelling", *Journal of Energy Storage*, vol. 5, pp. 10–24, 2016. Doi:10.1016/j.est.2015.11.001

[87] A. Hnatov, S. Arhun, O. Ulyanets and S. Ponikarovska, "Ultracapacitors electrobus for urban transport", In 2018 IEEE 38th International Conference on Electronics and Nanotechnology (ELNANO), 2018, pp. 539–543. IEEE.

[88] C. Woo-Young, "High-efficiency duty-cycle controlled full-bridge converter for ultracapacitor chargers", *IET Power Electronics*, vol. 9, no. 6, pp. 1111–1119, 2016. Doi:10.1049/iet-pel.2015.0090

[89] S. Vazquez, S. Lukic, E. Galvan, L. Franquelo and J. Carrasco, "Energy storage systems for transport and grid applications", *IEEE Transactions on Industrial Electronics*, vol. 57, no. 12, pp. 3881–3895, 2010. Doi:10.1109/tie.2010.2076414

[90] S. Jayasinghe and D. Vilathgamuwa, "Flying Supercapacitors as Power Smoothing Elements in Wind Generation", *IEEE Transactions on Industrial Electronics*, vol. 60, no. 7, pp. 2909–2918, 2013. Doi:10.1109/tie.2012.2233693

[91] M. Ucar and S. Ozdemir, "3-Phase 4-leg unified series—parallel active filter system with ultracapacitor energy storage for unbalanced voltage sag mitigation", *International Journal of Electrical Power & Energy Systems*, vol. 49, pp. 149–159, 2013. Doi:10.1016/j.ijepes.2013.01.005

16 Energy Scheduling and Flexibility Quantification in Buildings

Roya Ahmadiahangar, Andrei Blinov and Dimosthenis Peftitsis

CONTENTS

16.1 EMERGING POSSIBILITIES TOWARDS POSITIVE ENERGY BUILDINGS

The concerns associated with global climate change have instigated many actions to mitigate the footprint associated with human activity. Nowadays construction, maintenance and use of various buildings such as houses, offices, hospitals, public halls and warehouses account for around 40% of European Union (EU) energy consumption and 36% of CO_2 emissions [1, 2] (Figure 16.1). Therefore, many initiatives have been proposed to address this and increase the energy efficiency of existing and future buildings. Being an extremely multidisciplinary topic, with required expertise ranging from material and construction technologies to ergonomics and environmental sustainability, finding the optimal solution that is both energy and cost-efficient is a major challenge [3].

There are numerous projects launched all over the world showcasing the development of an energy-efficient building to a near-zero energy (nZEB) one [4–6]. Although the definition of nZEB varies by region [7, 8], the common understanding is that such building should have a reduced amount of energy used from fossil fuels, while its annual energy consumption is minimal or close to zero. In other words, the yearly energy consumption of the building should be close to its yearly production. At the same time, the architectonic quality and nZEB user experience/comfort should have minimal compromises when compared to a standard building.

The nZEB has to feature a well-insulated thermal envelope and incorporate high-performance heating, ventilation and air conditioning (HVAC), with local thermal energy production. Evidently, regardless of the insulation and HVAC system performance, to achieve zero consumption level, part of the energy required by the building needs to be produced locally in the form of electricity. Typically, this is achieved by photovoltaic (PV) generation, whereas the PV panels can be installed on the roof, integrated into various surfaces of the building or located nearby. Some of the studies demonstrate that combination of a ground

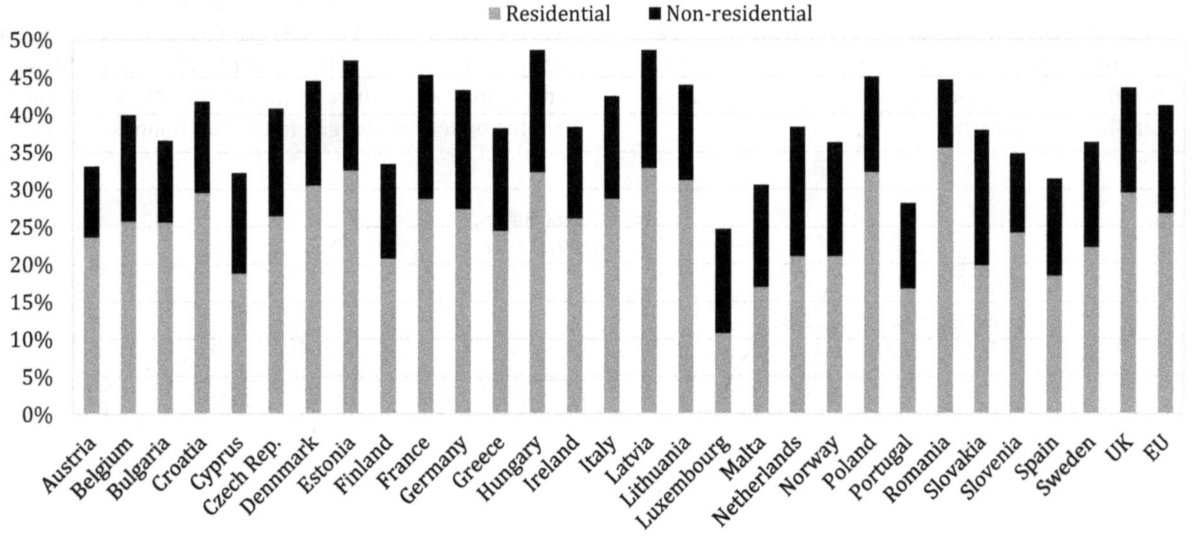

FIGURE 16.1 Share of buildings in total energy consumption of Europe (2016) [2].

DOI: 10.1201/9781003229124-16

source heat pump (geothermal) and PV is one of the most cost-optimal renewable systems, and one which can be considered widely and easily obtainable on-site [9]. Since electrical energy generated by a PV is one of the few types that can be easily exported to the grid and cover the building's energy demand, this technology is commonly considered crucial for nZEB.

Theoretically, with enough local energy generation and feed-in to the electric grid, a building can achieve net-zero consumption and even become energy-positive. However, buildings with high reliance on electrical energy export would have significant fluctuation in their input/output power profile (Figure 16.2), which can affect power grid stability, particularly in the case of large-scale applications of such nZEBs.

In the EU, the European Commission provided different requirements, depending on one of the four climate zones (see Table 16.1). It can be noticed that the recommendations are given considering both the energy efficiency of the building itself, as well as the local renewable regeneration. It is worth mentioning that most applications of nZEB in the EU directive place emphasis on new buildings. Nevertheless, there are recommendations and plans towards a revision of the current EU regulations to boost building stock renovation and improve the energy efficiency of existing building stock [10].

A further logical evolution of the nZEB concept is considered as a positive energy building (PEB). Although no official definitions exist on the properties of such buildings yet, certain estimations of properties and features can be outlined [12, 13]:

- A PEB should be capable of generating enough energy to not only completely cover its own demands, but also power consumer devices or even electric vehicles.
- A PEB should contribute to energy support of other buildings in the vicinity, with the possible aim of achieving energy neutrality or even positivity at the neighbourhood level.
- A PEB should feature active management of consumption, production and storage of the energy, aiming for maximisation of performance in a system-based approach.
- A PEB should involve a complex set of energy partnerships, offering new services to grid operators and utilities.
- Instead of focusing on the year-by-year energy balance, a PEB should aim towards full-time energy positivity, possibly using biomass as fuel during low-generation periods.

As follows, the PEB concept implicates a range of features that cannot be achieved only with the use of advanced materials and renewable sources. One of the most prominent properties of PEB is the capability of controlling its energy production and consumption at any point in time. This can be achieved by the use of local energy storage and demand-response programmes.

Typical characteristics of buildings with PV generation and battery energy storage systems (BESS) are shown in Figure 16.3. As observed, the BESS can be very effective at smoothing the consumption curve of the building, which formed a clear perspective for such systems. Although the market of the residential BESS is still forming, there is a range of available commercial BESS solutions from various manufacturers (Table 16.2).

On the other hand, despite price reduction and various governmental initiatives, in many cases, the economic feasibility of typical domestic storage, like the ones based on Li-ion batteries, is still not evident due to high investment costs and limited lifetimes. In many regions, the projected lifetimes of BESS is around 10 years under conventional operation [14]. Therefore, the development of sophisticated methods for controlling the energy flows and making the most use of the available resources is of major importance. Some studies considering a synergetic operation of both battery and HVAC systems demonstrated that the battery capacity requirement (and investment cost) can be reduced [15]. To reduce to impact of renewable generation on the grid and improve the flexibility of the neighbourhoods, a clustering-based approach was proposed in [16].

Nevertheless, the cost-effectiveness of BESS can be highly impacted by the scheduling approach and energy management [17–19]. The basic principle of BESS scheduling is that when there is higher on-site generation than load, the battery is charged to the maximum state of charge

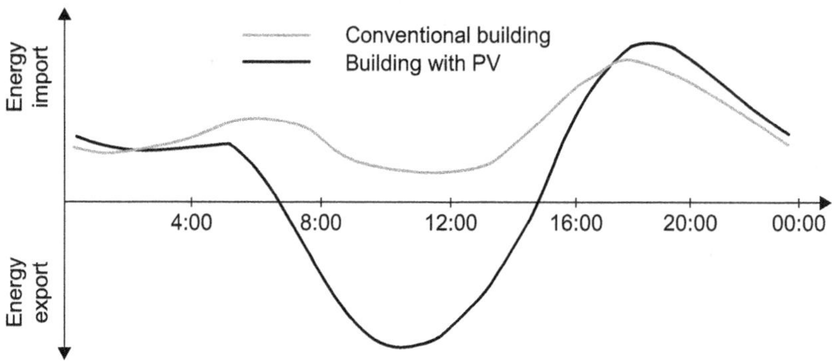

FIGURE 16.2 Typical 24-hour energy profile of conventional building and building with PV.

TABLE 16.1
EU Commission Recommendations on Energy Efficiency of Buildings [11]

Zone	Office, kWh/(m².y)			Single-Family House, kWh/(m².y)		
	Primary Energy	On-Site Sources	Net Primary Energy	Primary Energy	On-Site Sources	Net Primary Energy
Mediterranean	80–90	60	20–30	50–65	50	0–15
Oceanic	85–100	45	40–55	50–65	35	15–30
Continental	85–100	45	40–55	50–70	30	20–40
Nordic	85–100	30	55–70	65–90	25	40–65

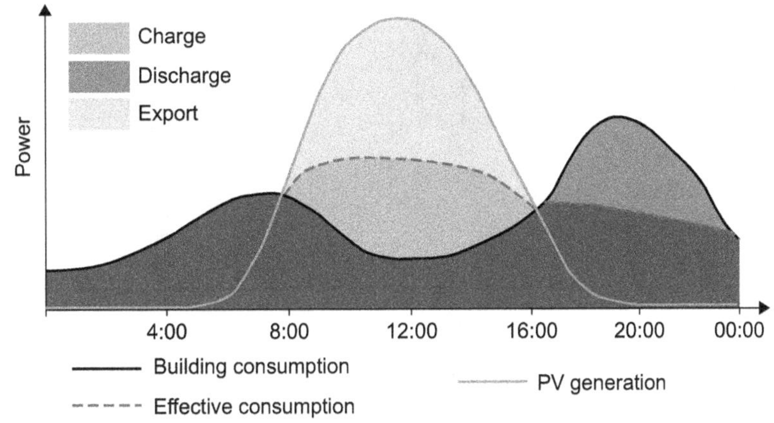

FIGURE 16.3 Typical 24-hour energy profile of a building with PV and battery storage.

TABLE 16.2
Examples of Commercially Available Residential Battery Energy Storage Systems

Manufacturer/Model	Maximal Energy Capacity (kWh)	Charge/Discharge Power (kW)	Battery Voltage (V)	Coupling
Tesla PowerWall	13.5	5	50	AC
Sonnen Batterie Eco	15	3.3	48	AC
SolarEdge+RESU10H	9.8	5	400	AC/DC
Enphase Encharge 3	3.5	1.3	67	AC
Enphase Encharge 10	10.5	3.8		
Nissan/Eaton xStorage	4.2 . . . 10	3.6 . . . 6	90	AC/DC
Varta Pulse/Pulse Neo 3	3.3	1.6/1.4	50	AC
Varta Pulse/Pulse Neo 6	6.5	2.5/2.3		
Sunny Boy Storage	External battery	3.7/5/6	360	AC
Victron Energy EasySolar	External battery	0.9/1.7/3.5	12.8–51.2	DC

(SoC) and discharged to the minimum when the reverse is the case. In this regard, several other objectives can be set, such as maximising self-consumption and minimising energy costs [20].

16.2 SCHEDULING APPROACHES FOR ENERGY MANAGEMENT OF BUILDINGS

Recently, residential-scale BESS has gained significant interest in both academia and industry under the new paradigm of renewable energy sources (RES) in many regions [21]. This interest has been driven by the rapid development of variable renewable generation, which substantially increases flexible resource requirements in power systems (PSs) [22, 23]. Another contributing factor is the declining capital costs for BESS technologies. With increasing amounts of BESS, it becomes an urgent problem deciding how to best operate BESS to meet different stakeholders' requirements. Different types of BESS that are utilised in gas, electricity and district heating networks provide support, in terms of flexibility of operation to the network that is utilising the storage device directly, as well as to all the network(s) coupled to this network through coupling components [24]. It is found that in smart energy systems, proper control strategies, which have large influences on the quasi-dynamic interactions, should be developed to use these interactions for increasing flexibility in different energy vectors [25, 26].

At the same time, The EU has set a goal that all buildings must be nZEBs by 2050 and made it compulsory to use certain amounts for the minimum share of RES in nZEBs [27]. In nZEB, the consumption is not homogenous. Use of appliances, which depend less on customer behaviour and habits, are more flexible regarding energy management [28]. The coordination between different energy management of nZEBs is vital to maintain grid efficiency. Otherwise, in the case of non-coordinated programmes, new peak hours are likely to be formed when electricity prices are low [29, 30]. The possible energy flexibility of different buildings and how to obtain this energy flexibility have been investigated in IEA-EBC Annex 67 Energy Flexibility Buildings [31]. The concluding phase of Annex 67 has, however, revealed that the most important area in increasing energy flexibility (EF) from the demand side is scaling up from single buildings to clusters of buildings (aggregation); this is the current focus for the next Annex 82 (2020–2024). Since the capacity of nZEB is typically insufficient to take part in the electricity market, aggregators act as mediators that monetise the flexibility of these nZEBs [32].

Figure 16.4 shows the data flow in optimal scheduling of nZEB [33], while the schematic of an energy management system for solar PV-BESS is depicted in Figure 16.5. The system consists of the following components.

- The nZEB Energy management unit allows the control of energy flow to support the concept of

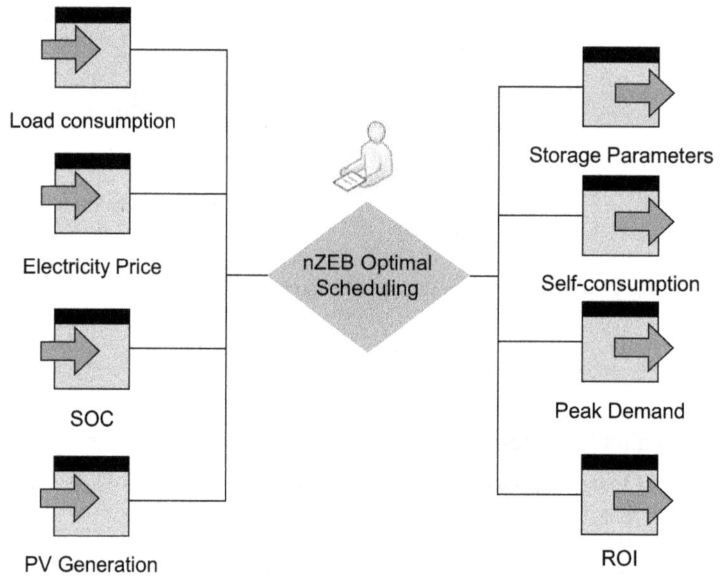

FIGURE 16.4 Data flow in optimal scheduling of nZEB.

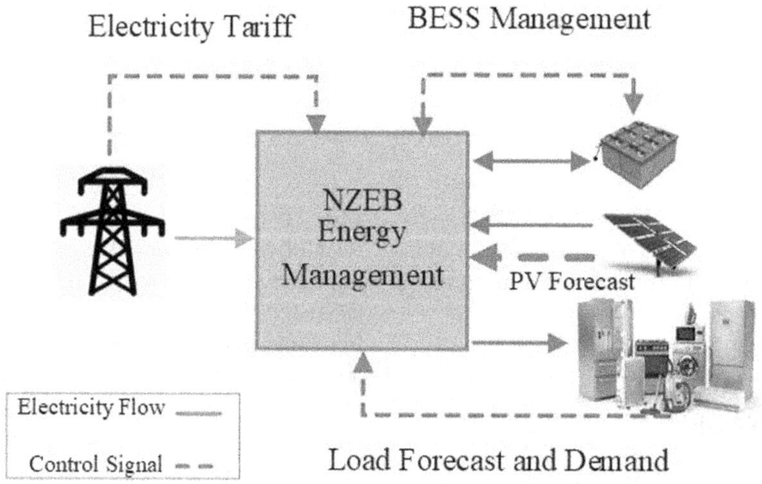

FIGURE 16.5 An energy management schematic for the nZEB.

PV-BESS; nZEB Energy management receives time of use (TOU) tariffs, BESS status and loads forecasts and decides how to optimise BESS operation.

- Energy from the rooftop PV panels will be delivered to the nZEB Energy management unit to meet the household's energy needs.
- BESS includes the ability to store energy from the PV panels or release it when needed.
- Residence-owned electrical appliances.

Through the weather and short-term load external forecasting systems, daily forecasts of PV and household loads can be obtained [34–36].

Several nZEBs in Estonia have been analysed using the proposed methodology. In this report, one residential nZEB in Estonia is used as a case study to evaluate the protocol. With an average elevation of just 61 metres above sea level, Estonia is a low-lying country with many buildings located close to the coast. The consumption profile is representative of a family with two adults and two children. A PV power generation chart reflecting the hourly generation of a 5kW unit is shown in Figure 16.6, from 1 January to 31 December 2019. Table 16.3 is based on BESS-PV system technical parameters in Estonia and electricity retail economic parameters in 2019.

Figure 16.7 shows the consumption profile of the understudy load for 2019. The total electrical energy consumption was 5239kWh.

A one-minute timeline was used in the second case study to determine the load. To design the BESS power electronic components in accordance with the maximum power of load and the frequency at which they occur, we studied in detail the maximum power of load and the frequency of their occurrence. Figure 16.8 shows the household load measured with one-minute accuracy in Estonia.

TABLE 16.3
Techno-Economic Parameters of the Case Study

Parameter	Value
PV peak power	1–5kWp
Battery energy capacity	5kWh
Battery investment cost	100 (Euro/kWh)
PV investment cost	1000 (Euro/kWp)
Electricity day price	0.039251 (Euro/kWh)
Electricity night price	0.025989 (Euro/kWh)
Peak tariff time range	07:00–23:59
Off-peak price time range	00:00–07:00

Using a simple analytical optimisation scheduling approach to maximise self-consumption, Figure 16.9 shows the calculated SoC and PB of the BESS. BESS charges during the day when PV is present and discharges at night when the load consumption is greater.

16.3 METHODS FOR QUANTIFICATION OF ENERGY FLEXIBILITY IN BUILDINGS

Flexibility is of prime importance for current and future PSs, with increasing grid integration of renewable energy sources (RES). In this regard, the main challenge is the management of the increased variability and uncertainty imposed by RES in the power balance [37].

Flexibility is a recent concept in PS that has been officially recognised by organisations like the International Energy Agency (IEA) [38], the North American Electric Reliability Corporation (NERC) [39], and the International Renewable Energy Association (IRENA) [40]. Flexibility can be defined as a PS's ability to respond to both expected and unexpected changes in demand and supply [41]. This

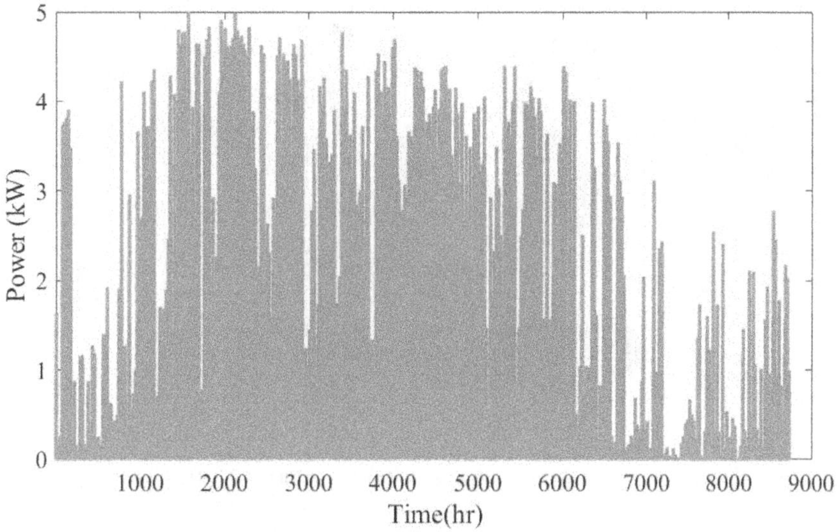

FIGURE 16.6 PV power generation by the hour in 2019.

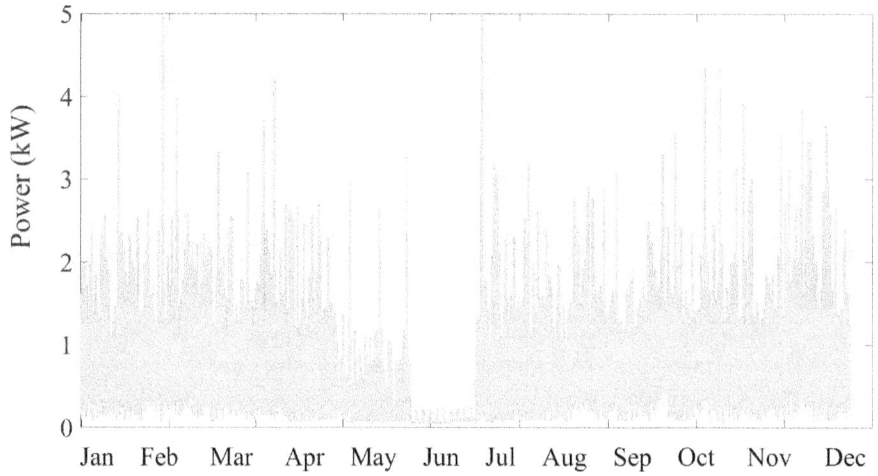

FIGURE 16.7 Load measurement for typical Estonian consumers.

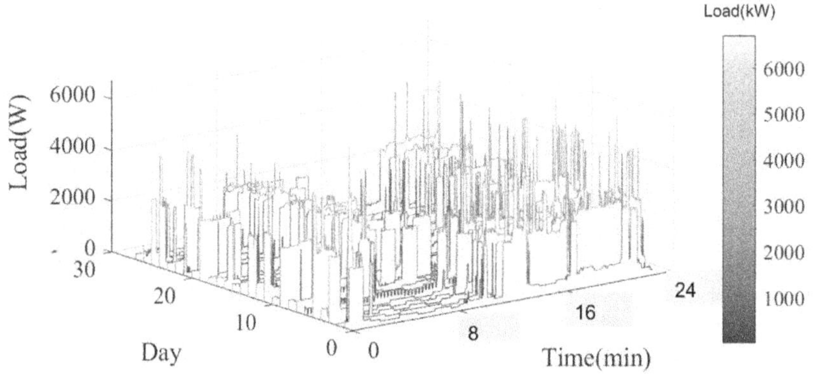

FIGURE 16.8 Accurately measured household load in Estonia.

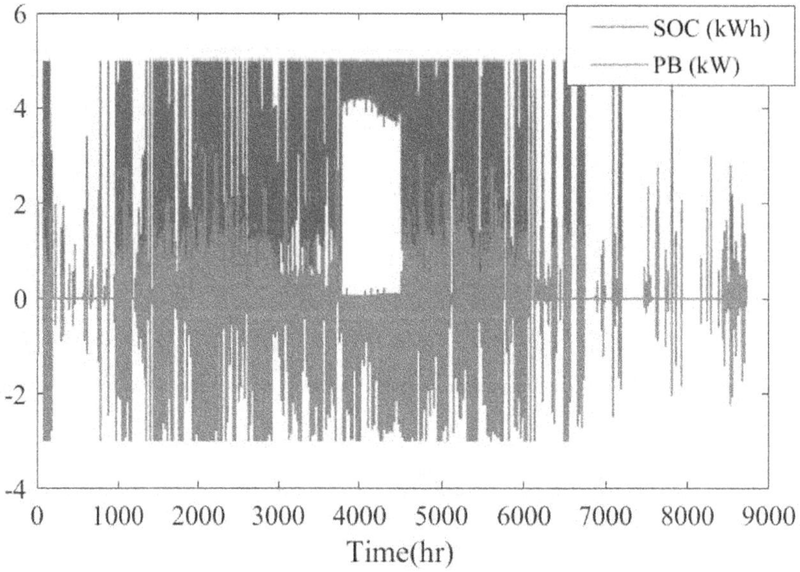

FIGURE 16.9 Calculated SoC and PB of the BESS.

concept can contribute to increasing the stability of the grid and the integration of RES. Intuitively, the increasing share of RES in PS is escalating the need for flexibility [42]. Important signs of inflexibility are known to be grid frequency fluctuations [43], area balance violations, negative market prices and price volatility [44]. Recent studies in this area are focusing on planning, scheduling and exploitation of flexibility mostly seen from the demand side [45–47], while later research has investigated generation units and large-scale ESS [48, 49]. DSF is defined as the capability of consumption modification in response to control signals. Possible sources of those control signals may be external signals (penalty) [50, 51] to smart metres from the aggregator, or internal control signals from the home energy management system (HEMS). The study in [52] quantifies the net revenues that can be captured by a flexible resource able to react to the short-term price variations on the day-ahead and intraday markets in Germany. One study [53] estimates flexibility of 12–23GW for the Northern European countries (Sweden, Denmark, Norway, Finland, Estonia, Latvia, and Lithuania) to be 15–30% of the peak load of the region.

The energy flexibility concept has been quantified in literature in several ways. The method of quantification is determined by the type of flexibility to be measured, as no one best method applies to all use cases. Using a flexibility envelope concept is one of the most applied methods; the authors in [54] propose a solution that captures energy flexibility by assessing the extent to which power can be increased or decreased for a given duration, e.g. user comfort or system constraints.

A building's or a device's flexibility depends on several factors, including the current energetic state, its distance from the constraints and the amount of time left until the constraints or a specific status are met. The shortcomings of this methodology include the assumption that the initial and final states of the system will be specified, as it is intended as an indication of flexibility potential, rather than as a scheduling method.

Energy consumption plays an essential role in the economy and has a great impact on environmental concerns, so the optimal use of energy has been considered an important goal in sustainable development. On top of this, progressively increasing deployment of smart meters and advanced data analysis techniques made it possible to facilitate a variety of applications, ranging from understanding energy consumption patterns to enabling utilisation of energy flexibility from the demand side (Figure 16.10).

Forecasting this flexibility, especially in nZEB and aggregated in nearly zero-energy districts (nZED), depends on several parameters, including weather, electricity generation and demand, grid constraints and market prices [55, 56], as well as the characteristics of the building's ventilation and heating system [57], control system, smart appliances, available capacity of ESS and the behaviour of occupants and their willingness to change their usage pattern and share of flexible loads.

Conventional approaches like computing the average of the data are not efficient in analysing this volume of data [58, 59]. Therefore, machine learning (ML) approaches are considered a powerful tool to deal with huge datasets [60]. The power consumption of flexible appliances and the

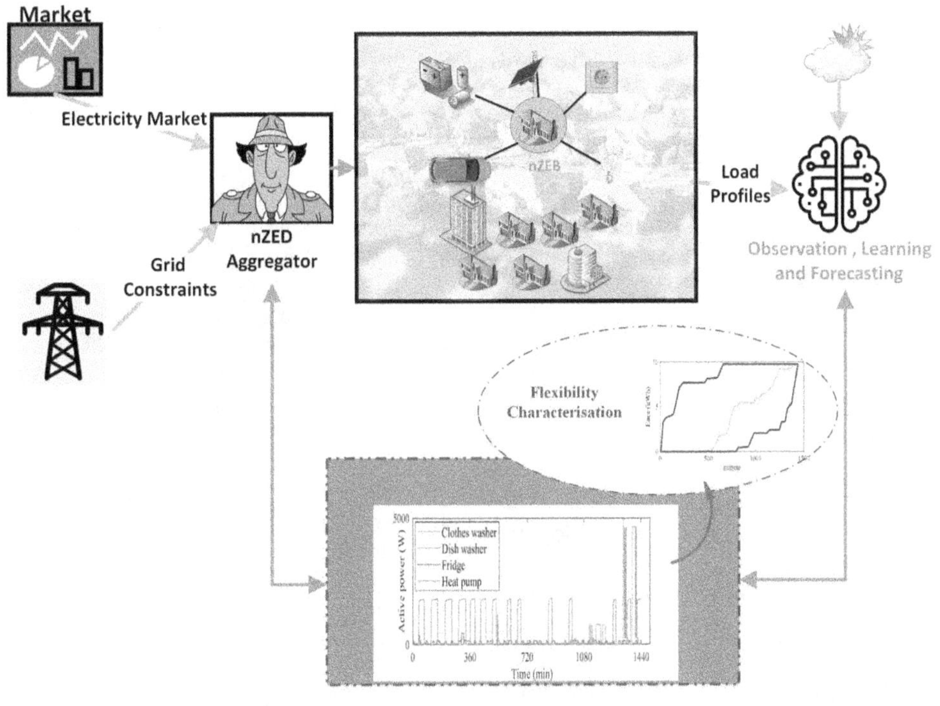

FIGURE 16.10 Energy flexibility quantification and utilisation.

usage behaviour of consumers is required in defining the EF of each nZEB [61]. The simplest way to get this information is the intrusive method; installing sensors on all appliances, which is costly, time-consuming and may cause discomfort for the tenants. To overcome the aforementioned challenges, non-intrusive load monitoring (NILM)—which extracts the consumption signal of appliances by analysing smart metre total active and reactive data—can be utilised [62, 63].

Another issue toward the utilisation of the flexibility in nZED is the lack of standard metrics for characterisation of flexibility. Some of the European Commission projects are concentrated on different aspects of PS flexibility, such as EMPOWER H2020 [64], INVADE H2020, REFLEX [65], Sim4Blocks [66], Flex4Grid [67], FEVER [68] and Ebalance+ [69]. Nevertheless, available metrics still do not address the variation in time of the overall demand aggregation [70, 71]. Although, while the common way to characterise the energy flexibility is by considering it as a static function at every time instant in all mentioned research, the validity of this approach is questionable because energy-based systems are never at a steady state. Despite extensive research, load forecasting and energy flexibility characterisation remain to be difficult problems.

Within the nZED, several energy technologies such as power electronics, HVAC, heat pumps, district heating, EVs and grids are connected via communication technology, which leads to a trans-disciplinary, multi-domain system. Civil engineers usually develop their models for residential nZEBs in IDA-ICE or Modelica software. In these models, meanwhile, there is not a good solution to integrate detailed models of building with electrical models, usually developed in MATLAB or RSCAD. The problem is that simulation packages for assessing system integration of components typically cover only one sub-domain, while terribly simplifying the others. Co-simulation overcomes this by coupling sub-domain models that are described and solved within their native environments, using specialised solvers and validated libraries [72–75].

16.4 CONCLUSIONS AND FUTURE PROSPECTS

More distributed renewable energy resources and consumption, and the increasing electrification of transportation, create new challenges in the power system and thus result in new needs for more flexible and cost-effective management of the electricity system. This, in turn, supports the need for a new approach and a vision for the development of the market for flexibility services.

The Clean Energy Directive, in turn, states that a supportive regulatory framework should be established that would provide an incentive for network operators to use flexibility services in network development and management. Also, the most uniform flexibility market products possible must be created between the transmission system operators (TSO) and distribution system operators (DSO), and through aggregators.

It has been suggested that different approaches can be taken to characterise and quantify the energy flexibility of each device. When it comes to residential buildings, they have drawbacks such as only considering smart appliances and using in-house systems. Extraction of the consumption patterns of flexible appliances, assuming a fixed power supply in the case of appliances with different modes of operation, and not taking into account the usage behaviour of customers and the uncertainties in the usage patterns all need to be covered in future research.

16.5 ACKNOWLEDGEMENT

This work was supported by the EEA Financial Mechanism Baltic Research Programme in Estonia—Grant EMP747.

REFERENCES

[1] Directive (EU) 2018/844 of the European Parliament and of the Council of 30 May 2018 amending Directive 2010/31/EU on the energy performance of buildings and Directive 2012/27/EU on energy efficiency, PE/4/2018/REV/1 [Online]. Available: https://eur-lex.europa.eu/

[2] Energy efficiency trends and policies in the household and tertiary sectors—an analysis based on the odyssee and mure databases [Online]. Available: https://www.odyssee-mure.eu/publications/archives/energy-efficiency-trends-policies-buildings.pdf

[3] L. Kranz et all, "ZEBRA 2020—NEARLY ZERO-ENERGY BUILDING STRATEGY 2020: strategies for a nearly zero-energy building market transition in the European Union," ZEBRA2020 Project Report, October 2016 [Online]. Available: https://www.bpie.eu/wp-content/uploads/2016/12/ZEBRA2020_Strategies-for-nZEB_07_LQ-double-pages.pdf

[4] Mengmeng Wang, Xiaojun Liu, Hanliang Fu, Baiyu Chen, "Scientometric of nearly zero energy building research: a systematic review from the perspective of co-citation analysis," Journal of Thermal Science 28, no. 6, pp. 1104–1114, 2019.

[5] Xu Wei, Zhang Shicong, "APEC nearly (net) zero energy building roadmap," Asia-Pacific Economic Cooperation (APEC) Energy Working Group, Project: EWG 15 2016A, November 2018 [Online]. Avaiable: www.apec.org/-/media/APEC/Publications/2018/12/

[6] Y. Saheb, S. Shnapp, D. Paci, "From nearly-zero energy buildings to net-zero energy districts—Lessons learned from existing EU projects," EUR 29734 EN, Publications Office of the European Union, Luxembourg, 2019 [Online].

[7] K. Peterson et all, "A common definition for zero energy buildings," The National Institute of Building Sciences, Project Report, September 2015 [Online]. Available: www.energy.gov/sites/default/files/2015/09/f26/bto_common_definition_zero_energy_build.

[8] "Commission Recommendation (EU) 2016/1318 of 29 July 2016 on guidelines for the promotion of nearly zero-energy buildings and best practices to ensure that, by

2020, all new buildings are nearly zero-energy buildings," vol. L 208, pp. 46–57. Available: http://data.europa.eu/eli/reco/2016/1318/oj

[9] I. Visa, M. D. Moldovan, M. Comsit, A. Duta, "Improving the renewable energy mix in a building toward the nearly zero energy status," Energy Buildings, vol. 68, pp. 72–78, 2014.

[10] Commission Recommendation (EU), 2019. 2019/786 of 8 May 2019 on building renovation, C/2019/3352. Off. J. Eur. Union. L127/34. Available: https://eur-lex.europa.eu/legal-content/EN/TXT/PDF/?uri=CELEX:32019H0786&rid=5

[11] Directive (EU) 2018/844 of the European Parliament and of the Council of 30 May 2018 amending Directive 2010/31/EU on the energy performance of buildings and Directive 2012/27/EU on energy efficiency, PE/4/2018/REV/1 [Online]. Available: https://eur-lex.europa.eu/

[12] Raymond J. Cole, Laura Fedoruk, "Shifting from net-zero to net-positive energy buildings," Building Research & Information, vol. 43, no. 1, pp. 111–120, 2015.

[13] Anna Magrini, Giorgia Lentini, Sara Cuman, Alberto Bodrato, Ludovica Marenco, "From nearly zero energy buildings (NZEB) to positive energy buildings (PEB): The next challenge – The most recent European trends with some notes on the energy analysis of a forerunner PEB example," Developments in the Built Environment, vol. 3, p. 100019, 2020.

[14] A. Dietrich, C. Weber, "What drives profitability of grid-connected residential PV storage systems? A closer look with focus on Germany," Energy Economics, vol. 74, no. 2018, pp. 399–416, 2018.

[15] M. B. Sanjareh, M. H. Nazari, G. B. Gharehpetian, R. Ahmadiahangar, A. Rosin, "Optimal scheduling of HVACs in islanded residential microgrids to reduce BESS size considering effect of discharge duration on voltage and capacity of battery cells," Sustainable Energy, Grids and Networks, vol. 25, p. 100424, 2020.

[16] A. Rosin, Roya Ahmadiahangar et al., "Clustering-based penalty signal design for flexibility utilization," in IEEE Access, vol. 8, pp. 208850–208860, 2020.

[17] A. B. Dayani, H. Fazlollahtabar, R. Ahmadiahangar, A. Rosin, M. S. Naderi, M. Bagheri, "Applying reinforcement learning method for real-time energy management," 2019 IEEE International Conference on Environment and Electrical Engineering and 2019, IEEE Industrial and Commercial Power Systems Europe (EEEIC/I&CPS Europe), Genova, Italy, 2019, pp. 1–5, doi:10.1109/EEEIC.2019.8783766.

[18] E. Azizi, A. M. Shotorbani, R. Ahmadiahangar, B. Mohammadi-Ivatloo, A. Rosin, R. Sadiq, K. Hewage, "Cost/comfort-oriented clustering-based extended time of use pricing," Sustainable Cities and Society, vol. 66, p. 102673, 2021.

[19] T. Häring, R. Ahmadiahangar, A. Rosin, H. Biechl, "Impact of load matching algorithms on the battery capacity with different household occupancies," IECON 2019–45th Annual Conference of the IEEE Industrial Electronics Society, Lisbon, Portugal.

[20] Mohammad Hosein Dadashi-Rad, Ali Ghasemi-Marzbali, Roya Ahmadi Ahangar, "Modeling and planning of smart buildings energy in power system considering demand response," Energy, vol. 118770, p. 213, 2020.

[21] A. Nikoobakht, J. Aghaei, M. Shafie-Khah, J. P. S. Cataláo, "Assessing increased flexibility of energy storage and demand response to accommodate a high penetration of renewable energy sources," IEEE Transactions on Sustainable Energy, vol. 10, pp. 659–669, 2019.

[22] A. Taşcıkaraoğlu, N. G. Paterakis, O. Erdinç, J. P. S. Catalão, "Combining the flexibility from shared energy storage systems and DLC-based demand response of HVAC units for distribution system operation enhancement," IEEE Transactions on Sustainable Energy, vol. 10, no. 1, pp. 137–148, 2019.

[23] H. Liao, J. V. Milanović, "Flexibility exchange strategy to facilitate congestion and voltage profile management in power networks," IEEE Transactions on Smart Grid, vol. 10, no. 5, pp. 4786–4794, 2019.

[24] Roya Ahmadiahangar, Argo Rosin, Ivo Palu, Aydin Azizi, "New approaches for increasing demand-side flexibility," in Demand-side Flexibility in Smart Grid, Singapore, Springer, 2020, pp. 51–62.

[25] Seyed Hamid Reza Hosseini, Adib Allahham, Sara Louise Walker, Phil Taylor, "Optimal planning and operation of multi-vector energy networks: a systematic review," Renewable and Sustainable Energy Reviews, vol. 133, p. 110216, 2020.

[26] Z. Pan, J. Wu, H. Sun, Q. Guo, M. Abeysekera, "Quasi-dynamic interactions and security control of integrated electricity and heating systems in normal operations," CSEE Journal of Power and Energy Systems, vol. 5, no. 1, pp. 120–129, 2019.

[27] Shady Attiaa, Polyvios Eleftherioub, Flouris Xenib, Rodolphe Morlotc, Christophe Ménézod, Vasilis Kostopoulose, Maria Betsie, "Overview and future challenges of nearly zero energy buildings (nZEB) design in Southern Europe," Energy and Buildings, vol. 155, p. 439–458, 2017.

[28] Hong Tang, Shengwei Wang, Hangxin Li, "Flexibility categorization, sources, capabilities and technologies for energy-flexible and grid-responsive buildings: state-of-the-art and future perspective," Energy, vol. 219, p. 119598, 2021.

[29] Cheng Fan, Gongsheng Huang, Yongjun Sun, "A collaborative control optimization of grid-connected net zero energy buildings for performance improvements at building group level," Energy, vol. 164, pp. 536–549, 2018.

[30] Yi Tang, Qian Chen, Jia Ning, Qi Wang, Shuhai Feng, Yaping Li, "Hierarchical control strategy for residential demand response considering time-varying aggregated capacity," Electrical Power and Energy System, vol. 97, pp. 165–173, 2018.

[31] "Annex 67," IEA-EBC [Online]. Available: www.annex67.org [Accessed March 2020].

[32] K. Bruninx, H. Pandžić, H. Le Cadre, E. Delarue, "On the interaction between aggregators, electricity markets and residential demand response providers," IEEE Transactions on Power Systems, vol. 35, no. 2, pp. 840–853, 2020.

[33] Roya Ahmadiahangar, Oleksandr Husev, Andrii Blinov, Hossein Karami, Argo Rosin, "Development of a Battery Sizing Tool for Nearly Zero Energy Buildings," in IECON 2020 The 46th Annual Conference of the IEEE Industrial Electronics Society, 2020.

[34] N. Shabbir, R. Ahmadiahangar, L. Kütt, A. Rosin, "Comparison of machine learning based methods for residential load forecasting," 2019 Electric Power Quality and Supply Reliability Conference (PQ) & 2019, no. Symposium on Electrical

Engineering and Mechatronics (SEEM), Kärdla, Estonia, 2019, pp. 1–4, doi:10.1109/PQ.2019.8818267.
[35] R. Ahmadiahangar, T. Häring, A. Rosin, T. Korõtko, J. Martins, "Residential load forecasting for flexibility prediction using machine learning-based regression model," 2019 IEEE International Conference on Environment and Electrical Engineering, no. and 2019 IEEE Industrial and Commercial Power Systems Europe (EEEIC/I&CPS Europe), Genova, Italy, 2019, pp. 1–4, doi:10.1109/EEEIC.2019.8783634.
[36] Noman Shabbir, Roya AhmadiAhangar, Lauri Kütt, Muhamamd N Iqbal, Argo Rosin, "Forecasting short term wind energy generation using machine learning," in 2019 IEEE 60th International Scientific Conference on Power and Electrical Engineering of Riga Technical University (RTUCON).
[37] Hussam Nosair, François Bouffard, "Flexibility envelopes for power system operational planning," IEEE Transactions on Sustainable Energy, vol. 6, no. 3, pp. 800–809, 2015.
[38] IEA, "Status of power system transformation 2018: advanced power plant flexibility," Paris, France, 2018.
[39] North American Electric Reliability Corporation, "Accomodating high levels of variable generation," NERC, Tech. Rep, Princeton, NJ, USA, 2009.
[40] International Renewable Energy Agency (IRENA), "Power system flexibility for the energy transition, part 1: overview for policy makers," Abu Dhabi, 2018.
[41] Jaquelin Cochran, Mackay Miller, Owen Zinaman, Michael Milligan, Doug Arent, Bryan Palmintier, "Flexibility in 21st century power systems," NREL, 2018. Available: https://www.nrel.gov/docs/fy14osti/61721.pdf
[42] Ana Fernández-Guillamón, Emilio Gómez-Lázaro, Eduard Muljadic, Ángel Molina-Garcí, "Power systems with high renewable energy sources: a review of inertia and frequency control strategies over time," Renewable and Sustainable Energy Reviews, vol. 109369, p. 115, 2019.
[43] M. Ayar, "A distributed control approach for enhancing smart grid transient stability and resilience," IEEE Transactions on Smart Grid, vol. 8, no. 6, pp. 3035–3044, 2017.
[44] Guochao Wang, Shenzhou Zheng, Jun Wang, "Fluctuation and volatility dynamics of stochastic interacting energy futures price model," Physica A: Statistical Mechanics and Its Applications, vol. 537, p. 122693, 2020.
[45] M. B. Anwar, H. W. Qazi, D. J. Burke, M. J. O'Malley, "Harnessing the flexibility of demand-side resources," IEEE Transactions on Smart Grid, vol. 10, no. 4, pp. 4151–4163, July 2019.
[46] F. Fanitabasi, E. Pournaras, "Appliance-level flexible scheduling for socio-technical smart grid optimization," IEEE Access, vol. 8, pp. 119880–119898, 2020.
[47] M. R. M. Cruz, D. Z. Fitiwi, S. F. Santos, S. J. P. S. Mariano, J. P. S. Cataláo, "multi-flexibility option integration to cope with large-scale integration of renewables," IEEE Transactions on Sustainable Energy, vol. 11, no. 1, pp. 48–60, 2020.
[48] Haoran Ji, Chengshan Wang, Peng Li, Guanyu Song, "Quantified analysis method for operational flexibility of active distribution networks with high penetration of distributed generators," Applied Energy, vol. 239, pp. 706–714, 2019.
[49] Jiahua Hu, Mushfiqur R. Sarker, Jianhui Wang, Fushuan Wen, Weijia Liu, "Provision of flexible ramping product by battery energy storage in day-ahead energy and reserve markets," IET Generation, Transmission & Distribution, vol. 12, no. 10, pp. 2256–2265, 2018.
[50] J. K. S. S. Balaji Dulipala, Sanjoy Debbarma, "Energy scheduling model considering penalty mechanism in transactive energy markets: a hybrid approach," International Journal of Electrical Power & Energy Systems, vol. 129, p. 106742, 2021.
[51] Seyed Morteza Ghorashi, Mohammad Rastegar, Soroush Senemmar, Ali Reza Seifi, "Optimal design of reward-penalty demand response programs in smart power grids," Sustainable Cities and Society, vol. 60, p. 102150, 2020.
[52] Stéphane Goutte, Philippe Vassilopoulos, "The value of flexibility in power markets," Energy Policy, vol. 125, pp. 347–357, 2019.
[53] Lennart Södera, Peter D. Lund, Hardi Koduvere, Torjus Folsland Bolkesjø, "A review of demand side flexibility potential in Northern Europe," Renewable and Sustainable Energy Reviews, vol. 91, pp. 654–664, 2018.
[54] R. D'hulst, W. Labeeuw, B. Beusen, S. Claessens, G. Deconinck, K. Vanthournout, "Demand response flexibility and flexibility potential of residential smart appliances: experiences from large pilot test in Belgium," Applied Energy, vol. 155, pp. 79–90, Oct. 2015, doi:10.1016/j.apenergy.2015.05.101.
[55] T. Morstyn, A. Teytelboym, M. D. McCulloch, "Designing decentralized markets for distribution system flexibility," IEEE Transactions on Power Systems, vol. 34, no. 3, pp. 2128–2139, 2019.
[56] I. Mamounakis, N. Efthymiopoulos, D. J. Vergados, G. Tsaousoglou, P. Makris, E. M. Varvarigos, "A pricing scheme for electric utility's participation in day-ahead and real-time flexibility energy markets," Journal of Modern Power Systems and Clean Energy, vol. 7, no. 5, pp. 1294–1306, 2019.
[57] Vahur Maask, Tobias Häring, Roya Ahmadiahangar, Argo Rosin, Tarmo Korõtko, "Analysis of ventilation load flexibility depending on indoor climate conditions," 2020 IEEE International Conference on Industrial Technology (ICIT), 2020, pp. 607–612, doi: 10.1109/ICIT45562.2020.9067153.
[58] Masoud Babaei, "Data-driven load management of stand-alone residential buildings including renewable resources, energy storage system, and electric vehicle," Journal of Energy Storage, vol. 28, 2020.
[59] M. Afzalan, F. Jazizadeh, "Data-driven identification of consumers with deferrable loads for demand response programs," IEEE Embedded Systems Letters, vol. 12, no. 2, pp. 54–57, June 2020.
[60] M. Sun, "Clustering-based residential baseline estimation: a probabilistic perspective," IEEE Transactions on Smart Grid, vol. 10, no. 6, pp. 6014–6028, 2019.
[61] Z. Wang Zhai, X. Yan, G. He, "Appliance flexibility analysis considering user behavior in home energy management system usings mart plugs," IEEE Transactions on Industrial Electronics, vol. 66, no. 2, p. 1391–1401, 2018.
[62] E. Azizi, A. M. Shotorbani, M.-T. Hamidi-Beheshti, B. Mohammadi-Ivatloo, S. Bolouki, "Residential household non-intrusive loadmonitoring via smart event-based optimization," IEEE Transactions on Consumer Electronics, vol. 55, no. 3, p. 233–241, 2020.
[63] S. Zhai, H. Zhou, Z. Wang, G. He, "Analysis of dynamic appliance flexibility considering user behavior via non-intrusive load monitoring and deep user modeling," CSEE

Journal of Power and Energy Systems, vol. 6, no. 1, pp. 41–52, March 2020.
[64] C. Cordobés, "EMPOWER H2020 project grant agreement 646476; esmart systems," Bellevue, WA, USA, 2017.
[65] "REflex, analysis of the european energy system," [Online]. Available: http://reflex-project.eu/ [Accessed March 2021].
[66] "Simulation supported real time energy management in building blocks," [Online]. Available: https://cordis.europa.eu/project/id/695965 [Accessed March 2021].
[67] "Flex4Grid, prosumer flexibility services for smart grid management," [Online]. Available: https://ec.europa.eu/inea/en/horizon-2020/projects/h2020-energy/grids/flex4grid [Accessed March 2021].
[68] "Flexible energy production, demand and storage-based virtual power plants for electricity markets and resilient DSO operation," [Online]. Available: https://cordis.europa.eu/project/id/864537 [Accessed March 2021].
[69] "ebalance," [Online]. Available: http://ebalance-project.eu/ [Accessed March 2021].
[70] Carlos Adrian Correa-Florez, Andrea Michiorri, George Kariniotakis, "Optimal participation of residential aggregators in energy and local flexibility markets," IEEE Transactions on Smart Grid, vol. 11, no. 2, pp. 1644–1657, 2020.
[71] S. Wang, X. Tan, T. Liu, D. H. K. Tsang, "Aggregation of demand-side flexibility in electricity markets: negative impact analysis and mitigation method," IEEE Transactions on Smart Grid, vol. 12, no. 1, pp. 774–786, Jan. 2021.
[72] Peter Palensky, Arjen A. van der Meer, Claudio David, "Cosimulation of intelligent power systems: fundamentals, software architecture, numerics, and coupling," IEEE Industrial Electronics Magazine, vol. 11, no. 1, pp. 34–50, 2017.
[73] B. P. Bhattarai et al, "Design and cosimulation of hierarchical architecture for demand response control and coordination," IEEE Transactions on Industrial Informatics, vol. 13, no. 4, pp. 1806–1816, 2017.
[74] Roya AhmadiAhangar, Argo Rosin, Ali Nabavi Niaki, Ivo Palu, Tarmo Korõtko, "A review on real-time simulation and analysis methods of microgrids," International Transactions on Electrical Energy Systems, vol. 29, no. 11, 2019, doi:10.1002/2050-7038.12106.
[75] L. Bottaccioli et al, "A flexible distributed infrastructure for real-time cosimulations in smart grids," IEEE Transactions on Industrial Informatics, vol. 13, no. 6, pp. 3265–3274, 2017.

17 Shade-Tolerant PV Microconverters

Vadim Sidorov, Abualkasim Bakeer, Hamed Mashinchi Maheri, Naser Hassanpour, Showrov Rahman, and Andrii Chub

CONTENTS

17.1 OVERVIEW OF A RESIDENTIAL PHOTOVOLTAIC SYSTEM

Depending on different configurations of photovoltaic (PV) converters, residential PV systems can be categorized broadly into two classes: 1) PV system with a central converter and 2) PV system with module-level power electronics (PV MLPE). The PV MLPE systems can be further sub-categorized into PV power optimizers and PV microinverters. Figure 17.1 demonstrates the classification of a residential PV system.

17.1.1 PV System with a Central Converter

In a conventional PV system, the solar modules are usually connected in series to form string(s), as shown in Figure 17.2. Often, multiple strings are connected in series for scaling up the ratings, in which case the total accumulated PV

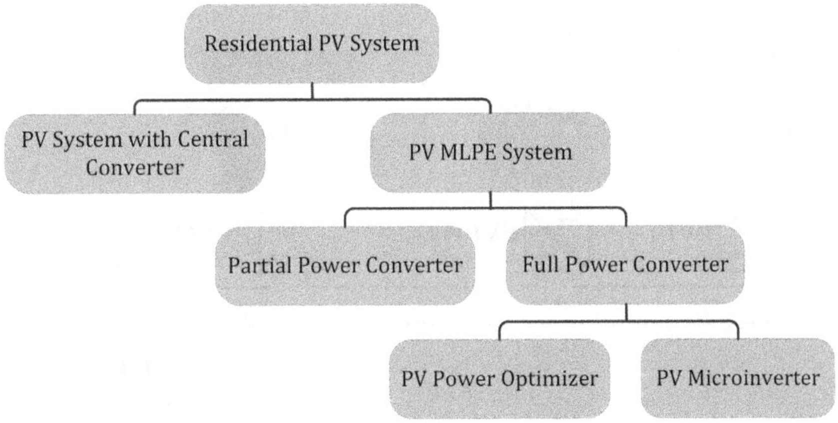

FIGURE 17.1 Classification of a residential PV system.

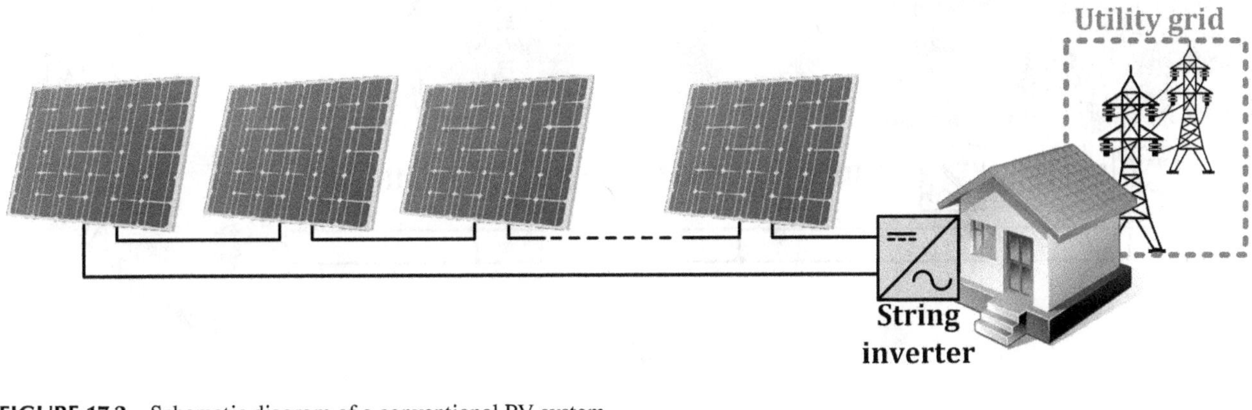

FIGURE 17.2 Schematic diagram of a conventional PV system.

DOI: 10.1201/9781003229124-17

system voltage is inverted to AC by a "central converter", which is also known as a "string inverter".

Although these types of PV systems have lower cost and component counts due to the panels connected in series, failure of a single solar module may jeopardize the operation of the whole string. Shading or partial shading problem is another serious drawback of the conventional PV system. If just a single panel is partially shaded and is producing 75% of its production capacity, then the entire string could produce only 75% of its full capacity. Therefore, in other words, the conventional PV system is what is called "as strong as its weakest point". Bypass diodes can sometimes mitigate this problem, but frequently, they are not entirely effective. Extensions of PV strings are also comparatively difficult in this type of the PV system since the string layout, module orientation, tilting, and the number of solar modules in a string should be identical.

17.1.2 PV System with Module-Level Power Electronics (MLPE)

Unlike the conventional PV system architecture based on central/string inverter(s), the application of module-level power electronics (MLPE) results in a better energy yield by reducing the mismatch in losses of the PV modules. The PV MLPE can be categorized into different sections depending on the criteria. For example, depending on the processed power, the PV MPLE systems can be divided into partial- or full-power conversion [1]. Between these two, the full-power conversion system is more favorable due to better versatility and performance in various operating scenarios of the PV installation. The full-power PV MLPE is mainly represented by the PV power optimizers (PVPO) and PV [2]. In fact, in many cases, PV MLPE is referred to as the collective term for a PV power optimizer and a PV microinverter.

PV MLPE offers several benefits over conventional PV systems, including higher system reliability, design flexibility, higher electricity yield, and shade tolerance. Module-level integration and power handing also make its use comparatively safe.

17.1.2.1 PV Power Optimizer (PVPO)

The PV Power Optimizer (PVPO) is basically a DC/DC converter (typically a non-isolated buck-boost DC/DC converter) added to the conventional series-connected PV string configuration, as shown in Figure 17.3(a) [3]. The PVPO sends optimized DC voltage to the string inverter, maximizing the overall electricity yield of the PV module. The main advantage of the PVPO is its ultra-wide input voltage operation range, and therefore they offer a higher degree of compatibility which can be paired with the vast majority of PV modules in the market.

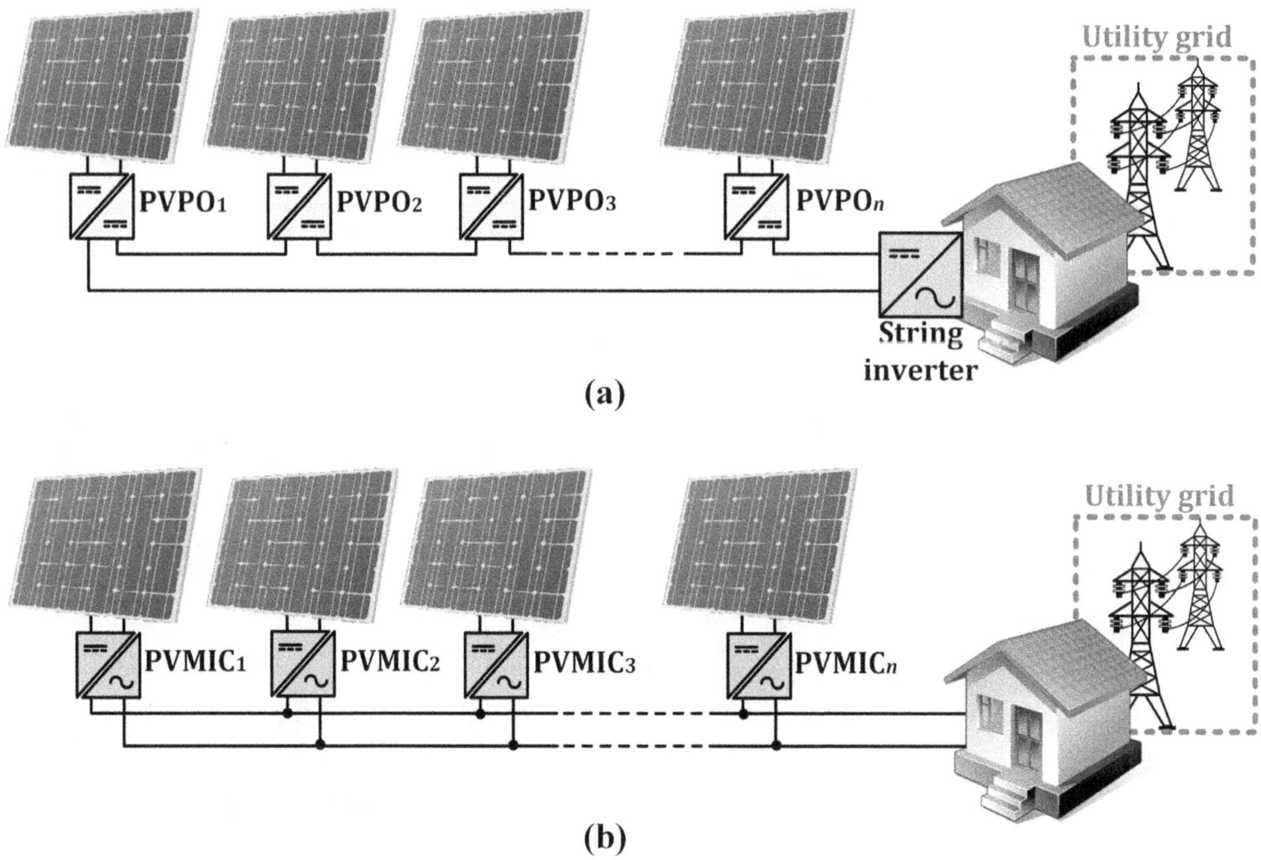

FIGURE 17.3 Generalized schematics of PV MLPE systems: (a) PVPOs; (b) PVMICs.

Nevertheless, the existence of the string inverter may still act like a "bottleneck" in the conversion system. Besides, due to the high DC voltage generation from the series-connected PV modules, arcing on the wires between the PV modules and the string inverter may occur.

17.1.2.2 PV Microinverter (PVMIC)

In the PV microinverter (PVMIC) system, the microinverters associated with the PV modules are connected in parallel, as shown in Figure 17.3(b). PVMICs maximize the energy harvest from the PV module; at the same time, they perform the DC-to-AC conversion. Therefore, direct AC connectivity with the grid is established, eliminating the necessity of a string inverter. Thanks to the parallel connection scheme, the PVMIC configuration can be easily scaled up/down by adding/removing the PV modules. Moreover, at any malfunction, module-level monitoring and/or shutdown can be performed. However, with its multistage power conversion and embedded control, protection, and communication systems, PVMICs have considerably higher component count. This, in turn, increases their cost, raising the cost per watt significantly. Furthermore, higher number of components could potentially lead to higher failure rates (especially at the high cell temperatures). This may affect the long-term reliability of the PVMIC system. Another issue with PVMICs is that they have a comparatively narrow maximum power point tracking (MPPT) voltage range, which is typically compromised by tradeoffs between cost, efficiency, and reliability.

17.2 CONVERTER EFFICIENCY ASSESSMENT

PV inverters are one of the key components influencing the overall energy yields and the efficiency of the PV system, provided that the installed PV modules are appropriately chosen. PV converter efficiency is often expressed as the weighted converter efficiency. Since the PV modules do not always operate at the same irradiance level, their outputs may also vary. That is why weighted efficiency, which applies weighting factors to provide an average efficiency of the converter, is typically used. The weighted converter efficiency is also recommended by the IEC 61683 and EN 50530 standards, and is based on the consideration that the PV modules are subjected to the continuous variation of the irradiance level. Also, according to the standards, the back panel temperature is assumed to be kept constant at 25° C with a varying irradiance profile.

Different versions of the weighted converter efficiency approximation method are followed in different regions. One of the most popular versions of the weighted converter efficiency uses hourly irradiance sampling to calculate the converter efficiency and can be expressed as follows:

$$\eta_{EURO} = 0.03\,\eta_{5\%} + 0.06\,\eta_{10\%} + 0.13\,\eta_{20\%} + 0.10\,\eta_{30\%} + 0.48\,\eta_{50\%} + 0.20\,\eta_{100\%} \tag{17.1}$$

The weighted converter efficiency η_{EURO} in Eq. (17.1) is formulated on the basis of six pre-determined levels of operations expressed in percentage—5, 10, 20, 30, 50, and 100%; and the weighting factors—0.03, 0.06, 0.13, 0.10, 0.48, and 0.20. The efficiency is calculated by multiplying the weighting factors to the converter level of operation. For example, the first term of the equation expresses that the weighting factor 0.03 or 3% multiplied with the converter efficiency is multiplied with 5% of its rated capacity with respect to the converter input DC power.

Typically, the method of efficiency calculation demonstrates higher accuracy for regions that have a moderate irradiance profile. Another popular weighted efficiency method is the California Energy Commission weighted efficiency, which has been proven as a more accurate efficiency calculation method for the regions with higher irradiance profile. It uses one second sampling data of irradiance, as well as different operation levels and weighting factors, as expressed in what follows:

$$\eta_{CEC} = 0.04\,\eta_{10\%} + 0.05\,\eta_{20\%} + 0.12\,\eta_{30\%} + 0.21\,\eta_{50\%} + 0.53\,\eta_{75\%} + 0.05\,\eta_{100\%} \tag{17.2}$$

However, the weighted converter efficiency method considers only the converter efficiency η_{EURO} or η_{CEC}, disregarding the efficiency of the converter in terms of MPPT, which refers to the algorithm that enables the converter to extract the maximum available power from the PV modules. Therefore, MPPT is also an influencing factor for determining the PV converter efficiency. The total efficiency or the overall efficiency of a converter can be expressed as a product of the converter efficiency and the MPPT efficiency.

Eq. (17.3) indicates the average output AC power and the average input DC power, and refers to the maximum available PV power. The definition of the overall efficiency is also described; P_{AC} and P_{DC} indicate the average output AC power and the average input DC power, respectively; and P_{MPP} refers to the maximum available PV power. The definition of the overall efficiency is also described in standard EN 50530 (2010) as: "Overall efficiency of photovoltaic inverters".

$$\eta_{TOTAL} = \eta_{CONV} \cdot \eta_{MPPT} = \frac{P_{AC}}{P_{DC}} \cdot \frac{P_{DC}}{P_{MPP}} \tag{17.3}$$

The converter efficiency η_{MPPT} becomes a key influencing aspect, especially in shaded conditions. Partial shading reduces the electricity generation and hence decreases the conversion efficiency. The obstruction of sunlight to the PV cell(s) or shading may occur in different ways, e.g., shade from a nearby tree, chimney, antennas, poles, etc. Tree leaves or bird droppings fallen on top of PV cell(s) may lead to the worst shading called "opaque shading" [4]. Figure 17.4(a) illustrates the partial shading caused by the shadow of a nearby pole and opaque shading due to tree leave(s) dropping onto the PV panel. Because of the shading, Cell numbers 30, 50, and 51 have varying irradiances.

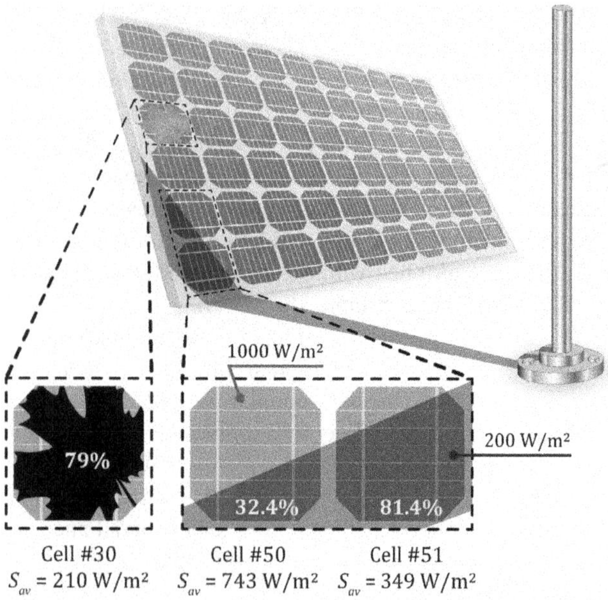

(A)

(B)

FIGURE 17.4 (a) Partial and opaque shading in a 60-cell PV panel; (b) structure of a 60-cell PV panel.

Even if a single module is shaded, the energy production of other modules on that PV system still suffers; due to that, the full power from the unshaded modules cannot be fully utilized. The shaded PV modules are usually bypassed by the integrated bypass diode(s), as shown in Figure 17.4 (b). However, the integration of bypass diodes generates multiple local maximum power point (LMPPT) peaks (shown in Figure 17.5), which makes it difficult for the converter to detect the global maximum power point tracking (GMPPT) to extract the maximum available power. Therefore, in the presence of partial shading, the overall efficiency of the converter highly depends on its ability to differentiate the GMPPT from the LMPPT.

The converter that can catch the GMPPT more accurately has higher η_{MPPT} efficiency. Likewise, since the overall efficiency depends on the product of η_{MPPT} and η_{CONV}, and the overall efficiency of the converter that catches the GMPPT accurately improves, as well. For example, in [5], three different PV microinverters are compared—a low-end microinverter (LoMI), a high-end microinverter (HiMI), and a shade-tolerant microinverter (STMI)—which are applied to building-integrated PV (BIPV) technology. The comparison considers different shading conditions and BIPV module combinations. According to [6], the STMI outperforms both LoMI and HiMI in all the test environments thanks to its greater η_{MPPT} efficiency.

17.3 PV MICROCONVERTERS

PV microconverters (PVMICs) can be divided to two types: the first type is DC/AC PVMICs, also known as PV microinverters, which include an internal inverter and connect directly to the grid; and the second type is DC/DC PVMICs, which connect to a centralized DC bus with a voltage of 350—400V and form a DC microgrid. In the second case, a centralized inverter is used for connecting the DC microgrid to the AC grid.

According to the characteristics of the PV module, PVMICs should have a wide input voltage range, at least from 20–45V for providing GMPPT and maximum operating power up to 350W. At the same time, PVMICs should operate under the grid voltage or under DC-bus voltage. It means that PVMICs should be high-boost or step-up converters. Regarding different standards and requirements of electricity, PVMICs can be with or without galvanic isolation.

The first type of the DC/AC PVMICs is a non-isolated boost converter. The main advantage of this type is few semiconductor devices are required and, as a result, cost is low. However, this converter should operate in a high-boost mode to achieve a high voltage gain. As a result, the semiconductor devices operate under high current stress. Due to these disadvantages, this first type of PVMICs is not popular in the market. One of the presented examples of non-isolated PVMICs is a modified Ćuk converter with an input voltage range from 40–100V (Figure 17.6) [7]. Experimental results show that the efficiency of the converter equals 93.5% at V_{in} = 50V, V_{out} = 110V, and P_{max} = 300W; the CEC efficiency equals 94.55%. According to the authors, the theoretical DC voltage gain of the converter can be up to 10V; however, in practice, the converter can operate with the maximum gain of up to 3V, a level gain which is not enough for operating under the AC grid of 230V. This is a serious disadvantage of the converter.

Due to the described disadvantages, isolated step-up PVMICs are more popular in the market. According to the structure of the topology, this type of PVMICs is divided into three categories: single-, two-, and three-stage converters. The main difference of the single-stage converter from the two- and three-stage converters is that it uses the full-bridge unfolder circuit for grid connection. The single-stage converter forms a half of sinusoidal voltage for the unfolder. The unfolder circuit operates at a double-line frequency and can be released based on transistors, as well as thyristors. Thyristors allow cost and size reduction of the converter, but they have a high level of the forward voltage drop that results in much higher conduction power losses than with

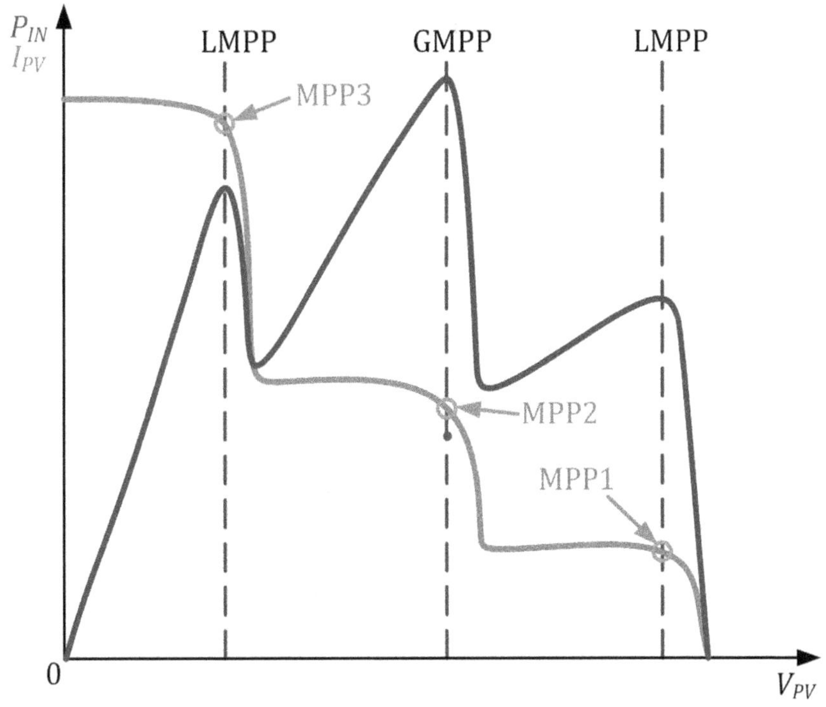

FIGURE 17.5 PV characteristics of the photovoltaic string under partial shading.

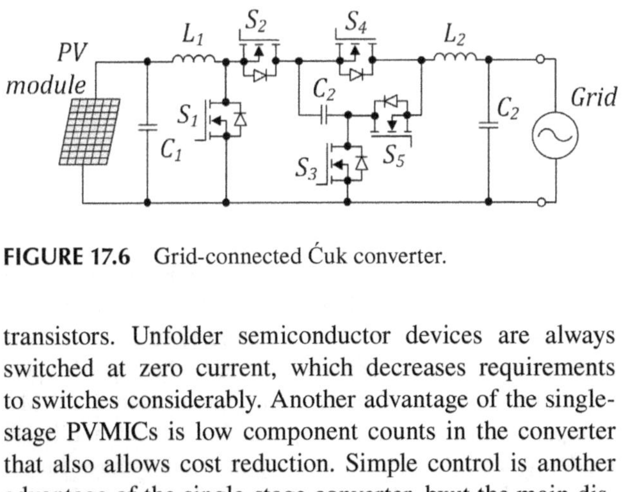

FIGURE 17.6 Grid-connected Ćuk converter.

transistors. Unfolder semiconductor devices are always switched at zero current, which decreases requirements to switches considerably. Another advantage of the single-stage PVMICs is low component counts in the converter that also allows cost reduction. Simple control is another advantage of the single-stage converter, bxut the main disadvantage of such single-stage PVMICs is the double-line–frequency voltage ripples at the input side. For filtering these ripples, large capacitors are required that increase the cost, weight, and size of a microconverter.

The popular topology of the single-stage PVMICs is the flyback microinverter (Figure 17.7). This converter can operate in continuous current mode (CCM), discontinuous current mode (DCM), or border current mode (BCM). DCM and BCM provide a zero-current switching for the main switch, which allows for reducing switching losses, but in comparison with the CCM, the DCM and BCM generate high RMS current in the converter and high current ripple in the primary side, through the primary capacitor C_1. As a result, the size and cost of the capacitor is increased. However, for providing the CCM in a wide power range, a large isolation transformer with a high magnetizing inductance is required. In this case, mixing different operation modes can be applied in the flyback microinverter for achieving high performance and small size and weight at the same time, as reported in different studies [8].

Many studies have also focused on improving the performance of the flyback topology. An active-clamped circuit or an active snubber are typical solutions for recycling the leakage energy and achieving soft-switching for the primary switch. These circuits include a capacitor that reduces the voltage overshoot and lowers the rising slope of the turn-off voltage. This additional capacitor does not affect the soft-switching of the main switch at turn-on. Another improvement for the scaling power of the flyback converter is using the interleaved topology. That allows spreading current and heat stress in a converter and utilizing low-cost transistors. Using an interleaved topology doubles equivalent switching frequency and the amplitude of current ripple in the primary side. As a result, interleaving gives flexibility in the design of the converter.

The topology of the interleaved flyback PVMICs with an active snubber and the unfolder based on transistors is shown in Figure 17.8. This topology is typical of commercial products. For example, the M250 from Enphase Energy and the solar microinverter from Microchip Technology were realized based on that topology [9, 10].

The next category of isolated PVMICs is two-stage microinverters, which consist of a step-up DC/DC circuit and a high-frequency inverter for connection to the grid. The first

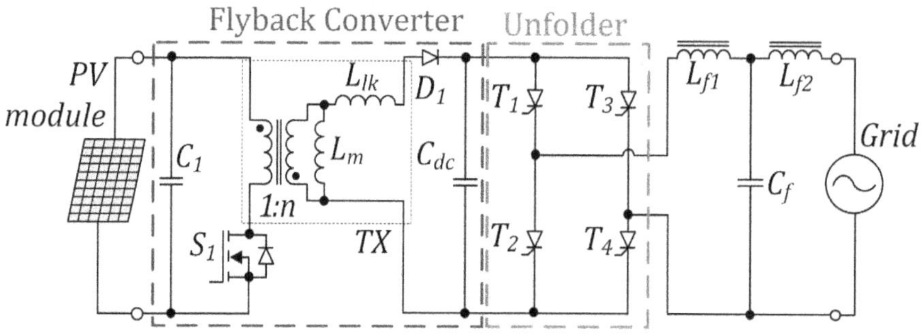

FIGURE 17.7 Flyback microconverter with the unfolder based on thyristors.

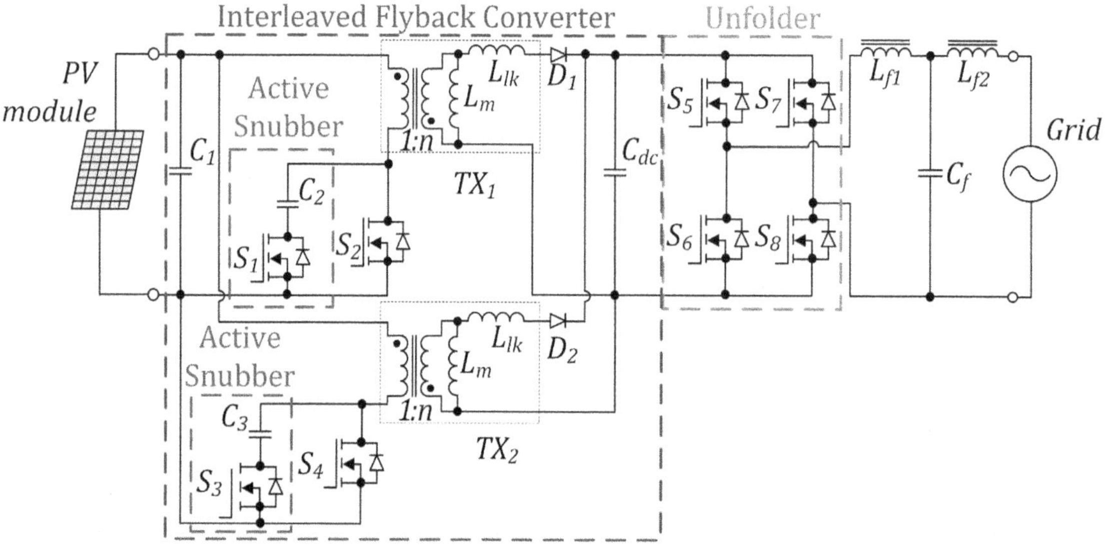

FIGURE 17.8 Interleaved flyback microconverter with an active snubber and the unfolder based on transistors.

stage forms DC voltage and implements MPPT, and the second stage usually operates at high frequency and forms sinusoidal current for the grid. One of the main advantages is the possible buffering of double-line–frequency voltage ripples in the DC-link capacitor, which operates under high voltage and low current stress. Although the two-stage PVMICs can be separately optimized, total efficiency is lower due to switching and conducting losses in both stages, and the system cost is higher than with single-stage microconverters. Each of the stages can be realized by applying a different topology.

The first example is a microinverter from Texas Instruments (TI), shown in Figure 17.9 [11]. The first stage of the topology consists of a step-up flyback DC/DC converter with an active clamping circuit and voltage-doubler rectifier (VDR). The active clamping circuit allows for achieving soft-switching of the main transistors S_2. The VDR was used in the converter to decrease the turn-ratio of the isolation transformer that allows for decreasing the size and cost of the transformer. At the same time, the VDR doubles the RMS current in the transformer, which increases conduction losses. The second stage was realized based on two DC/AC buck converters, which regulate the output voltage and current.

Another example of a two-stage topology is that based on the DC/DC push-pull converter, which is used as the first stage [12]. As can be seen in Figure 17.10, the stage was realized based on the conventional H-bridge, which operates under sinusoidal pulse-width modulation (SPWM). The push-pull converter can operate in the buck and boost mode, which allows for extending an input voltage range. This is a simple topology. However, in this converter, transistors operate with high switching current, which generates high switching losses. Another disadvantage of the converter is the charged energy in the leakage inductance of the transformer. This energy should be dissipated or discharged in the primary capacitor by an additional circuit.

A microinverter based on interleaved boost converters from STMicroelectronics is shown in Figure 17.11 [13]. Two buck converters apply pulse voltage to the isolation transformer and carry out MPPT. Unlike the previous example, there is no problem with the leakage inductance because the leakage current circuits through primary switches, but this topology consists of two inductors that increase the size, weight, and cost of the converter. Transistors also operate with high switching current. The DC/AC stage was realized

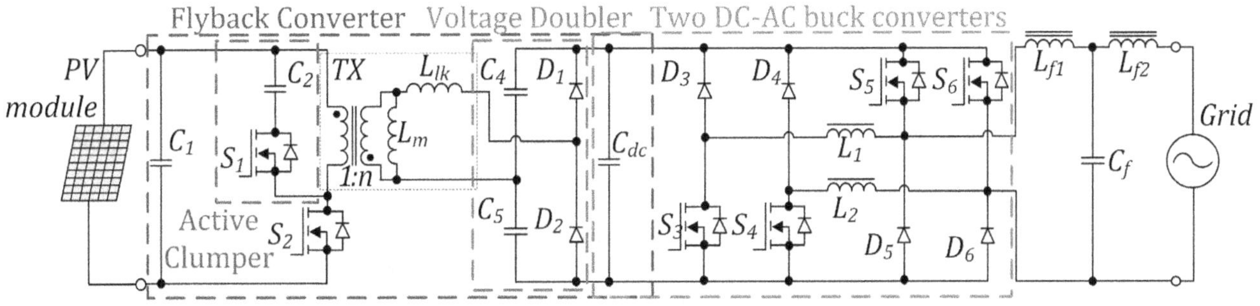

FIGURE 17.9 DC/DC flyback converter with a voltage-doubler rectifier and two DC/AC buck converters.

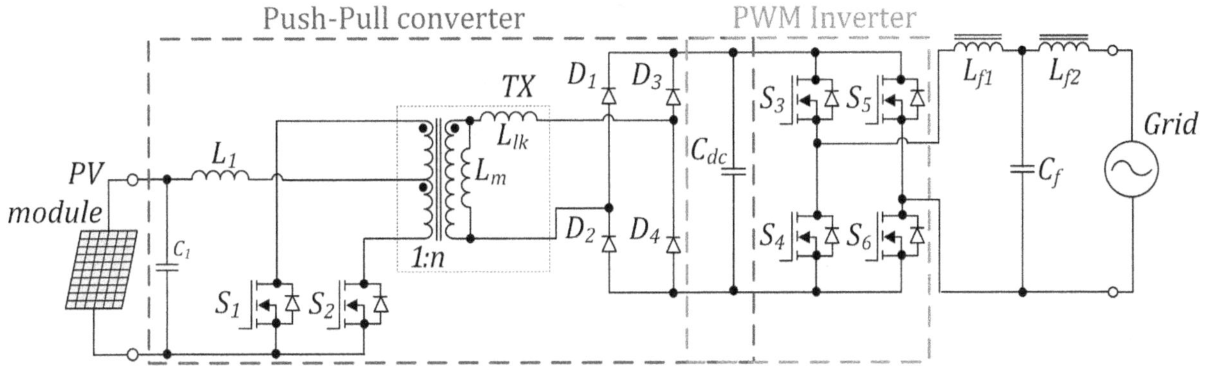

FIGURE 17.10 DC/DC flyback converter with a voltage-doubler rectifier and two DC/AC buck converters.

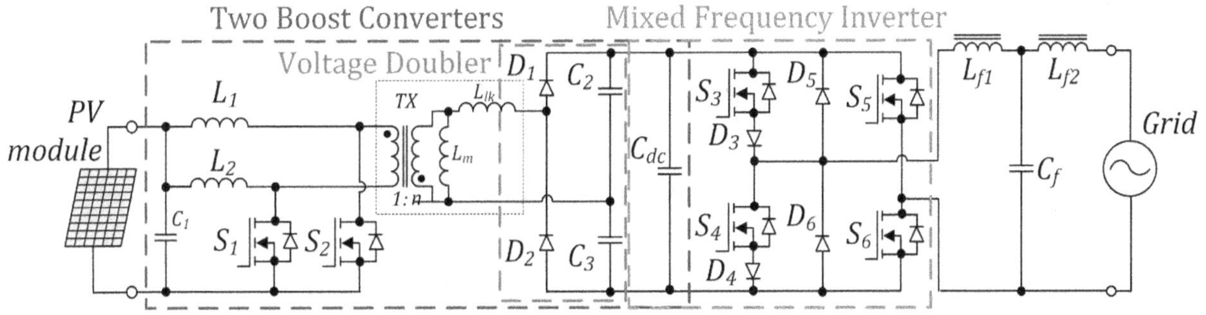

FIGURE 17.11 DC/DC flyback converter with a voltage-doubler rectifier and two DC/AC buck converters.

based on a mixed frequency inverter. In this inverter, only two transistors operate with high switching frequency, while the other two operate with a line frequency and have zero-current switching. This feature allows for a decrease in the requirements to the switching performance of switches S_5 and S_6; as a result, low-cost transistors can be selected for the second leg. Figure 17.11 shows that additional diodes were added in the inverter for blocking low-performance body diodes of switches S_1 and S_2. It increases conduction losses, but at the same time, decreases reverse recovery losses of the body diodes. To ensure circulating of current during dead times, additional anti-parallel Schottky diodes were added instead of body diodes in the inverter. This solution increases the size and cost of the inverter; however, it allows for achieving high performance.

A two-stage microinverter from Enecsys company has been realized based on the full-bridge LLC converter (Figure 17.12) [14]. The main advantage of the converter is soft-switching of transistors. It allows for a significant decrease in power losses. Nevertheless, the LLC converter features some disadvantages, such as frequency control methods and circulating power, and, as a result, narrowed operation ranges in practice. The reason is that to expand the input voltage range, in compliance with the frequency control, the operation frequency must be decreased or increased more than two times of a resonant frequency. Consequently, it requires a large input capacitor for filtering the input current at low frequency. But the main disadvantage of low operation frequency is the circulating energy in the primary side that produces conduction losses in the primary switches, the series capacitor, and the isolation transformer. A narrow input voltage range can be compensated by an additional controlled circuit. As shown in Figure 17.12, a current shape circuit allows for regulation of the input voltage and forms a half-sinusoidal current for the unfolder, which is utilized as a grid-connection circuit. Another disadvantage of the LLC converter is the series capacitor, which should provide maximum current of the transformer and have low deviation of capacitance for keeping the resonance frequency in the same level at different temperatures.

The next example of a two-stage microinverter is the Solar Optiverter from Ubik Solutions based on the hybrid quasi-Z–source (QZS) inverter with the series-resonant converter (SRC) and the PWM grid-connection inverter (Figure 17.13). This topology operates under the buck mode by applying the phase-shift modulation to control the SRC, and the boost control mode by applying short-through modulation to the QZS [15]. Operating in two control modes allows significant extension of the input voltage range. The QZS circuit has been realized based on the coupled inductor LQZS, integrated with the SRC. This feature of the topology allows reduction of elements and size of the Optiverter. The QZS circuit also operates as an input filter in the buck mode, as well as in the boost mode.

Another type of PVMIC is the DC/DC PV converter. From the theoretical view, all the described examples of microinverters can be applied for operation with a DC bus by removing a grid-connection inverter and changing the control method for operation with a DC output voltage. Since in the two-stage PVMICs, the first stage is DC/DC converters, also single-stage converters can be changed for operation with the DC voltage. In addition, another DC/DC converter can be described as a PVMIC. One example of DC/DC converters is a non-isolated single-stage DC/DC converter, which is shown in Figure 17.14. This topology has been utilized as a PV DC/DC converter from Femtogrid Energy Solutions [16]. The patent of this topology belongs to Tesla. The topology consists of only one controlled semiconductor device, which operates with soft-switching. As a result, the converter has low power losses and low cost. This topology can operate with high DC voltage gain, which allows for extension of the input voltage range.

The Optiverter based on the QSZ with SRC from Ubik Solutions is a good example of universal utilization of a PVMIC for the AC grid, as well as the DC grid. The first stage of the Optiverter is shown in Figure 17.15 [17]. This topology can be used for the DC grid because each stage has independent control, as previously stated; thus, stages of the topology can be separated. Also, the voltage range of the DC line is near the range of the DC grid. This feature allows for utilization of the same topology for the AC and DC grids. This approach unifies the production of the PVMICs and reduces the cost of the product.

Table 17.1 shows the comparison of different PV microinverters, as well as PV DC/DC microconverters. As can be seen, the typical input voltage ranges of the described microconverters equal about 20–45V, which is not

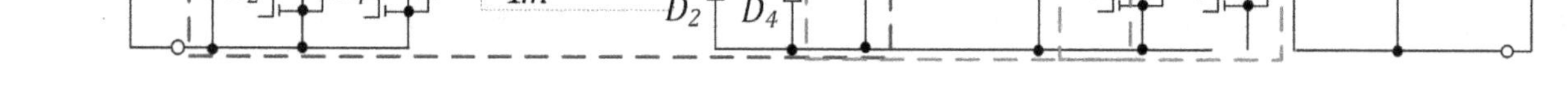

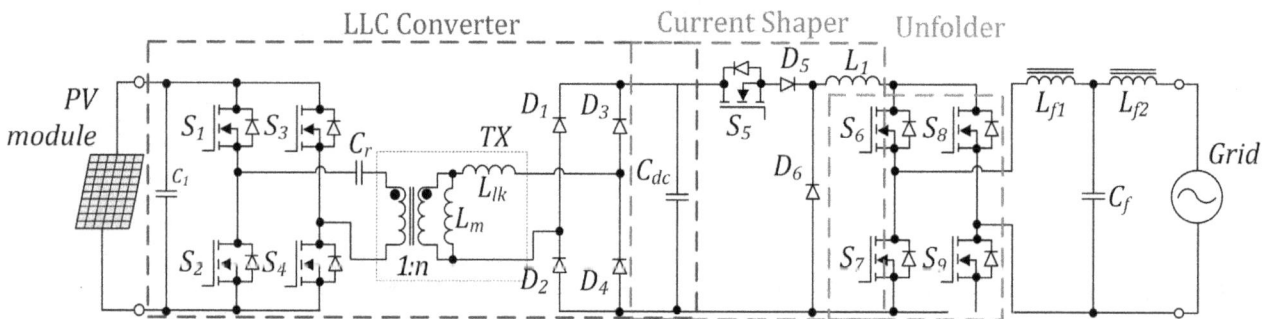

FIGURE 17.12 LLC DC/DC converter with a current shaper and unfolder.

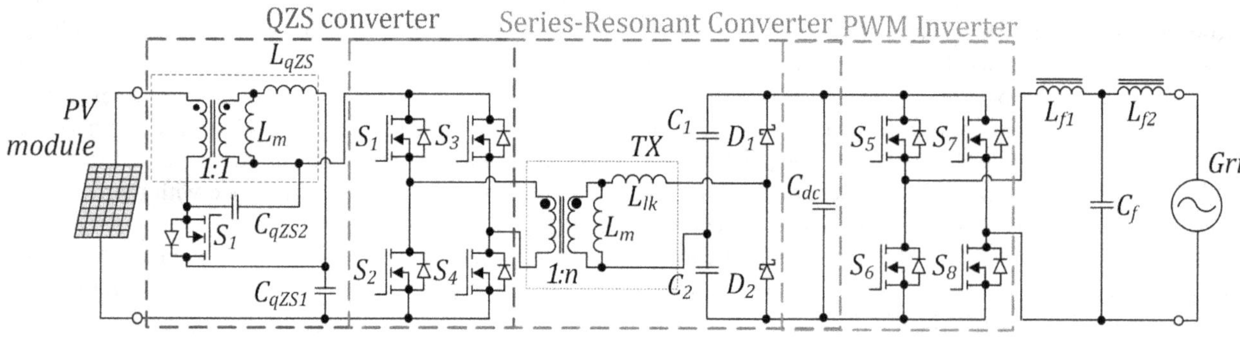

FIGURE 17.13 Hybrid quasi-Z–source with an SRC.

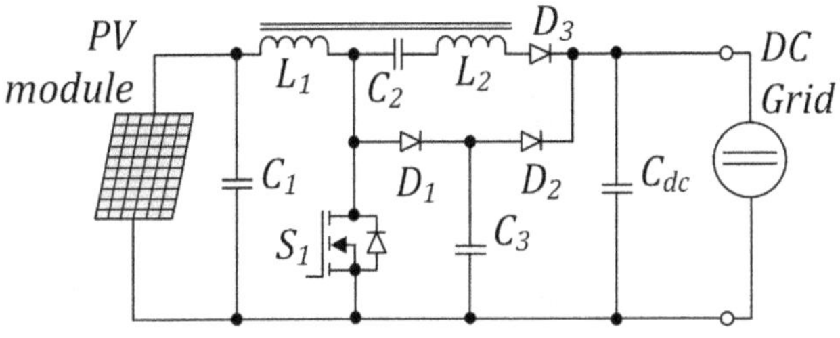

FIGURE 17.14 Non-isolated boost DC/DC converter.

enough for operating under partial shading conditions. Microconverters switch off when the input voltage drops. Only the Solar Optiverter can provide the required input voltage range for operating under partial shading conditions. At the same time, the converter has high efficiency. As the review of the PCMICs showed, non-isolated microinverters are not an appropriate solution for module-level PV converters. The reason is that this type of PVMICs cannot provide the typical input voltage range for operating with the majority of used PV modules and the AC output voltage of 230V, according to European standards. As a result, there are no industrial examples of module-level microinverters based on non-isolated topologies. The most popular solution for industrial PVMICs is the interleaved flyback topology with an active snubber or an active clamper.

Examples of the DC/DC PVMICs show the highest efficiency in comparison with the DC/AC converters. This is because this converter type does not include a grid-connection inverter, which has power losses. In any case, the DC/DC PV system requires one centralized interface inverter to connect the DC voltage bus to the utility AC grid. In summary, the PV system based on the DC bus and DC/DC PVMICs has higher efficiency than the system based on PV microinverters.

17.4 PV MODULE SUBSTRING-LEVEL SOLUTIONS

In this subsection, utilization of substring level approach for PV module microconverters is described. PV microinverters as the module-level power electronic approach was proposed to address the mismatch issue in the module level. The approach features a MPPT to increase the level of harvested energy with the benefit of increased reliability and a possibility of module-level performance

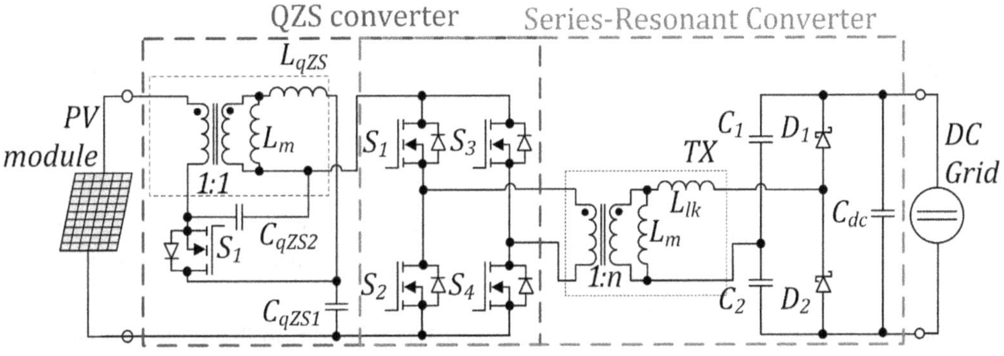

FIGURE 17.15 DC/DC QZS with SRC.

TABLE 17.1
Comparison of PV Microconverters

PVMIC	Input Voltage Range, V	Maximum Power, W	Output Voltage Range, V	Peak Eff., %	CEC Eff., %
Non-isolated Ćuk converter [7]	40–100	300	110	94.43	94.55
M250 Enphase [10]	16–60	250	189–264	96.5	96.5
Microchip microinverter [9]	25–45	250	90–140, 210–264	95	94.5
TI microinverter [11]	28–45	320	110–140, 220–280	93.34	92
Microinverter based on push-pull [12]	20–32.9	270	207–253	-	-
ST microinverter [13]	18–55	250	230–240	94.1	93.4
Enecsys microinverter [14]	20–44,	225	230	94.8	-
	27–54	340		95.4	—
Solar Optiverter [15]	8–60	300	230	96.2	95.3
Femtogrid [16]	8–42	310	360–400 DC	97	96.67
DC/DC QZS SR [17]	8–60	300	400 DC	97.4	96.75

monitoring. The approach shows significant benefits in the building level of PV systems, as most of the issues covering performance are related to this installation. Although the proposed approach was successful to track the MPP, the extracted power still is lower than the maximum available power of the PV module, which is due to the operation of the bypass diodes. As a solution for this issue, the concept of a substring level converter has been proposed. In the substring-level approach, the embedded bypass diodes in the junction box of each module are replaced by an active or passive topology, allowing harvesting of the maximum available power in the substring level. The concept is realized by processing the mismatched power between the shaded and the unshaded substring in an individual PV module or by the performance of the MPPT process for each substring separately by an individual operation of a DC/DC converter. The concept could be realized by different configurations of a power electronics converter. This study is focused on three emerged structures for this concept. The operation principle and possibilities of each structure are addressed. In order to show how the concepts could overcome the partial shading issues, the advantages and drawbacks of each approach are evaluated. Finally, the results of the assessment are provided in the conclusions.

17.4.1 Differential Power Processing (DPP)

Differential power processing approach is used to be a practical solution to overcome the issues related to the power mismatch in PV elements [18]. Here, it is focused on the substring level DPP concept as an efficient approach to address the power mismatch issues in the substring level. In the DPP approach, only a fraction of the overall power of the PV module is processed by DPP converters. Therefore, the candidate topologies for these converters require a compact and low-cost hardware implementation. The categorization of the DPP approach is realized by two terms of architecture and topology. The former refers to the configuration of the DPP converters with substrings, and the latter refers to the circuit topology of the DPP converter. From the aspect of architecture, the DPP system is divided into two categories of substring-to-substring and substring-to-bus. A variety of buck-boost and flyback topologies could be adopted as DPP converters. In this subsection, among different architectures of the DPP approach, the structure used successfully in the substring level with compatibility of the microinverter level is selected as a case study [19]. Figure 17.16 shows the main structure of the DPP concept. It is a substring-to-substring configuration adopted to a PV module with three substrings. In this approach, the embedded bypass diodes in the junction box are eliminated and the connection of the DPP converters that are bidirectional buck-boost converters is arranged such that the mismatch power between the substrings is processed. Figure 17.16(b) shows the possibility of the connection of DP converters to substrings.

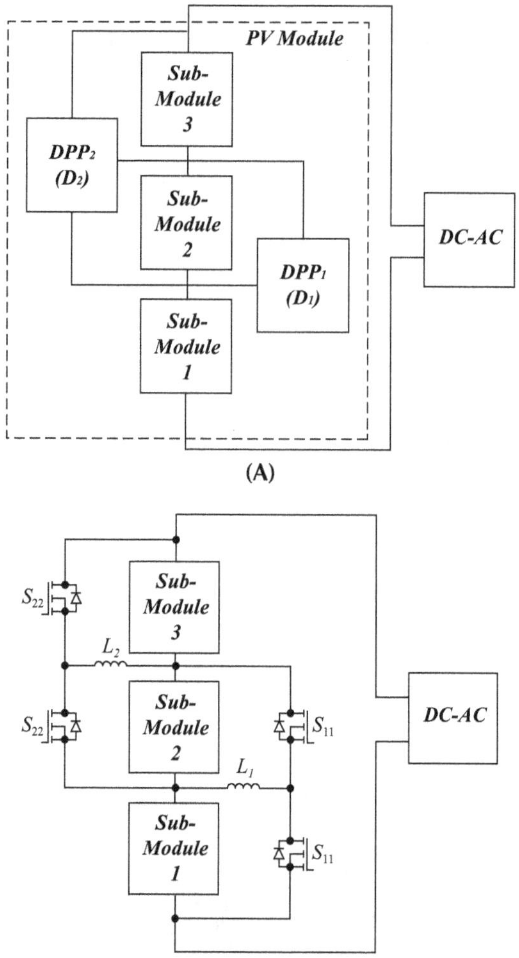

FIGURE 17.16 PV microinverter: (a) substring-to-substring DPP architecture; (b) possibility of connection of DPP converters.

17.4.1.1 Operation Principle and Design

The duty cycle of the switches of DPP converters is controlled and the mismatch power between adjacent substrings is processed to adjust the operating point of each substring to its maximum power. The switches are controlled as the proportional current is supplied by the inductors for operation of the substring in MPP. The average current of inductors is:

$$I_{L1} = I_{PV2} - I_{PV1} + (1 - D_2) I_{L2}, \tag{17.4}$$

$$I_{L2} = I_{PV3} - I_{PV2} + D_1 I_{L1}, \tag{17.5}$$

where I_L is the inductor current, I_{PV} is the current of each substring, and D is the duty cycle of each switch that depends on the voltage of substrings and is obtained by:

$$D_i = \frac{V_{PV,i}}{V_{PV,i+1} + V_{PV,i}}, \tag{17.6}$$

where $V_{PV,i}$ is the voltage of the *ith* substring

Considering Eq. (17.4), the average current of the inductors depends on each other. The overall power processed by the DPP converters is obtained by Eq. (17.8).

$$P_{DPP} = V_{PV1}\left|I_{L1}\right| + V_{PV2}\left|I_{L2}\right| \tag{17.7}$$

Analysis of Eqs. (17.4)–(17.5) shows that by increasing the difference between the current of substrings, the current of the DPP converter increases, resulting in a high processed power by the converter. The fraction of the overall power processed by the DPP approach is defined as:

$$\alpha = \frac{P_{DPP}}{P_{PV}} \tag{17.8}$$

where P_{PV} is the overall power produced by the substrings of the PV module

To implement the MPPT algorithm, it is required to sense the current of the PV module by the DPP converters; however, the microinverter must simultaneously measure the module current to track the maximum power of the PV module. As the DPP converter and the microinverter are installed in the backside of the PV module, the process of measurement is accurate in a fair way. However, this process is complicated for systems with a central inverter. In the design of the component as the DPP converter process, the fraction of the overall power of the PV system, the voltage, and the power rating of the component are low. This results in the reduction of cost and size. Operation at low power allows for selection of high switching frequency. As in most of the cases, a large part of the circuit is a magnetic component, operating at high frequency that results in significant reduction of the size of the inductors of the DPP converter. It enables designing a structure that will fit it in the junction box.

17.4.1.2 Discussion of Experimental Results

The operation of the DPP concept could be evaluated by the experimental test for a mismatch condition of 100% and 50% irradiance. The resulting P-V curve for both with and without operation of DPP converters is shown in Figure 17.17. As can be appreciated from Figure 17.17., the existing mismatch was compensated successfully. For the mismatch condition, operating of the bypass diodes results in different local maximum power points. Figure 17.17(b) shows that a microinverter is able to track the local MPP without a DPP converter. Once the DPP converter was activated, the limited power by the shaded substring was recovered and the PV module could provide the power in the unit MPP close to the maximum available power of the PV.

Evaluation of the results shows that the DPP approach features low power rating as it processes a small amount of the overall PV power. It has high system efficiency with a possibility of integration to the existing system. However, there is a question of how the approach operates at a high mismatch power such as opaque shading.

17.4.2 Integrated Power Optimizer (IPO)

Integrated power optimizer (IPO) is another approach of shading loss mitigation at the substring level. IPO is a passive structure with a possibility of integration with a boost converter used as an interface between the PV module and the DC bus. The combination could emerge as a unified PV microconverter to address the mismatch issue in the substring level. The mismatch current between the substrings is mitigated in the absence of any feedback control system. The power difference is processed by the IPO automatically, and MPP tracking is implemented by the boost DC/DC converter. Figure 17.18 shows the topology of the concept. In the following, the operating principle of the concept will be explained.

17.4.2.1 Operation Principle and Design

Typically, the IPO is a passive voltage multiplier (VM) network consisting of energy transfer capacitors and diodes coupled with an isolation step-down transformer. By eliminating the bypass diodes, the outputs of the VM network are connected to each substring in parallel. The operation of the VM network is based on the equalization of the voltage between substrings. For normal performance of the PV

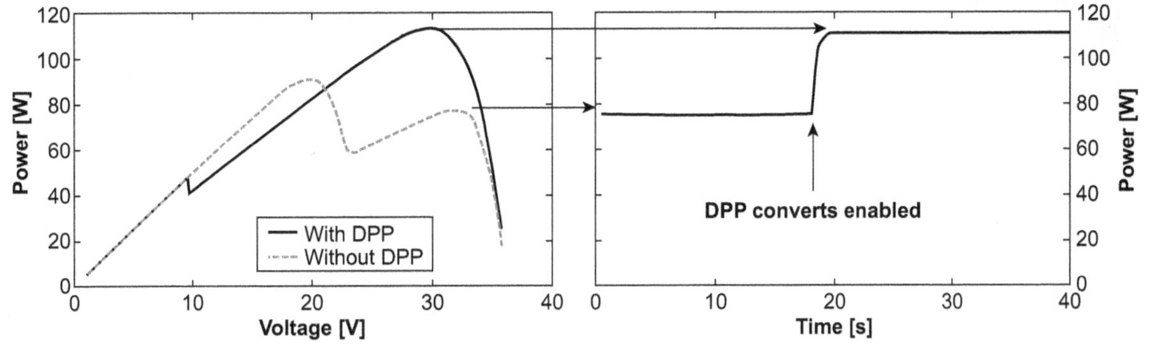

FIGURE 17.17 Experimental results for the mismatch condition of 100%, 50% irradiance: (a) P-V curve with and without a DPP converter; (b) maximum harvested power.

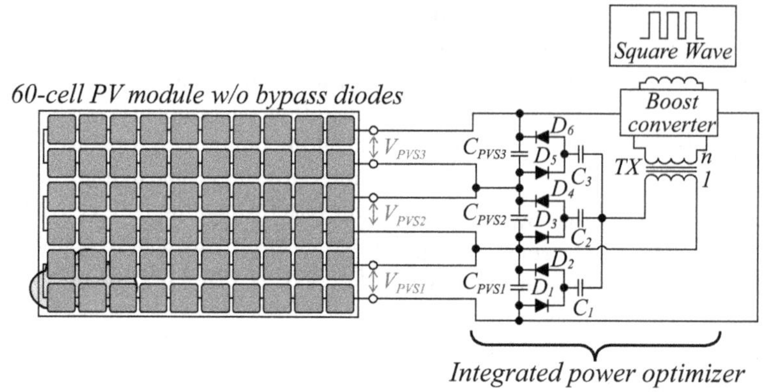

FIGURE 17.18 Generalized topology of the substring level microconverter with an IPO connected to the 60-cell PV module with one shaded substring.

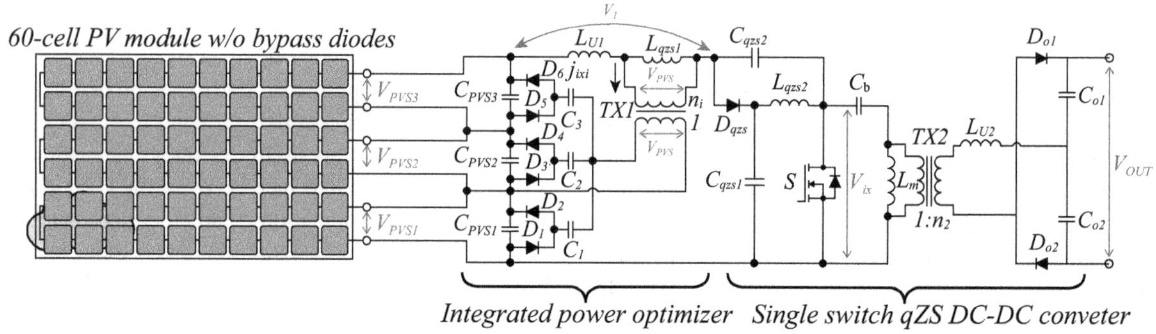

FIGURE 17.19 Generalized topology of SSqZSC with an IPO connected to the 60-cell PV module with one shaded substring.

module, there is equal voltage across the substrings and the IPO does not act. In the mismatch conditions, the mismatch current of substrings results in a voltage drop across the underperformance substring; the IPO acts in a way to compensate the voltage drop to equalize the voltage in all three outputs to ensure that the IPO injects the current to the PV module. For this operation, it is required to apply a square wave voltage to the primary side of the step-down transformer. The IPO could be combined with a boost converter to enable provision of that voltage. This concept was evaluated for a single-switch quasi-Z–source DC/DC converter (SSqZSC). For normal operation of SSqZSC, a square-wave voltage is generated across the inductors of the qZS network. Figure 17.19 shows an emerged topology of a substring PV microconverter that results from the combination of the IPO with SSqZSC. The input inductor of the qZS network is replaced with the primary side of the IPO transformer, and the magnetizing inductance of the transformer acts as an input inductor of SSqZSC. The IPO is driven by the voltage across the input inductor of qZS, which equals the DC voltage of SSqZSC and is obtained by Eq. (17.9):

$$V_{dc} = \frac{V_{PV}}{1-2D_S}, \tag{17.9}$$

where V_{PV} is the input voltage and D_S is the duty cycle of the switch of SSqZSC

For the mismatch conditions, the output diodes of the IPO parallel with shaded substrings conduct and the IPO supplies current to compensate the mismatch current. In order to evaluate the performance of the IPO concept, different shading scenarios could be considered. For example, if one of the substrings is shaded, the IPO will inject the current to the PV module when V_{dc} fulfills the following condition:

$$V_{dc} > n_1(V_{PVS} + 2V_D), \tag{17.10}$$

where V_D is the forward voltage drop of diodes and V_{PVS} is the voltage of the shaded substring

For the operation of the IPO in shading conditions, the turns ratio of the IPO should be designed based on Eq. (17.10).

17.4.2.2 Validation of the Concept

The SSqZSC with the IPO was simulated for the mismatch condition of 800/800/400W/m^2. The simulation results are shown in Figure 17.20 for the operation with and without IPO. For this condition, the maximum available power of the PV module equals 187.2W. As Figure 17.20 shows, the SSqZSC is able to track the local MPP with the power of 125.8W when the IPO does not act. Running the simulation when the IPO is enabled, the MPP is moved to (28.8V,

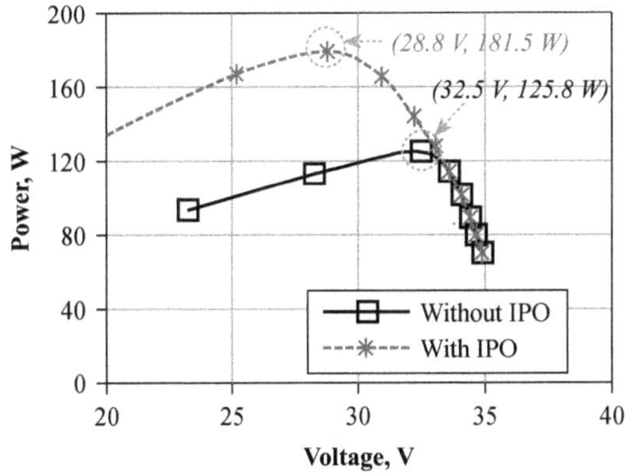

FIGURE 17.20 Localization of the MPP with and without the IPO.

181.5W). The simulation results show that the proposed approach enables harvesting of 97% of the available string power. However, the maximum power achieved for operating without the IPO corresponds to only 67% of the maximum available power.

Regarding the given discussion and simulation results, it is worth mentioning that the mismatch power between the substrings could be compensated by the IPO. The IPO could improve the maximum harvested power of the converter around 30%, against the operation without the IPO. For this operation, the converter is controlled by a simple MPPT algorithm and mismatched power is processed by the IPO automatically. As the VM network is free of magnetizing component, it could be designed in a compact PCB to be fitted in the junction box. The unified structure features high step-up voltage and galvanic isolation.

17.4.3 Substring-Level Microinverter with Parallel and Series DC/DC Converters

The other approach to overcome the power mismatch issues is substring level microinverters with parallel and series DC/DC converters. For this concept, the DC/DC stage of the conventional single input microinverter is reproduced to three individual ones. The input of each DC/DC converter is connected to the substring directly, and their output is combined in series or parallel configuration to the input of the DC/AC inverter. The generalized architecture of this concept is shown in Figure 17.21. Categorization of this concept is realized by the type of the inverter, power decoupling technique and the structure of the used transformer for galvanic isolation [20]. As compared with module-level counterparts, substring level microinverters feature increased component count and higher voltage-up ratio. The mentioned shortcomings would be helpful when selecting a topology for this concept. However, in the topology selection, it is worth taking into account that the emerged advantages of the concept should not be suppressed by the mentioned drawbacks. The topologies with reasonable semiconductor counts are acceptable, as the active and passive semiconductor devices play

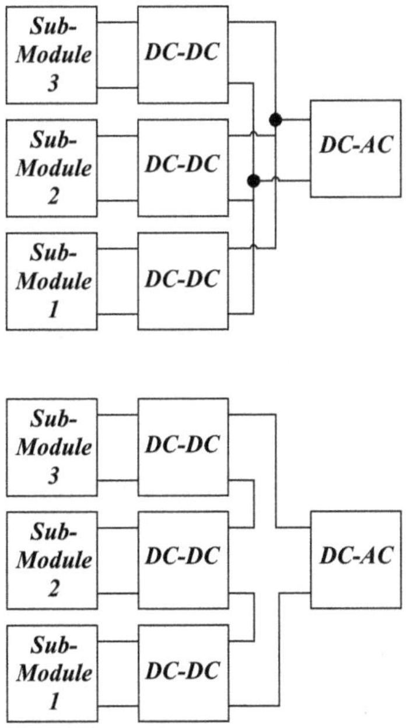

FIGURE 17.21 Possible architectures of the substring-level microinverter: (a) parallel DC/DC stage, (b) serial DC/DC stage.

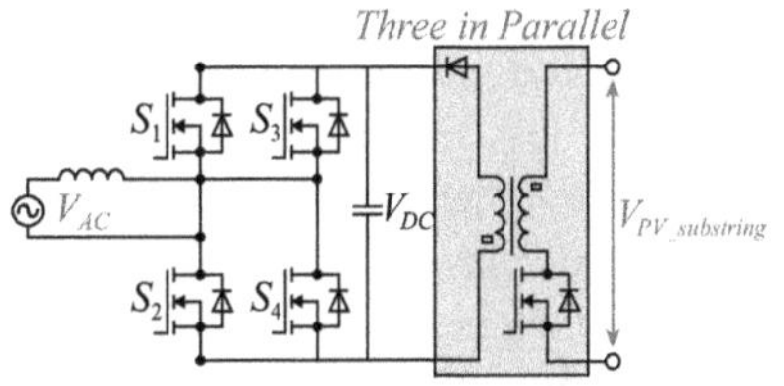

FIGURE 17.22 Three paralleled flyback converters and a full-bridge inverter.

the main role in power dissipation and control complexity. Considering the mentioned restrictions, the feasible topologies have been listed in [21]. In this subsection, two types of these topologies are addressed: pros and cons, along with the challenges, are clarified as guidelines for future studies on this topic.

17.4.3.1 Parallel Flyback DC/DC with PWM Full-Bridge Inverter

Figure 17.22 shows the topology of the parallel substring microinverter. For this topology, three paralleled flyback converters are used as the DC/DC stage, and the DC/AC stage consists of a full-bridge inverter. The output of DC converters is linked to the inverter by a DC-link, which is typically a capacitor. Also, an output low-pass filter is required to remove the high switching frequency component. The flyback converters operate in BCM Each flyback converter is controlled independently to extract the MPP from its substring and deliver it to the DC-link. The full-bridge inverter feeds this power to the grid. As the used converters in the DC/DC and DC/AC stage are well known from the point of operation and their operation is discussed in many sources, they will not be discussed here. For the design consideration of the converter, the inductors should be designed for operation in BCM. All the semiconductor devices are selected based on the rated power and voltage in each stage.

17.4.3.2 Serial Flyback DC/DC with PWM Full-Bridge Inverter

The configuration is shown in Figure 17.23. The input connection of flyback DC/DC converters is the same as in the parallel topology. Each substring has its individual connection with the converter. The output of flyback converters is configured in series and delivers the power to the DC/AC stage in the DC-link. As compared with the parallel configuration for the serial connection, the voltage step-up ratio could be low as the voltage of the DC-link is the sum of the output voltage of DC/DC converters. For both the series and parallel configuration, an output low pass filter is used to eliminate the high-frequency harmonics of the output current.

17.4.3.3 Topology Comparison

The performance and feasibility of the concepts could be assessed by the topology comparison. For the comparison, six different benchmarking factors were considered: total installed semiconductor factor, total stored capacitive energy, total stored inductive energy, transformer area product, conduction loss factor, and switching loss factor. The comparison was done for the same operation condition. The results are displayed in Table 17.2. As can be seen, the benchmarking factors for both of the structures are close together. However, the serial connection of DC/DC converters shows better performance than the paralleled one. This superiority is in the switching loss factor with a difference factor of 1.5. Regarding practical applications, the serial approach features control complexity [22] and requires a wider voltage step-up range than the paralleled configuration. Paralleled DC/DC configuration suffers from hard switching in the DC/AC stage, which results in high switching losses. This shortcoming could be relieved relatively by replacing GaN devices for optimization, leading to an achievable efficiency of 94.5%. As compared with the module-level microinverter, there is a common drawback of a higher step-up ratio of the DC/DC stage in the substring-level approaches. But still, the main advantages of the substring level approach is the increased energy yield for the mismatch condition [23]. The produced results show that the approach provides better performance at the deep mismatch conditions.

TABLE 17.2
Comparison of the Multi-Input Sub-Module Topology: Calculated Benchmarking Factors

	VA_{tot} (kVA)	E_{Cap}	E_{Ind}	A_{core}	λ_{cond}	λ_{sw}
Paralleled	4.88	149	9.49	1.68e-8	0.0641	2.29
Series	5.03	158	6.66	1.67e-8	0.0641	2.35

Three different substring-level shade-tolerant concepts are studied in this section. All the approaches could address the mismatch issues between the substrings. The mismatch power is mitigated in a substring-to-substring way by the DPP concept. while it is done in the string-to-substring configuration by the IPO. For the last concept, MPPT is done by each converter to harvest maximum power from each substring individually. From the control aspect, the first and the last approach have a closed-loop control to process the mismatch power. However, it is done by the IPO automatically. Implementation of the DPP and the series-parallel concept is realized by the DC/DC converters, resulting in increased component count and cost. Significant improvement in the compensation of the mismatch power is achieved by implementing three of the approaches, since 97% of the available power of the PV module could be harvested. This is around 1.5 times more than operation without the substring-level concept. Although the concept achieves a significant shade-tolerant ability as compared with the module-level microinverter, a detailed study is required to determine the capability of the approach for different mismatching conditions.

17.5 GLOBAL MAXIMUM POWER POINT TRACKING (GMPPT)

With the rapid rise of PV installed throughout the world, MPPT approaches are currently being researched to solve some of the obstacles associated with PV installation. The energy supply by the PV array would be lowered because of the problem of partial shading conditions (PSC) induced

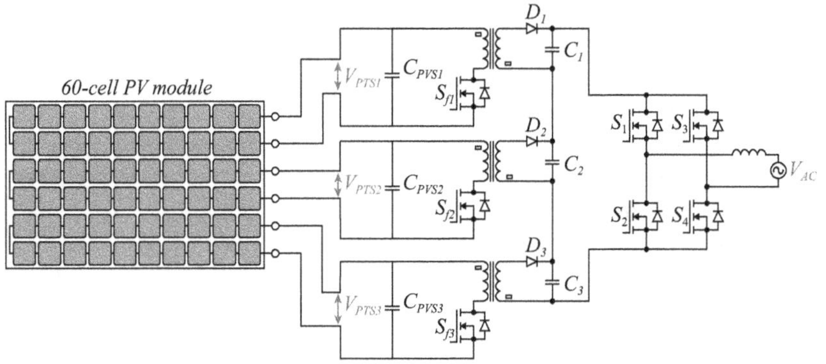

FIGURE 17.23 Three paralleled flyback converters and a full-bridge inverter.

by passing fog or natural clouds from the surrounding area. The anti-parallel bypass diode conducts when the PV array is shaded, disconnecting the shaded module from the PV string. Because the polarity of the voltage is flipped due to partial shading, the PV cell consumes rather than supplies power. In turn, the P-V characteristics of the PV string output will have multiple maxima points over the range of zero to an open-circuit voltage of the PV string, as shown, for instance, in Figure 17.24 for the 60-cell Si PV module. On the other hand, the system has one global power point at a voltage below the open-circuit voltage at approximately 10 V (i.e., MPP3). At this operating condition, the typical MPPT tracking methodology may be unable to capture the global power point. In these scenarios, the MPPT algorithm used should be able to sweep until it finds the best global maximum power point (GMPP), as well as offer sufficient accuracy independent of the PV array layout and the distribution of solar irradiance on the PV array surface. GMPPT would considerably improve PV system shade tolerance under hard shadowing situations, such as falling leaves, falling birds, or snow deposition on the PV module surface. One of the most important aspects of the GMPPT approach is the speed with which the peak point is reached, as well as the conversion efficiency.

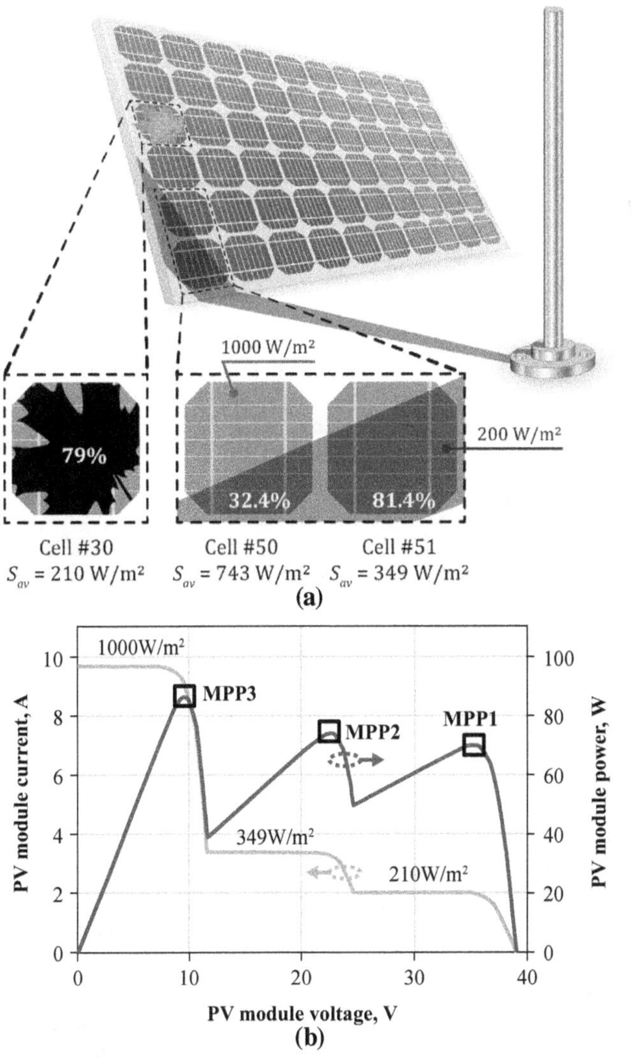

FIGURE 17.24 (a) Example of a shaded 60-cell Si PV module; (b) P-V and I-V characteristics.

The GMPPT can be classified into two categories: hardware and software approaches, as shown in Figure 17.25 [24, 25]. The modified MPPT techniques are an adaptation of well-known algorithms that try to tackle the PSC issue utilizing methods such as fuzzy logic controller, dividing rectangle, and load-line. Because of potential problems in limiting down the search range in the right position, this strategy is considered inapplicable. This might be due to improper assumptions regarding the dependency of the MPP position. The second GMPPT group includes the P-V curve sweeping methods, which are regarded as the most frequent research methods as no extra hardware is required. However, there is a tradeoff between power loss and sweeping time, since it is determined by the change in the operating point of the P-V curve at the open and short-circuit boundaries. In the third class, GMPP looks like an optimization problem, and hence metaheuristic approaches, such as particle swarm optimization (PSO) and ant colony optimization can be implemented, in which several conventional mathematical optimization issues and numerous engineering challenges are effectively utilized. The main drawback of these techniques is the time needed to locate the GMPP that demands a powerful implementation software. The last group includes the hardware methods that use additional circuits to sweep the P-V curve of the PV and simplify the operation of the power converter. The focus of this part is placed on the GMPPT based on the last GMPPT group (i.e., hardware methods).

17.5.1 Hardware GMPPT

The module for PV has nonlinear properties, and the impedance of the input of the converter should be matched to the maximum power depending on the level and temperature of the irradiance. The conventional MPPT cannot distinguish between local and global peak points, and will therefore remain at the local maxima.

One of the hardware approaches is called the generation control circuit; it is a tiny converter that acts as an active bypass in place of the bypass diode. It has an inconvenience of complexity and low power scale [26].

Another class uses external circuitry with the converter to change the impedance seen by the PV string. It is a more promising technique that can track the GMPP faster within a single P-V sweep. In the hardware methods, the P-V curve sweep is done using external hardware circuits, whereby the starting point of the P-V curve could be from either the open-circuit (V_{oc}, 0) or short-circuit (0, I_{sc}) boundary, as they are simple to apply. Then, it continues moving towards the GMPP. The open-circuit or short-circuit state of the PV string is applied for the t_{set} time duration, and then the algorithm begins to sweep the PV power and voltage until catching the GMPP during a time of t_{sweeep}, as shown in Figure

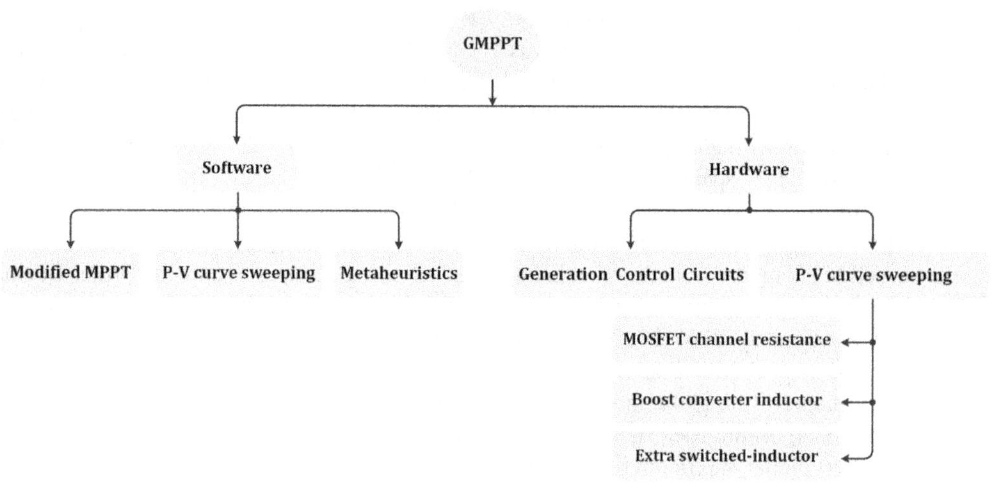

FIGURE 17.25 Classification of GMPPT for tolerance of partial shading conditions.

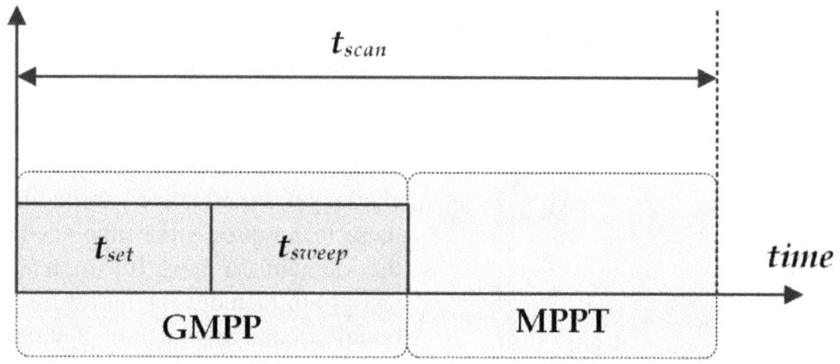

FIGURE 17.26 Timing diagram to scan the GMPP.

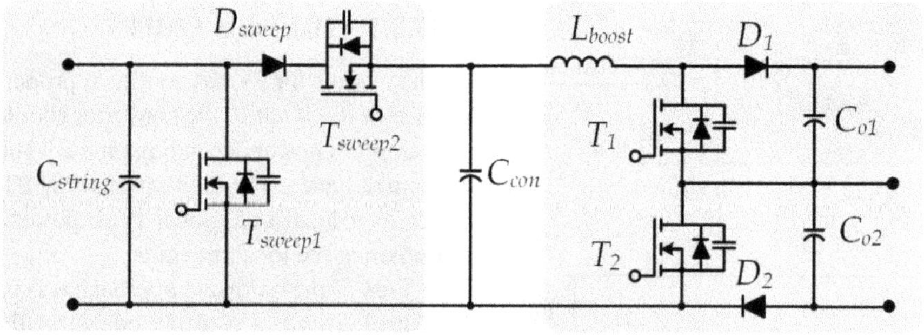

FIGURE 17.27 GMPPT hardware method based on the MOSFET channel resistance.

17.26. After identifying the GMPP window during the sum of t_{set} and t_{sweep}, a conventional MPPT—such as incremental conductance (INC) or perturb and observe (P&O)—can be applied to identify the GMPP exactly within a fine-tuning for the remaining part of the scan time t_{scan}. The three-level boost converter (TLB) is used as the interfaced topology as it has low voltage rating compared to the conventional boost converter and thus low power losses [27]. The scanning of the PV curve can be obtained by the following hardware methods: MOSFET channel resistance, boost converter inductor, and extra switched-inductor.

17.5.1.1 MOSFET Channel Resistance

This method has been proposed in [28], and its schematic diagram is shown in Figure 17.27. In this method, a short-current pulse (I_{SC}) is applied to the PV terminals and then the optimal operating current can be obtained by multiplying I_{SC} by the parameter value of k, as the optimal operating point of the PV system is proportional to the short-circuit current at the current environmental conditions. The value of the parameter k can be identified by the scan P-I curve. It is considered more concise to find the GMPP. The main

DC/DC converter is disconnected by turning off T_1 and T_2, so that the PV terminal looks like an open circuit. After the PV terminal voltage becomes lower than the DC-link voltage, the MOSFET T_{sweep1} is switched on in a linear region operation and the gate-source voltage is increased linearly for a time t_{sweep}. When the PV open-circuit voltage is larger than the DC-link voltage, the second switch T_{sweep2} should be turned on during the sweep period. With these MOSFET driving sequences, it will cross through several zones ranging from the cutoff to saturation and ohmic. After the t_{sweep} is finished and therefore the GMPP is defined, the converter comes back to its normal operation by turning on the switches T_1 and T_2 to apply the normal MPPT.

17.5.1.2 Boost Converter Inductor

In this method, the P-V features of the PV string are scanned based on the switching impedance circuit that incorporates a boost converter inductor [29]. The schematic diagram is shown in Figure 17.28. It can be applied only to active converters that can short the PV terminal, including the boost inductor like the three-level boost converter. It starts also by turning off the converter through switching off both T_1 and T_2. After the PV terminal is going to the open-circuit, the PV and the boost inductor are connected in series and shorted by turning on both T_1 and T_2. The value of the input capacitance C_{con} should be as small as possible to avoid measurement errors, where it will be added to the PV string capacitance. The sweeping time can be calculated as a function of the inductance value as well as the PV current at the current irradiation level:

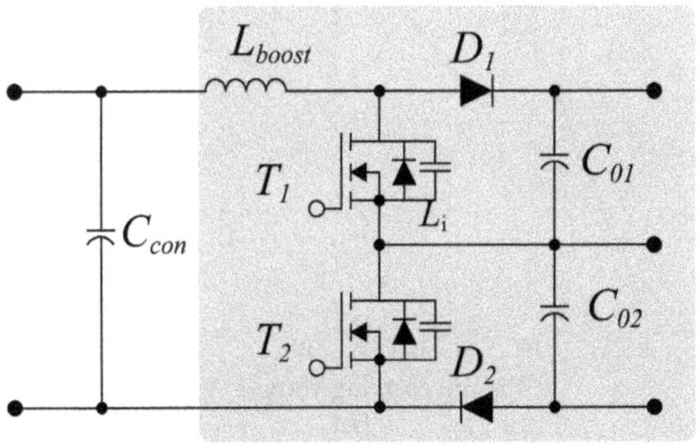

FIGURE 17.28 GMPPT hardware methods based on the inductor of the boost converter.

$$t_{sweep} \approx 2.L_{boost}\frac{I_{sc}}{V_{oc}} \tag{17.11}$$

where L_{boost} is the boost converter inductance, I_{sc} is the PV short-circuit current, and V_{oc} is the PV open-circuit voltage

17.5.1.3 Extra Switched-Inductor

This is similar to the previous method, but here an extra inductor L_{boost} with a switch T_{sweep1} and diode D_{sweep} are employed, as shown in Figure 17.29. This configuration allows decoupling between the TLB converter and the MPPT. The open-circuit is achieved for a time t_{set} similar to the previous method, then the PV terminal is shorted by the switch T_{sweep2} for the sweeping time t_{sweep}. The value of the external inductor L_{sweep} is added to the main boost inductor L_{boost} when the converter operates in the normal operation after identifying the GMPP.

17.5.2 Scanning of GMPP

The sweeping time to find the GMPP is of paramount importance at the expense of energy loss, and it should be as short as possible to increase the overall efficiency. It can consume several iterations to scan all the P-V until catching the GMPP. The effect of the junction capacitance of the PV modules should be minimized, as they are connected in a series configuration at increasing the string power. With the high scanning steps, more power oscillation is produced, resulting in more thermal cycles on the converter power devices. This, in turn, leads to an increase in the damage degradation of the converter, and so the probability of the converter failing is increased, which breaches the 25-year guarantee intended for PV systems. The accelerated GMPP is a tradeoff between the convergence speed and accuracy, as shown in Figure 17.30. The GMPP algorithm is executed on the digital platform, so the speed of the digital platform affects the selection of the scanning time, where for a time t_{sweep}, the number of N samples is given in Eq. (17.12), where T_s is the sampling interval of the digital controller:

$$N = \frac{t_{sweep}}{T_s} \tag{17.12}$$

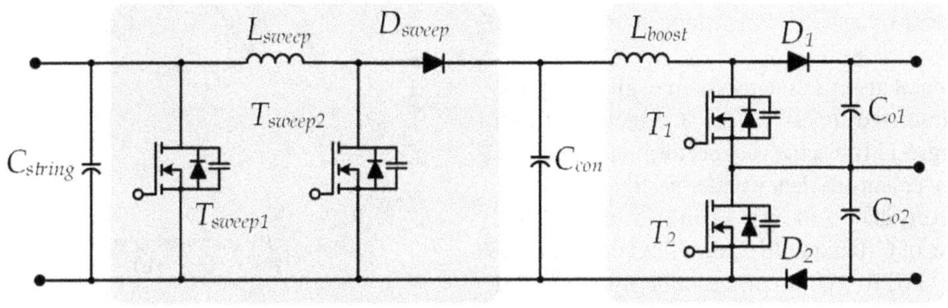

FIGURE 17.29 GMPPT hardware methods based on the extra inductor-switch cell.

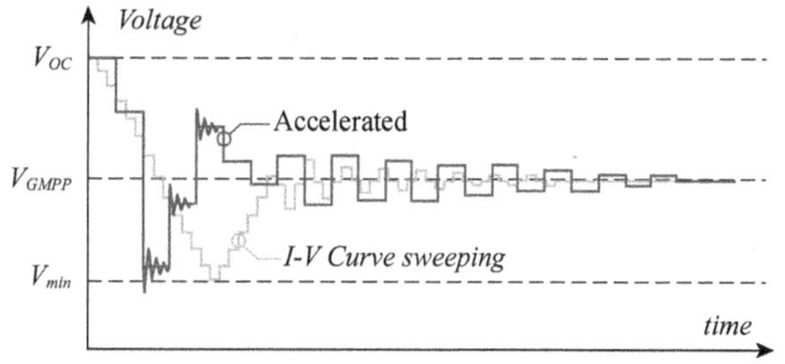

FIGURE 17.30 Sketch of the accelerated GMPPT.

17.5.3 Case Study

The building-integrated PV (BIPV) is used as an example of the commercial microinverter in the market, including an STMI, a LoMI, and a HiMI [30]. It was validated experimentally in different shaded and unshaded conditions, and evaluated in terms of maximum power point efficiency and power conversion efficiency. In this test, the microinverter is connected to a single-phase 230V utility grid at a nominal frequency of 50Hz. Also, the digital power analyzer based on Yokogawa WT1800 is employed to measure the efficiency, and the PV simulator of Agilent E4360A is used. In the experimental investigation, three modules for rooftop BIPV are considered:

1. Microstring of 8 × Gaia GS Integra Line SP 595
2. Microstring of 3 × SunTegra® Shingle
3. Microstring of 6 × GB-Sol PV Slate

The PV conversion system efficiency based on the EN 50530 (2010) "Overall Efficiency of Grid-Connected Photovoltaic Inverters" standard can be calculated as in Eq. (17.13), where η_{MPPT} is the efficiency and η_{MI} is the microinverter efficiency:

$$\eta_{PV} = \eta_{MPPT}\eta_{MI} \tag{17.13}$$

$$\eta_{MPPT} = \frac{P_{DC}}{P_{MPP}} \tag{17.14}$$

$$\eta_{MI} = \frac{P_{AC}}{P_{DC}} \tag{17.55}$$

The considered six PV power profiles are given in Figure 17.31, Figure 17.32, and Figure 17.33. The ST-1 profile shows the maximum power level at the minimum voltage (MPP3), whereas the second local MPP includes slightly lower power but two times greater voltage (MPP2), leading to much higher efficiency.

The experimental measurements for the given six PV profiles are summarized in Table 17.3. It is clear that in a BIPV microstring, STMI has the best performance in terms of PV energy conversion efficiency under partial shadowing scenarios. However, due to the use of the variable MPPT step, as in the case of G-1 and GBS-2, the HiMI occasionally can capture the GMPP, which can be done by the LoMI with a more basic control system. A STMI efficiency look-up table provides the control system with overall efficiency

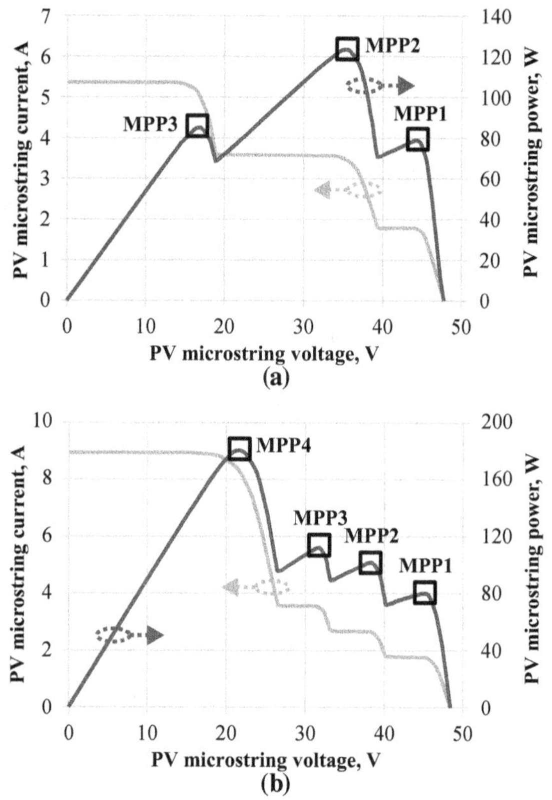

FIGURE 17.31 PV power profile for Gaia GS Integra Line SP 595 BIPV modules: (a) G-1; (b) G-2.

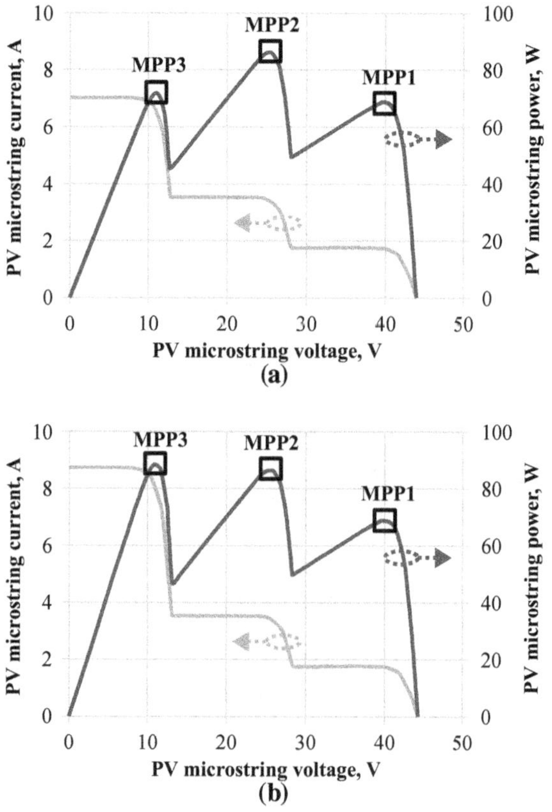

FIGURE 17.32 PV power profile for SunTegra® Shingle BIPV modules: (a) ST-1; (b) ST-2.

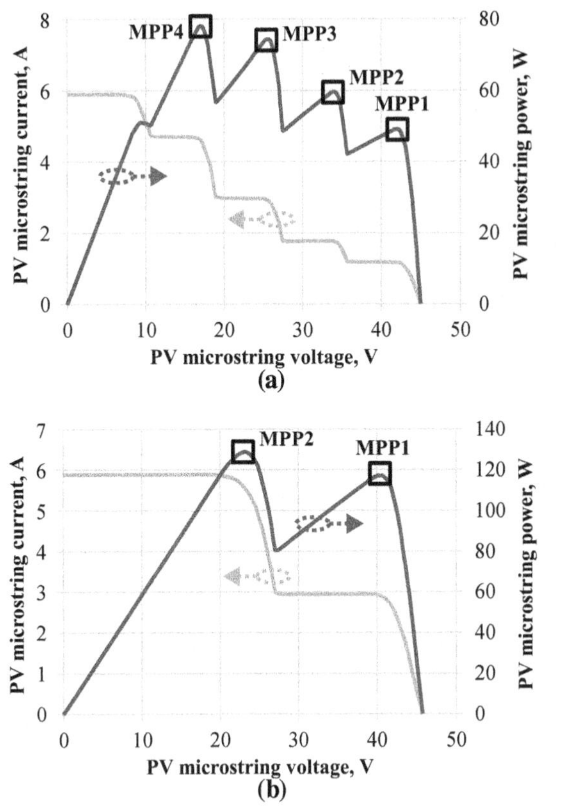

FIGURE 17.33 PV power profile for 6 × GB-Sol PV Slate BIPV modules: (a) GBS-1; (b) GBS-2.

optimization. As a result, in the instance of the ST-1 P-V profile, STMI works in MPP2 instead of MPP3 because it has a two-fold higher voltage and less than 2% higher than in MPP3 long-term trustworthiness.

As a result, the control system chose MPP2 as more favorable in terms of AC power pumped into the grid. This intelligent control results in a higher energy yield and less thermal cycling of the STMI, which should enhance long-term dependability. In addition, for the three types of BIPV modules chosen, further measurements of STMI efficiency were done using alternative configurations of the BIPV microstring. The results in Table 17.4 demonstrate the great scalability of the STMI (PV Optiverter), which can transmit PV power to the distribution grid with an efficiency of more than 80% even at BIPV microstring voltage levels three times lower than the lowest MPPT voltage of currently available microinverters.

17.6 FUTURE TRENDS

17.6.1 Compatibility with Different PV Modules (BIPV)

Considering material and shape, the PV modules are made in different sizes for a certain roof tile slate or shingle, with a variety of voltage ranges. The material is also changing day by day to reach higher efficiency and lower energy cost. Nowadays, the Si-based PV modules have reached a maturity and their cost reduction is decelerated. Concurrently, the thin-film technology is progressing to lower their cost

TABLE 17.3
EXPERIMENTAL MEASUREMENT OF THE PV ENERGY CONVERSION EFFICIENCY

Profile	Microinverter Type											
	LoMI				HiMI				STMI			
	MPP	η_{MPPT} [%]	η_{MI} [%]	η_{PV} [%]	MPP	η_{MPPT} [%]	η_{MI} [%]	η_{PV} [%]	MPP	η_{MPPT} [%]	η_{MI} [%]	η_{PV} [%]
ST-1	MPP1	77	92.3	71.1	MPP1	77	94.9	73.1	MPP2	97	90.5	**87.8**
ST-2	MPP1	79	92.9	73.4	MPP1	79	95.3	75.3	MPP2	99	91.6	**90.7**
G-1	MPP1	68	93.2	63.4	MPP2	99	95.7	**94.7**	MPP2	99	94	93.1
G-2	MPP1	44	93.0	40.9	MPP2	55	95.6	52.6	MPP4	99	91.5	**90.6**
GBS-1	MPP1	62	91.9	57	MPP2	75.5	95.3	72	MPP4	99	85.5	**84.6**
GBS-2	MPP1	91	94.1	85.6	MPP2	99	95.5	**94.5**	MPP2	99	91.5	90.6

TABLE 17.4
STMI EXPERIMENTAL EFFICIENCY WITH DIFFERENT BIPV MICROSTRINGS

Type of BIPV Roof Module	Number of Elements within a BIPV Microstring								
	1	2	3	4	5	6	7	8	9
Gaia GS Integra Line SP 595	-	82.9% 80W*	87.5% 120W	91% 160W	92.5% 200W	93.8% 240W	95.1% 280W	93% 320W	92% 360 W
SunTegra® Shingle	86% 105W	92.7% 210W	92.9% 315W	-					
GB-Sol PV Slate	-	85% 70W	89.9% 105W	92% 140W	94.6% 175W	94.2% 210W	94% 245W	-	

rapidly. Two dominant examples of them are cadmium telluride (CdTe) and copper indium (gallium) technologies. Series connection of the PV modules to string inverters or applying multilevel inverters can be a simple and conventional solution to all of these varying factors. The drawbacks of such connections are low reliability, unfavorable partial shading behavior, and restricted scalability. Furthermore, current microinverter technologies do not allow universal connection of PV modules made in various technologies and geometries with a wide voltage range. The newly introduced STMI called an Optiverter provides an ultra-wide input voltage regulation range, enabling application in modules with diverse characteristics. Performance of two conventional microinverters, LoMI and HoMI, has been compared with an Optiverter in terms of energy conversion capability for different PV modules. The comparison was done for 13 different BIPV modules in terms of all the voltage and current factors in standard test condition (STC). Then a microstring including 2–24 cells of these modules was made. The STMI provides a wider range in the selection of different modules due to its wide input voltage range. The result is having more freedom in implementing a rooftop BIPV system [31].

It can be seen from Figure 17.34 that the STMI provides high scalability in comparison with HoMI and LoMI.

17.6.2 Microconverter with Battery Operation Compatibility

Residential battery energy storage systems (RBESSs) are considered as the main contributor to energy performance and near zero energy building (NZEB) regulations. In case of having a DC bus in the building distribution system, RBESS can be connected to the bus by the same modularity of PV sources using microconverters. Having a modular energy storage system will increase flexibility and reduce weight and space, which are important factors in residential building applications. A versatile converter topology with a wide input voltage and output load regulation range are the main characteristics of a desired microconverter. The topologies developed for the PV-bus connection can be deployed for RBESS by adding bidirectional power flow capability to allow charge/discharge of batteries. Applying topology morphing control methods alongside a reconfigurable converter can be a significant modification to cope with charge/discharge profile of the battery. Furthermore, multimode control (MMC) method can be applied separately on charge/discharge operation modes to get the best performance. The qZS DC/DC converter with the reconfigurable qZS network [32] is one of the most promising candidates, which exhibits all the aforementioned requirements as a RBESS interface. Three-port microconverters are another trend to simultaneously support the connection of the PV-battery to the bus/grid. Compared with two-port microconverters, they are more complex, less reliable, and more difficult to implement as a modular system.

17.6.3 Ancillary Services to the Grid (Reactive Power Support)

Significant rise in the use of PV microinverters has raised interest in utilizing groups of them as a factor to support the distribution grid. They have potential to support reactive power to the distribution grid in order to regulate voltage and frequency of the grid by applying system-level control and communication with the distribution system operator (DSO). Furthermore, they can improve voltage profiles in the distribution feeders and can be used for power factor correction (PFC) in the level of single homes. The algorithms to control reactive power are the same as with string inverters. Two-stage microinverters are better candidates for reactive power support as the DC/DC stage can fully control PV voltage and MPPT. On the other hand, DC/AC stage can have more freedom to

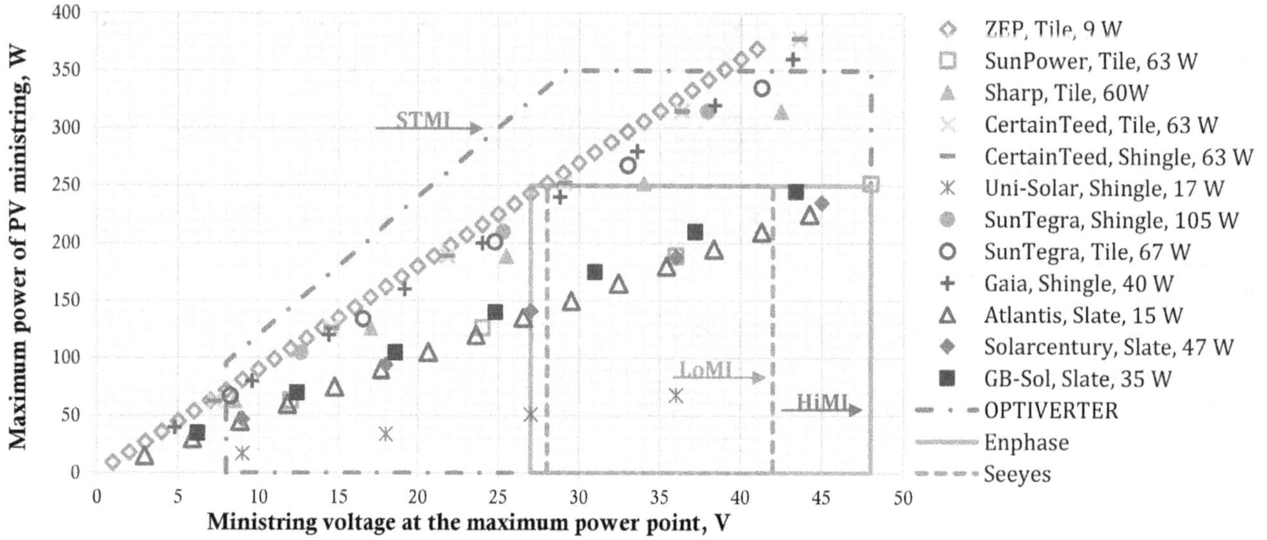

FIGURE 17.34 Possible configuration of a solar BIPV solar roof for various PV elements.

control active and reactive power injection by means of different control methods. A two-stage microinverter is implemented by connection of a partial power H-bridge DC/DC converter and an interleaved H-bridge inverter. The control scheme for the inverter stage includes an inner and outer loop, each containing a PI controller. The reactive power is controlled by the inner control loop and by a volt/voltage-ampere reactive command from the DSO [33]. AP Systems YC600 is a microinverter with a variety of grid support characteristics. It is a dual-module microinverter with independent MPPT (two-stage) topology. It is capable of providing 300VA peak power per channel for the input voltage range of 22–48V. The power factor of the converter is adjustable from 0.8 leading to 0.8 lagging.

17.6.4 Smart Grid Readiness (Remote Control by DSO)

A smart inverter must have digital architecture, bidirectional communications capability, and robust software infrastructure (Figure 17.35). The infrastructure allows the system to send and receive messages and commands. The benefits of such a software-controlled system include the ability to provide grid support options required by DSO. The options include ramp rate control, power curtailment, fault ride-through, and voltage profile support through voltage-ampere reactive power. A Hawaiian utility in cooperation with the company Enphase Energy implemented smart grid–ready inverters. To solve the problem of great voltage fluctuation caused by the growing number of solar installations in the area, the utility requested the inverters to expand their frequency trip limits from 59.6Hz to 57Hz. To create a cooperative effort between the inverter companies, utilities, and other third parties involved in the system, a variety of ongoing documentation and research is being done. UL/ANSI 1741, IEEE 1547, and IEC 61850 are the most important standards that are published to cover the requirement to implement communication interfaces for the inverters.

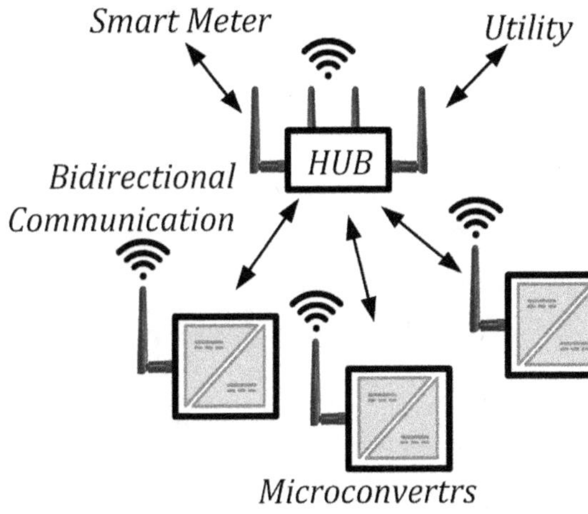

FIGURE 17.35 Communication network for a microinverter.

REFERENCES

[1] M. Kasper, D. Bortis and J. W. Kolar, "Classification and comparative evaluation of PV panel-integrated DC—DC converter concepts," IEEE Trans. Power Electron., vol. 29, no. 5, pp. 2511–2526, 2014.

[2] D. Vinnikov, A. Chub, E. Liivik, R. Kosenko and O. Korkh, "Solar optiverter—a novel hybrid approach to the photovoltaic module level power electronics," IEEE Trans. Ind. Electron., vol. 66, no. 5, pp. 3869–3880, May 2019.

[3] D. Vinnikov, A. Chub, E. Liivik, R. Kosenko and O. Korkh, "Solar optiverter—a novel hybrid approach to the photovoltaic module level power electronics," IEEE Trans. Ind. Electron., vol. 66, no. 5, pp. 3869–3880, May 2019.

[4] D. Vinnikov, A. Chub, E. Liivik, R. Kosenko and O. Korkh, "Solar optiverter—a novel hybrid approach to the photovoltaic module level power electronics," IEEE Trans. Ind. Electron., vol. 66, no. 5, pp. 3869–3880, May 2019.

[5] A. Chub, O. Korkh, R. Kosenko and D. Vinnikov, "Novel approach immune to partial shading for photovoltaic energy harvesting from building integrated PV (BIPV) solar roofs," in 2018 20th European Conference on Power Electronics and Applications (EPE'18 ECCE Europe), 2018.

[6] D. Vinnikov, A. Chub, E. Liivik, R. Kosenko and O. Korkh, "Solar optiverter—a novel hybrid approach to the photovoltaic module level power electronics," IEEE Trans. Ind. Electron., vol. 66, no. 5, pp. 3869–3880, May 2019.

[7] V. Gautam and P. Sensarma, "Design of Ćuk-Derived transformerless common-grounded PV microinverter in CCM," IEEE Trans. Power Electron., vol. 64, no. 8, pp. 6245–6254, Aug. 2017.

[8] M. A. Rezaei, K. Lee and A. Q. Huang, "A high-efficiency flyback micro-inverter with a new adaptive snubber for photovoltaic applications," IEEE Trans. Power Electron., vol. 31, no. 1, pp. 318–327, Jan. 2016.

[9] A. Dumais and S. Kalyanaraman, "Grid-connected solar microinverter reference design," Microchip Technology Inc., vol. 54, 2012.

[10] J. Yuan, F. Blaabjerg, Y. Yang, A. Sangwongwanich and Y. Shen, "An overview of photovoltaic microinverters: topology, efficiency, and reliability," in 2019 IEEE 13th International Conference on Compatibility, Power Electronics and Power Engineering (CPE-POWERENG), 2019, pp. 1–6.

[11] T. U. Guide, "Digitally controlled solar micro inverter design using C2000 piccolo microcontroller," June 2017, [Online]. Available: www.ti.com/lit/ug/tidu405b/tidu405b.pdf?ts=1624379226454&ref_url=https%253A%252F%252Fwww.google.com%252F

[12] Luis A. M. Barros, Mohamed Tanta, Tiago J. C. Sousa, Joao L. Afonso, J. G. Pinto, "New multifunctional isolated microinverter with integrated energy storage system for PV applications," Energies., vol. 13, no. 15, pp. 4016, 2020.

[13] R. Attanasio, "AN4070 Application Note: 250 W grid-connected microinverter," STMicroelectronics, 2012.

[14] P. Garrity, "Solar photovoltaic inverters," United States Patent US 8542512 B2, Sep. 24, 2013.

[15] D. Vinnikov, A. Chub, E. Liivik, R. Kosenko and O. Korkh, "Solar optiverter—a novel hybrid approach to the photovoltaic module level power electronics," IEEE Trans. Ind. Electron., vol. 66, no. 5, pp. 3869–3880, May 2019.

[16] R. Schaacke, "Electrically parallel connection of photovoltac modules in a string to provide a dc voltage to a dc

voltage bus," United States Patent US 8 432 143 B2, Apr. 30, 2013.

[17] D. Vinnikov, A. Chub, E. Liivik and I. Roasto, "High-performance quasi-Z-source series resonant DC—DC converter for photovoltaic module-level power electronics applications," IEEE Trans. Power Electron., vol. 32, no. 5, pp. 3634–3650, May 2017.

[18] P. S. Shenoy, K. A. Kim, B. B. Johnson and P. T. Krein, "Differential power processing for increased energy production and reliability of photovoltaic systems," IEEE Trans. Power Electron., vol. 28, no. 6, pp. 2968–2979, June 2013.

[19] S. Qin, A. J. Morrison and R. C. N. Pilawa-Podgurski, "Enhancing micro-inverter energy capture with sub-module differential power processing," 2014 IEEE Applied Power Electronics Conference and Exposition—APEC 2014, 2014, pp. 621–628.

[20] D. Leuenberger and J. Biela, "PV-module-integrated AC inverters (AC Modules) with subpanel MPP tracking," IEEE Trans. Power Electron., vol. 32, no. 8, pp. 6105–6118, Aug. 2017.

[21] D. Leuenberger and J. Biela, "PV-module-integrated AC inverters (AC Modules) with subpanel MPP tracking," IEEE Trans. Power Electron., vol. 32, no. 8, pp. 6105–6118, Aug. 2017.

[22] D. Leuenberger and J. Biela, "PV-module-integrated AC inverters (AC Modules) with subpanel MPP tracking," IEEE Trans. Power Electron., vol. 32, no. 8, pp. 6105–6118, Aug. 2017.

[23] R. Hernandez-Vidal, H. Renaudineau and S. Kouro, "Sub-module photovoltaic microinverter with cascaded flybacks and unfolding H-bridge inverter," 2017 IEEE 26th International Symposium on Industrial Electronics (ISIE), 2017, pp. 1035–1040.

[24] R. Kot, S. Stynski and M. Malinowski, "Hardware methods for detecting global maximum power point in a PV power plant," in proc. IEEE International Conference on Industrial Technology (ICIT), pp. 2907–2914, 2015.

[25] R. Kot, S. Stynski and M. Malinowski, "Experimental evaluation of hardware-based global maximum power point searching methods," in proc. 41st Annual Conference of the IEEE Industrial Electronics Society (IECON), pp. 000076–000081, 2015.

[26] R. Kot, S. Stynski and M. Malinowski, "Experimental evaluation of hardware-based global maximum power point searching methods," in proc. 41st Annual Conference of the IEEE Industrial Electronics Society (IECON), pp. 000076–000081, 2015.

[27] A. Bidram, A. Davoudi and R. S. Balog, "Control and circuit techniques to mitigate partial shading effects in photovoltaic arrays," IEEE J. Photovolt., vol. 2, no. 4, pp. 532–546, Oct. 2012.

[28] T. Noguchi, S. Togashi and R. Nakamoto, "Short-current pulse-based maximum-power-point tracking method for multiple photovoltaic-and-converter module system," IEEE Trans. Ind. Electron., vol. 49, no. 1, pp. 217–223, Feb. 2002.

[29] R. Kotti and W. Shireen, "Fast converging MPPT control of photovoltaic systems under partial shading conditions," in proc. IEEE International conference on Power Electronics, Drives and Energy Systems (PEDES), pp. 1–6, Dec. 2012.

[30] A. Chub, O. Korkh, R. Kosenko and D. Vinnikov, "Novel approach immune to partial shading for photovoltaic energy harvesting from building integrated PV (BIPV) solar roofs," in proc. 20th European Conference on Power Electronics and Applications (EPE'18 ECCE Europe), 2018, pp. P. 1–P.10.

[31] A. Chub, O. Korkh, R. Kosenko and D. Vinnikov, "Novel approach immune to partial shading for photovoltaic energy harvesting from building integrated PV (BIPV) solar roofs," in proc. 20th European Conference on Power Electronics and Applications (EPE'18 ECCE Europe), 2018, pp. P. 1–P.10.

[32] Andrii Chub, et al., "Bidirectional DC—DC converter for modular residential battery energy storage systems," IEEE Trans. Ind. Electron., vol. 67, no. 3, pp. 1944–1955, 2019.

[33] Dong Dong, et al., "A PV residential microinverter with grid-support function: design, implementation, and field testing," IEEE Trans. Ind. Electron., vol. 54, no. 1, pp. 469–481, 2017.

18 LCL-Filter Design and Application

Oleksandr Husev and Oleksandr Matiushkin

CONTENTS

18.1 GRID-CONNECTED INVERTERS AND TYPES OF OUTPUT FILTERS

In recent years, renewable energy has been the center of attention. In turn, it has raised attention to the DC/AC converters connected to the grid (Figure 18.1). Passive magnetic elements that are required for output filtering performance, along with boost capabilities, are one of the most expensive and traditionally bulky components. An output filter is a necessary component of pulse width modulation (PWM) converters. It is design is described in many papers [1–26].

The Google Little Box Challenge (GLBC) demonstrated a big interest in the power electronics converters as a part of photovoltaic (PV) systems, and also demonstrated the significance of high-power density inverters for the future development of PV systems [27–29]. At the same time, the main limitation in the application of the GLBC project outcomes consists in the narrow range of input voltage regulation.

Works [30–34] present a good overview of conventional solar single-phase inverters. More complex solutions, like high step-up inverters [35] and common-ground inverters [36], can also be regarded as competitive solutions.

Research directions in the area consist of the single-stage alternatives. Inverters with active boost cells are described in [37–40]. They can provide a very high boost of the input voltage, but they suffer from high current spikes in passive elements and semiconductors. Impedance-source (IS) networks are reported as a promising single-stage solution. Z-source inverters (ZSIs), quasi-Z–source inverters (qZSIs) and all other derivatives are proposed for grid integration. There are several review papers [41–45], along with many others addressing different relevant issues [46–51]. Also,

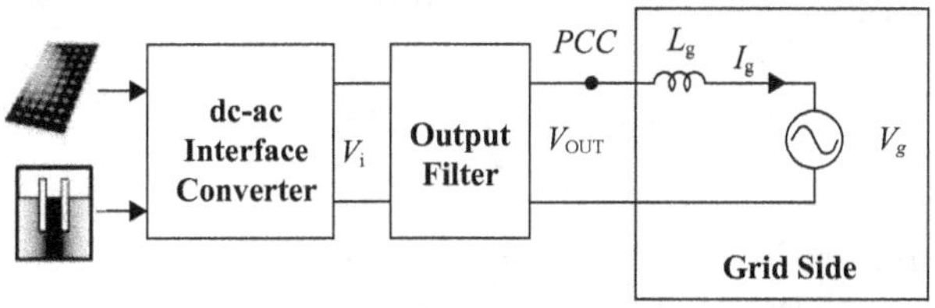

FIGURE 18.1 Grid-connected renewable energy system.

recent research papers have revealed evident drawbacks of the IS-based converters in terms of power density and efficiency [52–54].

Split-source inverters (SPIs) [55, 56] are proposed as an alternative solution to IS-based inverters. In this case, the improved performance can be achieved in terms of passive component counts, but it is accompanied by higher voltage and current stresses at lower voltage gains. Also, there is no ST immunity. The need for the inverter with a wide input voltage regulation with acceptable efficiency and high-power density pushes the academia and the industry to searching for new solutions. Several interesting other buck-boost single-stage inverters are proposed in [57–60].

An Aalborg inverter is an inverter that combines buck and boost functionalities [61–66]. Two independent buck-boost stages are utilized to achieve the output sinusoidal voltage shape. The first stage is responsible for positive output voltage generation, and the other is responsible for negative output voltage generation. Despite the main advantage of the solution which consists in the minimum voltage drop across the passive and active components in the power loop, this solution uses a double number of semiconductors and passive elements in the buck-boost stage. This drawback is a main obstacle to practical implementation. Also, two power sources are required. A quite similar idea is discussed in [67] and called a dual-buck–structured inverter. It works in a similar way, but has a single input voltage source.

The solution based on the input boost and buck converters, along with a line frequency unfolding circuit, is proposed in [68, 69] and can be interesting for the industrial applications. The operation principle consists of modulation of the DC-link voltage by means of boost stage if the input voltage is below required output voltage. Buck stage performs further modulation if the input voltage is higher than required output voltage. Unfolding circuit provides positive and negative output voltage generation. A modified solution which is based on buck-boost stage allows reducing the number of inductors is described in [70, 71].

Figure 18.2 shows a few representatives of these discussed converters. Figure 18.2(a) shows the conventional solution based on the boost cell and VSI; Figure 18.2(b) shows the qZS-based inverter as a single-stage solution.

DOI: 10.1201/9781003229124-18

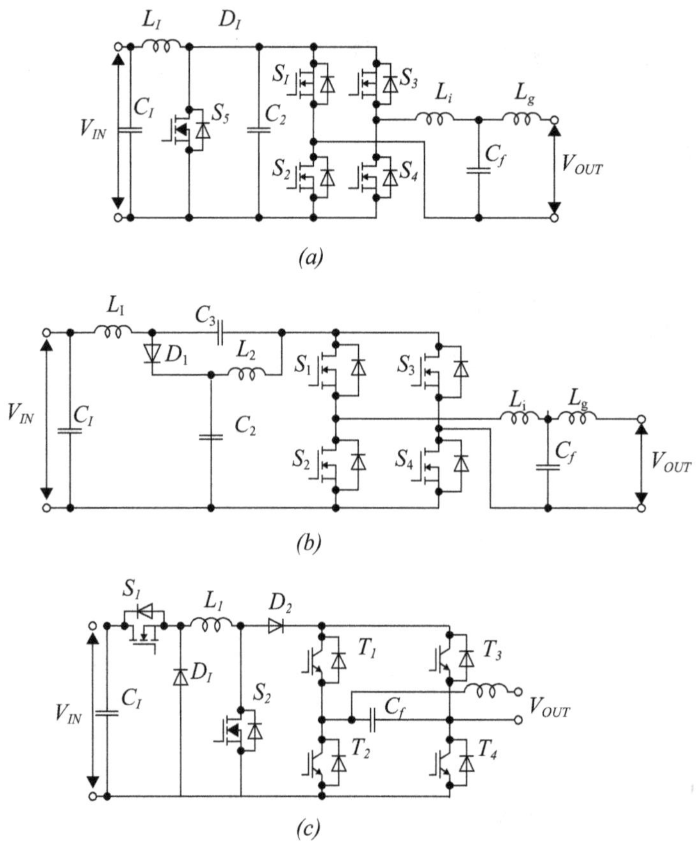

FIGURE 18.2 Grid-connected DC/AC representatives: (a) conventional two-stage solution; (b) qZS-based inverter; (c) buck-boost inverter with unfolding circuit.

Finally, Figure 18.2(c) shows the buck-boost converter with unfolding circuit. All these solutions can provide tolerance to the wide input voltage range and have an output filter to provide acceptable quality of the grid current.

In the case of the voltage source inverters (VSIs), the semiconductor stage is represented by a voltage generator. It corresponds to a conventional solution (Figure 18.2a) and qZS inverters (Figure 18.2b). In the case of the buck-boost converter with unfolding circuit, the situation is more complicated and a more complex approach is required to calculate the optimal passive filtering components.

The conclusion from this chapter is that despite the similar output filter stage, the calculation approach of the output filter can be different and will be covered in the next subsections.

There are many possible filter structures, such as L, LC, LCL, LCCL, LLCL, LCL-LC, etc. [P. Channegowda et al. (2010); M. Liserre et al. (2005)], [72, 73]. A high-order output filter may have smaller size but requires resonance damping and control that is more sophisticated. Figure 18.3 shows several output topologies, along with sensing signals that are considered for DC and AC application.

In the first approximation (Figure 18.3[a]), the simple inductance can be considered as an output filter. It can be used in the VSIs. The LC-filter (Figure 18.3[b]) can be considered as a further derivation of the L-filter suitable for the VSIs, as well. The grid side current measurement is required for the power flow control in both cases.

LCL-filters can be used as an alternative solution for the size reduction of an output filter (Figure 18.3[c]). Further derivation of the high order filters is shown in Figure 18.3(d)–(e). Figure 18.4(d) shows a conventional LLCL-filter, while Figure 18.4(e) shows an LCLCL-filter. The high-order output filters can provide good filtering features along with small size. At the same time, complex control—including active damping—has to be applied.

The grid-side voltage and current sensors are required for all solutions. In the case of high-order filters, an additional voltage or current sensors are required for an active damping control. At the passive damping, the additional converter-side current sensing can be eliminated.

18.2 GENERAL DESIGN PRINCIPLES OF THE OUTPUT FILTER OF A GRID-CONNECTED VOLTAGE SOURCE INVERTER

In order to provide the L-filter selection, several criteria have to be considered. For example, the ripple criterion consist in the error between the injected grid current and reference current.

Another criterion consists in the current ripple calculation on the switching harmonic. A typical harmonic spectrum of the inverter before the filter is shown in Figure 18.4. Taking into account that the current ripple is contributed only by the switching frequency, it is possible to define the converter voltage harmonic as following:

$$THD_I \approx \sqrt{\frac{I_{SW}^2}{I_1^2}} = \frac{I_{SW}}{I_1} \quad (18.1)$$

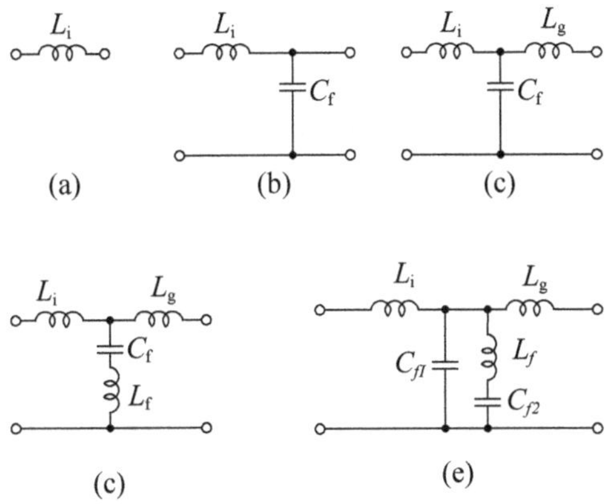

FIGURE 18.3 Possible output filter configurations: (a) L-filter; (b) LC-filter; (c) LCL-filter; (d) LLCL-filter; (e) LCLCL-filter.

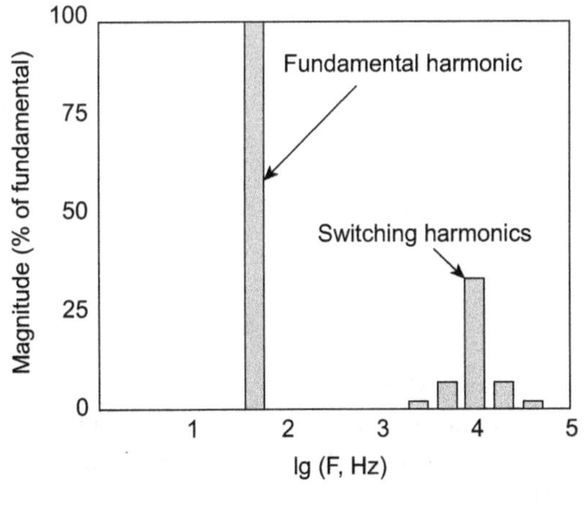

FIGURE 18.4 Harmonic spectrum of the injected grid current.

where I_{SW} is the grid root mean square (RMS) value of the harmonic current at the switching frequency, and I_1 is the RMS value of the fundamental current harmonic of the grid

Taking into account high switching frequency, the grid can be represented as a short circuit while the converter is a harmonic generator [Gonzalez-Castrillo et al. (2011); M. Liserre et al. (2005)]. In this case, the current ripple is passing from the converter side to the grid which is depicted in Figure 18.5 and can be expressed as:

$$G_L(h_{SW}) = \frac{I_g(h_{SW})}{V_i(h_{SW})} = \left| \frac{-j}{w_1 \cdot h_{SW} \cdot (L_i + L_{g2})} \right| \quad (18.2)$$

where w_1 is the fundamental harmonic, h_{SW} is the number of switching harmonic, and total filter inductance $L_f = L_i + L_{g2}$

From Eq. (18.2), the current ripple is expressed as:

$$I_{SW} = I_g(h_{SW}) = V_i(h_{SW}) \cdot \frac{1}{w_1 \cdot h_{SW} \cdot L_f} \quad (18.3)$$

The switching frequency harmonic component of the output voltage of the inverter is defined as $V_i(h_{SW})$. As a result, taking into account Eqs. (18.2) and (18.3) and the power factor is equal to 1, the value of inductance can be defined as:

$$L_f \geq \frac{V_i(h_{SW}) \cdot V_g}{w_1 \cdot h_{SW} \cdot P \cdot THD_I} \quad (18.4)$$

where P is nominal output power

If the switching frequency component of the voltage harmonic is unknown, the current ripple can be calculated using the inverter voltage waveform. As shown in P. Channegowda et al. (2010), such approach gives the same result since there is a proportional dependence between the ripple magnitude and harmonics content.

LCL-Filter Design

The equivalent circuit of the converter is very similar to the previous case (Figure 18.6). The grid is considered as a short circuit, while converter is a harmonic generator. The same condition is used for transfer function calculation:

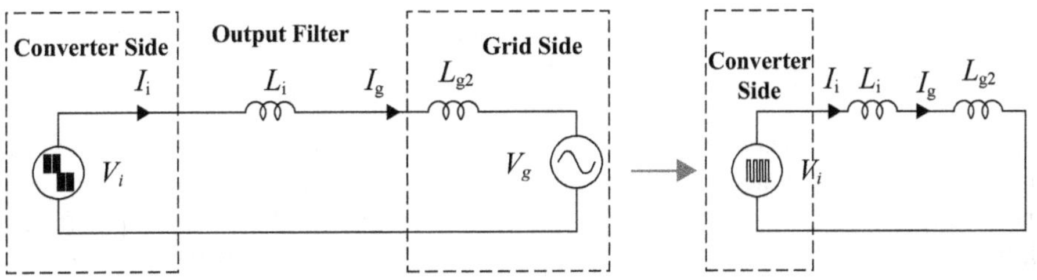

FIGURE 18.5 Equivalent circuit of the grid-connected converter with an L output filter.

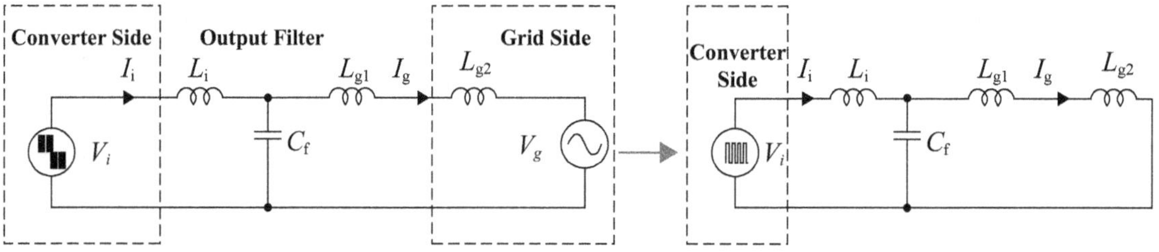

FIGURE 18.6 Equivalent circuit of the grid-connected inverter with LCL output filter.

$$G_{LCL}(h_{SW}) = \frac{I_g(h_{SW})}{V_i(h_{SW})} = \left| \frac{-j}{w_1 \cdot h_{SW} \cdot L_i - L_i \cdot L_g \cdot w_1^3 \cdot h_{SW}^3 \cdot C_f + w_1 \cdot h_{SW} \cdot L_g} \right| \tag{18.5}$$

where it is assumed that $L_g = L_{g1} + L_{g2}$

Based on Eq. (18.5), the current ripple is defined as following:

$$I_{SW} = I_g(h_{SW}) = \left| \frac{V_i(h_{SW})}{w_1 \cdot h_{SW} \cdot L_i - L_i \cdot L_g \cdot w_1^3 \cdot h_{SW}^3 \cdot C_f + w_1 \cdot h_{SW} \cdot L_g} \right| \tag{18.6}$$

In order to get the inductance value from Eq. (18.6), it is necessary to find the harmonic component of the inverter voltage at the switching frequency $V_i(h_{SW})$. Based on Eqs. (18.1) and (18.6), it is possible to obtain the following:

$$THD_I = \frac{V_i(h_{SW}) \cdot V_g \cdot G_{LCL}(h_{SW})}{P} \tag{18.7}$$

The resonance frequency f_{RES} is defined as:

$$f_{RES} = \frac{1}{2\pi} \sqrt{\frac{L_g + L_i}{L_g \cdot L_i \cdot C_f}}. \tag{18.8}$$

The Bode plots of the LCL-filter transfer function is shown in Figure 18.7. In order to avoid resonance problems, the resonance frequency f_{RES} has to be in a range between ten times the line frequency f_0 and one-half of the switching frequency f_{SW} [M. Schweizer et al. (2011); Yi Tang et al. (2012)]. Based on this:

$$10 \cdot f_0 \le f_{RES} \le \frac{f_{SW}}{2} \tag{18.9}$$

The relation between two inductances can be represented using the index r for grid inductance L_g as a function of inverter side inductance L_i, [Yi Tang et al. (2012); M. Liserre et al. (2005)].

$$L_g = r \cdot L_i \tag{18.10}$$

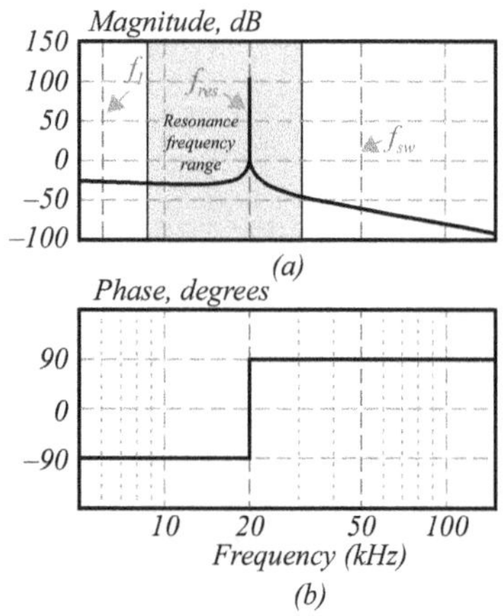

FIGURE 18.7 Requirements for resonance frequency selection.

Assuming that f_{RES} is determined, from Eq. (18.8), the value of inductance from the inverter side can be defined as following:

$$L_i = \frac{1+r}{r} \cdot \frac{1}{4\pi^2 \cdot C_f f_{RES}^2} = \frac{1+r}{r} \cdot L \tag{18.11}$$

where L is defined as the value of the weighted inductance:

$$L = \frac{1}{4\pi^2 \cdot C_f f_{RES}^2} \tag{18.12}$$

The value of the capacitor C_f is limited by the capacitive reactive power in ratio to nominal output power which is predetermined by its relative value Δ [W. Wu et al. (2012)]:

$$C_f \le \frac{\Delta \cdot P}{V_g^2 \cdot w_1} \tag{18.13}$$

where P is the nominal output power and V_g is the grid RMS voltage

On the one hand, the larger value of the capacitor leads to the better filtering capability, but on the other hand, the current stress on the semiconductors and passive elements along with reactive power oscillation is larger, as well.

Finally, from Eqs. (7), (10) and (11), the relative index r can be derived solving of a quadratic equation:

$$r^2 + r \cdot \left(2 - \frac{\dfrac{V_i(h_{SW}) \cdot V_g}{2 \cdot P \cdot THD_U \cdot w_1 \cdot h_{SW} \cdot L \cdot h_{SW}}}{\left(C_f \cdot L \cdot w_1^2 \cdot h_{SW}^2 - 1\right)} \right) + 1 = 0. \quad (18.14)$$

According to this approach, the weighted inductor, the value of the capacitor and the index r can be defined. It provides definition of the output filter.

The relative index r in Eq. (18.14) may have several solutions. These solutions define the ratio between the converter side and the grid side inductances.

For example, the total inductance $L = L_i + L_g$ is assumed to be constant. From Eqs. (18.10) and (18.11), it can be expressed as:

$$L_g = \frac{r}{1+r} \cdot L, \; L_i = \frac{1}{1+r} \cdot L \quad (18.15)$$

The voltage and current transfer functions for the high-frequency ripples can be expressed as the function of r which has extreme points. Figure 18.8 shows general dependences for both approaches. It is seen that the function has maximum and minimum values where minimum value r_{MIN} corresponds to the maximum attenuating factor of the high-switching ripples. Mathematically, two maximum values that correspond to the one resonance frequency can be obtained. In order to find a certain value of index r, extreme points must be found:

$$\frac{d}{dr} G_{LCL}(r) = 0, \; \frac{d}{dr} K_{LCL}(r) = 0 \quad (18.16)$$

One of the value corresponds to the optimal solution In a very general case, the equation has several solutions, but only range $0 < r \leq 1$ can be accepted. This means that the converter-side inductance has to be larger to the grid-side inductance. In the opposite case, high current spikes in the semiconductors will be observed.

As an intermediate conclusion in order to define the output filter, the resonance frequency and output current ripple value have to be predefined. After this step, the capacitor value calculation should be performed. The final step, which consists of inverter-side and grid-side inductors calculations, can be processed. If the values of inductors do not have a proper solutions, the process can go through several iterations, changing the value of capacitor or resonance frequency.

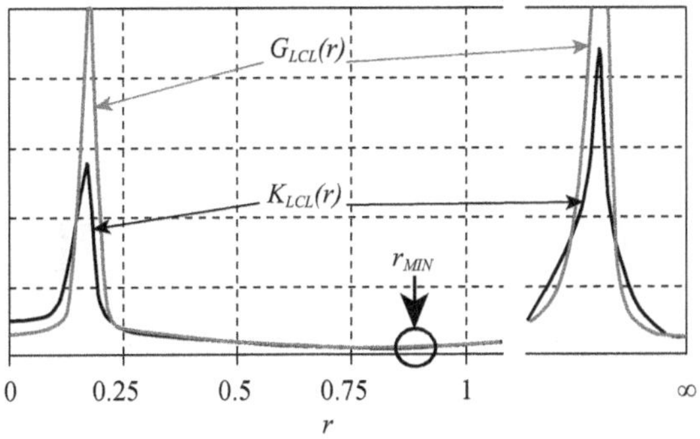

FIGURE 18.8 Voltage and current transfer functions as a function of the inductance ratio.

18.3 OUTPUT FILTER DESIGN OF THE GRID-CONNECTED INVERTER WITH UNFOLDING CIRCUIT

The second selected case study example is the buck-boost DC/DC cell with an unfolding circuit. Buck-boost cell provides the necessary shape modulation, while the unfolding circuit provides the sign changing of the output voltage.

Figure 18.9 shows the equivalent circuit of this converter. It demonstrates the output filter stage and semiconductor stage represented by a voltage generator or equivalent simple switch. In the case of the buck mode, the equivalent circuit depicted in Figure 18.9(a) is valid. It remains the unite LCL-filter structure during the switching period with the corresponding method described earlier.

Figure 18.9(b) shows the equivalent circuit of the same converter in the boost mode. The boost cell has well-known gain factor. At the same time, due to the output voltage modulation, the gain depends on the instantaneous value of the output voltage, which in turn depends on the phase:

$$D(\varphi) = \frac{\left|V_M \cdot \sin(\varphi)\right| - V_{IN}}{\left|V_M \cdot \sin(\varphi)\right|} \quad (18.17)$$

where V_{IN} is the input voltage

The design guidelines of the inductor and the capacitor are similar to the simple boost cell. The difference consists in the additional output inductance as part of the LC-filter. The steady-state analysis allows for the passive components that are required to keep the voltage and current ripple in the predefined range to be defined. The differential equations derived for the first circuit along with the inductor current and capacitor voltage ripples are as follow.

$$L_1 \cdot \frac{di_L}{dt} = V_{IN}, \; C_1 \cdot \frac{dv_C}{dt} = -i_g \quad (18.18)$$

$$\Delta i_L = \frac{V_{IN}}{2 \cdot L_1} \cdot D(\varphi) \cdot T_{SW}, \; \Delta v_C = \frac{v_{OUT}}{2 \cdot R \cdot C_1} \cdot D(\varphi) \cdot T_{SW} \quad (18.19)$$

where i_g is the output current, T_{SW} is the switching period, v_{OUT} is an instantaneous value of the output voltage and R is the equivalent load resistance

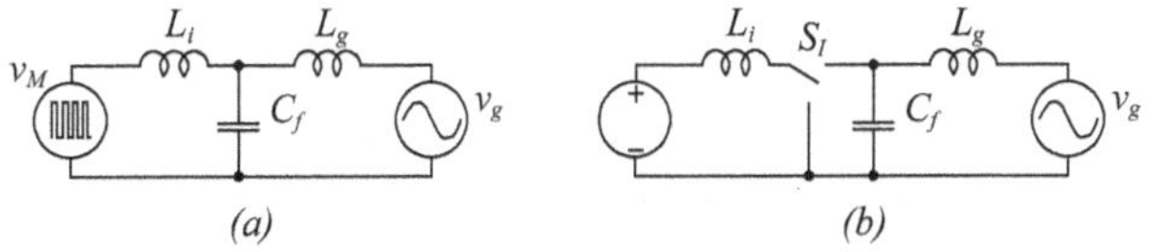

FIGURE 18.9 Equivalent circuits of the grid-connected inverter based on the buck-boost cell and unfolding circuit: (a) buck mode; (b) boost mode.

The output inductance design requires more sophisticated approach. The current ripple across output inductance depends only on the ripple across the capacitor. It is known that magnetic flux deviation is proportional to the inductance and current deviation, which in turn is proportional to the voltage applied across inductance. As a result, the output current ripple can be expressed by as following:

$$L_g \cdot \left(2 \cdot \Delta i_g\right) = \Delta\Psi,\ \Delta i_g = \frac{\Delta v_C}{8 \cdot L_g} \cdot T_{SW} \quad (18.20)$$

where $\Delta\Psi$—deviation of the magnetic flux

The ripple current and voltage factors for each component are introduced for convenience. The peak values of the currents across inductances:

$$I_{LMAX} = \frac{2 \cdot P}{V_{IN}},\ I_{gMAX} = \frac{2 \cdot P}{V_M}, \quad (18.21)$$

$$K_L = \frac{2 \cdot \Delta i_L}{I_{LMAX}},\ K_C = \frac{2 \cdot \Delta v_C}{V_M},\ K_g = \frac{2 \cdot \Delta i_g}{I_{gMAX}} \quad (18.22)$$

As a conclusion, taking into account that the current and voltage ripples across the passive components depend on the phase of the output sinusoidal voltage, the final expressions for value of inductances and the capacitor are the following:

$$L_i = \frac{V_{IN}^2 \cdot \left(\left|V_M \cdot \sin(\varphi)\right| - V_{IN}\right)}{2 \cdot P \cdot K_L \cdot f_{SW} \cdot \left|V_M \cdot \sin(\varphi)\right|} \quad (18.23)$$

$$C_1 = \frac{2 \cdot P \cdot \left(V_M \cdot \sin(\varphi) - V_{IN}\right)}{V_M^3 \cdot K_C \cdot f_{SW}} \quad (18.24)$$

$$L_g = \frac{V_M^2 \cdot K_C}{16 \cdot P \cdot K_g \cdot f_{SW}} \quad (18.25)$$

where f_{SW} is the switching frequency

18.4 EXAMPLES OF OUTPUT FILTER DESIGN

To demonstrate an example, the output filter was calculated for a three-level (3L) grid-connected inverter with 1kW nominal power, as depicted in Figure 18.10. The preliminary calculations performed for 3L topology reveal that output harmonic component at the switching frequency $V_i(h_{SW})$ is about 0.45% V_g. It also depends on the modulation index and belongs to the worst case. The RMS value of the output grid voltage V_g is assumed to be 230V.

According to this described approach, the value of the output inductances of the L-filter can be calculated according to Eq. (18.4). For example, for the previously mentioned parameters, and predefined THD_I= 3% and selected switching frequency of 25kHz, the output inductance can be defined as:

$$L_f \geq \frac{0.45 \cdot 230^2}{2\pi \cdot 50 \cdot 500 \cdot 1000 \cdot 0.03} = 5.05\,mH \quad (18.26)$$

In case of LCL-filter calculation, the filtering capacitor value C_f is limited by the reactive power oscillation at the rated power. Assuming the reactive power is less than 2%, the capacitance can be calculated using Eq. (18.13):

$$C_f \leq \frac{1\% \cdot P}{V_g^2 \cdot w_1} \leq \frac{1\% \cdot 1000}{230^2 \cdot 2\pi \cdot 50} \leq 0.6\mu F \quad (18.27)$$

The resonance frequency of the LCL-filter can be found after the capacitance calculation. As previously mentioned, according to Eq. (18.9), the resonance frequency has to be selected in a range from ten times the line frequency f_0 up to one-half of the switching frequency f_{SW}. Also, the resonance frequency has to be calculated in order to satisfy the $0 < r \leq 1$ condition.

For example, if the calculated value of the capacitor is C_f = 0.47μF, using Eq. (18.12) along with Eq. (18.14) and solving this system relative to the index r, it is possible to obtain the certain index r for particular resonance frequency f_{RES}. If there are no real solutions or $r > 1$, the resonance frequency must be changed in order to satisfy the condition

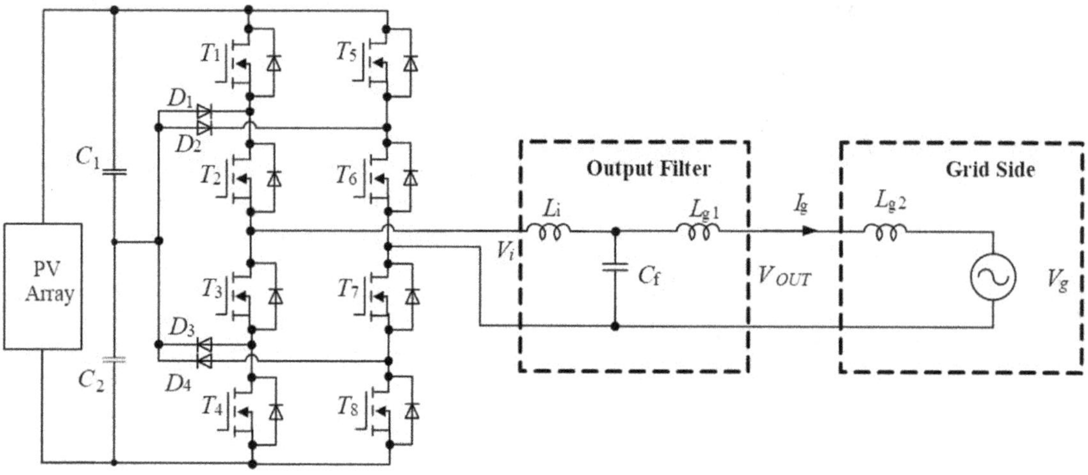

FIGURE 18.10 3L NPC VSI as a case study example.

$0 < r \le 1$. After several iterations, in this particular example the value of $r = 0.97$ for $f_{RES} = 10.2$kHz is obtained. This resonance frequency value is very close to its maximum possible value taking into account the condition $0 < r \le 1$.

Finally, after finding the weighted inductance value L and considering Eqs. (11)–(10), the values of the filter inductors L_g and L_i can be expressed:

$$L = \frac{1}{4\pi^2 \cdot 0.47 \cdot 10^{-6} \cdot 10200^2} = 0.51\,mH \quad (18.28)$$

$$L_i = \frac{1+0.97}{0.97} \cdot 0.51 = 1.1\,mH \quad (18.29)$$

$$L_g = (1+0.97) \cdot 0.51 = 1.02\,mH \quad (18.30)$$

Table 18.1 sums up the results of the output filter calculation for different switching frequencies. In advance to the LCL-filter, the value of inductance for a simple inductor filter is shown, as well. Assuming that all inductors are rated for almost the same current, the overall size of magnetics in an LCL-filter is at least two times smaller as compared to a simple L-filter.

It should be noticed that the value of the filtering capacitor is a quite flexible parameter. At the same time, its maximum value is limited by current ripple. The very large capacitors and correspondingly small inductors lead to the high current ripple, which in turn leads to increased losses.

An alternative approach based on the voltage distortion approach is described in [74]. As compared to the classical approach, it is based on the voltage transfer function for both grid-connected and off-grid inverters. The output current quality of the inverter is compared with the obtained different output filter parameters. The THD_I of the output current of the inverter is the main criterion for comparison.

Figure 18.11 shows the dependences of the THD_I of the output current of the inverter versus its output power. Figure

TABLE 18.1
Calculated Values of the Output Filters Components for 1kW Grid-Connected Inverters at Different Switching Frequencies

Switching	L	LCL			
Frequency, kHz	L_f, mH	L_i, mH	L_g, mH	C_f, mF	f_{RES}, kHz
25	5.05	1.1	1.02	0.47	10.2
		0.69	0.68	1	8.6
50	2.52	0.36	0.35	0.47	17.4
		0.25	0.24	1	14.5
100	1.26	0.13	0.12	0.47	29.5
		0.09	0.08	1	24.5

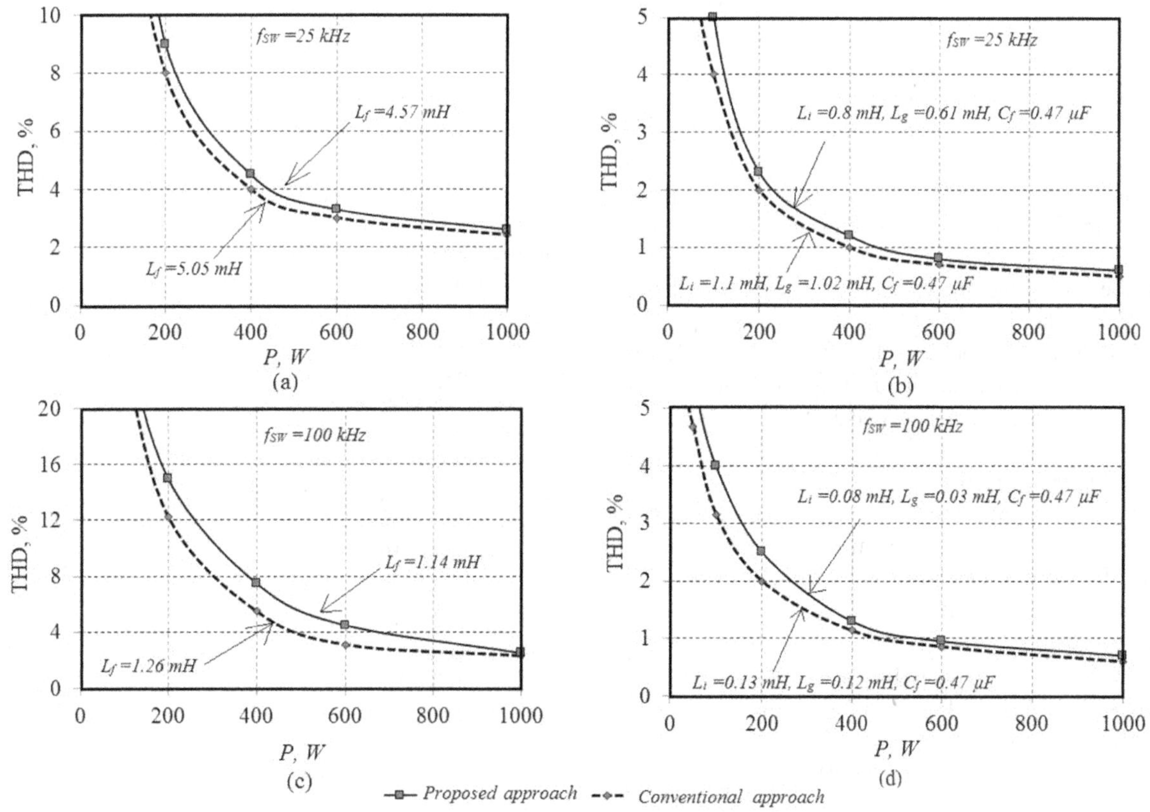

FIGURE 18.11 THD_I as a function of the output power: (a) L-filter with 25kHz; (b) LCL-filter with 25kHz; (c) L-filter with 100kHz; (d) LCL-filter with 100kHz.

18.11(a) and Figure 18.11(c) illustrates the THD_I as a function of the output power in case of a grid-connected inverter with L-filter. The upper figure corresponds to 25kHz of the switching frequency, while the lower to 100kHz. The values of the filtering components are shown in Table 18.1. It can be seen that a slightly lower inductor value leads to a worse output current quality, but within the predefined range.

Figure 18.11(b) and Figure 18.11(d) illustrate the THD_I of the LCL-filter as a function of the output power. Figure 18.11(b) corresponds to 25kHz of the switching frequency, and Figure 18.11(d) to 100kHz. The main conclusion from these described figures is that despite on the significant difference in the values of inductor, the difference in the output current total harmonic distortion (THD) is not significant. Difference can be noticed at the low power points that correspond to the inverter idle mode. Simulation and experimental verification are described in [75].

The second case study example shows the buck-boost inverter with unfolding circuit (Figure 18.2[c]). The LCL-filter parameters depend on the resonance frequency and on the mode of operation. Table 18.2. shows the main parameters that belong to the buck mode of the buck-boost inverter with unfolding circuit. The example of calculation is performed using before defined Eqs. (23)–(25).

One of the main goals of the iterations of calculation is to estimate the size of the overall filter by changing the value of the capacitor. Usually, the inductors occupy the largest area within the overall size of the output filter. Taking into account that the size of the each inductor coil depends on accumulated energy, the maximum current across inductor has to be taking into consideration, as well.

Figure 18.12 is derived in order to demonstrate the dependence of the inductance values on the value of the capacitor. It shows this dependence in case of constant power, value of the output current ripple, and the switching frequency.

Figure 18.12a demonstrates that the size of the converter side inductor is increasing, with the capacitor value increasing. At the same time, the value of the grid side inductor is decreasing. It should be noted that the current ripple in the grid side inductance should be very small, while the current ripple in the converter side inductor is much higher (up to 20%). It means that the overall size of the magnetics elements is reducing with capacitance value increasing. It is shown in Figure 18.12(b).

Converter-side inductance can be reduced keeping the same grid current ripples by means of resonance frequency growing. In the opposite case, higher values of the filtering capacitor may cause additional zero crossing distortions of the grid current.

It should be noticed that in the case of reducing the grid-side inductor to a value less than 50μH, the impact of the grid impedance becomes very remarkable. It is known that grid inductance may reach up to 40μH. In this case, an LCL-filter can be replaced by a simple LC-filter. Table 18.2 shows comparisons of the injected to grid current ripple in both LCL- and LC-filters.

Finally, to summarize the theoretical statements, Figure 18.13(a)–(f) demonstrates the simulation results of the buck-boost converter with unfolding circuit in the buck mode in case of different filtering capacitors.

The boost requires different approach. In this case, the LCL-filter can not be considered as classical LCL-filter

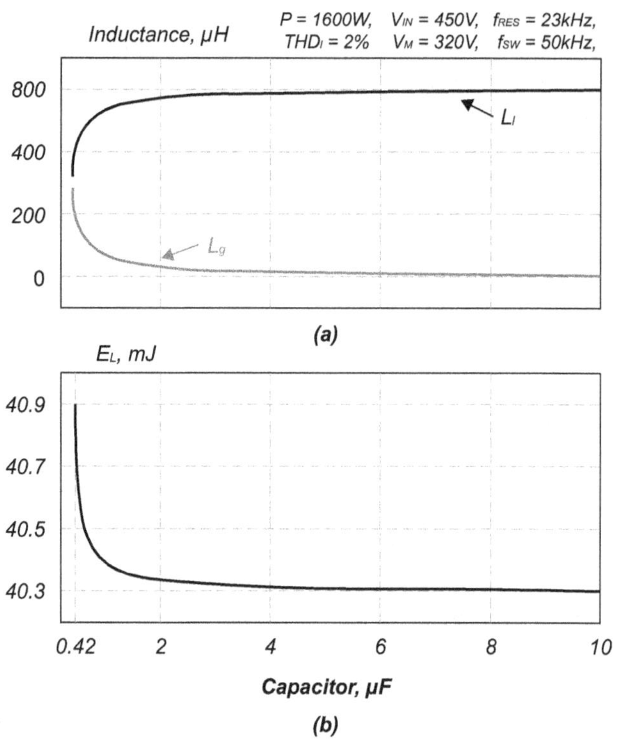

FIGURE 18.12 Magnetics elements estimation as a function of the capacitance value: (a) converter-side and grid-side inductance values as a function of the filtering capacitor; (b) energy accumulated in the magnetics elements as a function of the filtering capacitor.

TABLE 18.2
Parameters of the Buck-Boost Inverter with Unfolding Circuit in the Buck Mode

	C_1= 0.42μF		C_1= 0.84μF		C_1= 1.26μF	
Parameter	**LCL**	**LC**	**LCL**	**LC**	**LCL**	**LC**
Current change of the grid inductor Δi_g, A	0.2	-	0.19	-	0.21	-
Current change of the input inductor Δi_L, A	1.5	1.4	1.3	1.25	1.25	1.2
Voltage change of the output capacitor Δv_C, V	8.5	8.3	4	3.7	2.9	2.4
Grid inductance L_g, μH	138		62		40	
Input inductance L_i, μH	665		741		763	
Frequency of switching f_{SW}, kHz	50					
Sample frequency, kHz	50					
Input voltage V_{IN}, V	450					
Grid amplitude voltage V_M, V	320					
Grid frequency voltage f_{SINE}, Hz	50					
Input power P, W	1600					

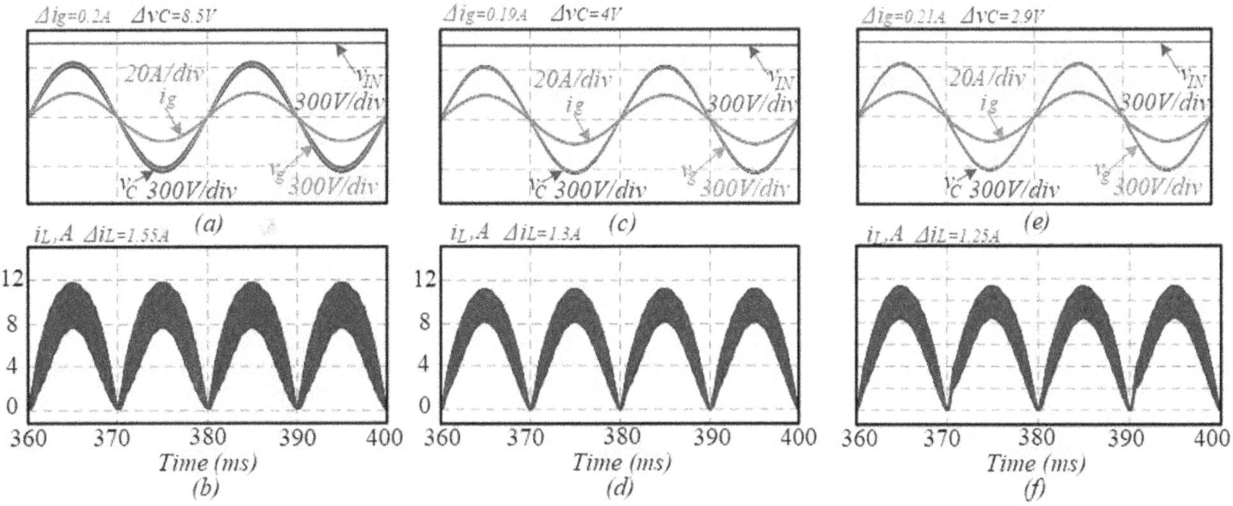

FIGURE 18.13 Simulation results of the buck-boost converter with unfolding circuit in the buck mode: (a) the output current and voltage, along with the grid voltage; (b) the inductor current; (c) the output current and voltage, along with the grid voltage with a double capacitor; (d) the inductor current with a double capacitor; (e) the output current and voltage along with the grid voltage with a triple capacitor; (f) the inductor current with a triple capacitor.

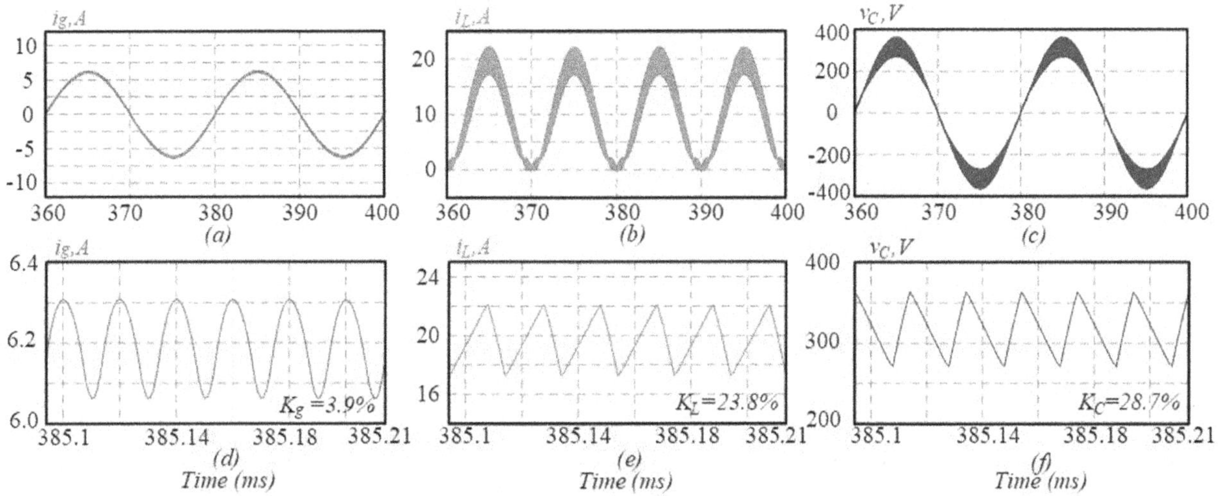

FIGURE 18.14 Simulation results of the buck-boost converter with unfolding circuit in the boost mode: (a) the output current; (b) the inductor current; (c) the output voltage; (d) the ripple across output current; (e) the ripple across the input inductor; (f) the ripple in the output voltage.

because of different equivalent circuits. The ripple of the grid current depends only on the ripple of the voltage across the filtering capacitor. Due to this, the higher value of the filtering capacitor can be required in order to keep the output capacitor voltage at the same level. Table 18.3 shows simulation parameters of the buck-boost converter with unfolding circuit in the boost mode. Figure 18.14 shows correspondent simulation results.

18.5 CONCLUSIONS

This chapter presents a detailed output filter design of any grid-connected or off-grid inverter. Several conventional approaches are discussed. General approach for the L- and LCL-filter design is disclosed, step by step. Analytical expressions allow obtaining the lowest output filter size of the L- or LCL-filter. The simulation, along with experimental results in the reference research papers, confirmed the analytical expressions outcomes. The smallest size of the output LCL-filter corresponds to the higher resonance frequency that applies limitation to the range of the possible switching frequency.

In case of inverters with unfolding circuit, the calculation approach has to be modified. On the one hand, it is very similar to the LCL-filter calculation in the buck mode, but on the other hand, the boost mode requires another approach.

TABLE 18.3
Parameters of the Buck-Boost Inverter with Unfolding Circuit in Boost Mode

Parameter	Value
Output current ripple factor K_g, %	4
Output voltage ripple factor K_C, %	30
Input current ripple factor K_L, %	24
Output inductance L_g, μH	960
Input inductance L_i, μH	287
Output capacitor C_l, μF	0.9
Input voltage V_{IN}, V	100
Input power P, W	1000
Frequency of switching f_{SW}, kHz	50
Grid amplitude voltage V_M, V	320

REFERENCES

[1] A.A. Rockhill, M. Liserre, R. Teodorescu, P. Rodriguez, "Grid-filter design for a multimegawatt medium-voltage voltage-source inverter," IEEE Trans. Ind. Electron., vol. 58, no. 4, pp. 1205–1217, Apr. 2011.

[2] P. Gonzalez-Castrillo, E. Romero-Cadaval, M. Milanes-Montero, F. Barrero-Gonzalez, M. Guerrero-Martinez, "A new criterion for selecting the inductors of an active power line conditioner," Compatibility and Power Electronics (CPE), 2011 7th International Conference-Workshop, pp. 167–172, 1–3 June 2011.

[3] P. Channegowda, V. John, "Filter optimization for grid interactive voltage source inverters," IEEE Trans. Ind. Electron., vol. 57, no. 12, pp. 4106–4114, Dec. 2010.

[4] E. Romero-Cadaval, M.I. Milanés-Montero, F. Barrero-González, "A modified switching signal generation technique to minimize the RMS tracking error in active filters," IEEE Trans. Power Electron., vol. 20, no. 5, pp. 1118–1124, Sept. 2005.

[5] P.C. Loh, S.W. Lim, F. Gao, F. Blaabjerg, "Three-level Z-source inverters using a single LC impedance network," IEEE Trans. Power Electron., vol. 22, no. 2, pp. 706–711, 2007.

[6] T.G. Habetler, R. Naik, T.A. Nondahl, "Design and implementation of an inverter output lc filter used for DV/DT reduction," IEEE Trans. Power Electron., vol. 3, no. 17, pp. 327–331, 2002.

[7] V. Golubev, "Calculation and optimization of the LC-filter of the pulse converter of an AC voltage," Tekhnichna Elektrodynamika, no. 1, pp. 33–37, 2012.

[8] H. Kim, S.-K. Sul, "A nove l-filter design for output LC filters of PWM inverters," JPE, vol. 22, no. 2, pp. 706–711, 2007.

[9] Y. Sozer, D.A. Torrey, S. Reva, "New inverter output filter topology for PWM motor drives," IEEE Trans. Power Electron., vol. 15, no. 6, pp. 1007–1017, 2000.

[10] M. Schweizer, R. Blattmann, J.W. Kolar, "Optimal design of LCL harmonic filters for three-phase PFC rectifiers," 37st Annual Conference of IEEE Industrial Electronics Society, IECON 2011, pp. 1503–1510, 2011.

[11] M. Liserre, F. Blaabjerg, S. Hansen, "Design and control of an LCL-filter-based three-phase active rectifier," IEEE Trans. Ind. Appl., vol. 41, no. 5, pp. 1281–1291, Sept.–Oct. 2005.

[12] Yi Tang, P.C. Loh, P. Wang, F.H. Choo, F. Gao, F. Blaabjerg, "Generalized design of high performance shunt active power filter with output LCL filter,"IEEE Trans. Ind. Electron., vol. 59, no. 3, pp. 1281–1291, Mar. 2012.

[13] M. Liserre, F. Blaabjerg, R. Teodorescu, "Grid impedance estimation via excitation of LCL-filter resonance," IEEE Trans. Ind. Appl., vol. 43, no. 5, pp. 1401–1407, Sept.–Oct. 2005.

[14] M. Liserre, A. Dell'Aquila, F. Blaabjerg, "Genetic algorithm based design of the active damping for a LCL-filter three-phase active rectifier," IEEE Trans. Power Electron., vol. 19, no. 1, pp. 76–86, Jan. 2004.

[15] E. Figueres, G. Garcera, J. Sandia, F. Gonzalez-Espein, J.C. Rubio, "Sensitivity study of the dynamics of three-phase photovoltaic inverters with an LCL grid filter," IEEE Trans. Ind. Electron., vol. 56, no. 3, pp. 706–717, Mar. 2009.

[16] G. Shen, X. Zhu, J. Zhang, D. Xu, "A new feedback method for PR current control of LCL-filter-based grid-connected inverter," IEEE Trans. Ind. Electron., vol. 57, no. 6, pp. 2033–2041, Jun. 2010.

[17] K. Jalili, S. Bernet, "Design of LCL-filters of active-front-end two level voltage source converters," IEEE Trans. Ind. Electron., vol. 56, no. 5, pp. 1674–1689, May 2009.

[18] D.-E. Kim, D.-C. Lee, "Feedback linearization control of grid-interactive PWM converters with LCL FILTERS," JPE., vol. 9, no. 2, pp. 288–299, 2009.

[19] M. Malinowski, M.P. Kazmierkowski, W. Szczygiel, S. Bernet, "Simple sensorless active damping solution for three-phase PWM rectifier with LCL filter," 31st Annual Conference of IEEE Industrial Electronics Society, IECON 2005, pp. 987–991, 2005.

[20] M. Malinowski, S. Bernet, "A simple voltage sensorless active damping scheme for three-phase PWM converters with an LCL-filter," IEEE Trans. Ind. Electron., vol. 55, no. 4, pp. 1876–1880, Apr. 2008.

[21] J.S. Lee, H.G. Jeong, K.B. Lee, "Active damping for wind power systems with LCL filters using a DFT," JPE, vol. 12, no. 2, pp. 326–332, 2012.

[22] H.G. Jeong, D.-K. Yoon, K.B. Lee, "Design of an LCL-filter for three-parallel operation of power converters in wind turbines," JPE, vol. 13, no. 3, pp. 437–446, 2013.

[23] X.Q. Guo, W.Y. Wu, H.R. Gu, "Modeling and simulation of direct output current control for LCL-interfaced grid-connected inverters with parallel passive damping," Simulat. Model. Pract. Theor., vol. 18, pp. 946–956, 2010.

[24] W. Wang, X. Ma, "Transient analysis of low-voltage ride-through in three-phase grid-connected converter with LCL filter using the nonlinear modal series method," Electr. Power Syst. Res., vol. 105, pp. 39–50, 2013.

[25] W. Wu, Y. He, F. Blaabjerg, "An LLCL power filter for single-phase grid-tied inverter," IEEE Trans. Power Electron., vol. 27, no. 2, pp. 782–789, 2012.

[26] W. Wu, Y. He, T. Tang, F. Blaabjerg, "A new design method for the passive damped LCL- and LLCL-filter based single-phase grid-tied inverter," IEEE Trans. Ind. Electron., accepted for publication, vol. 99, 2012.

[27] D. Bortis, D. Neumayr, J.W. Kolar, "ηρ-Pareto optimization and comparative evaluation of inverter concepts considered for the GOOGLE Little Box Challenge," in Proc. of IEEE 17th Workshop on Control and Modeling for Power Electronics (COMPEL), pp. 1–5, 2016.

[28] R. Ghosh, Miao-xin Wang, S. Mudiyula, U. Mhaskar; R. Mitova, D. Reilly, D. Klikic, "Industrial approach to design a 2-kVa inverter for Google little box challenge," IEEE Trans. Ind. Electron., vol. 65, no. 7, pp. 5539–5549, July 2018.

[29] A. Morsy, P. Enjeti, "Comparison of active power decoupling methods for high-power-density single-phase inverters using wide-bandgap FETs for Google little box challenge," IEEE J. Emerg. Sel. Top. Power Electron., vol. 4, no. 3, pp. 790–798, 2016.

[30] S.B. Kjaer, J.K. Pedersen, F. Blaabjerg, "A review of single-phase grid-connected inverters for photovoltaic modules," IEEE Trans. Ind. Appl., vol. 41, no. 5, pp. 1292–1306, Sept.–Oct. 2005.

[31] D. Meneses, F. Blaabjerg, O. Garcıa, J.A. Cobos, "Review and comparison of step-up transformerless topologies for photovoltaic AC-module application," IEEE Trans. Power Electron., vol. 28, no. 6, pp. 2649–2663, 2013.

[32] K. Zeb, Saif Ul Islam, W. Uddin, I. Khan, M.A. Khan, Sajid Ali, T.D.C. Busarello, H.J. Kim, "An overview of transformerless inverters for grid connected photovoltaic system" in proc. of 2018 International Conference on Computing, Electronic and Electrical Engineering.

[33] S. Kouro, J.I. Leon, D. Vinnikov, L.G. Franquelo, "Grid-connected photovoltaic systems: an overview of recent research and emerging PV converter technology," IEEE Ind. Electron. Magazine, vol. 9, no. 1, pp. 47–61, 2015.

[34] M. Noman, H. Khan, M. Forouzesh, Y.P. Siwakoti, Li Li, T. Kerekes, F. Blaabjerg, "Transformerless inverter topologies for single-phase photovoltaic systems: a comparative review," IEEE J. Emerg. Sel. Top. Power Electron., vol. 8, no. 1, pp. 805–835, 2020.

[35] K. Alluhaybi, I. Batarseh, H. Hu, "Comprehensive review and comparison of single-phase grid-tied photovoltaic microinverters," IEEE J. Emerg. Sel. Top. Power Electron., vol. 8, no. 2, pp. 1310–1329, June 2020.

[36] F.B. Grigoletto, "Five-level transformerless inverter for single-phase solar photovoltaic applications," IEEE J. Emerg. Sel. Top. Power Electron., vol. 8, no. 4, pp. 3411–3422, Dec. 2020.

[37] J. Kikuchi, T.A. Lipo, "Three-phase PWM boost-buck rectifiers with power-regenerating capability," IEEE Trans. on Ind. Applic., vol. 38, pp. 1361–1369, Sept./Oct. 2002.

[38] F. Gao, R. Teodorescu, F. Blaabjerg, P.C. Loh and D.M. Vilathgamuwa, "Performance Evaluation of Buck-Boost Three-Level Inverters with Topological and Modulation Development," in Proc. Power Electronics and Applications, 2007 European Conference on, pp. 1–10, 2007.

[39] F. Gao, R. Teodorescu, F. Blaabjerg, P.C. Loh, D.M. Vilathgamuwa, "Topological design and modulation strategy for buck—boost three-level inverters," IEEE Trans. Power Electron., vol. 24, no. 7, pp. 1722–1732, July 2009.

[40] R.L. Mathew, K.S. Jiji, "A three level neutral point clamped Inverter with buck-boost capability for renewable energy sources," in Proc. of International Conference on Sustainable Energy and Intelligent Systems (SEISCON 2011), pp. 201–206, 2011.

[41] F.Z. Peng, "Z-source inverter," IEEE Trans. Ind. Appl., vol. 39, no. 2, pp. 504–510, 2003.

[42] Y.P. Siwakoti, F. Peng, F. Blaabjerg, P.C. Loh, G.E. Town, "Impedance source networks for electric power conversion part-I: a topological review," IEEE Trans. Power Electron., vol. 30, no. 2, pp. 699–716, Feb. 2015.

[43] Y.P. Siwakoti, F. Peng, F. Blaabjerg, P. Loh, G.E. Town, "Impedance source networks for electric power conversion part-II: review of control method and modulation techniques," IEEE Trans. Power Electron., vol. 30, no. 4, pp. 1887–1906, Apr. 2015.

[44] A. Chub, D. Vinnikov, F. Blaabjerg, F.Z. Peng, "A review of galvanically isolated impedance-source DC—DC converters" IEEE Trans. Power Electron., vol. 31, no. 4, pp. 2808–2828, Apr. 2016.

[45] L. Yushan, H. Abu-Rub, G. Baoming, "Z-source/quasi-Z-source inverters: derived networks, modulations, controls, and emerging applications to photovoltaic conversion" IEEE Ind. Electron. Mag., vol. 8, no. 4. pp. 32–44, Dec. 2014.

[46] O. Husev, A. Chub, E. Romero-Cadaval, C. Roncero-Clemente, D. Vinnikov, "Hysteresis current control with distributed shoot-through states for impedance source inverters," Int. J. Circuit Theory Appl., vol. 44, no. 4, pp. 783–797, 2015.

[47] C. Roncero-Clemente, E. Romero-Cadaval, M. Ruiz-Cortés, O. Husev, "Carrier level-shifted based control method for the PWM 3L-T-Type qZS Inverter with capacitor imbalance compensation," IEEE Trans. Ind. Electron., vol. 65, no. 10, pp. 8297–8306, Oct. 2018.

[48] O. Husev, R. Strzelecki, F. Blaabjerg, V. Chopyk, D. Vinnikov, "Novel family of single-phase modified impedance-source buck-boost multilevel inverters with reduced switch count," IEEE Trans. Power Electron., vol. 31, no. 11, pp. 7580–7591, Now. 2016.

[49] T. Shults, O. Husev, F. Blaabjerg, C. Roncero, E. Romero-Cadaval, D. Vinnikov, "Novel space vector pulse width modulation strategies for single-phase three-level NPC impedance-source inverters," IEEE Trans. Power Electron., vol. 34, no. 5. pp. 4820–4830, 2019.

[50] T. Shults, O. Husev, F. Blaabjerg, J. Zakis, K. Khandakji, "LCCT-derived three-level three-phase inverters," IET Power Electronics, vol. 10, no. 9. pp. 996–1002, 2017.

[51] Y.P. Siwakoti, F. Blaabjerg, V.P. Galigekere, A. Ayachit, M.K. Kazimierczuk, "A-source impedance network," IEEE Trans. Power Electron., vol. 31, no. 12, pp. 8081–8087, 2016.

[52] O. Husev, F. Blaabjerg, C. Roncero-Clemente, E. Romero-Cadaval, D. Vinnikov, Y.P. Siwakoti, R. Strzelecki, "comparison of impedance-source networks for two and multilevel buck-boost inverter applications," IEEE Trans. on Power Electron., vol. 31, no. 11, pp. 7564–7579, Nov. 2016.

[53] R. Burkart, J.W. Kolar, G. Griepentrog, "Comprehensive comparative evaluation of single- and multi-stage three-phase power converters for photovoltaic applications," in Proc. of Intelec, pp. 1–8, 2012.

[54] D. Panfilov, O. Husev, F. Blaabjerg, J. Zakis, K. Khandakji, "Comparison of three-phase three-level voltage source inverter with intermediate dc-dc boost converter and quasi-Z-source inverter," IET Power Electr., vol. 9, no. 6, pp. 1238–1248, 2016.

[55] A. Abdelhakim, P. Mattavelli, G. Spiazzi, "Three-phase split-source inverter (SSI): analysis and modulation," IEEE Trans. Power Electron., vol. 31, no. 11, pp. 7451–7461, Nov. 2016.

[56] A. Abdelhakim, P. Mattavelli, F. Blaabjerg, "Performance evaluation of the single -phase split -source inverter using an alternative DC -AC configuration," IEEE Trans. Power Electron., vol. 65, no. 1, pp. 363–373, Jan. 2018.

[57] H. Ribeiro, F. Silva, S. Pinto, and B. Borges, "Single stage inverter for PV applications with one cycle sampling technique in the MPPT algorithm," in Proc. 35th Annu. Conf. IEEE Ind. Electron., pp. 842–849, 2009.

[58] Z. Fedyczak, R. Strzelecki, G. Benysek, "Single-phase PWM AC/AC semiconductor transformer topologies and applications," in Proc. of Power Electronics Specialists Conference, pp. 1–6, 2002.

[59] A. Kumar, V. Gautam, P. Sensarma, "A SEPIC derived single stage buck-boost inverter for photovoltaic applications," in Proc. of IEEE International Conference on Industrial Technology (ICIT), pp. 403–408, 2014.

[60] T.M. Nishad, K.M. Shafeeque, "A novel single stage buck boost inverter for photovoltaic applications," in Proc. of 2016 International Conference on Electrical, Electronics, and Optimization Techniques (ICEEOT), pp. 3067–3071, 2016.

[61] W. Wu, J. Ji, F. Blaabjerg, "Aalborg inverter—a new type of 'buck in buck, boost in boost' grid-tied inverter," IEEE Trans. on Power Electron., vol. 30, no. 9, pp. 4784–4793, Sept. 2015.

[62] S. Zhang, W. Wu, H. Wang, M. Huang, N. Gao, F. Blaabjerg, "Single-stage MPPT control realization for Aalborg inverter in photovoltaic system," in IECON 2017–43rd Annual Conference of the IEEE Industrial Electronics Society, pp. 4233–4238, 2017.

[63] W. Wu, S. Feng, J. Ji, M. Huang, F. Blaabjerg, "LLCL-filter based single-phase grid-tied aalborg inverter," in Pro. International Power Electronics and Application Conference and Exposition, Shanghai, China, pp. 658–663, Nov. 5–8, 2014.

[64] Z. Liu, H. Wu, Y. Liu, J. Ji, W. Wu and F. Blaabjerg, "Modelling of the modified-LLCL-filter-based single-phase grid-tied Aalborg inverter," IET Power Electron., vol. 10, no. 2, pp. 151–155, 2017.

[65] W. Wu, Z. Wang, J. Ji and F. Blaabjerg, "Performance analysis of new type grid-tied inverter-aalborg inverter," in Proc.of 16th European Conf. on Power Electronics and Applications, Lappeenranta, Finland, pp. 1–10, Aug. 26–28, 2014.

[66] H. Wang, W. Wu, H. Shu-hung Chung, F. Blaabjerg, "Coupled-inductor-based aalborg inverter with input DC energy regulation," IEEE Trans. on Ind. Electron., vol. 65, no. 5, pp. 3826–3836, May 2018.

[67] A. Ali Khan, H. Cha, "Dual-buck-structured high-reliability and high-efficiency single-stage buck—boost inverters," IEEE Trans. on Ind. Electron., vol. 65, no. 4, pp. 3176–3187, Apr. 2018.

[68] Z. Zhao, M. Xu, Q. Chen, J.-S. (Jason) Lai, Y. Cho, "Derivation, analysis, and implementation of a boost—buck converter-based high-efficiency PV inverter," IEEE Trans. on Power Electron., vol. 27, no. 3, pp. 1304–1313, Mar. 2012.

[69] Z. Zhao, J.-S. (Jason) Lai, Y. Cho, "Dual-mode double-carrier-based sinusoidal pulse width modulation inverter with adaptive smooth transition control between modes," IEEE Trans. on Ind. Electron., vol. 60, no. 5, pp. 2094–2103, Mar. 2013.

[70] M. Jang, M. Ciobotaru, V.G. Agelidis, "A compact single-phase bidirectional buck-boost-inverter topology," in Proc. of 2012 International Conference on Renewable Energy Research and Applications (ICRERA), pp. 1–6, 2012.

[71] O. Husev, O. Matiushkin, C. Roncero, F. Blaabjerg, D. Vinnikov, "Novel family of single-stage buck-boost inverters based on unfolding circuit," IEEE Trans. Power Syst., vol. 34, no. 8. pp. 7662–7676, 2019.

[72] D. Pan, X. Ruan, X. Wang, F. Blaabjerg, X. Wang, and Q. Zhou, "A highly robust single-loop current control scheme for grid-connected inverter with an improved LCCL filter configuration," IEEE Trans. Power Electron., vol. 33, no. 10, pp. 8474–8487, Oct. 2018.

[73] F. Li, X. Zhang, H. Zhu, H.Y. Li, C.Z. Yu, "An LCL-LC filter for grid-connected converter: topology, parameter, and analysis," IEEE Trans. on Power Electron., vol. 30, no. 9, pp. 5067–5077, 2015.

[74] O. Husev, A. Chub, E. Romero-Cadaval, C. Roncero-Clemente, D. Vinnikov, "Voltage distortion approach for output filter design for off-grid and grid-connected PWM inverters," JPE., vol. 15, pp. 278–287, 2015.

[75] O. Matiushkin, O. Husev, D. Vinnikov, C. Roncero-Clemente, "Optimal LCL-filter study for buck-boost inverter based on unfolding circuit," in proc. of 14th International Conference on Compatibility, Power Electronics and Power Engineering (CPE-POWERENG), 2020 IEEE.

Index

For Product Safety Concerns and Information please contact our EU
representative GPSR@taylorandfrancis.com
Taylor & Francis Verlag GmbH, Kaufingerstraße 24, 80331 München, Germany

www.ingramcontent.com/pod-product-compliance
Lightning Source LLC
Chambersburg PA
CBHW080936220326
41598CB00034B/5799
9781032134253